Richard C. Powell

Physics

A Non-Calculus Based Course for STEM Students

Richard C. Powell
Tucson, AZ, USA

ISSN 2192-4791 ISSN 2192-4805 (electronic)
Undergraduate Lecture Notes in Physics
ISBN 978-3-032-09360-8 ISBN 978-3-032-09361-5 (eBook)
https://doi.org/10.1007/978-3-032-09361-5

This Springer imprint is published by the registered company Springer Nature Switzerland AG
The registered company address is: Gewerbestrasse 11, 6330 Cham, Switzerland

Undergraduate Lecture Notes in Physics

Undergraduate Lecture Notes in Physics (ULNP) publishes authoritative texts covering topics throughout pure and applied physics. Each title in the series is suitable as a basis for undergraduate instruction, typically containing practice problems, worked examples, chapter summaries, and suggestions for further reading.

ULNP titles must provide at least one of the following:

- An exceptionally clear and concise treatment of a standard undergraduate subject.
- A solid undergraduate-level introduction to a graduate, advanced, or non-standard subject.
- A novel perspective or an unusual approach to teaching a subject.

ULNP especially encourages new, original, and idiosyncratic approaches to physics teaching at the undergraduate level.

The purpose of ULNP is to provide intriguing, absorbing books that will continue to be the reader's preferred reference throughout their academic career.

Acknowledgements

The author gratefully acknowledges the help he received from Doug Powell in the preparation of this manuscript. In addition, some of the experiments and problems were initially published by Apologia Ministries in "*Advanced Physics in Creation*" and the author appreciates Apologia granting him the rights to publish them in this text.

Introduction

Abstract

Physics is the science that allows us to explain how things work in the world around us. It describes the properties of matter and light as well as phenomena such as gravity, electricity and magnetism. It is the basis for all of the sophisticated technology we enjoy today. Understanding physics is critical to our decision making in a high-tech society. It gives us the knowledge we need to use new technology for the betterment of all mankind without damaging our ecological surroundings. For example, it impacts our choice of the source of energy to provide the power we need for the future. If you want to be a good steward of the Earth, you should make every effort to understand physics.

Main Topics in the Introduction

- Review of Vector Math
- Review of Units
- Review of Significant Figures

This text is designed for a comprehensive one-year course in physics. It would be ideal for high school senor STEM students or college freshmen taking a non-calculus based physics course. It would be helpful for students to have completed an introductory physics course or an advanced physical science course with a strong physics component before taking this course. However, these prerequisites are not required. This is a self-contained course with all the information required for a student to have a complete understanding of modern physics and should be accessible to any serious student of science. Each module focuses on a specific area of physics and contains example problems, homework problems, and suggested experiments. Each chapter also contains discussions of special topics related to that area of physics plus a section of a next level discussion that is optional depending on the student.

Along with advancing a student's scientific knowledge, it should also improve their problem solving and critical thinking skills. The book is divided into sixteen modules each of which address a different area of physics. Each module begins with a brief review of the fundamental concepts of that area of physics. The rest of the module extends these concepts to a more advanced level.

Because physics is an observational science, doing the experiments in each module will help students develop their observational skills. Learning physics does not mean memorizing important concepts and laws. You demonstrate your knowledge of physics by explaining observations and working through problems. There is a specific procedure to follow in working problems. The first step is to draw a sketch of the problem to visualize the data. Then write lists of the known and unknown parameters. You should then use the equations you have learned with the known parameters to solve for the unknown parameters. It's that simple! If you follow this technique of problem solving, you will be able to clearly communicate your solution to others. The problems you will be asked to work may seem sometimes to be trivial or irrelevant. However, physicists like to develop simple models to simulate important real-world events. For example, the problem of a block sliding down an incline plane might be a model of a car skidding to a stop on a hill. The same physics is used to analyze these two situations. Understand the basics and you will understand your world.

Because of the observational nature of physics, different observers in different locations will describe the same event differently. Even though the laws of physics are independent of the point of observation, the way an observer sees an event does change. Therefore, it is important when describing an event or working a problem that you define your point of observation. This is referred to as a reference frame. A reference frame includes the point of origin and the directions relevant to the situation. A typical reference frame is shown in Fig. 1. The o where the three axes cross is the origin. The arrows designated as x, y, and z are called Cartesian coordinate directions. They are perpendicular to each other and can be used to locate any point in space. The orientation of the three axes is arbitrary and can be chosen to best fit the problem. However, for a conventional "right-handed" coordinate system, if the fingers of your right hand are pointed along the x-axis and rotated toward the y-axis, your thumb will point in the direction of the z-axis.

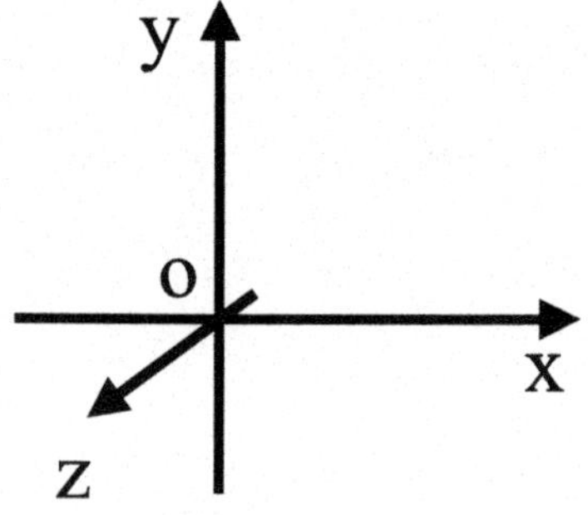

Fig. 1 Reference frame

In this course, we will usually have to work in only two dimensions. The x and y coordinates may represent horizontal and vertical directions, or they may represent compass directions on the ground such as east and north.

Review of Vectors

As a review from your mathematics courses, we deal with two types of parameters, scalars and vectors. A scalar has only a magnitude whereas a vector has both a magnitude and a direction associated with it. It is important when giving a vector parameter as a solution to a problem to give both its magnitude and its direction. The basics of working with vectors warrants a brief review. A vector is designated in one of two ways. The first is to give its magnitude and its angular position with respect to a reference frame such as the one shown in Fig. 1. The second is to give its directional components with respect to the reference frame. In this case it is common to use unit vectors **i**, **j**, and **k** to designate the x, y and z directions. These are vectors with unit magnitude. A general convention is to write vectors in bold type. So a vector might be written as $\mathbf{v} = 5\mathbf{i} + 3\mathbf{j} + 4\mathbf{k}$ where $v_x = 5$, $v_y = 3$, and $v_z = 4$.

Figure 2 shows a generic two-dimensional vector with length (magnitude) R and direction θ. The angle θ designating the direction of a vector is always measured in a counterclockwise direction from the positive x-axis. Many times, it is useful to break a vector down into its components. In the two-dimensional case shown in Fig. 2,

$$\mathbf{R} = \mathbf{R_x} + \mathbf{R_y} \tag{1}$$

where

$$R_x = R\cos\theta,\ R_y = R\sin\theta,\ \tan\theta = R_y/R_x,\ \text{and}\, R^2 = R_x^2 + R_y^2. \tag{2}$$

If the angle θ designating the direction of the vector is greater than 90 °, the angle in the right triangle used to break the vector into its components is a different, related angle.

It is important to do mathematical operations with vectors. You will recall that the addition of two vectors can be done graphically by putting the tail of one vector

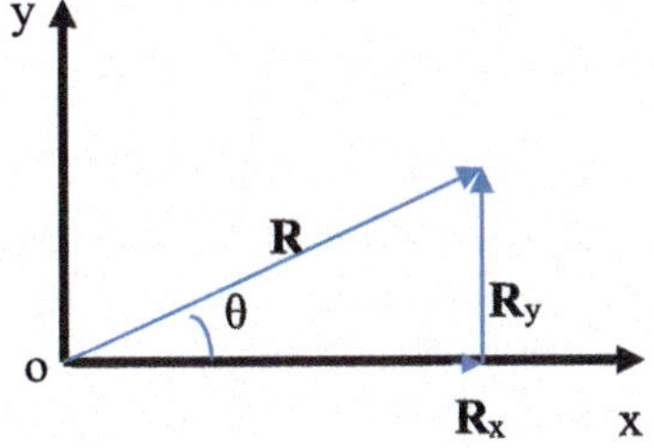

Fig. 2 Vector components

at the head of the other vector and drawing the resultant vector from the tail of the first to the head of the last. Instead, this addition can be done by breaking each vector into its x and y components. Then all the x components can be added together to obtain the x-component of the resultant vector and all the y components can be added together to get the y-component of the resultant vector. At this point the Pythagorean Theorem and the other equations given in Eq. 2 can be used to determine the magnitude and direction of the resultant vector.

Multiplication with vectors is more complex. In some cases, the product of two physical properties represented by vectors results in a scalar physical property. For this situation the multiplication is the sum of the products of the components of the vectors. This type of multiplication is called a scalar product because the result is a scalar. It is also called a dot product because it is represented by a dot, $\mathbf{A} \bullet \mathbf{B}$. The result can be found using the Cartesian components of the vectors or using the angle between the vectors,

$$\mathbf{A} \bullet \mathbf{B} = A_x B_x + A_y B_y = AB\cos\theta. \tag{3}$$

Note that the angle θ in this case is not designating the direction of one of the vectors. It is the angle between the two vectors. The relationship between the two parts of Eq. 3 can be seen by orienting the reference frame so vector **B** is along the x-axis. Then $A\cos\theta$ is A_x while B_x equals B and B_y is zero. The two parts of the equation are then equivalent. Another way of looking at this is that $A\cos\theta$ is the component of **A** in the direction of **B**.

As an example of when we need to use a dot product, "work" is the product of the component of force in the direction of displacement times the displacement. In other words, the scalar quantity work is the dot product of the vectors of force and displacement.

Note that if the two vectors are orthogonal, the dot product is zero. Also, the dot product operation is commutative. You can see from Eq. 3 $\mathbf{A} \bullet \mathbf{B} = \mathbf{B} \bullet \mathbf{A}$.

The second type of vector multiplication occurs when the physical processes being multiplied result in a physical entity represented by a vector. This is called a vector product because the result is a vector. It is also called a cross product because it is represented by a ×, A × B. The magnitude of the resultant vector can again be found either from a combination of the products of the vector components or from the angle between the two vectors. However, in this case, using the vector components is more complicated so we will just use the simpler expression,

$$|\mathbf{A} \times \mathbf{B}| = AB\sin\theta. \tag{4}$$

In this case, $B\sin\theta$ is the component of **B** perpendicular to the direction of **A**. If the vectors are parallel their cross product is zero.

The direction of the resultant vector of a cross product is found by using the right-hand rule: point the fingers of your right hand in the direction of the first vector and curl them toward the direction of the second vector. Your thumb will point in the direction of the resultant vector. It is always perpendicular to the directions of both original vectors. Using this right-rule, it is easy to see that the cross product operation is NOT commutative, **A x B = – B x A**. The magnitude of the resultant vector is the same but the direction is reversed.

As an example of when we need to use a cross product, the "torque" needed to cause rotational motion is the product of the lever arm times the component of force perpendicular to the direction of the lever arm. In other words, the torque vector is the cross product of the vectors of the lever arm and force.

As an example of vector multiplication, study Example 1.

Example 1

Find the dot product and the cross product of the following two vectors: $\mathbf{A} = 3\mathbf{i} + 4\mathbf{j}$ and $\mathbf{B} = 5\mathbf{i} + 0\mathbf{j}$.

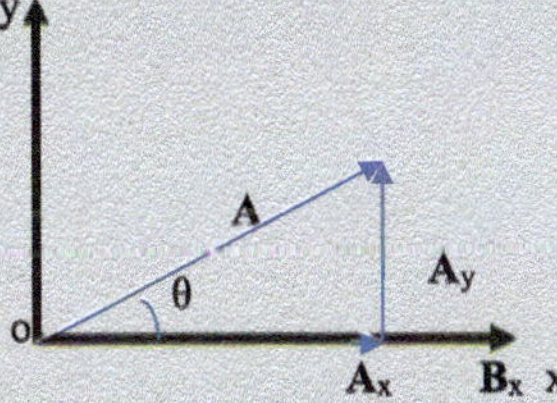

Knowns: $A_x = 3$, $A_y = 4$, $B_x = 5$, $B_y = 0$

Unknowns: $\mathbf{A} \bullet \mathbf{B}$, $\mathbf{A} \times \mathbf{B}$

From Eq. 2, the angle between the two vectors is given by

$\theta = \tan^{-1} A_y/A_x = 4/3 = 53\,^{\circ}$

and the magnitude of **A** is

$$A = (A_x^2 + A_y^2)^{1/2} = (9 + 16)^{1/2} = 5.$$

Then from Eq. 3,

$\mathbf{A} \bullet \mathbf{B} = AB\cos\theta = (5)(5)\cos 53\,^{\circ} = 15.$

The result of the dot product of these two vectors is a scalar with a magnitude of 15.

The magnitude of the cross product is given by Eq. 4

$|\mathbf{A} \times \mathbf{B}| = AB\sin\theta = (5)(5)\sin 53\,^{\circ} = 20.$

The direction of the cross product is found by putting the fingers of your right hand in the direction of **A** and rotating them toward **B**. When you do this your thumb points into the page. Thus the result of the cross product of these two vectors is a vector of magnitude 20 pointing into the page.

Units and Numbers

Every quantity in physics has units associated with it. When you solve a problem in physics, it is not enough to give the solution as some number, you must also say what units the number represents (and the direction if the answer is a vector quantity). Unfortunately, physicists around the world have not always agreed on what to use for units so several different systems of units have been established. However, at this time there is general agreement for everyone working in physics to use the international scientific unit system (SI system) based on the following seven quantities: length in meters (m); mass in kilograms (kg); time in seconds (s); temperature in Kelvin units (K); electrical current in Amperes (A); luminous intensity in candelas (cd); and the amount of a substance in moles (mol). This is what we will use in this course. Appendix II shows how to convert these SI units to other common unit systems and also shows how other derived units are related to these basic units. Sometimes it is convenient to express your results in terms of other units. For example, you might want to know your speed in miles per hour instead of meters per second. However, unless expressly asked to used other units, always give your answer in SI units.

There are of course many quantities other than these seven basic ones that we deal with in physics and they each have their own units. However, each of these other units can be derived from the seven basic units listed above. For example, the unit of force in the SI system is a Newton. This is equivalent to a ($kg \cdot m \cdot s^{-2}$). Similarly, the unit of energy in the SI system is a Joule which is a ($kg \cdot m^2 \cdot s^{-2}$). These derived units will be defined when we have need to use them.

When you work a problem, it is helpful to do a unit analysis. To do this, you check the units you have in each step of the solution. The units of the final answer must be consistent with what that quantity is. For example, if the final result is the speed of an object and your unit analysis comes out to be kilograms, there is an error somewhere in your calculations. Unit analysis is an important way to catch errors.

To refresh your memory about units, consider the following examples.

Example 2

The speed limit on many highways in the United States is 65 miles per hour. What is the speed limit in centimeters per second? (There are 1609 meters in a mile.)

$$\mathrm{v} = \frac{65\ \cancel{miles}}{\cancel{hour}} \times \frac{1609\ \cancel{m}}{1\ \cancel{mile}} \times \frac{1\ cm}{0.01\ \cancel{m}} \times \frac{1\ \cancel{hour}}{3600\ s} = 2900\ \mathrm{cm/s}.$$

Note how all of the units cancel out except cm and s and cm/s is a correct unit for speed.

Example 3
How many ergs are in 151 Joules of energy?

$$E = 151\,J = 151\,\frac{kg\text{-}m^2}{s^2} \times \frac{1000\,g}{kg} \times \frac{1\,cm^2}{0.0001\,m^2} = 1.51 \times 10^9\,\frac{g\,cm^2}{s^2}$$

$$= 1.51 \times 10^9\ \text{ergs}.$$

In this calculation you need to know the basic units that define the derived units of Joules and ergs. These can be found in Appendix II. Also you need to be sure and square the cm to m conversion.

Next, we need to discuss how to express the numerical part of our answer. This is determined primarily by how accurately we can make the measurements or calculations involved in the problem. It is conventional to express the number in however many digits we can confidently measure or calculate and round off the remaining digits. For example, if our calculator gives us an answer that reads 2365.67 m but we know that the data we used to obtain this is only accurate to a tenth of a meter, we have no confidence in the last number to the right of the decimal point. Therefore, we can round off this answer to 2365.7. This number has five significant figures. When dealing with a set of numbers to be added or subtracted, multiplied or divided, they should all be rounded off to the number of significant figures equal to the smallest number of significant figures of any member of the set. The answer will also have this number of significant figures.

When we deal with very large or very small numbers it is customary to use exponential notation. This means our answer could be written as 2.3657×10^3. Preceding the exponent, we have a number between one and ten. The exponent indicates how many steps the decimal point should be moved to the right (if it is positive) or to the left (if it is negative). The Appendix III lists a table giving the names we use for common positive and negative exponential numbers. These can apply to any different unit such as a picosecond or a picometer and indicates an exponential factor of 10^{-12} in either time or length.

It also should be pointed out that there are some special numbers in physics and some of the important ones are listed in Appendix I. These are important constants such as the speed of light, the charge on an electron, or Avogadro's number. These are not things you need to memorize but you do need to know where to find them listed since you will have to use them quite a lot.

With this brief review, you should be ready to begin your study of physics. The most difficult part will probably be associated with the math needed to work problems. If possible you should have handy a book of math tables that includes trig identities, common equations like the Pythagorean theorem, and relationships between angles.

Contents

List of Figures

Kinematics

1

Chapter Summary

The creation of the universe occurred with a "big bang" that started everything in motion. The universe has continued to be in motion since that beginning. Within our universe, the earth we live on is revolving around the sun and rotating on its axis. Thus having something in motion is a very common situation. Because if this, it is important that we know how to understand and describe the properties of motion. Some of the themes of the chapter are represented in Fig. 1.1.

Main Concepts in This Chapter

- Kinematics
- Kinematic Equations of Motion
- Motion in One Dimension
- Motion Graphs
- Motion in Two Dimensions

1.1 Introduction

The first six chapters of this book discuss an area of physics known as classical mechanics. This area of physics helps us understand why things move the way they do in time and space and how to use mathematics to describe this motion. To put this in perspective, think about moving objects you see around you such as cars, runners, footballs, or anything else that interests you. The motion might be

R. C. Powell, *Physics*, Undergraduate Lecture Notes in Physics,
https://doi.org/10.1007/978-3-032-09361-5_1

Fig. 1.1 Races on a straightaway are excellent examples of linear motion. *Credit* R. Henrik Nilsson, Creative Commons Share Alike 4.0

in a straight line (one dimension) like a hundred-yard dash in a track meet or in the shape of an arc (two dimensions) like a golf shot. There is also an enormous amount of motion that you *cannot* see. For example, the molecules which make up the air around you are speeding around the room, colliding into one another and also into you. In fact, the speed at which they move and the violence with which they collide with your body determines whether you feel hot, warm, cool, or cold. Being able to accurately describe and predict the motion of all of these objects is an important knowledge for us to have.

In this chapter we will learn about how to describe the motion of an object. This is called kinematics. In the second chapter we will learn about how forces can be used to change the state of motion of an object. Later we will learn how an object's position and state of motion are related to its energy.

1.2 Fundamental Concepts of Kinematics

To describe the motion of an object, we need to know three things. The first is the change in the object's position. The second is the object's velocity which tells us how fast the object's position is changing. The third is acceleration which tells us how fast the object's velocity is changing. Our view of these things depends on our position as an observer. If you are looking out the window of a moving train and see a car driving down the highway, the motion of the car will appear to be different to you than it appears to be for a person sitting on a bench in the train station. Thus, to make our description of the motion of an object meaningful, we must define the frame of reference we are using to make our observations. A frame of reference includes a position point from which measurements are made and a definition of directions, called a coordinate system. Coordinate directions in three dimensions are generally designated as x, y, and z, and these, along with time, are the dimensions we use to describe problems in kinematics. Note that the units of length in the SI system are meters while time is a scalar quantity with units of seconds.

There are several important concepts to remember about these basic definitions of kinematics. The first deals with the description of the position of an object with respect to a specific reference frame. It is a vector with its tail at the origin of the coordinate system of the reference frame and directional components (x, y, z). However, most motion can be separated into three separate motions, one in each dimension direction. The object's distance is how far away it is which is the magnitude of the position vector. The coordinate system can be oriented so that this vector is in the x-direction. Position is then a vector, $\mathbf{x}$, in one dimension. It can be treated as a scalar distance $\pm x$ where its direction is given by a plus or minus sign with respect to the origin.

Displacement is the amount that an object's position changes. In one dimension, this is designated by $\Delta \mathbf{x}_{\mathrm{d}}$ where the Greek symbol delta is commonly used in physics to designate the change of an entity. Both distance and displacement have units of length such as feet or meters. However, distance is a scalar quantity which means that it only has a magnitude associated with its measurement while displacement is a vector quantity which means that it has both a magnitude and a direction. For motion in one dimension, the directional part of a displacement measurement can be defined as being in a positive or negative direction from the origin. In two and three dimensions an arrow is used to point the direction of the displacement vector. If we are interested in the total distance a moving object travels, we don't care which direction it travels so we simply add all the displacement magnitudes without regard to whether they are in the plus or minus direction. Remember it is possible to have motion resulting in a large distance traveled but with zero displacement if the object ends up at the same place where it started.

The second important concept to remember is the difference between speed and velocity. Velocity is defined as how fast the object is changing position. This is a vector that has both magnitude and direction. Speed is a scalar quantity which tells how fast an object is changing its position but ignores the direction of the change. It is the magnitude of the velocity vector. If the velocity of a moving object changes during the time interval of its displacement, its instantaneous velocity can be defined as its velocity at a specific instant in time while its average velocity is its total displacement divided by the time interval of the trip. The mathematical expression for average velocity v_{a} is

$$\mathbf{v}_{\mathrm{d}} = \frac{\Delta \boldsymbol{x}_{\mathrm{d}}}{\Delta t}\ \mathrm{m/s}. \tag{1.1}$$

The units of velocity are units of length over time. Since the displacement is a vector and the time interval a scaler the average velocity is a vector. In one dimension the direction of the velocity vector is given by a plus or minus sign. A plus sign designates an object moving away from the origin point of the reference frame while a negative sign designates an object moving toward the origin point.

The average speed of a moving object during its trip, v_a, is defined as the distance traveled by the object divided by the time interval of the travel.

$$v_{\mathrm{a}} = \frac{\Delta x_{\mathrm{t}}}{\Delta t}\ \mathrm{m/s}. \quad (1.2)$$

Speed is a scaler quantity since both the distance traveled and the time interval are scalers. We can also define instantaneous speed, v_i, as the speed of an object at any instant in time. This is just the magnitude of the instantaneous velocity of the object at that instant in time.

From the above definitions, it is clear that velocity is independent of the path taken during the motion of an object while speed depends on the path taken. Speed is a scalar that is always positive with no dependence on direction while velocity is a vector that can be positive or negative depending on its direction.

The third important concept to consider is acceleration. Acceleration tells us how quickly the velocity of an object is changing over some interval in time. It is a vector quantity (since velocity is a vector). For motion in one dimension, acceleration can be designated as positive if the vector is pointed away from the origin of the reference frame and as negative if it is pointed toward the origin of the reference frame. For motion in two or three dimensions, acceleration can change the direction that an object is traveling as well as changing its speed.

The mathematical expression for acceleration a is

$$\mathbf{a} = \frac{\Delta \mathbf{v}}{\Delta t}\ \mathrm{m/s^2} \quad (1.3)$$

with the units of meters per second squared in the SI system.

If the velocity vector and the acceleration vector are pointed in the same direction, the moving object is speeding up. If the velocity vector and acceleration vector are pointed in opposite directions, the moving object is slowing down. Sometimes the word deceleration is used for negative acceleration implying an object is slowing down. However, this is ambiguous since it does not account for the situation when the velocity is also in the negative direction, so a negative acceleration means the object is speeding up in the negative direction. Thus, it is best to not use the word deceleration. Remember that an object with zero acceleration is not necessarily at rest. It may have a constant, nonzero velocity.

1.3 Kinematic Equations of Motion

For an object moving with a constant acceleration, its motion can be described mathematically by using what are called equations of motions.

Kinematic Equations of Motion

$v_d = \frac{\Delta x_d}{t}$	Eq. 1.1
$a = \frac{\Delta v}{t}$	Eq. 1.3
$v_f^2 = v_i^2 + 2a\Delta x$	Eq. 1.4
$v_f = v_i + at$	Eq. 1.5
$\Delta x = v_i t + \frac{1}{2}at^2$	Eq. 1.6

The first two of these equations are just the definitions of velocity and acceleration given in the previous section where we have set the initial time to zero on our clock. Since we are using the expression for the average velocity of the motion, we can also express $\mathbf{v}_d$ in terms of the usual way for calculating an average value in terms of the initial and final values,

$$v_d = (v_f + v_i)/2.$$

Substituting this expression into Eq. 1.1 and solving for time gives

$$t = 2\Delta x/(v_f + v_i).$$

Then solving for time in Eq. 1.3 gives

$$t = (v_f - v_i)/a.$$

If we equate these two expressions for the time elapsed in the event and solve for the final velocity, we have the third equation of motion,

$$v_f^2 = v_i^2 + 2a\Delta x. \quad (1.4)$$

The fourth equation of motion is derived from Eq. 1.3

$$a = \frac{\Delta v}{t} = (v_f - v_i)/t.$$

Solving this equation for the final velocity gives

$$v_f = v_i + at. \quad (1.5)$$

To develop the fifth equation of motion we need rely on the graph method of motion description described in your introductory physics course and discussed in detail later in this chapter. For a velocity–time graph of motion with a constant acceleration, the area under the curve is the displacement. With the assumption that the motion starts at time equals zero, the displacement is the sum of the areas of a rectangle (if the initial velocity is not zero) and a triangle

$$\Delta x = v_i t + \frac{1}{2}(v_f - v_i)t = \frac{(v_i + v_f)}{2}t.$$

Substituting Eq. 1.5 for v_f then gives

$$\Delta x = v_i t + \frac{1}{2}at^2 \qquad (1.6)$$

which is the last equation of motion.

These five equations of motion are the tools we need to mathematically describe motion in one dimension with a constant acceleration. They relate the parameters of position, velocity, acceleration, and time. Which equations we need to use for a specific problem depends on which of these parameters we know and which ones we need to determine. It is important to emphasize that they only work in situations when there is a constant acceleration.

1.4 Applications of the Equations of Motion in One Dimension

To demonstrate the use of these kinetic equations, let us consider some examples.

Example 1.1

An object travels in one dimension with a constant acceleration of 0.56 m/s^2 to the west.

If it starts with an initial velocity of 3.2 m/s east and travels for 15.0 s, what are its displacement and final velocity?

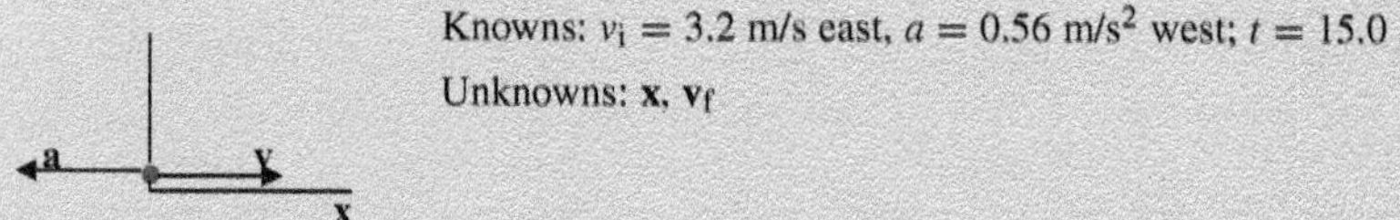

The kinetic equation that relates x, v, a, and t is Eq. 1.6. Now remember, we must be *very careful to consider direction* in this situation. Notice that the initial velocity is directed to the east but the acceleration is directed to the west. That means that the acceleration *opposes* the initial velocity. Thus, the object is initially slowing down. Mathematically, we take these directions into account with positive and negative signs. First, we define our directions with the coordinate system that is shown above. The motion to the west is negative and motion to the east is positive. The definition could be reversed. As long as we *stick to the definition*, which way is positive and which way is negative is irrelevant. Taking this definition into account, we can plug into the equation:

$$\begin{aligned}\Delta x &= v_i t + \frac{1}{2}at^2 \\ &= (3.2\,\text{m/s})(15\,\text{s}) + \left(-0.56\,\text{m/s}^2\right)(15.0\,\text{s})^2/2 \\ &= -15\,\text{m}.\end{aligned}$$

The displacement, then, is − 15 m, which means 15 m to the west. The object, therefore, actually started out moving east, but it ended up west of its initial position. Thus, we know that the object eventually slowed to a halt and then started traveling west. At that point, the acceleration and velocity were pointed in the same direction, so the object began speeding up. That continued, and by the end of the 15.0 s, the object was 15 m west of its initial position.

To find the final velocity, either Eq. 1.4 or Eq. 1.5 would work, since we have all of the variables in each equation except for the final velocity $\mathbf{v}_f$. Using Eq. 1.4,

$$v_f = v_i + at = 3.2\,\text{m/s} + \left(-0.56\,\text{m/s}^2\right)(15.0\,\text{s}) = -5.2\,\text{m/s}.$$

The final velocity, then, is 5.2 m/s to the west. We knew that it had to be pointed to the west, since the displacement told us that in the end, it was moving west. The negative sign on the velocity simply confirms this fact. Remember, since velocity is a vector quantity, the answer must include the direction.

It is also possible to have one-dimensional motion in the vertical direction with constant acceleration. Objects near the surface of the earth experience the same constant, acceleration due to gravity and if this is the only relevant force, they undergo free fall. Whether the object weighs an ounce or a million pounds, the acceleration it experiences due to gravity does not change. In the absence of air resistance, then, all objects fall with exactly the same acceleration. This is called "free fall." Of course, air resistance reduces the acceleration due to gravity. A feather falls much more slowly than a rock not because of gravity being different but because of air resistance being different. A feather experiences the same acceleration due to gravity as does the rock. However, since air resistance affects the feather more strongly than the rock, the feather falls more slowly than the rock. In many cases it is possible to neglect air resistance as being negligible with respect to a specific problem. In these cases, it is possible to treat the object as being in free fall.

The acceleration due to gravity is $g = 9.81\ \mathrm{m/s^2}$ downwards in metric units and $g = 32.2\ \mathrm{ft/s^2}$ downwards in English units. You will be required to know those values from memory. Since the acceleration due to gravity is constant while an object is near the surface of the earth, free fall is a real-world example of motion under constant acceleration, and it is an excellent situation in which to use the kinematic equations of motion.

Example 1.2

A person standing on a bridge holds a ball at a height of 25.0 m above a river. He throws the ball straight up in the air with an initial velocity of 10.0 m/s. Ignoring air resistance, what will the velocity of the ball be when it strikes the water below? What is the maximum height (relative to the river) that the ball reaches in its path? How long does the ball stay in the air before hitting the river?

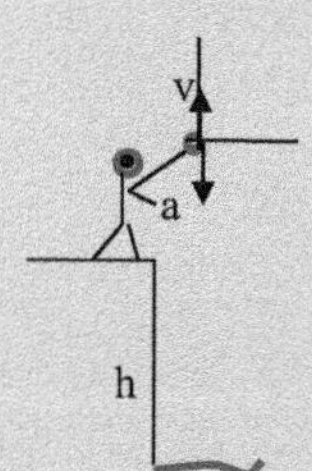

Knowns: $h = 25.0$ m; $v_i = 10$ m/s up; $a = -g$

Unknowns: v_f; x_{max}; t

Note that the choice of the coordinate system as shown gives the positive direction as up and the negative direction as down.

Again, you must be keenly aware of direction and use the proper signs.

You know the acceleration is in the downward direction. Because the object is in free fall, its acceleration is the acceleration due to gravity. The problem wants us to calculate the final velocity, so Eq. 1.4 is the equation to use with the known and unknown parameters we have,

$$\begin{aligned} v_f^2 &= v_i^2 + 2a\Delta x = (10.0\,\mathrm{m/s})^2 + 2(-9.81\,\mathrm{m/s^2})(-25.0\,\mathrm{m}) \\ &= 1.00\times 10^2\,\mathrm{m^2/s^2} + 491\,\mathrm{m^2/s^2} = 491\,\mathrm{m^2/s^2}. \end{aligned}$$

So,

$$v_f = -24.3\,\mathrm{m/s}\ \text{in the downward direction.}$$

The river is 25.0 m *below* the point at which the ball was thrown. Thus, the displacement is $-$ 25.0 m.

Notice two things about the solution. First, we had to report 100 as 1.00×10^2 because we needed three significant figures. The number 100 has only one significant figure, so we had to use scientific notation to write the answer to the proper number of significant figures. Also, notice that the final velocity has a ± sign because of the square root. However, we can reason out which is right. The ball is moving downwards when it hits the water. Since upwards is the positive direction, we know that the ball's velocity must therefore be negative. Thus, the answer is – 24.3 m/s, or 24.3 m/s downwards.

Next we can calculate the maximum height (relative to the river) that the ball reaches in its path. As you learned in your introductory course, the velocity of the ball will be zero when it reaches its maximum height. Thus, we know that the final velocity at that point is zero. We also know the initial velocity and the acceleration, so Eq. 1.4 is the equation to use again but this time the displacement is the unknown. Keeping the direction definitions the same:

$$v_f^2 = v_i^2 + 2a\Delta x = (10.0\,\mathrm{m/s})^2 + 2(-9.81\,\mathrm{m/s^2})(x) = 0$$

So,

$$x = (-1.00 \times 10^2\,\mathrm{m/s^2})/(2)(-9.81\,\mathrm{m/s^2}) = 5.10\,\mathrm{m}.$$

This is 5.10 m up from the thrower's hand. Since the thrower's hand is 25.0 m above the river, the answer is that the maximum height is 30.1 m above the river. Notice here that because we are adding, we do not count significant figures. Instead, we go by precision. Since the least precise number goes out to the tenths place, the answer goes out to the tenths place.

The final question is how long does the ball stay in the air before hitting the river? Because of the variables that we now know, we could use two equations to solve for this. We know the final velocity when it strikes the river (– 24.3 m/s), we know the acceleration (– 9.81 m/s^2), we know the initial velocity (10.0 m/s), and we know the displacement (– 25.0 m). Thus, *either* Eqs. 1.5 or 1.6 will work. Let us use Eq. 1.5. Once again, remember that the signs must be right!

$$v_f = v_i + at - 10.0\,\mathrm{m/sed/} - 9.8$$

So,

$$t = (v_f - v_i)/a = (-24.3\,\mathrm{m/s} - 10.0\,\mathrm{m/s})/(-9.81\,\mathrm{m/s^2})$$
$$= 3.50\,\mathrm{s}.$$

Now you should do the following experiment to demonstrate the use of kinetic equations to describe motion.

Experiment 1.1

Measuring Your Vertical Pitching Speed

Supplies:

- A ball that is easy to throw but not too light (Use a baseball or golf ball or something that is heavy enough so that air resistance is not a factor.)
- A tape measure or meter stick
- A stopwatch
- Someone to help you

Introduction: In order to test the speed at which a baseball pitcher throws, baseball teams usually employ radar guns, which use the Doppler effect to determine the speed of the ball.

However, you can get a pretty good estimate of the speed at which you throw by simply using the kinematics equations in a situation which involves *mostly* one-dimensional motion.

Procedure:

1. Find an area outdoors where you can throw the ball around but not risk breaking things. The area should be level within a circle of at least 1.5 m from where you are standing.
2. Stand the way that you normally would stand when you throw a ball, and hold your arm straight up as high as it will go.
3. Have your helper measure the distance from the center of your palm to the ground. Record that distance.
4. Hold the ball in one hand and the stopwatch in the other. Do not adjust your stance.
5. Now throw the ball *straight up* in the air, extending your arm fully in the release. When the ball leaves your hand, start the stopwatch. When it hits the ground, stop the stopwatch. In order for this experiment to work, the ball must land pretty near where you are standing, telling you that it traveled mostly in one dimension (straight up and straight back down). Ideally, you should have to jump out of the way to avoid being hit. However, as long as the ball lands within 1.5 m of where you are standing, the motion is *mostly* one-dimensional, and the error associated with the slight motion in the other dimension is not large. If the ball landed within 1.5 m of where you are standing, record the time. If not, ignore the time.
6. Repeat step #5 until you have 10 times recorded. Thus, you have 10 trials in which you have measured the time it takes for the ball to travel up and then back down to the ground.
7. Average your 10 times to get the average time it took the ball to travel. The process of averaging, as you should already know, compensates for random experimental errors such as not starting the stopwatch at precisely the time the ball leaves your hands or not stopping the stopwatch at precisely the time that the ball hits the ground.
8. Now you have all of the information you need in order to calculate the initial velocity, which is the velocity at which you can throw a ball straight up in the air. The displacement is the distance measured from the center of the palm to the ground. Remember, however, that the sign is important. Define upwards motion as positive. With that definition, the displacement is negative. In addition, you know the acceleration: $-9.81\ \text{m/s}^2$ or $-32.2\ \text{ft/s}^2$, depending on the units with which you want to work. Finally, you know the time (the average of your 10 trials). If you put those numbers into Eq. 2.5, you have everything except the initial velocity, so you can solve for it.
9. So that the number has some meaning for you, convert it to miles per hour. You can use the fact that 1 m = 0.000621 miles and 1 h = 3600 s. Your answer will probably be somewhere between 15 and 70 miles per

hour. When I did this experiment, my throwing speed was 28 miles per hour.

Now please understand that there are errors associated with this experiment. Most people can throw faster horizontally than vertically, because the mechanics are easier and the whole body can aid in the throw. Also, the fact that the motion is not perfectly one-dimensional adds error as well. Air resistance also plays a very small role. Overall, however, this experiment gives you a good estimate of the motion characteristics.

To demonstrate your ability to use the kinematic equations to describe motion in one dimension, you should work the following Student problems.

Student

1.1 A baseball player throws a ball into the air with an initial velocity of 22.0 m/s upwards. It travels up and then right back down, hitting the ground 4.58 s later.
   a. From what height was the ball thrown?
   b. How long did it take the ball to reach its maximum height?
   c. What was the ball's acceleration at its maximum height?

1.2 An object is given an initial shove which provides it with an unknown initial velocity. Its acceleration is measured to be a constant 1.5 m/s^2. At a time of 3.0 s after the shove, its velocity is measured to be − 12.0 m/s. How fast will it be traveling at a time of 5.0 s after the shove?

1.5 Motion Graphs

In some cases, it is useful to show how an object is moving by constructing a graph of the data describing the motion. We can make measurements of the position of a moving object at several different points and determine its velocity and acceleration. Then we can plot position, velocity, or acceleration on the y-axis of the graph versus time on the x-axis of the graph. (Note that we call the horizontal axis the x-axis but it represents time, not distance. The distance of the object is given by x but plotted along what is conventionally called the y-axis.) If the object is not moving, the graphs are straights lines in the y-direction with the velocity and acceleration being zero and the position being a constant displacement from the origin of the reference frame.

For movement at a constant velocity (called uniform motion) the motion graphs look like the example in Fig. 1.2. Acceleration is zero so it is plotted as a straight line along the x-axis. Velocity is constant so it is plotted as a straight line parallel

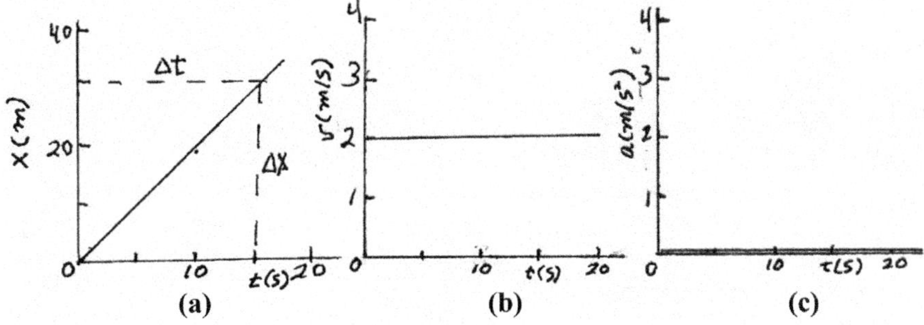

Fig. 1.2 Graphs for motion with a constant velocity

to the x-axis at the value equaling the magnitude of the velocity. The position is a straight line increasing in both the x- and y-directions. The slope of a straight line on a generic x–y graph is given by

$$\text{slope} = \frac{\Delta y}{\Delta x}. \text{ With the parameters for this graph, slope} = \frac{\Delta x}{\Delta t}$$

In this case, the slope is the change in distance divided by the change in time which is the definition of velocity. Remember this is the average velocity of the given time interval. In this case with constant velocity, it is also the instantaneous velocity at any point in this time interval. In Fig. 1.2 the slope of the distance-time graph equals 2 which is the value of the constant velocity in the velocity–time graph. The slope of the line on this graph is zero which is consistent with the value shown in the acceleration-time graph. The area under the velocity–time plot is the product of the constant velocity and the time duration which is the displacement. In this example it is 2 m/s × 20 s = 40 m which is consistent with the position-time graph.

As a different example, consider motion with a constant acceleration that is not zero. The motion graphs are shown in Fig. 1.3. The acceleration-time graph is a horizontal line at the value of the acceleration $a = 2$ m/s^2. The area under the line at $t = 2$ s is the velocity at that time which is 4 m/s. The velocity is a straight line increasing with time on the velocity–time graph. The slope of the velocity–time graph is equal to the acceleration. The position-time graph is not a straight line but a curve whose slope increases with time. As you should have learned from mathematics, the slope of a curve at a given point is defined as the slope of a line tangent to the curve at that point. Thus, at any point in time the tangent to the position-time curve gives the instantaneous velocity at that time.

In the general case when neither velocity nor acceleration is constant over the time interval of interest, a position versus time curve might look something like Fig. 1.4. It is still true that the average velocity of an object can be determined by Eq. 1.1 using the values from this graph and that the slope of the graph at any

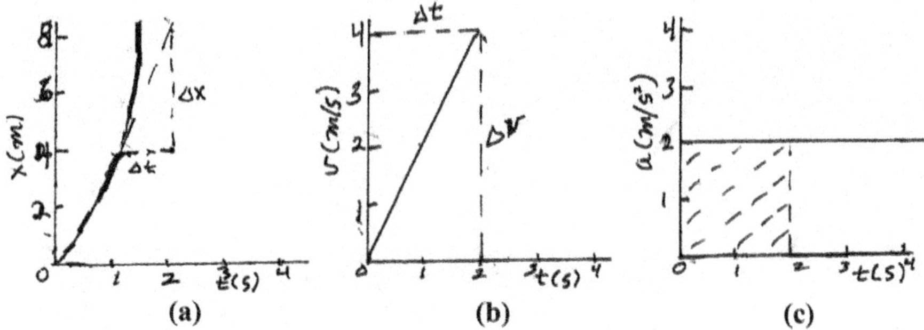

Fig. 1.3 Graphs for motion with a constant acceleration

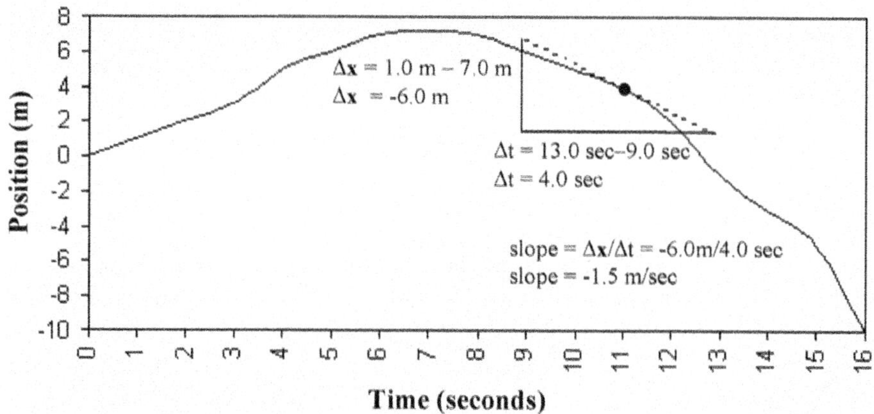

Fig. 1.4 Typical position versus time graph

position in time gives the **instantaneous velocity** of the object. We can *approximate* the instantaneous velocity with Eq. 1.1, as long as we choose a very short time interval.

In Fig. 1.4, the position being positive or negative denotes the direction of displacement from its starting position. The slope of the graph being positive indicates a velocity in the positive direction while a negative slope indicates a velocity in the opposite direction. Every change in slope indicates a change in velocity. When the graph runs horizontal, the velocity is zero. In Fig. 1.4, the velocity is positive from 0 to 7 s. It goes to zero at 7 s and becomes negative from 7 to 16 s.

For example, suppose we wanted to determine the velocity of the object in Fig. 1.4 at a time of 11 s. We could do that by drawing a line that is tangent to the curve at 11 s. The slope of that line would then be the instantaneous velocity of the object at 11 s. The fact that the slope is negative means that the object is traveling in a direction that is opposite to the direction we defined as positive in our coordinate system.

Note that the graph between 0 and 3 s is a straight line. This means that between zero and 3 s, the velocity is constant. Since it does not change during that time interval, the average velocity is also the instantaneous velocity for that period.

The important point is:

When the position versus time curve is a straight line, the velocity is constant and thus here is no difference between instantaneous and average velocity.

To derive a velocity versus time curve from a position versus time curve, it is necessary to determine the slope of the first curve whenever it changes value. If we do this for the graph in Fig. 1.4, the result looks like the graph in Fig. 1.5. It is obvious from Fig. 1.4 that the slope of the graph is constant for the first 3 s and then increases for the next second. After that it becomes less until $t = 7$ s where it goes to zero. From then on it is negative and changes to a greater negative value for the final second. You should make some measurements of the slopes to confirm this.

Now we can switch our discussion to velocity versus time graphs. As you learned in your first-year physics course, the slope of a velocity versus time graph gives us the **acceleration**.

$$a = \frac{\Delta v}{t}.$$

As was the case with velocity, using the data from the graph with this equation gives us the *average* acceleration over a particular time interval. The smaller the time interval, the closer the equation approximates the instantaneous acceleration at a time within that time interval. To get the true instantaneous acceleration at a given time, however, we would have to calculate the slope of a line tangent to the velocity versus time curve at that point. All of the skills that you just learned

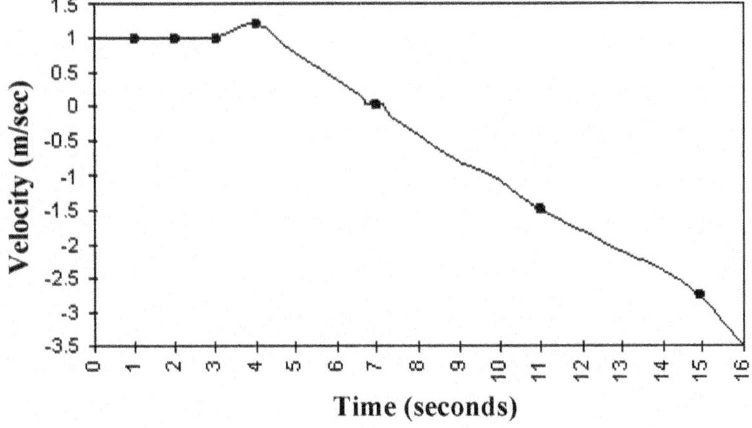

Fig. 1.5 Velocity versus time graph

about analyzing a position versus time curve can be applied to a velocity versus time curve as well. Thus, you could analyze the slope of a velocity versus time curve at several points and sketch the resulting acceleration versus time curve. One important fact is that **the area under an object's velocity versus time curve is the total displacement that object experiences during its motion.**

To demonstrate this, considering the following example.

Example 1.3

The behavior of an object's one-dimensional velocity with respect to time is given below. What is the total displacement of the object over the entire time represented by the graph?

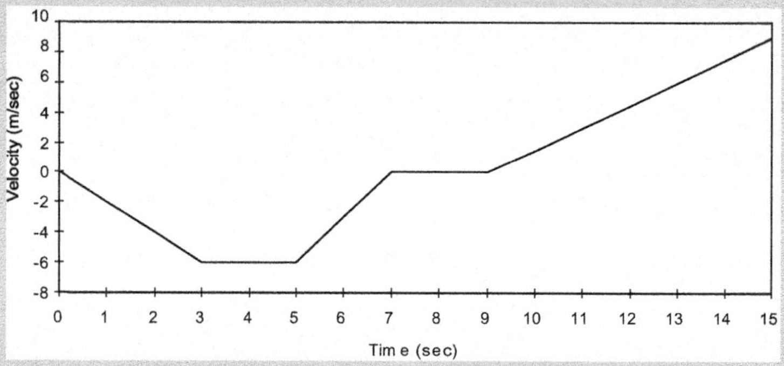

The answer to the question of total displacement is the area between the graph and the $x = 0$ axis. To see this explicitly, we can draw the horizontal line at $v = 0$ and shade the areas of interest.

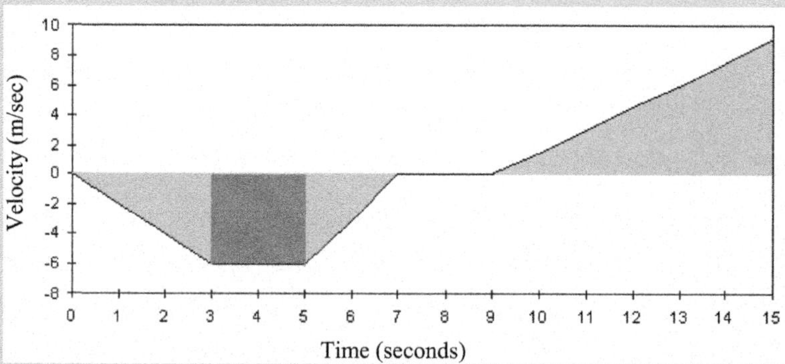

As seen in the figure, we need to calculate the areas of three triangles and one rectangle. Going from left to right, this is

$$\Delta x = (1/2)(3\,s)(-6\,m/s) + (2\,s)(-6\,m/s)$$
$$+ (1/2)(2\,s)(-6\,m/s) + (1/2)(6\,s)(9\,m/s)$$
$$= -9\,m - 12\,m - 6\,m + 27\,m = 0\,m.$$

This shows that the total displacement in this example is 0 m. Thus, the motion of the object represented by the graph starts out in the negative direction and then turns around and goes in the positive direction and ends up exactly where it began. Note that the units come out to be meters as they should for distance.

Now work the following Student problems to be sure you understand how to use the motion graphs.

Student

1.3 A position versus time graph for an object moving in one dimension is given below. Determine the time or times at which the object stops and changes direction.

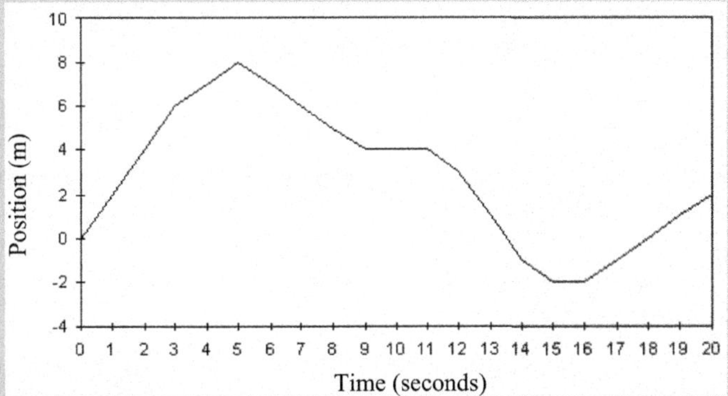

1.4 An object traveling in one dimension has the velocity versus time graph given below. What is the total displacement of the object?

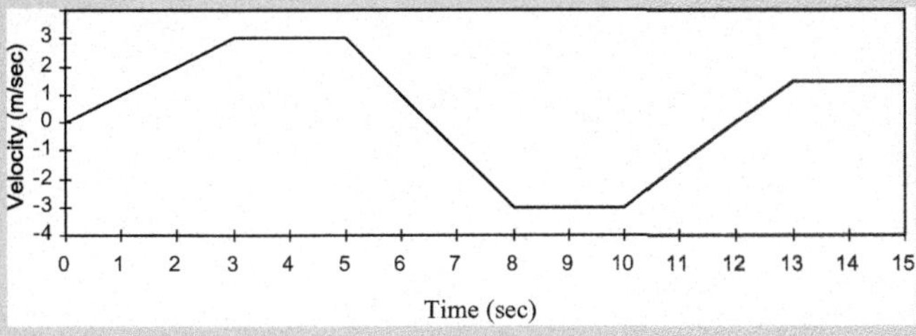

1.6 Kinematics in Two Dimensions

In many cases the motion of an object is not in a straight line and it has to be described in two dimensions. In these cases, the variables we have been dealing with such as displacement, velocity, and acceleration must be treated as vectors instead of scalars. This means that they each have a specific direction associated with them as well as a magnitude and a set of units. All three entities (magnitude, direction, and units) must be specified for an answer to any problem to be correct. Remember, a bold symbol such as **v** indicates a vector quantity.

Just as in one-dimensional motion, motion in two dimensions is relative and a reference frame and reference point must be defined to make the description of the motion meaningful. The reference frame is generally defined with axes in the *x*- and *y*-directions and the reference point at the origin where the axes cross. These directions will have specific meanings for different problems such as directions on a map or horizontal versus vertical. In many cases the best technique for solving a problem is to break it up into two one-dimensional problems, one in the *x*-direction and one in the *y*-direction. After solving these problems as we did in the last section, the *x*- and *y*-components of each variable can be combined to give the final vector quantity. In general, there are two types of problems to consider for motion in two dimensions. The first are navigation problems where the motion is on a flat plane. The second deals with projectiles which involve a combination of horizontal and vertical motion. These are each discussed below.

Let's begin by considering navigation problems.

Example 1.4

A boat is crossing a river that is 100 m wide. If the river runs from east to west with a current of 1 m/s and the boat is crossing from south to north with a constant velocity of 10 m/s in the north direction, how long does it take to cross the river and how far downstream will the boat be when it lands? What is the direction of travel?

The first step in solving this problem is to draw a picture of it. This is shown in below. In addition, the reference frame and reference point are shown in the figure. The *x*-axis is

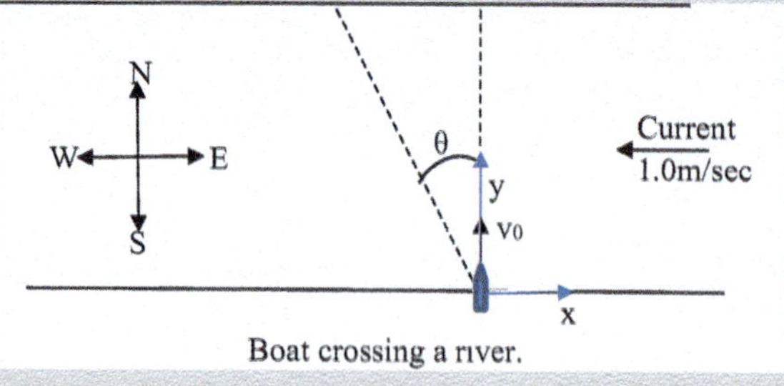

Boat crossing a river.

chosen to point in the direction of east and the y-axis is pointing in the direction north. They cross at the reference point which is located at the starting position of the boat on the south bank of the river. The knowns and unknowns are:

Knowns: $x_0 = y_0 = 0$ m; $y_f = 100$ m; $v_x = -1$ m/s; $v_y = 10$ m/s; $\mathbf{a} = 0$

Unknowns: t; x_f

To calculate the time it takes to cross the 100 m river at a constant velocity of 10 m/s, we use the equation

$$y_f = v_y t \text{ or solving for time } t = y_f/v_y.$$

Substituting in the known quantities gives the time crossing the river as

$$t_f = 100\,\text{m}/10\,\text{m/s} = 10\,\text{s}.$$

Now we can solve for the final x-position of the boat after this amount of time. For this we use the equation

$$x_f = v_x t_f.$$

Substituting the known values of the velocity in the x-direction and the time of travel gives

$$x_f = (-1\,\text{m/s})10\,\text{s} = -10\,\text{m}.$$

In other words, the boat lands on the north side of the river 10 m west of its starting point on the south side of the river.

The direction of travel can be found by forming a right triangle with either the x- and y-components of distance traveled or the x- and y-components of velocity. So

$$\tan\theta = 10\,\text{m}/100\,\text{m} = 0.1.$$

So $\theta = 5.7°$. The vector direction of travel is 5.7° to the west of north.

This example is a good model for how to work problems involving navigation. Note that we worked this as two one-dimensional problems, one in the y-direction and one in the x-direction.

Next let us consider a projectile problem in two dimensions. A projectile is anything that is launched into the air and allowed to travel uninhibited for a certain amount of time. (For these problems, air resistance can be neglected. We will take it into account in the next chapter.) This includes golf balls, baseballs, footballs, cannonballs, etc. The problem can be broken up into its horizontal and vertical components, and each of these can be treated separately. They are related by having the same time-of-travel in both directions. In your introductory physics course, you learned that the vertical component can be treated like a free-fall problem as discussed in the last chapter. The acceleration acting on the projectile is gravity in the downward direction. This causes the projectile to start with an upward velocity, reach a maximum height where its velocity is zero, and go in the downward direction until something stops it. The trajectory that the projectile travels is an arc the shape of a parabola that is symmetric about the point of maximum height. When it returns to its initial height in the vertical direction, it will be traveling with a velocity having the same magnitude as its initial velocity but its direction rotated 90° in the downward direction. The horizontal can be treated like motion in one dimension with a constant velocity. As always, the motion is relative so to describe it accurately it is necessary to define a point of reference. This is best demonstrated by working through an example.

Example 1.5
A ball launched at an angle of 66° above the horizontal at an initial velocity of 4.5 m/s. Find the maximum height of the ball's trajectory, the distance it travels before it hits the ground, and the time it takes to hit the ground.

As always, the first steps are to sketch the problem, chose a coordinate system with a reference point, and list the knowns and unknowns. The sketch of the problem with the coordinate system is shown in the figure below:

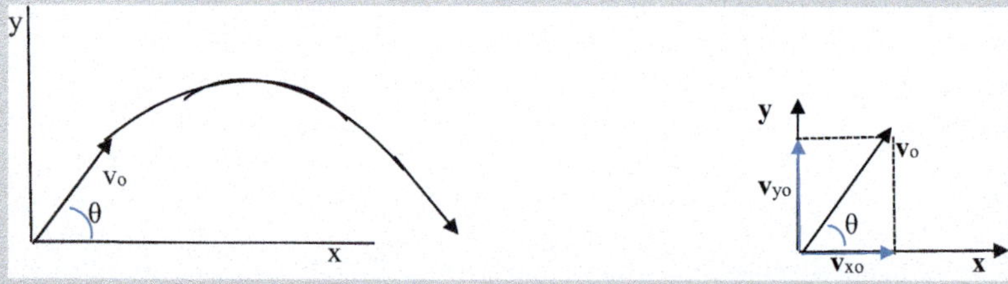

Knowns: $x_o = y_o = 0$ m; $\theta = 66°$; $v_0 = 4.5$ m/s; $a_y = -\,g$; $a_x = 0$

Unknowns: y_{max}; t_f; x_{max}

The first step is to find the x- and y-components of velocity. This is done by using trigonometry of right triangles:

$$v_{0y} = v_0 \sin 66^\circ = 4.5(0.9135) = 4.1\,\mathrm{m/s},$$
$$v_{0x} = v_0 \cos 66^\circ = 4.5(0.467) = 2.1\,\mathrm{m/s}.$$

Next we use the kinetic equation of motion for velocity in the vertical direction and solve it for time:

$$v_y = v_{0y} + a_y t = v_{0y} - gt \quad \text{or} \quad t = \left(v_{0y} - v_y\right)/g.$$

At the position of maximum height, the velocity in the y-direction is zero so

$$t_{\max} = v_{0y}/g.$$

The acceleration due to gravity in the downward direction has compensated for the initial impetus given the ball in the upward direction at launching. Using this fact in the kinetic equation of motion for position in the vertical direction gives:

$$\begin{aligned} y_{\max} &= y_0 + v_{y0} t_{\max} + a t_{\max}^2/2 \\ &= 0 + (4.1\,\mathrm{m/s})(4.1\,\mathrm{m/s})/g - (g/2)\left[(4.1\,\mathrm{m/s})/g\right]^2 \\ &= 4.1(4.1/9.81) - (9.81/2)[4.1/9.81]^2 = 0.86\,\mathrm{m}. \end{aligned}$$

So the ball gets to a height of 0.86 m before it falls back to the ground.

Now we can find how long it takes to hit the ground after launch by using the fact that this will just be twice the time to reach its maximum height:

$$t_{\mathrm{total}} = 2t_{\max} = 2v_{0y}/g = 2(4.1\,\mathrm{m/s})/\left(9.81\,\mathrm{m/s^2}\right) = 0.83\,\mathrm{s}.$$

This ball is in the air slightly less than a second.

Finally, we can salve for the distance the ball travels in the horizontal direction using the kinetic equation of motion for distance traveled with constant velocity:

$$x_{\max} = v_{x0} t_{\mathrm{total}} = (2.1\,\mathrm{m/s})(0.83\,\mathrm{s}) = 1.74\,\mathrm{m}.$$

Note that once again we have solved a two-dimensional motion problem by breaking it up into two one-dimensional problems so we can use our kinetic equations.

You should remember from your introductory physics course that there is a special case for projectile problems when the end of the projectile's travel is at the same height as its starting height. The total distance traveled in this case is called the range of the projectile. An expression for the range of a projectile can be found by combining the last two equations of the previous example problem with the expressions for the *x*- and *y*-components of the initial velocity:

$$R = x_{\max} = v_{x0}t_{\text{total}} = 2v_{x0}v_{0y}/g = 2v_o^2 \sin\theta \cos\theta/g$$

So,

$$R = \frac{v_o^2 \sin(2\theta)}{g}. \tag{1.7}$$

In the final step of this derivation we used the trig identity: 2 sin θ cos θ = sin 2θ. This can be a very helpful expression to use when it is applicable. To illustrate this, work through the following example.

Example 1.6

A ship is firing its guns (v_o = 750.0 ft/s) at another ship which has run out of ammunition. To avoid the guns, the other ship is trying to hide. It gets behind an island that has a large mountain (2000.0 ft high) on it. That way, the mountain is between the two ships. The ship that is firing is 2600.0 ft from the tip of the mountain. How far from the tip of the mountain must the hiding ship stray before the firing ship can hit it? The firing ship can fire its guns at any angle.

Begin by sketching the problem and listing the known and unknown parameters.

Knowns: v_o = 750.0 ft/s; h = 2000.0 ft; x_0 = 2600.0 ft

Unknowns: x_1

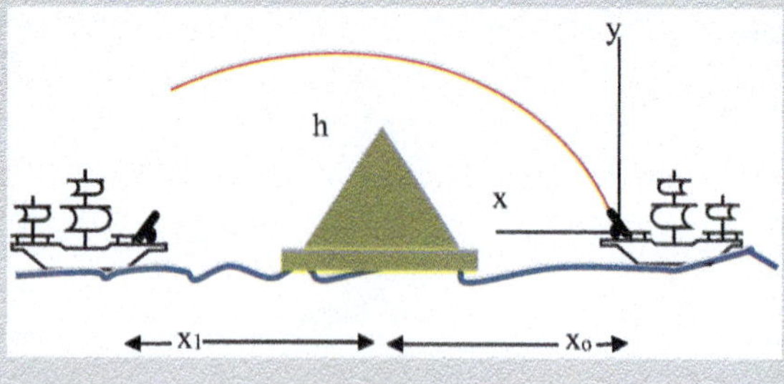

In order for the projectile to get over the mountain, it must be at a height of at least 2000.0 ft after it goes a distance of 2600.0 ft. To calculate the time to reach its maximum height we can use Eq. 1.6 for the horizontal direction:

$$\Delta x = v_{ix}t + \frac{1}{2}at^2 = 2600.0\,\text{ft} = (750.0\,\text{ft/s})(\cos\theta)t + 0$$

So,

$$t = 3.467/\cos\theta\ \text{s}.$$

Now we can use this time in the vertical direction of motion for the time it takes to reach 2000.0 ft.

Again using Eq. 1.6,

$$\begin{aligned}\Delta y = v_{iy}t + \frac{1}{2}at^2 &= 2000.0\,\text{ft} \\ &= (750.0\,\text{ft/s})(\sin\theta)(3.467/\cos\theta)\ \text{s} \\ &\quad + \frac{1}{2}(-32.2\,\text{ft/s}^2)(3.467/\cos\theta)^2\ \text{s}^2\end{aligned}$$

So,

$$2000\,\text{ft} = \left(2.600 \times 10^3\,\text{ft}\right)\tan\theta - (194\,\text{ft})\sec^2\theta,$$

where we have used two well-known trig identities. We can simplify this further using the relationship:

$$\sec^2\theta = 1 + \tan^2\theta.$$

This results in a quadratic equation with tan θ as the variable

$$(194\,\text{ft})\tan^2\theta - \left(2.600 \times 10^3\,\text{ft}\right)\tan\theta + 2194\,\text{ft} = 0.$$

Using the quadratic formula to solve this for tan θ gives

$$\tan\theta = \frac{2.600 \times 10^3\,\text{ft} \pm \sqrt{\left(2.600 \times 10^3\,\text{ft}\right)^2 - 4(194\,\text{ft})(2194\,\text{ft})}}{2(194\,\text{ft})}$$

The plus or minus sign gives us two possible answers for the launch angle,

$$\tan\theta = 351/388 \text{ so } \theta = 42.1^\circ \text{ or } \tan\theta = 4849/388 \text{ so } \theta = 85.4^\circ.$$

These represent the minimum and maximum launch angles for getting the projectile over the mountain. The maximum angle will result in the projectile landing closest to the mountain, so this is the angle we need to use in answering the question. We can put this into the range equation to determine the shortest distance the projectile will travel if it just clears the mountain,

$$R = \frac{v_o^2 \sin(2\theta)}{g} = (750.0\,\text{ft/s})^2 \sin\left(2 \times 85.4°\right)/\left(32.2\,\text{ft/s}^2\right) = 2.80 \times 10^3\,\text{ft}.$$

Since the question is how far from the mountain must the ship be to not be hit, we must subtract the distance the first ship is from the mountain from the total range of the projectile. This gives

$$x_1 = R - x_o = 2800\,\text{ft} - 2600\,\text{ft} = 200\,\text{ft}.$$

In other words, the target ship must stay closer to the mountain than 200 ft in order not to be hit by the projectile.

If you were able to follow the above example in detail, you should be able to work these next two Student problems.

Student

1.5 A person is shooting cannonballs at a castle. The cannon is 150.0 m from the castle wall, and the wall is 45.0 m high. It can shoot its projectiles with a speed of 65.0 m/s, and it is currently aimed at an angle of 25.0°. Will the projectiles pass over the wall? If not, how high will they be when they hit the wall?

1.6 For the situation above, what range of firing angles must be used in order to get the projectiles over the wall?

1.7 Any projectile following a parabolic path will pass through every height (except for its maximum height) twice. Develop an equation that gives the time in between the first time it reaches any height, h, and the second time it reaches the same height. The equation should be given in terms of h, the initial speed (v_o), the launch angle (θ), and gravitation acceleration (g).

1.8 Consider a golf hole with an elevated tee hitting to a hole on a green 50 m below it and 100 m distant. If the initial velocity of the ball is 50m/sec only in the horizontal direction, will the ball land near the hole?

Special Topic

The range Eq. 1.7 gives us two possible solutions for sending a projectile from its origin to its target. We can zing it low and fast or lob it high and slow. These choices are shown in the following figure. For example, we can start it with a launch angle of 30° and a speed of 33.7 m/s or with a launch angle of 45° at a speed of 31.3 m/s. Both of these will hit a target 100 m distant. The first one will reach a maximum height of 14.5 m in 1.72 s while the second one will reach a maximum height of 24.9 m in 2.4 s. Thus an artillery gunner with a cannon or a quarterback throwing a pass must decide if they need a high trajectory to get over a defensive obstacle or a low trajectory to deliver the projectile quickly.

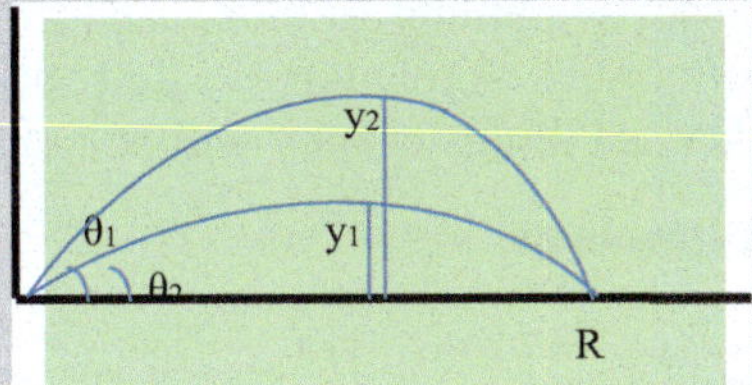

Next Level

In real-life situations, the time dependence of position may be expressed by a complicated mathematical equation. To determine the velocity at any instant in time requires the use of calculus. A simple example of this is position expressed as a polynomial like

$$x(t) = 5.0t + 0.2t^3$$

Velocity at any time t is given by $v(t) = \frac{dx(t)}{dt}$ where the right side of the equation is called the differential of x with respect to t. In introductory calculus you will learn the rule for taking the differential of a polynomial: if $x(t) = At^n$ then the differential is $\frac{dx(t)}{dt} = nAt^{n-1}$. Applying this rule to our example expression for $x(t)$ gives

$$v(t) = 5.0 + 0.6t^2$$

Now that we have an expression for $v(t)$ we can determine the instantaneous velocity at any time. For example, at a time of 4 s, $v(4\ \text{s}) = 14.6$ m/s.

In your calculus class you will learn the rules for taking the differentials of other types of mathematical expressions so you can follow this procedure for any mathematical relationship between position and time. The same procedure can be followed for determining instantaneous acceleration since $a(t) = \frac{dv(t)}{dt}$. In the example above, $a(t) = 1.2t$ so $a(4\ \text{s}) = 4.8\ \text{m/s}^2$.

Summing Up

In this chapter you learned how to describe motion using the Equations of Motion and motion graphs. In future courses you may learn how to extend these concepts using the more sophisticated mathematics of calculus. Being able to describe motion is critically important to understanding the creation we live in as demonstrated by our discussion of projectiles and navigation. In the next chapter we will learn about what causes something to move, and in future chapters we will study special types of motion such as rotational motion, periodic motion, and wave motion. What you have learned in this chapter will be used in these future studies.

Answers to the Student Problems

1.1

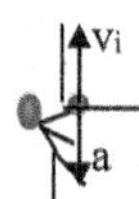

Knowns: $v_i = 22.0$ m/s; $a = -9.81$ m/s^2; $t = 4.58$ s

Unknowns: h; t_{max}; a_{max}

As indicated by the coordinate system in the figure, the upward direction is positive and downward is negative.

a. In this part, we need to calculate the displacement. The only equation which relates displacement to the quantities we have is 1.6.

$$\Delta x = v_i t + \frac{1}{2}at^2 = (22.0\,\text{m/s})(4.58\,\text{s}) + (-9.81\,\text{m/s}^2)(4.58\,\text{s})^2/2$$
$$= 101\,\text{m} - 103\,\text{m} = -2\,\text{m}$$

We can have only one significant figure in the end because the final thing we do is subtract the two numbers. Since they both go out to the ones place, the answer can go only to the ones place.

Notice that the displacement is negative. That means the ground is 2 m *down* from the initial starting place. Thus, the initial height of the ball is 2 m above the ground.

b. At its maximum height, the velocity of the ball is zero. We know its initial velocity, and we want the time, so the easiest equation to use is Eq. 1.5.

$$v_f = v_i + at = 0\,\text{m/s} = 22.0\,\text{m/s} - \left(9.81\,\text{m/s}^2\right)t$$

So,

$$t = (-22.0\,\text{m/s})/\left(-9.81\,\text{m/s}^2\right) = 2.24\,\text{s}.$$

Notice that we do not worry about the significant figures associated with the final velocity. That's because the final velocity is *exactly* 0 m/s, since it is stopped.

c. When the ball reaches its maximum height, its velocity is zero. However, the acceleration is a constant $-$ 9.81 m/s^2 throughout its motion, even when its velocity is zero.

1.2

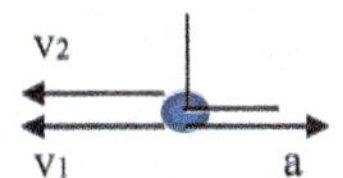

Knowns: $a = 1.5$ m/s^2; $t_1 = 3.0$ s; $v_1 = -$ 12.0 m/s; $t_2 = 5.0$ s

Unknowns: v_2

The known parameters tell us that the object is slowing down because the acceleration opposes the velocity. We need to know the velocity at 5.0 s and we know that at $t = 3.0$ s it is traveling at $-$ 12.0 m/s. Let's just assume the problems *starts* at $t = 3.0$ s. Then the *initial velocity* is $-$ 12.0 m/s. We need to know the velocity at 5.0 s. That's 2.0 s after the time that we know the velocity. Thus, in the situation we are defining, the time is 2.0 s. Now we can use Eq. 1.5,

$$v_f = v_i + at = -\,12.0\,\text{m/s} + \left(1.5\,\text{m/s}^2\right)(2.0\,\text{s}) = -9.0\,\text{m/s}$$

Now, of course, the way we redefined things would not have worked had we wanted to know something about the situation *prior to* $t = 3.0$ s. However, we did not need to know anything about that time frame, so our redefinition worked fine.

1.3 The object stops when the slope of the curve is zero. That happens in three places on the graph: at $t = 5$ s; between 9 and 11 s; and between 15 and 16 s. Although the velocity is zero in three regions of the graph, the question asked at what times did the object stop *and change direction*. That happens at 5 s, because the slope goes from positive before 5 s to negative after 5 s. However, between 9 and 11 s, the velocity is zero, but the slope before 9 s is negative as is the slope after 11 s. Thus, the object did not change directions. Finally, from 15 to 16 s, the velocity is zero and the slope changes from negative prior to 15 s to positive after 16 s. In the end, then, the object stopped and changed directions twice: once at 5 s and once from 15 to 16 s.

1.4

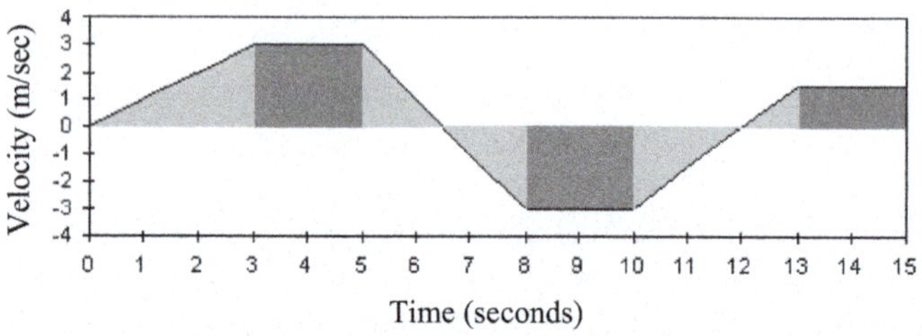

To determine the area under the curve, we must split the graph into regions for which we can calculate the area. In this graph, then, we have 5 triangles (areas 1, 3, 4, 6, and 7) and 3 rectangles (areas 2, 5, and 8). That's a lot of areas to figure, but it's not really difficult:

Area of region #1 (a triangle) = (1/2)(3.0 s)(3.0 m/s) = 4.5 m.
Area of region #2 (a rectangle) = (2.0 s)(3.0 m/s) = 6.0 m.
Area of region #3 (a triangle) = (1/2)(1.5 s)(3.0 m/s) = 2.3 m.
Area of region #4 (a triangle) = (1/2)(1.5 s)(− 3.0 m/s) = − 2.3 m.
Area of region #5 (a rectangle) = (2.0 s)(− 3.0 m/s) = − 6.0 m.
Area of region #6 (a triangle) = (1/2)(2.0 s)(− 3.0 m/s) = − 3.0 m.
Area of region #7 (a triangle) = (1/2)(1.0 s)(1.5 m/s) = 0.75 m.
Area of region #8 (a rectangle) = (2.0 s)(1.5 m/s) = 3.0 m.

The total displacement, then, is simply the sum of the areas:
Total displacement = 4.5 m + 6.0 m + 2.3 m− 2.3 m− 6.0 m− 3.0 m + 0.75 m + 3.0 m = 5.3.

1.5 To get over the wall, the projectile must be at a height of 45.0 m once it travels 150.0 m from the cannon as shown in the sketch. Thus, we need to determine the height of the projectile when it is that far from the cannon.

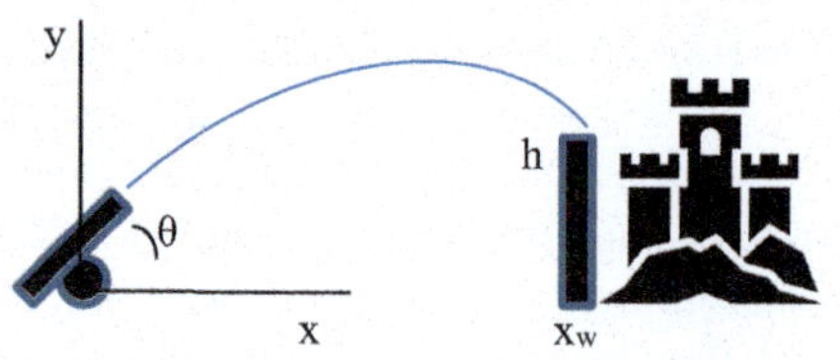

Knowns: $x_w = 150.0$ m; $h = 45$ m; $v_o = 65.0$ m/s; $\theta = 25.0°$

Unknowns: y_w

We start in the horizontal dimension to get the time for the projectile to travel to the wall.

$$\mathbf{x_w} = \mathbf{v_{ox}}t + \frac{1}{2}\mathbf{a}\mathbf{t}^2$$

Since the acceleration in the x-direction is zero, and the x-component of the initial velocity is $v_o \cos \theta$, the time is given by

$$t = x_w/v_o \cos\theta = 150.0\,\text{m}/(65.0\,\text{m/s})\cos 25.0^\circ = 2.55\,\text{s}$$

Next we work in the y-direction and ask how high the projectile will be after traveling for 2.55 s. Using the same kinetic equation in the y-direction gives

$$\mathbf{y_w} = \mathbf{v_{oy}}t + \frac{1}{2}\mathbf{a}\mathbf{t}^2$$

In this direction the acceleration is $-g$ and the initial component of velocity is $v_o \sin \theta$. Knowing time we can solve for the height at the wall

$$\begin{aligned} y_w &= (v_o \sin\theta)t - \frac{1}{2}gt^2 \\ &= (65.0\,\text{m/s})(0.42)(2.55\,\text{s}) - \left(9.81\,\text{m/s}^2\right)(2.55\,\text{s})^2 \\ &= 38.2\,\text{m} \end{aligned}$$

Since $y_w < h$ will not make it over the wall. It will hit 38.2 m up the 45.0 m wall.

1.6 The sketch is the same as the last problem but the knowns and unknowns are different.

Knowns: $x_w = 150.0$ m; $h = 45$ m; $v_o = 65.0$ m/s; $y_w = h$

Unknowns: θ

We know that the height must be 45.0 m when the projectile is 150.0 m from the cannon. Using the expression we derived for the time it takes for the projectile 150.0 m in the x-direction,

$$t = x_w/v_o \cos\theta = 150.0\,\text{m}/(65.0\,\text{m/s})\cos\theta = 2.31/\cos\theta$$

At that time, the projectile *must* be at least 45.0 m high. Thus, we can take that time and put it into the equation for the vertical dimension we derived in the previous problem

$$y_{\mathrm{w}} = (v_{\mathrm{o}} \sin\theta)t - gt^2 = 45.0\,\mathrm{m}$$
$$= (65.0\,\mathrm{m/s})\sin\theta(2.31\,\mathrm{s})/\cos\theta$$
$$-\left(9.81\,\mathrm{m/s^2}\right)(2.31\,\mathrm{s}/\cos\theta)^2$$

Now we can use the trig identities tan θ = sin θ/cos θ and $1/\cos^2\theta = 1 + \tan^2\theta$ to generate a quadratic equation for tan θ,

$$(26.2\,\mathrm{m})\tan^2\theta - \left(1.50\times10^2\,\mathrm{m}\right)\tan\theta + 71.2\,\mathrm{m} = 0.$$

Using the quadratic formula gives

$$\tan\theta = \frac{1.50\times10^2\,\mathrm{m} \pm \sqrt{\left(1.50\times10^2\,\mathrm{m}\right) - 4(26.2\,\mathrm{m})(71.2\,\mathrm{m})}}{2(26.2\,\mathrm{m})}$$
$$= \frac{1.50\times10^2\,\mathrm{m} \pm 123\,\mathrm{m}}{52.4\,\mathrm{m}}$$

So the two possible answers are $\theta = 27°$ and $\theta = 79.1°$.

The cannon must be aimed between these two angles to get the projectiles over the wall.

1.7

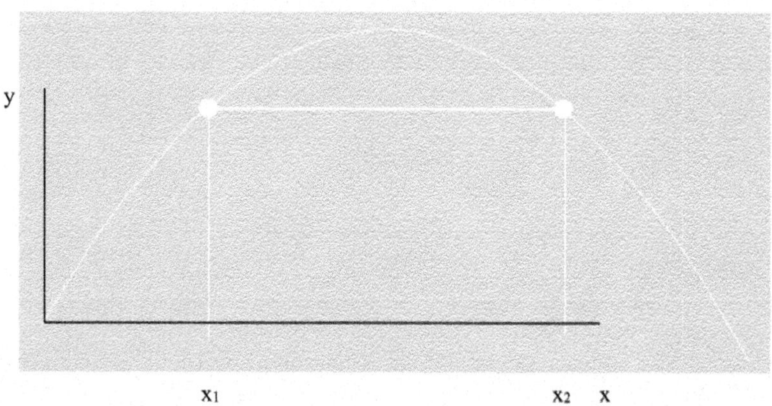

Knowns: v_{o}; θ; h; x_1; x_2

Unknowns: t_{h}

For this problem, we need to work only in the vertical dimension where the initial velocity is v_{o} sin θ, the height is h, and the acceleration is $-\,g$.

Thus, the time can be solved for as follows:

$$\mathbf{y_h} = \mathbf{v_{oy}} t_h + \frac{1}{2}\mathbf{a} t_h^2 = h = v_o \cdot \sin\theta t_h - \frac{1}{2} g t_h^2$$

This is a quadratic equation in t_h,

$$\frac{1}{2} g t_h^2 - v_o \cdot \sin\theta t_h + h = 0.$$

We can solve this using the quadratic formula

$$t_h = \frac{v_o \sin\theta \pm \sqrt{(v_o \sin\theta)^2 - 4\frac{1}{2}gh}}{2\left(\frac{1}{2}g\right)} = \frac{v_o \sin\theta \pm \sqrt{(v_o \sin\theta)^2 - 2gh}}{g}.$$

This expression gives the two times the projectile is at a height h during its travel. We want to calculate the *difference* in these two times.

$$\Delta t_h = \frac{2\sqrt{(v_o \sin\theta)^2 - 2gh}}{g}$$

This formula uses only v_o, θ, g, and h, so it is the equation we wanted.

1.8

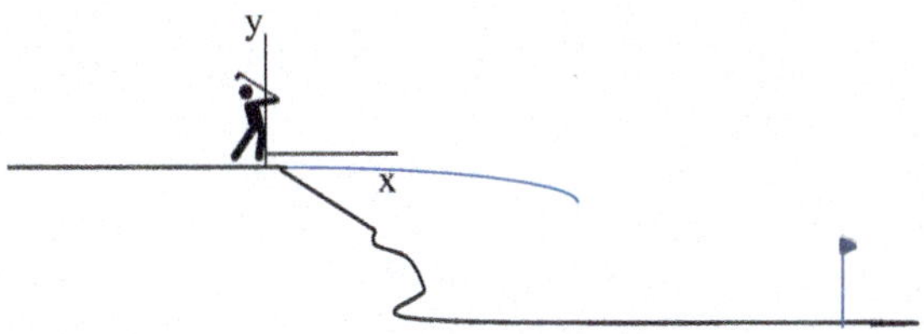

The sketch for this problem is shown above. The lists of knowns and unknowns are:

Knowns: $v_0 = 50$ m/s $= v_{0x}$; $v_{0y} = 0$; $\theta_0 = 0°$; $a_x = 0$; $a_y = g$; $y_0 = x_0 = 0$ m; $y_f = -50$ m; tee to hole distance is $x_h = 100$ m

Unknowns: x_f

First use the kinetic equation of motion for the vertical direction to get the time of flight:

$$y_f = v_{y0}t + at^2/2$$
$$-\ 50\,\mathrm{m/s} = 0 - 9.8\,\mathrm{m/s^2}\,t^2/2$$
$$t^2 = 2(50)/9.8 \text{ or } t = 3.194\,\mathrm{s}.$$

Using this time of flight in the kinetic equation of motion for distance traveled in the horizontal direction gives:

$$x_f = v_{x0}t + at^2/2 = (50\,\mathrm{m/s})(3.194\,\mathrm{s}) + 0 = 159.7\,\mathrm{m}.$$

Thus the ball overshoots the hole by 159.7 – 100 = 59.7 m. Note that this result is quite different from the one you would get if you incorrectly used the range equation.

Study Guide for This Chapter

1. At a certain time, an object's velocity vector is given by the following arrow:

   One second later, the velocity vector is given by this arrow:

   Draw the vector which represents the object's average acceleration during that one second.
2. A student observes several instances of motion with constant acceleration and then makes the following statement: "If the acceleration of an object is constant, it must travel in a straight line." Is the statement true? Why or why not?
3. Under what specific condition of acceleration will instantaneous velocity and average velocity be the same at all times?
4. One student throws a ball up into the air and determines the time it takes for the ball to hit the ground. A second student, at the same instant, drops a ball and times how long it takes for his ball to hit the ground. Compare the times measured by the two students. Compare the acceleration of the two balls.
5. Consider a projectile launched at an angle of 45° relative to the ground. Describe the relative directions of the acceleration vector and velocity vector in both the horizontal and vertical dimensions over the entire path of the projectile, assuming that it eventually lands on the ground again.
6. A projectile is launched from level ground with a velocity whose magnitude is v and whose direction is θ. What are the magnitude and direction of the velocity when the projectile reaches the ground again, providing that it hits at the same height as the height from which it was launched?

7. Which of the following would follow a parabolic trajectory:
   a. A baseball thrown horizontally from the roof of a building.
   b. An airplane flying from one city to another.
   c. A rocket leaving the launchpad, firing its engines.
   d. A rocket after the engines have burned out but still under the influence of earth's gravity.
8. A student throws a ball horizontally at 30.0 m/s. A second student drops a ball from exactly the same height at the same instant. Compare the time at which the balls hit the ground.
9. An object travels in one dimension with the following position versus time graph.
   a. Over what time interval is the object moving with the greatest speed?
   b. At what times is the object stopped?
   c. At what times does the object change directions?
   d. Sketch the velocity versus time graph for the same time interval.

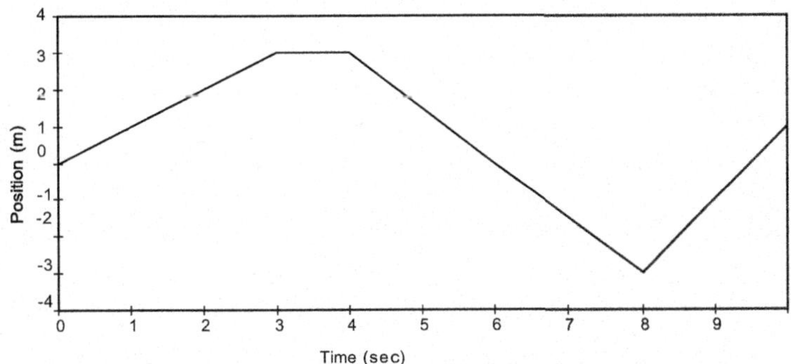

10. An object travels in one dimension according to the following graph.
    a. During which time intervals is the object speeding up?
    b. At what times is the object at rest?
    c. Sketch the acceleration versus time graph.

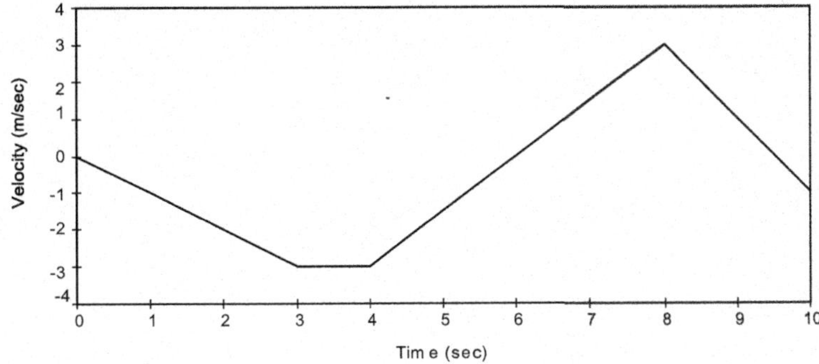

11. A rock is dropped off of a cliff. It takes 1.5 s for it to fall half the distance to the ground. What is the total time it will take to fall from the top of the cliff to the ground? Ignore air resistance.
12. Starting from rest, an object travels with an acceleration of 2.0 m/s^2 for 25 s and, at the end of that time period, it begins traveling with an acceleration of – 2.0 m/s^2 for another 25 s.
    a. What is the final velocity?
    b. What is the maximum speed attained during the trip?
    c. What is the total displacement over the entire trip?
13. Two bikers start at the same point and travel in opposite directions at constant speed. The speed of the first biker is twice that of the second biker. At the end of 30.0 min, they are 9401 m apart from one another. What is the speed of each biker?
14. A boy stands on a ladder and throws a ball straight up into the air with an initial velocity of 5.0 ft/s. If the boy releases the ball 15.0 ft above the ground and it falls straight back down, how fast will it be going when it impacts the ground? Ignore air resistance.
15. A gun has a maximum range of 500.0 m on level ground. If the gun is fired at an angle of 30.0° relative to the ground, what is its range?
16. A cannon shoots a projectile horizontally off the edge of a cliff. It is aiming at a ship that sits on the ocean, 500.0 m below and 2500.0 m west of the cliff. At what speed must the projectile be fired in order to hit the ship?
17. Suppose in problem number 8 that the cannon can only shoot its projectiles with one speed: 300.0 m/s. At what angle must the cannon aim to hit the ship?
18. A football player just scored a touchdown and wants to throw the ball to a fan that's wearing his jersey. The fan is 300.0 ft (horizontally) from the player, but he is also up in the stands, which are 50.0 ft above the point where the football player will release the ball. If the football player aims the ball at 30.0° relative to the horizontal, at what speed will he have to release the ball so that the ball hits his fan?

Next Level

19. The position of an object as a function of time is given by the polynomial:

$$x(t) = 2.0 + 10t - 3.1t^2$$

What is the equation for the velocity and what is the value of the velocity at 5 s?

What is the equation for acceleration and what is the value of acceleration at 5 s?

20. Consider the position versus time graph shown in problem 9 above. According to the graph, what is the average velocity during the first segment of the trip? What is the polynomial expression that describes position as a function of time during this segment? Using simple calculus, what is the velocity during this segment of the trip and how does it compare with your number from the graph? What is the acceleration during this segment of the trip?

2 Dynamics of Motion

Chapter Summary

This chapter deals with forces that can change an object's state of motion. There are six simple machines that allow us to change the magnitude or direction of forces. These are: a lever; a wheel; a pulley; a screw, an inclined plane; and a wedge. All of these are important in how we control the motion of moving objects. Some of the themes of the chapter are represented in Figs. 2.1, 2.3 and 2.4.

Main Ideas for This Chapter

- Newton's Laws
- Equilibrium
- Natural Forces—Gravity, Friction, Air Resistance
- How to apply these concepts to the world around us

2.1 Introduction

In the previous chapter, we have been discussing the way to describe how something moves. Now we will learn about why something moves the way it does. The additional concept we need in order to do this is force. Force is an interaction that can change the motion of an object in either speed or direction. It can be either a push or a pull and is a vector quantity with magnitude and direction. In the SI unit system the unit of force is a Newton with dimensions of kg m/s^2. In general, there are two kinds of forces: contact forces and field forces. Contact forces require that the entity initiating the force touch the object to which it is applying the force. Field forces act at a distance and do not require the objects to touch.

R. C. Powell, *Physics*, Undergraduate Lecture Notes in Physics,
https://doi.org/10.1007/978-3-032-09361-5_2

An example of this is the force of gravity. The earth exerts a gravitational force on the moon but does not touch it.

Much of what we know about forces and how they affect the motion of an object dates back to the work of Sir Isaac Newton in the 1600s. As a result of studying God's Creation, Newton published a monumental work entitled *The Mathematical Principles of Natural Philosophy*. Later on, it became known simply as *The Principia*. Most historians consider the publication of this work in the year 1687 as the beginning of the scientific discipline we call physics. In *The Principia*, Newton discussed his views on dynamics. He laid out three basic laws of motion and discussed the consequences of those laws. These laws contradicted some long-held beliefs in the scientific community, but Newton used rigorous experiments to provide evidence for his laws. The remainder of *The Principia* is divided into three books in which Newton applied his laws of motion to several situations. In the first book, he concentrated on how his laws of motion affected the motion of bodies in different situations. The second book covered the motion of bodies in fluids. That book also explored the behavior of the fluids themselves. Finally, the third book dealt with the motion of the planets, moons, and comets in the solar system. In that book, he developed his universal law of gravitation, which we will study in a future chapter.

The genius of Newton is best demonstrated by the fact that his laws still form the basis of most of the physics that we study today. Certainly there is more to modern physics than Newton's Laws. For example, the foundation of modern physics is a theoretical framework known as quantum mechanics, which we will discuss in the final two chapters of this course. Quantum mechanics has been successful in describing most of the features of the atomic and subatomic world, which are situations in which Newton's Laws do not work. However, when you move out of the atomic and subatomic worlds and observe the macroscopic world, you find out that Newton's Laws work very well.

2.2 Newton's Laws

The important concepts of dynamics are summarized in three laws of motion articulated by Sir Isaac Newton:

Newton's Laws

Newton's First Law: A body at rest will remain at rest unless acted on by an external force. A body in motion will continue that motion unless acted on by an external force.

Newton's Second Law: The sum of all the forces acting on an object equals the object's mass times its acceleration.

Fig. 2.1 A satellite in orbit around the earth. *Credit* NASA, public domain

Newton's Third Law: For every action there is an equal and opposite reaction.

Newton's First Law is sometimes called the law of inertia since inertia is defined as the resistance of an object to changing its state of motion. You experience this when you are riding in a vehicle that suddenly either brakes or speeds up so you are thrown either forward or backwards. The Law of Inertia, was probably considered the most controversial statement in *The Principia.* It squarely contradicted the teachings of Aristotle, a Greek philosopher who lived from 384 to 322 BC. Aristotle was revered by scientists throughout the ages, and he said that the "natural state" of an object is at rest. Thus, an object would "like" to be at rest, and a force has to act on an object to keep it from being at rest. Newton's First Law says that there is no "natural state" for an object. Instead, objects simply move (or don't move) until acted on by an outside force. Thus, if an object is moving, it will continue to move *forever* unless a force acts to stop it. Even though this law was controversial, rigorous experiments demonstrated that it was true.

Newton's Second Law can be written as an equation

$$\mathbf{F} = m\mathbf{a} \tag{2.1}$$

where $\mathbf{F}$ is the vector sum of all the forces acting on the object and $\mathbf{a}$ is the resulting acceleration vector. In your introductory physics course you used this equation extensively to work problems. When the sum of the external forces on an object is zero, the object has zero acceleration and is said to be in equilibrium. Later in this chapter we will discuss static and dynamic equilibrium.

Using Newton's Second Law it is possible to show the difference between the concepts of mass and weight. The mass of an object is how much matter it possesses. It is a scalar quantity. The weight of an object is the force of gravity acting on its mass. It is a vector quantity. From Eq. 2.1,

$$\mathbf{F} = m\mathbf{a} \text{ becomes } \quad w = mg \tag{2.2}$$

where w denotes weight and g is the acceleration due to gravity. Although w and g are both vectors, we can just work with their magnitudes since we know their direction is always down. The mass of an object is always the same while its weight can vary if it is in an environment with a gravitational acceleration different from the usual force of earth's gravity.

In considering Newton's Laws of Motion, there are two concepts of mass. The first is that of inertial mass. Inertial mass is measure of an object's resistance to a change in its state of motion. This means that when you try to change an object's velocity (either get it moving from rest or change its speed and/or direction while it is moving), it will resist that change. The amount by which it resists depends on its inertial mass. In fact, that's really what the second law says. Equation 2.1 says that the force required to accelerate (change the velocity of) an object depends on the mass. If one object has twice the mass of a second object, the second object will require only one-half of the force to change its velocity to the same degree.

The second concept of mass is a measure of the amount of *matter* in an object. This concept of mass is often referred to as gravitational mass, because the more matter an object contains, the more gravitational force it experiences when exposed to another object.

What is the relationship between inertial and gravitational mass? They are numerically equal. If an object has a gravitational mass of 10.0 g, it will have an inertial mass of 10.0 g. Why bother to make the distinction, then? Inertial mass *resists* changes in motion. However, gravitational mass *encourages* changes in motion. The larger the gravitational mass of an object, the larger the gravitational force it experiences when exposed to another massive object. The larger the force, the more the change in velocity. These two different concepts of mass, then, actually lead to two different behaviors when it comes to different situations. Einstein used the fact that gravitational mass and inertial mass are numerically equal as a major point in developing his theory of general relativity, which we will discuss near the end of this course. We will revisit the concept of weight versus mass later in the chapter.

For working problems in dynamics, we follow the same procedure we used in kinematics, sketch the problem with reference point and coordinate system, list the knowns and unknowns, and use the appropriate equations to obtain the solution. Now it is necessary to add to our sketch a diagram showing all the forces relevant to the problem.

Before going through example problems, we need to remember the importance of a reference frame for our analysis. Remember from your introductory course that velocity is *relative*. It must be defined relative to a reference frame. Suppose, for example, that two spaceships are flying in space and are approaching one another such that they get 10,000 m closer to one another every second. You could analyze their motion in several different ways. You could say that the first ship is approaching the second ship with a velocity of 10,000 m/s and that the second ship is standing still in space. You could also say that the second ship is approaching the first ship with a velocity of − 10,000 m and the first ship is standing still in space. You could also say that both ships are moving, one with a velocity of 5000 m/s and the other with a velocity of − 5000 m/s. Each of these answers is correct. We must choose a reference frame and define velocity relative to that reference frame. Then the velocity of each ship can be determined relative to the chosen reference frame. The description of the motion is correct within the chosen reference frame.

The surface of the earth is convenient to use for a reference frame for problems when we can neglect the motion of the earth in space. A satellite in a geosynchronous orbit never moves relative to some fixed point on the earth. If you were to look at a satellite in geosynchronous orbit, you would say that the satellite isn't moving. However, relative to the sun, the other planets, the moon, and any non-geosynchronous satellites, that satellite is moving like crazy. Thus, in the reference frame of the earth's surface, the satellite is stationary. In the sun's reference frame, however, it is moving. Velocity, then, depends on how you define your reference frame.

There is a special class of reference frames called inertial reference frames. An inertial reference frame is one in which an object that is subject to no force travels at a constant velocity. Thus, an inertial reference frame is any reference frame in which Newton's First Law of motion is true. Remember in the statement of the law of inertia, the motion of an object does not change unless it is acted on by some outside force. If we choose a reference that is accelerating, it must have an outside force acting on it. It is usually best to choose an inertial reference frame when solving a problem in dynamics. It could be argued that because of the earth's rotation and its motion around the sun that we are always in motion with acceleration so nothing can be defined as a truly inertial reference frame. Although this is true, for most problems the earth's motion can be neglected and an inertial reference frame can be attached to the earth's surface. Just remember:

we will treat the earth as an inertial reference frame because it is close to one. It is not a true inertial reference frame, however. Any reference frame moving

with a constant velocity with respect to the earth, then, will also be considered an inertial reference frame.

Newton was able to discover his laws because the earth is close to an inertial reference frame.

Thus, for all practical purposes, Newton's First Law works on the earth.

Newton's Third Law can be demonstrated by considering a baseball pitcher throwing a fastball that is hit by the batter. When the hit occurs, the ball exerts a force on the bat and the bat exerts an equal and opposite force on the ball. Note that the initial force and the reaction force act on two different objects (bat and ball) so they do not cancel each other out. Newton's Third Law tells us that forces always occur in pairs. There is no such thing as an isolated force. An important example of this is the normal force that accounts for the effective weight of an object and determines the magnitude of friction.

To demonstrate the effect of forces, let us go through the following example.

Example 2.1

As an example, consider a skier who starts from rest and accelerates down a hill that is at an angle of 30° to the horizontal. If the coefficient of kinetic friction between the snow and the skis is 0.1, how fast is the skier going after 5 s? (Neglect any air resistance.)

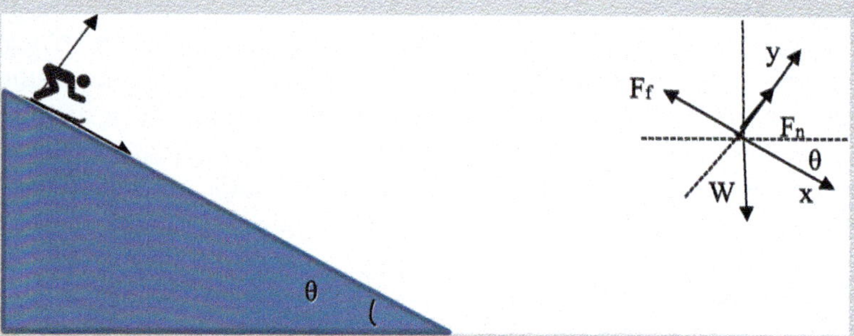

The sketch of this problem is shown above along a reference frame and a force diagram. The knowns and unknowns are:

Knowns: $v_0 = 0\,\text{m/s}$; $\theta = 30°$; $\mu_k = 0.1$; $t = 5\,\text{s}$

Unknowns: v_f

Since the skier stays on the ground where $y = 0\,\text{m}$, there is no velocity or acceleration in the y-direction. This means the sum of all the forces in this direction are zero. Therefore,

$$F_N - F_{gy} = 0 \text{ or } F_N = F_{gy} = mg\cos\theta.$$

In the x-direction, the net forces are given by

$$F_x = F_{gx} - F_f = mg\sin\theta - \mu_k F_N = mg\sin\theta - \mu_k mg\cos\theta = ma_x.$$

Solving this for acceleration down the slope gives

$$a_x = g(\sin\theta - \mu_k\cos\theta) = 9.81\,\mathrm{m/s^2}[0.5 - (0.1)(0.866)] = 4.06\,\mathrm{m/s^2}.$$

Now that we know the acceleration in the downhill direction we can calculate the velocity at any time using

$$v_f = v_0 + at = 0 + \left(4.06\,\mathrm{m/s^2}\right)(5\,\mathrm{s}) = 20.3\,\mathrm{m/s}.$$

Thus after 5 s, the skier will have accelerated from zero to a velocity of 20.3 m/s. By waxing his skis to reduce friction, he could be going faster than this. Remember we are able to choose the coordinate system in our reference frame to simplify the geometry of the problem.

An Atwood's machine consists of two unequal masses joined together by a rope. The rope is hung vertically over a light pulley, and the motion of the system can then be analyzed in terms of the acceleration of the masses and the tension in the rope. You know what will happen in a situation like this. The heavier mass will begin to accelerate downward while the lighter mass will accelerate upwards. In order to analyze this situation, you need to realize two things. First, since the masses are connected by one rope, the tension is the same throughout the rope. Thus, the rope pulls on m_1 with the same force as that with which it pulls on m_2. Second, since the rope stays tight and does not "bunch up," the masses must be moving with the same acceleration. Thus, the acceleration of each mass is equal in magnitude. We can look at *each mass separately* and sum up the forces, setting the sum equal to the mass times the acceleration (Fig. 2.2).

Look, for example, at the first mass. There are two forces acting on it. Gravity pulls it down (the negative direction) with a strength of $m_1 g$, and the tension on the rope pulls it up. Thus, we can sum up the forces as:

$$T - m_1 g = m_1 a$$

Fig. 2.2 Atwood's machine

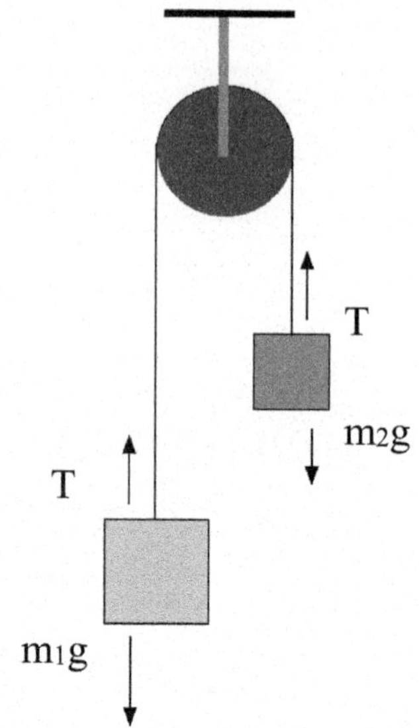

What are the unknowns in the equation? Remember, we want to determine the acceleration in terms of the masses and the acceleration due to gravity. Thus, we can treat m_1, m_2, and g as givens. That means there are two unknowns, T and a. However, if we solve for T in terms of a, we might then learn something else by looking at the other mass.

$$T = m_1 a + m_1 g$$

Now let's look at the other mass. Once again, there are two forces acting on it. Gravity pulls it down with a strength of $m_2 g$, while the tension in the string pulls it up. Pulleys, however, present a problem. They *change the direction.* Mass 1 is moving downward. When that happens, mass 2 moves upward. As a result, *downward motion for mass 1 is upward motion for mass* 2. In terms of mass 2, then, upward motion must be defined as *negative* and downward motion must be defined as *positive.* Be sure you understand this point. We defined negative motion when we dealt with mass 1. We said that negative motion for mass 1 is downward motion. However, when mass 1 moves in the negative direction, mass 2 moves *upwards.* Thus, for mass 2, *upwards motion is negative.* This is very important. When we define the sign for direction, it must be consistent. The pulley makes negative motion for mass 1 upward motion for mass 2. Thus, for mass 1, downward motion is negative. That means for mass 2, upward motion is negative.

$$m_2 g - T = m_2 a$$

Now, we can substitute the equation we derived for T into this equation, which will reduce this equation to one unknown:

$$m_2 g - (m_1 a + m_1 g) = m_2 a.$$

Solving this for acceleration gives

$$a = \left(\frac{m_2 - m_1}{m_2 + m_1}\right) g \tag{2.3}$$

Notice how this equation works out to give you the direction of the acceleration. If m_2 is larger than m_1, then the acceleration is positive. According to our definition of direction, positive motion is upward motion for mass 1 and downward motion for mass 2. Thus, the equation tells us that if m_2 is larger than m_1, mass 1 rises while mass 2 falls. That is, of course, what one expects. If, on the other hand, m_1 is larger than m_2, the acceleration is negative. That means mass 1 will fall and mass 2 will rise, as you would expect. That's why it is important to understand how to do sign definitions with pulleys. Since pulleys reverse the direction of motion from one side to the other, you have to be careful in determining the signs associated with the forces at play. If you know the masses involved, Eq. 2.3 gives us the resulting acceleration, and this can be used in either of the two expressions for T to get the tension in the rope.

To give you a little practical experience with Atwood's Machine, perform the following Experiment.

Experiment 2.1

Building and Using Atwood's Machine

Supplies:

- Two Ziplock bags (The smaller the bag, the better.)
- A string that is about 1.5 m in length
- Sand (Kitty litter, dirt, or small gravel will work as well.)
- A mass scale (It can be one of the scales used to measure food, which cost about $6 at any grocery store. It should be marked off in grams, and it should have a range of no more than 500 g.)
- A pulley (You can get these at any hardware store. They are plastic and are typically called "clothesline" pulleys. Get one that is at least 2.5 inches in diameter.)
- A stopwatch
- Meter stick or another metric ruler
- A high place to which you can attach the pulley (I used the shower curtain rod in my shower.)

Introduction: Atwood's machine can be used for many different applications. It is sometimes used to measure the acceleration due to gravity. We will use it to get an idea of the friction associated with the pulley.

Procedure:

1. Place one of the bags on the scale and fill it with sand until it has a mass of 50 g. Remember, the scale can be read to one decimal place more than what it is marked off. Thus, your reading should really be 5.0×10^1 g.
2. Take that bag off and put the other bag on the scale. Fill it with sand until it has a mass of 6.0×10^1 g.
3. Zip both bags sealed.
4. Tie one bag to one end of the string. Probably the easiest way to do this is to punch a hole through the bag near the top and then thread the string through the hole.
5. Tie the pulley to a rod or other structure that is at least 1.5 m above the ground. I used the curtain rod in my shower.
6. Thread the string through the pulley, and then attach the other bag to the free end of the string.
7. You now have a working version of Atwood's machine. The bags of sand are the masses, and when you release them, the lighter mass will rise and the heavier mass will fall.
8. Now we want to measure the acceleration of your Atwood's machine. To do this, hold the stopwatch in one hand and the lighter bag in the other.
9. Pull the lighter bag down so that the heavier bag is all the way at the top, touching the pulley.
10. Release the bag and, at the same time, start the stopwatch. Stop the stopwatch when the lighter mass reaches the pulley.
11. While the system is in this configuration, measure the length of the string from the pulley to the heavier bag. This is the distance that the lighter mass traveled from the time you released it to the time that it touched the pulley.
12. Repeat steps 8–10 nine more times. If one or two of the times are significantly (30% or so) different than the others, discard that trial and try again.
13. Average your 10 time measurements to get one average time.
14. You now have the distance that your masses traveled and the time it took them to travel that distance. Using kinetic Eq. 1.6, you can calculate the acceleration of the masses. Remember, the initial velocity is zero.

15. The acceleration you calculated in #14 is the experimental acceleration. Now, use the equation we just derived for the acceleration of Atwood's machine Eq. 2.3 to calculate the *theoretical* acceleration of your Atwood's machine.
16. Compare the two accelerations. The experimental number should be *lower* than the theoretical number. Why? Because of three things that were ignored in the example. First, friction was ignored. Second, it actually takes energy to turn the pulley. You will learn more about that later. Also, the mass of the string was ignored. After all, it has to accelerate as well. All of these effects will slow down the system, reducing the acceleration.
17. Take the experimental acceleration and divide it by the theoretical acceleration. This gives you an idea of how much friction plays a role in the machine. For example, my experimental acceleration was 82% of the theoretical value. That tells me that the factors I ignored in the calculation reduced the acceleration by roughly 18%.
18. So that you can take a shower again, remove the Atwood's machine from the shower, and clean everything up as well.

You now should be ready to do some Student problems.

Student

2.1 If an object is at rest in an inertial reference frame, can we conclude that no forces are acting on it?

2.2 Two observers move with constant velocity relative to one another. Are they both in inertial reference frames?

2.3 In Example 2.1, we split up the force due to gravity in terms of components that were parallel to and perpendicular to the surface of the incline. The force diagram shows that the angle opposite of the parallel component is the angle of the incline, θ. Use geometry to prove this fact.

2.4 In the following situation, $m_1 = 40.0\,\text{kg}$ and $m_2 = 20.0\,\text{kg}$. Neglecting friction, the mass of the string, and the mass of the pulley, what is the acceleration?

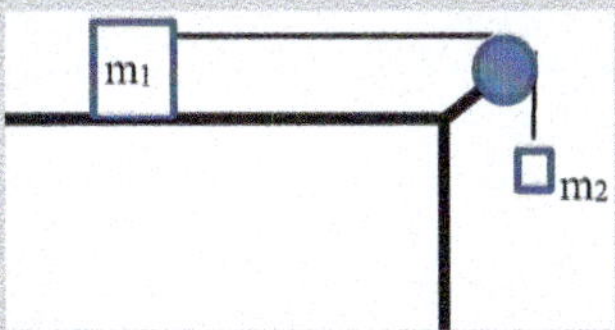

2.5 Now take the friction between m_1 and the table into account in Problem 2.4. If the coefficient of sliding friction is 0.25, what is the acceleration?

2.3 Equilibrium

Newton's Second Law leads to a special situation called equilibrium. An object is in equilibrium when the sum of the forces acting on the object is zero. Under those conditions, Newton's Second Law says that the object has no acceleration. It might be at rest, or it might be moving at a constant velocity. Either situation is considered a state of equilibrium. An object is in static equilibrium if it is at rest and dynamic equilibrium if it is moving at a constant velocity. The following example is a problem involving static equilibrium.

Example 2.2

A 900.0 g picture is hung according to the diagram shown below. Calculate the tension in each of the three strings.

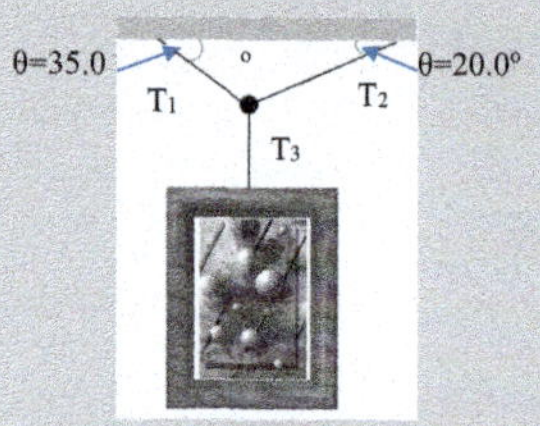

Knowns: $m = 900$ g; $\theta_1 = 35.0°$; $\theta_2 = 20.0°$

Unknowns: T_1; T_2; T_3

To do this problem, we first have to realize that the picture is hanging still. Thus, its acceleration is zero. Next, we have to look at the forces in the problem. First, we look at the forces acting on the picture. There are only two. Gravity is pulling the picture straight down. It has a strength equal to the weight of the picture, which is mg. The string is exerting **tension** $\boldsymbol{T_3}$, pulling the picture up.

Since the picture is not moving, we know it is not accelerating. This means that the sum of the forces on the picture is zero. However, we need to sum up the forces as *vectors*. They are each in the same dimension (the vertical dimension), so we use signs to indicate direction. Let us define down as negative, so the force due to the weight is $-mg$. This means that the sum of the forces, which equals zero is:

$$T_3 - mg = 0$$

Since we know the mass, we can calculate T_3

$$T_3 = mg = (0.9000\,\text{kg})\left(9.81\,\text{m/s}^2\right) = 8.83\,\text{N}$$

Note that to get the answer in Newtons of force, the mass in grams had to be converted to kilograms.

Now that we know the tension in the bottom string, the force diagram for the problem looks like

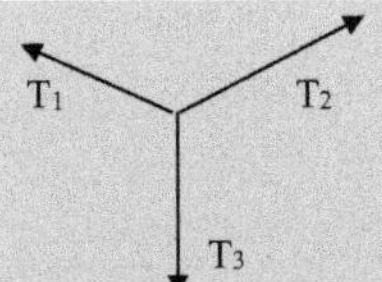

The sum of these three forces must equal zero. Since they are in different directions, we must work with the x- and y-components. For the horizontal direction this becomes

$$T_2 \cos 20.0^\circ - T_1 \cos 35.0^\circ = 0.$$

So,

$$T_2 = T_1 \cos 35.0^\circ / \cos 20.0^\circ = 0.872 T_1.$$

Now let's go to the vertical dimension. That should give us more information. In that dimension, the point connecting the strings is not moving. Thus, acceleration is zero. That means the sum of the forces is zero. T_3 is pulling straight down (the negative direction), while the vertical components of T_2 and T_1 are pulling straight up.

$$T_2 \sin(20.0) + T_1 \sin(35.0) - 8.83\,\text{N} = 0$$

Now let's take the equation we have for T_2 from the horizontal dimension and stick it into this

$$0.872 T_1 \sin(20.0) + T_1 \sin(35.0) - 8.83\,\text{N} = 0$$

Solving this for T_1 gives wires.

$$T_1 = 10.1\,\text{N}$$

To get the final tension we can use the expression we derived for T_2 in terms of T_1

$$T_2 = 0.872 T_1 = 0.872(10.1\,\text{N}) = 8.81\,\text{N}.$$

At this point we know the tension in all three.

See if you can work through the following Student problem.

Student

2.6 **Consider hanging a painting in an art museum with two wires as shown in the figure below. If the painting has a mass of 1.0 kg and the**

wires are at angles of $\theta_1 = 45°$ and $\theta_2 = 30°$ to the horizontal, what is the tension in each wire?

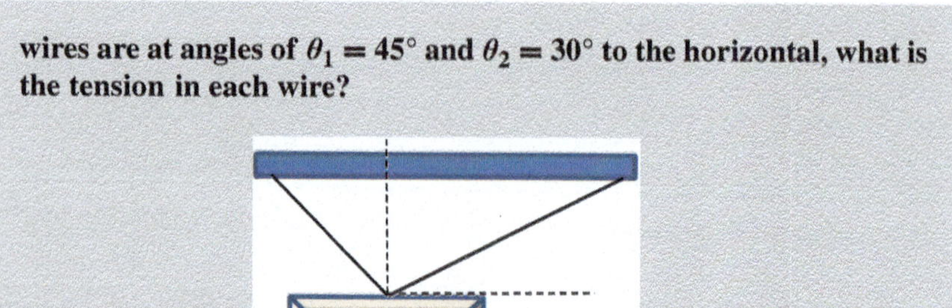

2.4 Natural Forces—Gravity, Friction, Air Resistance

Along with man-made forces, there are several different types of forces in nature that we routinely encounter. These include air resistance, friction, and gravity. We dealt with the last of these previously in our study of free fall and projectile problems. Remember, gravity is the mutual attraction of two objects that each have mass. For the situations discussed in this chapter, one of these objects is the earth and the other is some moving object. The force of the earth's gravitational attraction on an object always pulls it toward the center of the earth. We will see below how important this is in generating a normal force that determines effective weight and friction. Air resistance and friction are similar in that they both occur because an object is interacting with its environment. Both types of force oppose the direction of motion of an object. In both cases the magnitude of the force depends on the state of motion of the object. These details are discussed more thoroughly below.

Air resistance arises when an object is moving through the atmosphere and has to push the air molecules out of the way. This is a force that acts opposite to the direction of motion and thus tends to slow the moving object down. Engineers usually try to design the shape of an object that will move through the air to minimize air resistance. In the kinematic descriptions of motion in the previous chapter, we always neglected air resistance. In this chapter we will learn how to account for the effects of air resistance.

The force of air resistance on a projectile is called drag. It increases with the velocity of the projectile and can limit the velocity of an object undergoing free fall. When air resistance is not negligible, we need an equation for the drag force acting on the moving object. From doing experiments, the famous scientist Lord Rayleigh derived the expression

$$F_{\text{drag}} = C_{\text{D}}\rho A v^2/2 \tag{2.4}$$

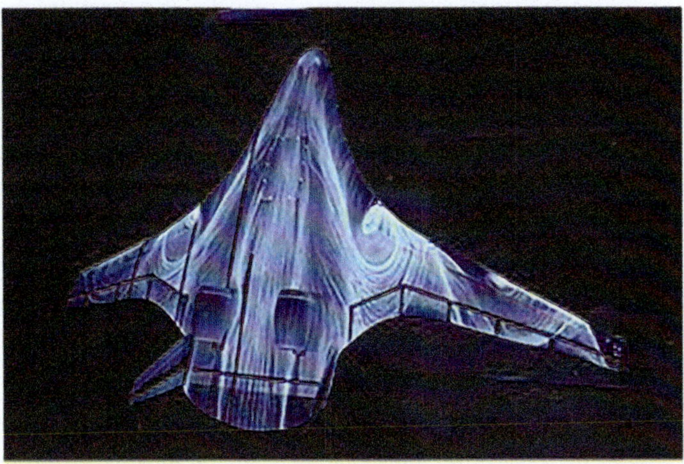

Fig. 2.3 Wind tunnel test of air resistance. *Credit* NASA Langley/Preston Martin, public domain

where ρ is the density of the air, v is the velocity of the object, A is the cross-sectional area of the object perpendicular to the direction of motion, and C_D is a dimensionless constant called the drag coefficient, which depends on the shape of the object and the skin friction of the material it is made of. The magnitude of C_D is 0.012 for a large, sub-sonic aircraft. For an object with a shape like , $C_D = 1.14$, while an object with the same value of A and made out of the same material but with a shape that has a rounded front has a $C_D = 0.045$. To minimize drag, airplanes are designed to minimize A and C_D, by giving them aerodynamic designs and making them from materials with low "skin friction."

Ignoring "C_D," what does the equation tell us about the air resistance an object experiences? First, it depends on the density of the air. The more dense the air, the more air resistance there is. This should make sense. After all, if there are a *lot* of molecules in the air through which you are traveling, there will be a lot of resistance. The fewer molecules, the lower the resistance. This is why airplanes like to fly at high altitudes. The higher the altitude, the less dense the air. Thus, the higher a plane flies, the less air resistance it experiences. This means that the higher a plane flies, the less fuel it consumes, because the less it has to fight air resistance. Now, of course, it takes extra fuel to climb to a high altitude, since while the airplane climbs, it is working against gravity. As a result, the trip has to be long enough to be worth getting to a higher altitude. Nevertheless, in general, airplanes like to fly as high as is realistic to reduce the density of the air through which they are traveling in order to reduce drag.

The force of air resistance also depends on the size of the object. The larger the cross section that is perpendicular to the direction of travel, the more the air resistance. This should also make sense. If an object travels through the air, the part of the object that must "shove" air out of the way is the part that will cause the drag. Thus, the cross section of the object which is perpendicular to its velocity vector will be the one that affects drag. The larger the cross section, the more air that must be "shoved" out of the way so as to allow the object to move. Thus, the more air resistance it will experience. You can see the effect that the cross section of an object has on air resistance by performing this simple experiment.

Experiment 2.2
The Effect of Cross Section on Air Resistance

Supplies:

- Two round balloons (They need to both be the same size before they are inflated.)
- A stepladder or other platform on which to stand

Introduction: The force of air resistance depends on the cross-sectional area of an object. This experiment demonstrates that effect.

Procedure:

1. Take one balloon and inflate it mostly full. Tie it off so that it stays inflated.
2. Take the second balloon and inflate it only slightly. Make sure you have inflated it enough so that it has the proper shape, but it should be no more than half the size of the other balloon.
3. Climb up on the ladder with the two balloons. **BE CAREFUL!!!**
4. Hold one balloon in each hand at the same height, and drop them at the same time.
5. Note which one hits the ground first.
6. Repeat the experiment once or twice to make sure your results make sense.

What happened in the experiment? The small balloon should have hit the ground noticeably sooner than the large balloon. Why? They were roughly the same shape, so the "C" for each object was the same. They were traveling through the same air, so "ρ" was the same for each balloon. Their initial velocities were the same so, at least initially, "v" was the same in both cases. However, the cross section perpendicular to the motion (A) for each balloon was different. The "A" for the large balloon was significantly larger than the "A" for the small balloon. As a result, the large balloon experienced more drag and was thus slower at reaching the ground.

Finally, the force of air resistance also depends on the speed of the object as it travels. In fact, it depends on the speed of the object more than it depends on the other factors. The force of air resistance increases as the *square* of the speed. Thus, if the object moves three times as fast, it experiences *nine times* the air resistance! Supersonic jets such as the Concorde airplane fly at very high

Fig. 2.4 A T-38 supersonic jet, used to train astronauts. *Credit* NASA picture, public domain

altitudes—much higher than that of slower jets. The main reason for this is that the effect of velocity on air resistance is so high that a supersonic jet must do anything it can to counteract air resistance. About the only thing it can do is climb to very high altitudes where the density of air is as low as possible. In general, then, the faster a jet travels, the higher it travels so as to try and counteract the strong effect that its speed has on the air resistance.

Now that you have learned the basic factors which affect air resistance, we can apply Eq. 2.4 to the concept of **terminal velocity**. Since air resistance depends on speed, the faster an object travels, the larger the air resistance. Well, consider what this means for an object which is falling. As it falls, its velocity increases. This means the air resistance increases. What does that do to the acceleration of the object? It *decreases* the acceleration. Thus, the farther an object falls, the *less* it accelerates, because air resistance increases with increasing speed. At some point, the object will be traveling so fast that the air resistance becomes strong enough to *completely counteract the force of gravity*. At that point, the object will still fall because it has velocity in the downward direction. However, since there is no more acceleration, the velocity will not change. No matter how far the object falls after that point, its speed will not increase. That speed is called the object's "terminal velocity."

Returning to the free-fall problem discussed in Chap. 1, the total force on an object is gravity minus drag,

$$F = mg - C_{\mathrm{D}}\rho A v^2/2 = ma.$$

The terminal velocity v_{T} is when $a = 0$. This leads to the expression

$$v_{\mathrm{T}} = \sqrt{\frac{2mg}{C_{\mathrm{D}}\rho A}}. \tag{2.5}$$

Example 2.3

An object with a cross-sectional area of 2×10^{-3} m^2 and a mass of 5.0×10^{-2} kg, free falls in air with a density of 1.290 kg/m^3. What is its terminal velocity if it has a shape with $C_D = 1.14$ and what is its terminal velocity if it has a shape with $C_D = 0.045$?

F_{drag}

mg

Knowns: $A = 2 \times 10^{-3}$ m^2; $m = 5.0 \times 10^{-2}$; $\rho = 1.290$ kg/m^2; $C_D = 1.14$ or 0.045

Unknowns: v_T

Using Eq. 2.5,

$$v_T = \sqrt{\frac{2mg}{C_D \rho A}} = \sqrt{\frac{2 \times 5 \times 10^{-2}\,\text{kg} \times 9.81\,\text{m/s}^2}{C_D 1.290\,\text{kg/m}^2 \times 2 \times 10^{-3}\,\text{m}^2}} = \sqrt{\frac{380}{C_D}}$$
$$= 18.25\,\text{m/s for } C_D = 1.14$$
$$= 91.89\,\text{m/s for } C_D = 0.045.$$

During the previous discussion of free fall, we were told to ignore air resistance. In many cases, that introduces only a small error in the calculation, since air resistance is negligible in many situations. However, air resistance *can* be a factor, especially when an object is large, has a small mass, or is traveling very quickly. For these cases the drag force given by Eq. 2.4 can be non-negligible in the dynamics of the problem.

Before we leave this section, we need to note two quick points about Eq. 2.5. First of all, this equation is only valid for objects that are either large or moving at reasonably high speeds. This equation applies to most objects in free fall, since free fall produces large speeds.

For small objects traveling slowly, we need to develop a completely different equation. Equation 2.5 is, by far, the most widely applicable equation for objects and situations with which we are familiar. Just be sure and remember that it is not applicable in *all* situations. Second, this equation actually applies to object moving through *any* fluid. Thus, if you substitute the density of water for the density of air in Eq. 2.5, you will end up calculating the terminal velocity of the object as it falls in a deep pool of water.d3.

Next Level

When an object is falling in a liquid it will also reach a terminal velocity. In Chap. 14 we will learn that an object in a liquid will experience a force of buoyancy equal to the weight of the liquid it displaces. The equation for terminal velocity must be modified to include this force.

For an object of weight W falling in a liquid, terminal velocity is reached when $W = F_B + F_D$ where F_B and F_D are the forces of buoyancy and drag, respectively. For a spherical object, these three forces can be expressed as: $W = (\pi d^3 \rho_s/6)g$; $F_D = C_D \rho v_t^2 A/2$; $F_B = (\pi d^3 \rho/6)g$. Here A is the projected area of the sphere of diameter d, ρ is the density of the liquid, and ρ_s is the density of the sphere. With these expressions, the terminal velocity is

$v_t = \pm\sqrt{\frac{4gd}{3C_D}\left(\frac{\Delta\rho}{\rho}\right)}$ where $\Delta\rho$ is the difference between ρ_s and ρ.

Generally, $\rho_s > \rho$ and the terminal velocity is positive in the downward direction. However, if $\rho_s < \rho$ like an air bubble in a glass of soda the buoyancy causes the bubble to rise so the terminal velocity is in the negative upward direction. If $\rho_s = \rho$ the buoyancy force is equal and opposite to the object's weight so there is no motion and $v_t = 0$.

Before we leave this topic, try your hand at the following Student problem.

Student

2.7 A 0.22-caliber rifle (a standard rifle used for target practice) fires a bullet whose diameter is 0.22 inches and whose mass is 38.0 g. Given that the density of air is 1290 g/m^3 and $C = 0.50$, what is the terminal velocity of a bullet shot from a 0.22-caliber rifle fired vertically in the air?

Next we want to consider some special aspects of gravity. One important natural force that we deal with in essentially every problem is gravity. You may have learned that any two objects that have mass experience a gravitational attraction toward each other. We will revisit this general gravitational force in Chap. 4 focusing on rotational motion such as the planets orbiting around the sun. For this chapter, we are interested in the force of the earth's gravitational attraction on an object near its surface. We refer to this as the weight of the object and through Newton's Second Law express it as

$$W = mg$$

where g is the earth's gravitational acceleration equal to 9.81 m/s^2.

According to Newton's Third Law there must be an equal and opposite force acting on the object to offset the force of its weight. This is called the normal force, F_N. It acts in the upward direction perpendicular to the surface on which the object is resting. The normal force is important for two reasons. The first is in determine the effective weight of an object and the second in calculating the friction between the object and the surface on which it is sitting.

The first if these is associated with having an object in a non-inertial reference frame. Let's suppose that, for some insane reason, you decided you wanted to weigh yourself in an elevator. Thus, you take your bathroom scale into the elevator, set it on the floor, and jump on the scale. The elevator is still and, in a moment, the scale reads that you weigh 122 pounds. Now let's suppose that someone calls the elevator from a floor high above the floor you are on. Thus, the elevator begins to move upwards. If you watched the scale while the elevator began to move upward, you would see your weight as read by the scale *increase*. Then, as the elevator reached a constant velocity on the ride upward, the scale would go back to reading 122 pounds. Then, as the elevator began to slow down, you would see your weight as read by the scale *decrease*. When the elevator stopped, your weight would go back to 122 pounds. Why does your weight fluctuate like that? During parts of the trip in the elevator, you are not in an inertial reference frame. Why? Because during parts of the trip, the elevator was *accelerating*. Since you were not in an inertial reference frame, your weight appeared to fluctuate. To understand why your weight appeared to fluctuate, you must analyze the situation from an inertial reference frame. Then, you can reason out what happened physically. What would be a proper reference frame from which to analyze the situation? If an observer stood outside of the elevator and observed the motion, that observer would be in an inertial reference frame, because he or she would be at rest relative to the earth. Thus, let's analyze the situation from *that* perspective.

Example 2.4

A man weighs himself on an elevator while the elevator is at rest. The weight is 122 pounds. Then, the elevator begins to move upward with an acceleration of 8.00 ft/s². What weight does the scale read then?

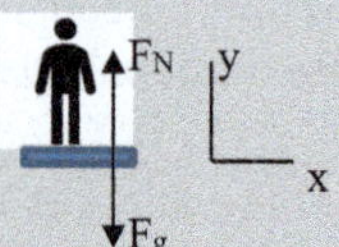

Notice that now we are analyzing things relative to the inertial reference frame of the earth. Inside the elevator, there is no observed acceleration. The man and the bathroom scale are at rest relative to the elevator. Relative to the earth, however, they are accelerating at 8.00 ft/s². If we view the situation from this perspective, it is easy to understand it.

Knowns: $W = 122\,\text{lbs}$; $a = 8.00\,\text{ft/s}^2$

Unknowns: W_a

Remember that the scale actually reads the normal force exerted on it. According to Newton's Second Law, the sum of the forces equals the mass times the acceleration,

$$F_\text{N} - F_\text{g} = ma.$$

Solving for the normal force,

$$F_N = F_g + ma.$$

When the elevator is at rest or moving with a constant velocity, the acceleration is zero so the normal force equals the man's gravitational weight,

$$F_N = 122\,\text{lbs}.$$

However, when the elevator is accelerating the normal force on the scale is

$$\begin{aligned} F_N &= F_g + ma = 122\,\text{lbs} + (W/g)a \\ &= 122\,\text{lbs} + \left(122\,\text{lbs}/32.2\,\text{ft/s}^2\right)\left(8.00\,\text{ft/s}^2\right) \\ &= 152\,\text{lbs}. \end{aligned}$$

As the elevator accelerates upward, then, the scale reads a weight *larger* than the man's true weight. When the elevator slows down, the acceleration becomes negative. If you plug a negative acceleration into the problem, you will find that the scale reads a weight *smaller* than the man's true weight.

What if the elevator cable were cut, and the elevator began to free fall? At that point, the man's acceleration would be $-\ g$. If you put that into the equation we developed above, you would see that the scale would read zero. Thus, in that situation, the man would appear to be weightless. This is, in fact, the way we *simulate* weightlessness here on earth. If we put an experiment in a container that falls freely, the experiment can be considered to occur in weightless conditions, because in that reference frame, the experiment has no weight. The National Aeronautics and Space Administration (NASA), for example, has a 430 ft deep shaft called the "drop tower" which is used to simulate weightlessness.

There is a common misconception that there is no gravity in space. That's because we see pictures of astronauts floating around while in spaceships. They are *not* floating around because there is no gravity in space. If there were no gravity in space, the planets would not orbit the sun nor would the moons orbit the planets! The reason you see astronauts float in spaceships is that those spaceships are in orbit, and while in orbit, the spaceship is accelerating downwards. We will revisit this in Chap. 4.

The other important aspect of the normal force is that it is needed to calculate the force of friction between two objects. Friction is a force that occurs whenever the surfaces of two objects rub together. For example, friction occurs when an object slides along a table. It acts opposite the direction of motion and therefore slows down a moving object. The origin of friction is the attractive forces between the atoms on the surfaces of the two objects. By making the surfaces as smooth as possible, the force of friction can be minimized. The force of friction is given by the product of a coefficient of friction, μ, and the normal force that the surface exerts on the object. For a horizontal surface the normal force is equal and opposite to the weight of the object. If an object is already moving the friction is called kinetic friction. If the object is at rest and the applied force is trying to make it move, the friction is called static friction. In general, the coefficient of static friction is greater than the coefficient of kinetic friction because the attractive forces between the surface molecules is greater when the object is not moving.

$$F_f = \mu_k F_n; \text{ and } F_f \leq \mu_s F_n. \tag{2.6}$$

As seen from Eq. 2.6, the force of static friction varies from just enough to stop the object from moving up to its maximum amount.

Before leaving the topic of friction, do the following lab experiment.

Experiment 2.3

Measuring the Coefficient of Kinetic Friction

Supplies:

- A Ziplock bag
- A block of wood or some other object with a flat surface (I used a small cardboard box with some sand inside.)
- A board that is as wide as the block and about 1.5 m long
- A string that is as long as the board
- Sand (Kitty litter, dirt, or small gravel will work as well.)
- A mass scale
- A pulley
- Hammer and nails
- A stopwatch
- Meter stick or other metric ruler
- A table or desk (The closer the height is to 1.5 m, the better.)

Introduction: In your first-year course, you should have learned how to measure the coefficient of static friction between a board and a block by tilting the block. That, however, measures only the coefficient of static friction. This method, by contrast, will measure the coefficient of kinetic friction. You will do this using a setup that looks like the drawing for "student" problem 2.4.

Procedure:

1. Take the board and attach the pulley to the board with nails. This can be the tricky part. You want the pulley to be at the right height. Make

sure that the top of the pulley is not higher than the top of the block as it rests on the board.

2. You will need to attach the string to the block of wood. If the block has no place to attach the string, drive a nail into it so that you can attach the string to the nail. Place the nail so that the string will stretch *in a straight line which is parallel to the board* from the block to the pulley. This will ensure that the string exerts a force on the block which is parallel to the board.
3. Measure the mass of the block.
4. As you did in the previous experiment, fill a Ziplock bag with sand. You want the mass of the bag and sand to be about equal to the mass of the block.
5. Seal the bag and punch a hole through it near the top so that you can thread the string through it.
6. Attach the string to the block, place it over the pulley, and attach the other end of the string to the bag.
7. Now you have a system that looks much like the drawing for "student" problem 3.4.
8. Lay the board/pulley system on the desk so that the bag hangs over the edge of the table and the pulley turns freely.
9. Pull the block back along the board so that the bag is lifted to where it is just touching the pulley.
10. Release the block. If the block slides across the board, your setup is working. If not, the friction between the board and the block is too large, and you need to add more mass to the bag.
11. Once the system allows the block to slide across the board when it is released, you are ready to begin the experiment. Repeat steps (9)–(10) ten times, each time using the stopwatch to measure the time from when you released the block to when the bag touches the ground. Average those 10 results so that you have a good idea of the time that it took for the bag to fall.
12. Pull the bag up so that it is just touching the pulley as you did each time in step (9).
13. Measure the distance between the bottom of the bag and the floor. That is the distance over which the bag dropped.
14. Now you can calculate the coefficient of sliding friction between the board and the block. First, use the time and distance you measured in Eq. 1.6 to determine the acceleration of the system.
15. Next, analyze the two masses in the system independently. If you are unsure how to do this, review the solution to "student" problem 2.4. The best way to begin the analysis is with the bag. There are only two forces on the bag: gravity and tension. Since you have measured the acceleration, you can get a value for the tension in the string.

16. Now look at the block. There are two forces acting horizontally on the block: the tension on the string (you determined its value in the previous step) and friction. Once again, since you know the acceleration, you can determine the size of the frictional force.
17. Since the frictional force is just the coefficient of sliding friction times the normal force, and since the normal force is just the mass of the block times g, you can now solve for the coefficient of sliding friction. The number should be less than one. Mine was 0.58.
18. You can try this with different boards. The smoother the board, the lower your coefficient should be.
19. Clean everything up.

Many times man-made forces are so much greater than natural forces that air resistance and friction can be neglected. In working problems, you will be told if you can neglect air resistance and friction or not. If these effects are not negligible, you should now know how to include them in analyzing problems.

2.5 Applications and Examples

You should now be able to use Newton's Laws of Motion plus the kinetic equations you learned in the first chapter to analyze almost any motion. There are two techniques that we used in working the previous examples that you should think about using when you are faced with a problem to solve. First, if there is more than one object in the problem, we treated each object separately. Second, we analyzed the components of the forces on an object separately in the x- and y-directions. Let's illustrate this with a couple of more samples.

Example 2.5

Two blocks ($m_1 = 40.0$ kg, $m_2 = 25.0$ kg) are touching one another and sitting on the floor ($\mu_s = 0.450$, $\mu_k = 0.250$). A person pushes the blocks with a force of 200.0 N. What is the acceleration of the system? What force pushes on the 25.0 kg mass?

Knowns: $m_1 = 40.0\,\text{kg}$; $m_2 = 25.0\,\text{kg}$; $F = 200.0\,\text{N}$; $\mu_s = 0.450$; $\mu_k = 0.250$

Unknowns: a; F_2

The 40.0 kg object is the one that actually experiences the force of the person pushing.

What are all of the forces acting on it? Well, gravity pulls the object downwards, and the floor exerts a normal force up on the block. As a result of the contact between the block and the floor, there is also a friction force which opposes the motion. In addition, the 200.0 N force acts on the block. There is, however, one more force. The 40.0 kg block pushes against the 25.0 kg block. Newton's Third Law says that the 25.0 block must push against the 40.0 kg block with an equal and opposite force. We will call this force "*P*," because it results from the fact that the blocks push against each other. This force is often referred to as a **contact force**. The forces working on this object are shown below:

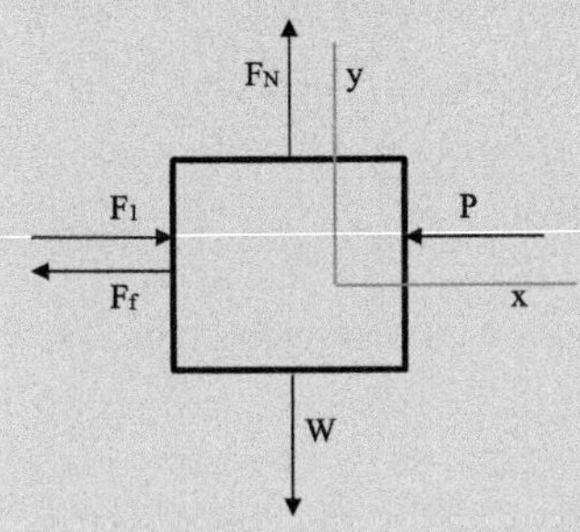

We are interested in the horizontal direction, because that's where all of the action takes place. However, we do need to look at the vertical dimension since friction is involved. To determine the force due to friction, we must determine the normal force.

Therefore, let's start in the vertical dimension. There is no acceleration in the vertical. Thus, the sum of the forces is zero. As a result:
$F_N - W = 0$
So
$F_N = (40\,\text{kg})(9.81\,\text{m/s}^2) = 392\,\text{N}.$

Now that we know the normal force, we can calculate the frictional force. Since the system is moving, we will use μ_k:

$$F_f = \mu_k F_N = (0.250)(392\,\text{N}) = 98.0\,\text{N}$$

We now have a number for the frictional force. Thus, we can look at the horizontal dimension. Using the convention that motion to the right is positive, we get the following equation from Newton's Second Law:

$$F_1 - F_f - P = ma$$

So,

$$200.0\,\text{N} - 98.0\,\text{N} - P = (40.0\,\text{kg})a.$$

Solving for *P* gives

$$P = 102.0\,\text{N} - (40.0\,\text{kg})a.$$

Now let's look at the other mass. The 25.0 kg mass is also affected by gravity, the normal force, and friction. In addition, it is being pushed with a force of "*P*" by the 40.0 kg block. Remember, this "*P*" is equal in magnitude to the "P" we just solved for, because of Newton's Third Law. It is simply pointed in the opposite direction.

Notice that the normal force, frictional force, and weight are *different* than the values for the normal, frictional, and weight forces acting on the first mass. The contact force is the same, however. Once again, to get a number for the frictional force we must first look at the vertical dimension:

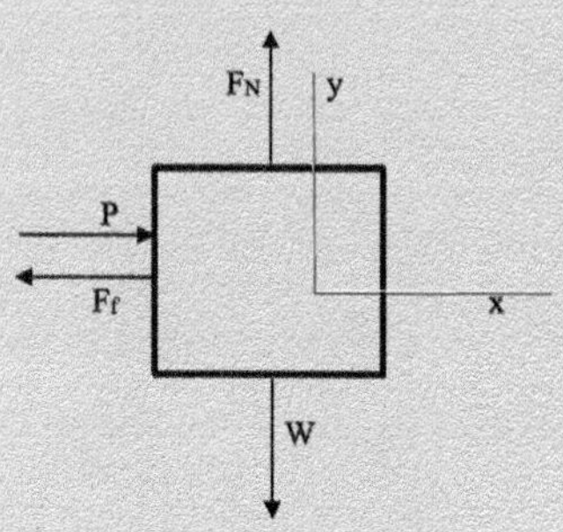

$$F_N - W = 0$$

So

$$F_N = (25.0\,\text{kg})(9.81\,\text{m/s}^2) = 245\,\text{N}.$$

Now we can calculate the frictional force:

$$F_f = \mu_k F_N = (0.250)(245\,\text{N}) = 61.3\,\text{N}.$$

Next, we sum up the forces in the horizontal dimension and set them equal to mass times acceleration:

$$P - F_f = ma$$

Since we know an expression for P in terms of acceleration, we can now solve for acceleration:

$$P - 61.3\,\text{N} = (25.0\,\text{kg})a$$
$$102.0\,\text{N} - (40.0\,\text{kg})a - 61.3\,\text{N} = (25.0\,\text{kg})a,$$

or

$$a = (40.7\,\text{N})/(65.0\,\text{kg}) = 0.626\,\text{m/s}^2.$$

The acceleration is actually much easier to find. Because the coefficient of friction is the same for both masses, all you have to do is add the masses together and treat the entire system as one block of mass 65.0 kg. Thus, there is only one equation to solve. However, there would be no way to answer the second part of the question. The second part asks what force pushes on the 25.0 kg block. Many students would say a force of 200.0 N, because the 200.0 N force is somehow "transmitted" through the first block and to the second. However, that is not the case. The first block pushes on the second block with a contact force. We have an equation for that contact force, so we can use it:

$$\begin{aligned} P &= 102.0\,\mathrm{N} - (40.0\,\mathrm{kg})a \\ &= 102.0\,\mathrm{N} - (40.0\,\mathrm{kg})\left(0.626\,\mathrm{m/s^2}\right) = 77.0\,\mathrm{N}. \end{aligned}$$

The force pushing on the second block, then, is nowhere near 200.0 N. It is 77.0 N.

Example 2.6

As another example, consider the situation shown in the figure below. Two blocks are attached by a rope thrown over a pulley. One has a mass of 40 kg and slides on an incline plane with a coefficient of kinetic friction of 0.2, while the other has a mass of 60 kg and undergoes a vertical fall. Find the tension in the rope and the acceleration of the blocks.

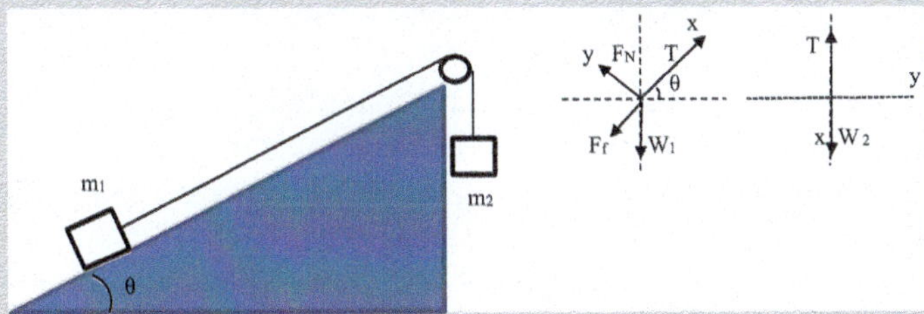

The sketch and the force diagrams for each of the two blocks is shown above. This can be treated as two problems, one a block sliding on an incline plane and the other the motion of a suspended block. The tension in the in the rope and the common acceleration of the blocks are the variables that connect the two problems. The knowns and unknowns are:

Knowns: $m_1 = 40\,\mathrm{kg}$; $m_2 = 60\,\mathrm{kg}$; $\theta = 45°$; $\mu_k = 0.2$

Unknowns: a, T

Note that we have to be careful about the directions of the coordinate systems in our two-block problem. For block 1, we chose sliding up the plane as the positive x-direction and perpendicular to the plane as the positive y-direction. The pulley changes the direction of these axes so for block 2 the positive x-axis is in the downward vertical direction while the positive y-axis is in the horizontal direction to the right. Then the tension in the rope and the acceleration of the blocks will be the same for both blocks as they must be.

We can begin with an analysis of block two. The sum of the forces on it are tension in the upward direction and its weight in the downward direction. According to Newton's Second Law the sum of these forces equals its mass times its acceleration,

$$-T + w_2 = m_2 a \text{ or } T = m_2 g - m_2 a = 60\,\text{kg}\left(9.81\,\text{m/s}^2 - a\right)$$
$$= 589\,\text{kgm/s}^2 - (60\,\text{kg})a.$$

Next consider the x-component of forces on block 1. These include the rope tension in the positive direction, friction in the negative direction, and the x-component of its weight in the negative direction,

$$T - F_\text{f} - W_1 \sin\theta = m_1 a$$

or

$$T = \mu_\text{k} w_1 \cos\theta + w_1 \sin\theta + m_1 a$$

where we used the fact that the normal force on the block is found from the y-component of its weight. According to Newton's Third Law, the normal force is equal to this force in the opposite direction. Multiplying the coefficient of friction by the normal force gives the force of friction. This equation can be solved for the tension in the rope,

$$T = \mu_\text{k} w_1 \cos\theta + w_1 \sin\theta + m_1 a$$
$$= (0.2)(40\,\text{kg})\left(9.81\,\text{m/s}^2\right)\cos 45^\circ$$
$$+\ 40\,\text{kg}\left(9.81\,\text{m/s}^2\right)\sin 45^\circ + 40\,\text{kg a}$$
$$= 332.96\,\text{kg m/s}^2 + 40\,\text{kg a}.$$

We can now equate these two expressions for the tension in the rope,

$$589\,\text{kg m/s}^2 - 60\,\text{kg a} = 332.96\,\text{kg m/s}^2 + 40\,\text{kg a}$$

So,

$$100\,\text{kg a} = 256\,\text{kg m/s}^2.$$

Then

$$a = 2.56\,\text{m/s}^2.$$

Now that we know the acceleration of the blocks, we can substitute this into either of the expressions for the tension on the rope

$$T = 589\,\mathrm{kg\,m/s^2} - (60\,\mathrm{kg})\left(2.56\,\mathrm{m/s^2}\right) = 435\,\mathrm{N}.$$

Thus, the tension of 435 N in the rope allows the blocks to accelerate at 2.56 m/s^2 in the positive direction.

Now try the following Student problems.

Student

2.8 In the drawing to the right, the block on the incline has a mass of 150.0 kg, and the block hanging from the pulley has a mass of 100.0 kg. Neglecting friction, what value of θ will allow the 150.0 kg mass to move up the incline at a constant velocity?

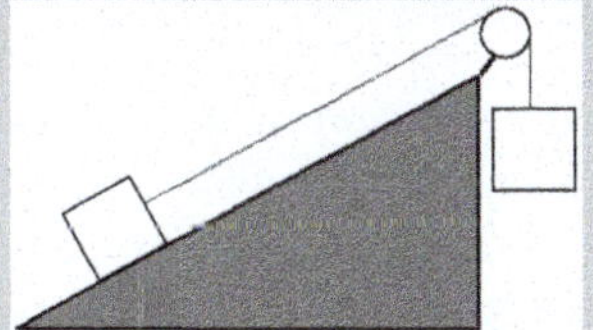

2.9 Consider the situation depicted in the drawing to the right with the angle of inclination being 35.0°. If the coefficient of kinetic friction is 0.250 and the system is moving, what is the acceleration of the system and the tension in each string?

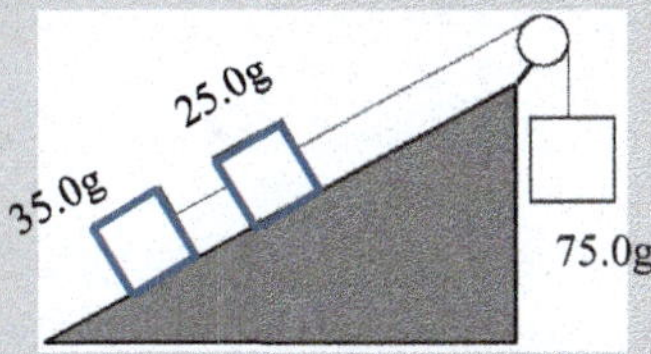

2.10 Three masses are pushed along a floor that has a coefficient of kinetic friction equal to 0.350. The force used to push the masses is 375.0 N
a. What is the acceleration of the system?
b. List the values of all forces which act on the 15.0 kg block.

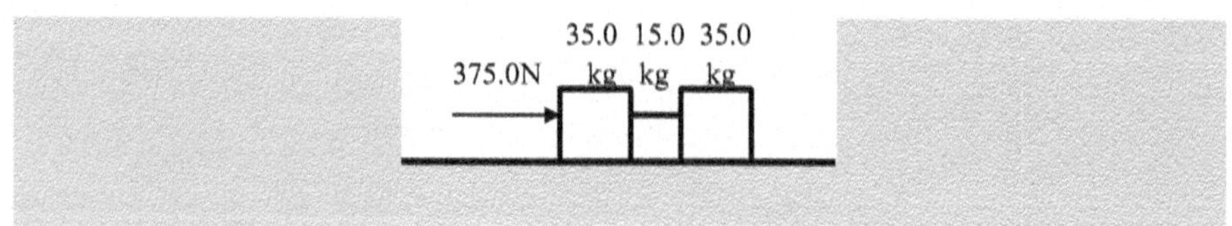

Special Topic

Here are six "simple machines" that have been extremely beneficial to mankind. Each of these allows us to change the magnitude or direction of an applied force. By doing this, the machine increases the usefulness of the force. One is the wheel and axle. A rotation of a small axle is translated to a rotation of a large wheel. The second one is a lever. With appropriate lever arms, a small force can lift a heavy object. Next is a screw that changes a rotational motion to a linear motion. An inclined plane allows us to change the height of an object without directly lifting it. A pulley changes the direction of an applied force and can be rigged to increase the effectiveness of the force. Finally, a wedge changes a downward force to a horizontal force. We will discuss some of these in more detail in the following chapters. The increase in the effectiveness of a force by using these simple machines is called their mechanical advantage. It is interesting to think about how such simple items can play such an important role in our life.

Summing Up

In this chapter you learned about how forces change the state of motion of objects. In addition, you learned that some forces have to touch an object to be effective while other forces can act at a distance. The major natural force we discussed was gravity. We will return to the discussion of gravity in Chaps. 5 and 16. There are three other natural forces we will learn about in future chapters: electromagnetic force; strong nuclear force; and weak nuclear force. Together these four forces provide the basic properties of the creation we live in.

Answers to the Student Problems

2.1 No, we cannot. If it is at rest (or moving with a constant velocity), we can conclude that the *sum of the forces* acting on the object is zero. It is entirely possible that two or more forces are acting on the object. They simply cancel each other out.

2.2 Not necessarily. If they move with constant velocity *relative to each other*, that means either they have no acceleration (and thus are in inertial reference frames) or have *the same acceleration*. Two cars each accelerating at 3.5 m/s^2 in the same direction will move at constant velocity with respect to each other. However, they are not inertial reference frames.

2.3 Let's start by drawing a horizontal line where the gravitational vector and the incline meet. When we do that, we can use the geometric law that says when two parallel lines are cut by a transversal, corresponding angles are congruent.

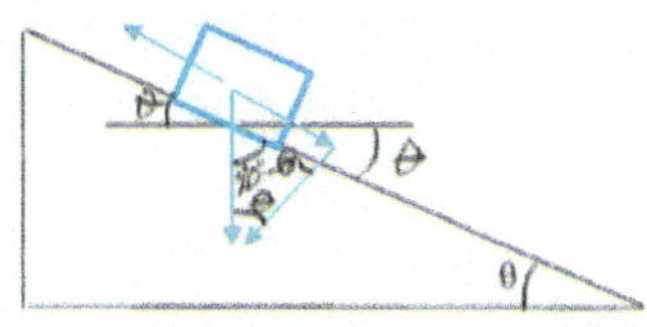

When we split the gravitational force vector into components parallel to and perpendicular to the incline, a right triangle is formed. Since the internal angles of a triangle must add to 180, that means the remaining angle must be θ.

2.4 Knowns: $m_1 = 40.0$ kg and $m_2 = 20.0$ kg
Unknown: a

In all problems like this, we treat the masses independently. The second mass is the easiest to analyze first. It has two forces acting on it. Gravity pulls it down, and the tension in the string pulls it up. Using the normal definition of down being negative motion:

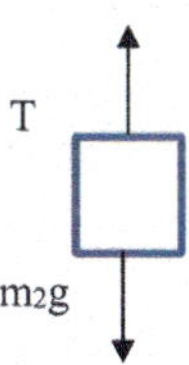

$$T = m_2 \cdot g = m_2 \cdot a$$
$$T = m_2 \cdot a + m_2 \cdot g$$
$$T = (20.0\,\text{kg}) \cdot (a) + 196\,\text{N}$$

With that equation for T, we can move to the first mass to calculate the acceleration. Since friction is not being considered, the only horizontal force acting on the mass is tension. There are vertical forces acting on the mass, but we need not consider them since we are ignoring friction. Now, we *must* worry about the signs used. We defined negative as m_2 moving down. When m_2 moves downward, however, m_1 moves to the right. Thus, the motion of m_1 to the right is negative. This means:

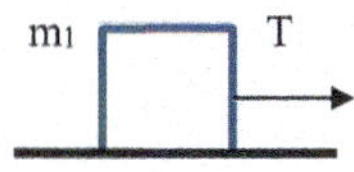

$$-T = m_1 \cdot a$$

We can plug the equation we had previously for T into this equation and solve for a:

$$(-20.0\,\text{kg}) \cdot a - 196\,\text{N} = (40.0\,\text{kg})a$$

$$a = (-196\,\text{N})/(60.0\,\text{kg}) = -3.27\,\text{m/s}^2$$

Notice that the acceleration is negative, which means that m_1 is moving to the right. There is a simpler way to do this problem and get this answer. However, you need to know how to do problems this way, because the simple method will only work in very simple cases.

2.5 This is the same as the problem above except that we now have the force of friction acting to the left on block 1. The coefficient of kinetic friction is 0.25.
The new force diagram for block 1 is shown on the right.
Our analysis of m_2 stays the same. Thus, from m_2 we get the same equation:
$T = (20.0\,\text{kg}) \cdot (a) + 196\,\text{N}$

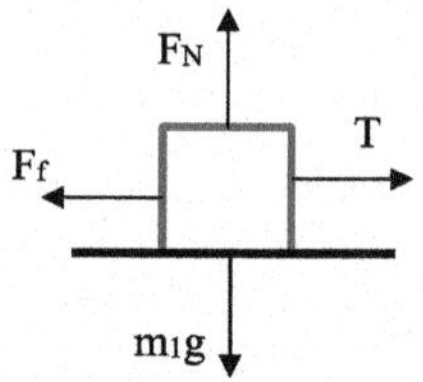

When we look at m_1, however, things change.
Remember, friction depends on the normal force, which is a vertical force. Thus, we have to consider vertical forces now. In the vertical dimension, gravity pulls the mass down and the normal force pushes it back up. Since there is no motion in that dimension, we know that the sum of these forces is zero.

$$F_N - m_1 \cdot g = 0 \text{ so } F_N = 392\,\text{N}.$$

Now we can use F_N to calculate the frictional force:

$$F_f = \mu \cdot F_N = (0.25) \cdot (392\,\text{N}) = 98\,\text{N}.$$

Now we can sum up the forces in the horizontal dimension. Remember, however, that in analyzing m_2, we already defined the direction of motion. When m_2 moves downward the motion is negative. Thus, when m_1 moves to the right, motion is negative. This means:

$$F_f - T = m_1 \cdot a$$
$$98\,\text{N} - T = (40.0\,\text{kg}) \cdot a$$

Now we can plug in the equation we got for T from m_2 and solve for a:

$$98\,\text{N} - (20.0\,\text{kg})a - 196\,\text{N} = (40.0\,\text{kg})a.$$

So,

$$a = (98\,\text{N} - 196\,\text{N})/(60.0\,\text{kg}) = -1.6\,\text{m/s}^2.$$

Once again, the negative simply means that m_1 is moving to the right and m_2 is moving downwards.

2.6

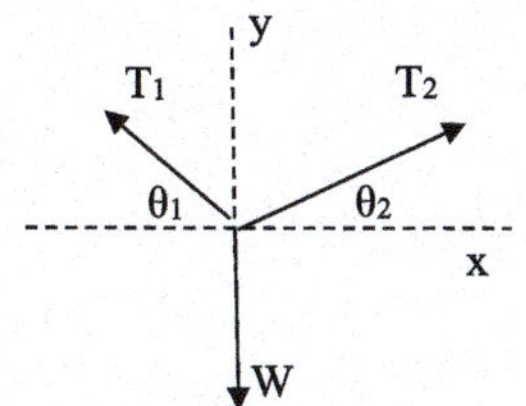

The force diagram for the problem is shown in the figure. The knowns and unknowns are:

Knowns: $\theta_1 = 45°; \theta_2 = 30°$, $m = 1.0\,\text{kg}$

Unknowns: $T_1; T_2$

Because it is in static equilibrium, the sum of the forces in the x-direction must equal zero,

$$-T_1 \cos 45° + T_2 \cos 30° = 0 \text{ or } -0.707T_1 + 0.866T_2 = 0$$

In the y-direction the sum of the forces also has to equal zero,

$$T_1 \cos 45° + T_2 \cos 30° - mg = 0 \text{ or } 0.707T_1 + 0.5T_2 = \left(9.8\,\text{m}/sec^2\right)(1.0\,\text{kg}).$$

If we add these equations for the x- and y-components we get,

$$0 + 1.366T_2 = 9.8\,\text{N or } \underline{T_2 = 7.17\,\text{N}}.$$

Substituting this value into the equation for the x-component forces and solving for T_1 gives

$$\underline{T_1 = 12.42\,N}.$$

Note that when the wires are attached at different angles, the tension in each wire is significantly different.

2.7 Knowns: $d = 0.22$ inches; $m = 38.0$ g; $\rho = 1290\ \text{g/m}^3$; $C_\text{D} = 0.50$

Unknowns: v_T

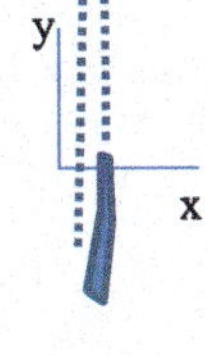

A bullet is cylindrical. However, it travels so that the circular part of the cylinder is perpendicular to the direction of motion. Thus, the area of the cylinder's circle (πr^2) is the cross-sectional area we need to calculate.

However, we need the unit to be m^2, so we must convert 0.11 inches (half the diameter) to m. That works out to 0.0028 m.

$$\text{Area} = \pi r^2 = (3.1415) \cdot (0.0028\,\text{m})^2 = 2.5 \times 10^{-5}\,\text{m}^2.$$

Now we can plug the numbers into 2.5 for terminal velocity:

$$v_T = \sqrt{\frac{2mg}{C_D \rho A}} = \sqrt{\frac{2(38.0\,g)\,m/9.81\,m/s^2)}{(0.50)(1290\,g/m^3)(2.5\times 10^{-5}\,m^3)}} = 210\,m/s.$$

Thus, if a 0.22-caliber rifle were shot directly into the air, it would fall back down to the earth with a maximum speed of 210 m/s.

2.8 Knowns: $m_i = 150.0$ kg; $m_p = 100.0$ kg; $a = 0$ m/s^2

Unknowns: θ

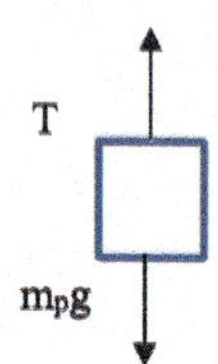

In this situation, we must once again analyze the masses individually. Let's start with the 100.0 kg mass. There are two forces, tension and gravity. If the system moves at *constant velocity*, that means *acceleration is zero*. Thus, the sum of the forces is zero. Let's say that motion of the 100.0 kg mass downwards is negative. That means:

$$T - (100.0\,\text{kg}) \cdot g = 0$$
$$T = (100.0\,\text{kg}) \cdot g = 981\,\text{N}$$

Instead of getting an equation for T this time, we actually got a number. That should make the next step even easier!

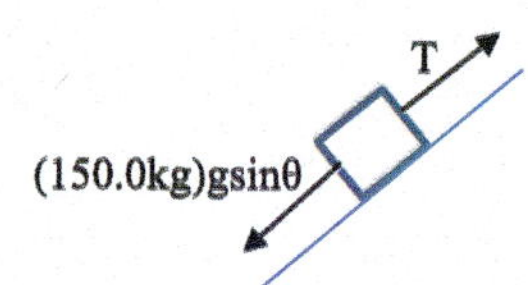

Now we look at the other block. This block has two forces working on it in the dimension parallel to the incline. The tension of the string is pulling it up the incline, and the component of the gravitation force that is parallel to the incline is pushing it down the incline.

There are forces in the dimension vertical to the incline but, since we are ignoring friction, we can ignore them.

Since motion of the 100.0 kg mass downwards is negative motion, that means motion up the incline is negative. This means that the sum of the forces (which equals zero) is:

$$(150.0\,\text{kg}) \cdot g \cdot \sin\theta - T = 0$$

We know what T is, so we can plug that number in and solve for theta.

$$(150.0\,\text{kg})(9.81\,\text{m/s}^2)\sin\theta - 981\,\text{N} = 0$$

$$\Theta = \sin^{-1}\left\{\frac{981\,\text{N}}{(150.0\,\text{kg})(9.81\,\text{m/s}^2)}\right\} = 41.8^\circ$$

If the angle of the incline is 41.8°, the masses will move at a constant velocity.

2.9 Knowns:
$\theta = 35^\circ$;
$m_1 = 35.0\,\text{g}$;
$m_2 = 25.0\,\text{g}$;
$m_3 = 75.0\,\text{g}$

Unknowns: T_1; T_2; a

Let's analyze the easiest mass first: the 75.0 g mass. There are only two forces acting on it, and they sum to give the mass times the acceleration.

Note that mass is not in the standard units, so we must convert to kg. If downward motion is negative:

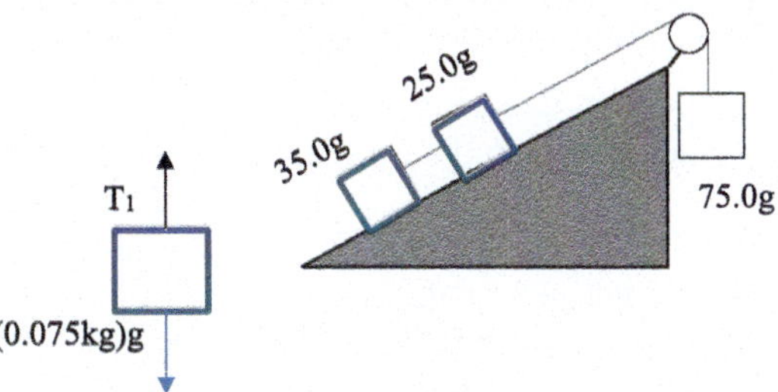

$$T_1 - (0.0750\,\text{kg}) \cdot g = (0.075\,\text{kg}) \cdot a$$
$$T_1 = (0.0750\,\text{kg}) \cdot a + 0.736\,\text{N}$$

This gives us an equation for the tension in the *first* string. However, there are two strings, so that complicates matters. Now let's look at the 25.0 g mass:

The tension in the second string, the component of the gravitational force parallel to the incline, and friction all pull on the mass. Only the tension of the first string pulls upward on the mass. Since we are dealing with friction we have to determine the normal force.

FN y
T1
x
(0.0250kg)gsinθ
T2
Ff
(0.0250)gcosθ

That's easy, however. The sum of the forces perpendicular to the incline tells us that $F_{\text{normal}} = (0.0250\,\text{kg}) \cdot g \cdot \cos\theta$. Since we know g and θ we can evaluate this. The normal force is 0.201 N. This means the kinetic frictional force is 0.0503 N. Now can sum up the forces parallel to the incline that work on this block:

$$T_2 + F_f + (0.0250\,\text{kg}) \cdot g \cdot \sin\theta - T_1 = (0.0250\,\text{kg}) \cdot a$$

Plugging in what we know (F_{friction}, g, θ, and the equation for T_1):

$$T_2 + 0.0503\,\text{N} + 0.141\,\text{N} - (0.0750\,\text{kg}) \cdot a - 0.736\,\text{N} = (0.0250\,\text{kg}) \cdot a$$

We can't solve this equation because it has two unknowns. However, we can solve for T_2 in terms of a and hope that the last mass gives us what we need.

$$T_2 = 0.545\,\text{N} + (0.100\,\text{kg}) \cdot a$$

The last mass is a bit easier. It has T_2 pulling it up the incline, and friction plus the gravitational force component parallel to the incline pulling it down. Since we are dealing with friction, we must also use the dimension perpendicular to the incline to determine the normal force.

The vertical dimension tells us that $F_{\text{normal}} = (0.0350\,\text{kg}) \cdot g \cdot \cos\theta$. This tells us that the frictional force is 0.0703 N. Summing up the forces parallel to the incline gives us:

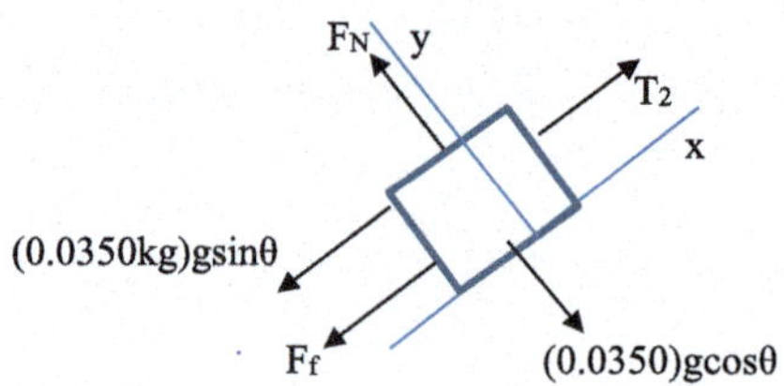

$$(0.0350\,\text{kg}) \cdot g \cdot \sin\theta + F_f - T_2 = (0.0350\,\text{kg}) \cdot a$$

Now we can plug in what we know from the other equations we derived and solve for a:

$$0.197\,\mathrm{N}\,0.0703\,\mathrm{N}\,0.545\,\mathrm{N}\,(0.1000\,\mathrm{kg})a(0.0350\,\mathrm{kg})a$$

$$a = (-0.278\,\mathrm{N})/(0.135\,\mathrm{kg}) = -2.06\,\mathrm{m/s^2}.$$

Then with the equations we derived for T_2 and T_1 we can calculate the tensions in the strings,

$$T_1 = (0.0750\,\mathrm{kg}) \cdot a + 0.736\,\mathrm{N} = 0.581\,\mathrm{N}$$
$$T_2 = 0.545\,\mathrm{N} + (0.1000\,\mathrm{kg}) \cdot a = 0.339\,\mathrm{N}$$

2.10 Knowns: $m_1 = 35\,\mathrm{kg}$; $m_2 = 15\,\mathrm{kg}$; $m_3 = 35\,\mathrm{kg}$; $P_1 = 375$; $\mu_k = 0.350$

Unknowns: a; forces on m_2

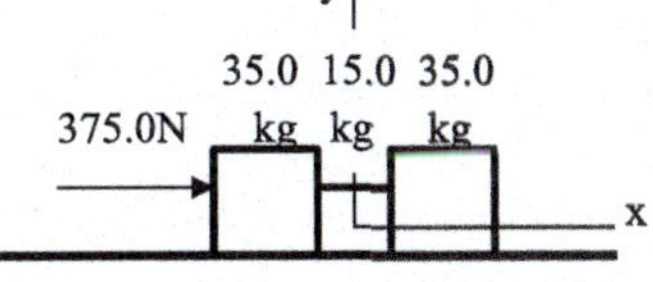

You could get the acceleration of the system by just adding up the masses and treating the three as one large mass. However, to answer the second part of the question, you have to do a full analysis. Thus, you might as well do it right away. The first block has the $P_1 = 375.0\,\mathrm{N}$ force pushing to the right, which we will define as the positive direction. It pushes the second block, which in turn pushes back to the left. Finally, friction pushes to the left. Since we are dealing with friction, we have to determine the normal force.

The vertical dimension tells us that the normal force is 343 N, which tells us the frictional force is 1.20×10^2 N. Summing up the horizontal forces, then:

$$375.0\,\mathrm{N} - P_1 - 1.20 \times 10^2\,\mathrm{N} = (35.0\,\mathrm{kg}) \cdot a$$

So,

$$P_1 = 255\,\mathrm{N} - (35.0\,\mathrm{kg}) \cdot a$$

Now we can move on to the next mass:

In this mass, the force with which the first mass pushes is pointed to the right. However, this mass pushes against the third mass, and in turn that third mass pushes back with an equal but opposite force, which I will call P_2. The vertical dimension tells us that the normal force is 147 N, which means the frictional force is 51.5 N.

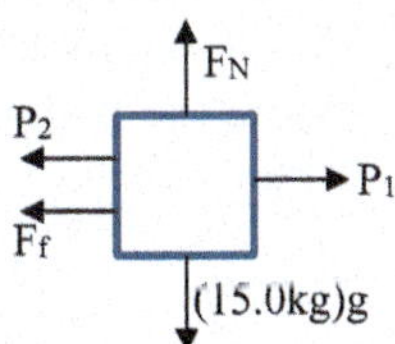

$$P_1 - F_f - P_2 = (15.0\,\text{kg}) \cdot a$$
$$P_2 = 204\,\text{N} - (50.0\,\text{kg}) \cdot a$$

Finally, we reach the last block. The vertical dimension tells us the normal force is 343 N, which tells us the frictional force is 1.20×10^2 N. Summing up the horizontal forces, then:

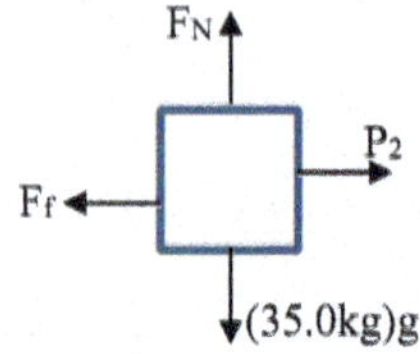

$$P_2 - F_f = (35.0\,\text{kg})a$$

So,

$$a = (84\,\text{N})/(85.0\,\text{kg}) = 0.99\,\text{m/s}^2$$

Now that we know a, we can figure out P_1 and P_2, which will tell us all of the forces acting on the 15.0 kg block.

$$P_1 = 255\,\text{N} - (35.0\,\text{kg}) \cdot a = 221\,\text{N}$$
$$P_2 = 203\,\text{N} - (50.0\,\text{kg}) \cdot a = 154\,\text{N}$$

Note that to have the correct answer, we must show the directions of each force as well as its magnitude and correct unit.

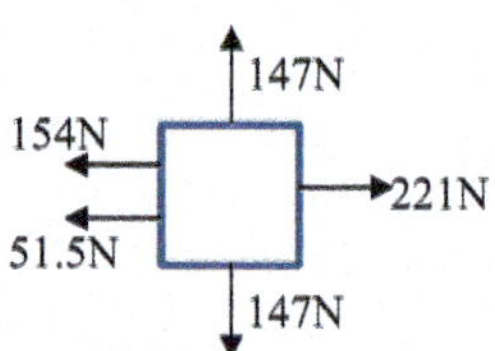

Study Guide for This Chapter

1. You are inside a large room, and there is a ball hanging on a string that is attached to the ceiling. The ball hangs straight down. As you watch, the ball seems to swing to the right of its own accord. Slowly, the ball begins to drift back to the point from which it started, and it once again hangs straight down from the ceiling. Throughout this entire time, there was no wind or other motion in the room.
   a. While the ball is moving, are you in an inertial reference frame? Why or why not?

b. What was the net direction of the sum of forces acting on the room while the ball was swinging to the right?
c. When the ball once again hung straight down from the ceiling, were you in an inertial reference frame?

2. Suppose you are in a car with a ball that hangs from a string attached to the ceiling. As you move with constant speed, the ball hangs straight down. Suppose you were to follow a curve in the road, never altering your speed. Which way would the ball swing?
3. Consider Atwood's machine. Suppose the string and pulley were weightless and there was no friction. If the masses on the machine move with constant velocity, what is the mass of the second compared to the first?
4. Suppose you were sitting on a car seat that could read the force with which your back pushed against the seat. When the car is at rest, the car seat reads 20 pounds, indicating that you are leaning back in the seat. Assume that you do not adjust the way you are sitting.
   a. As the car started to accelerate, would the reading go up or down?
   b. When the car reaches a constant speed, what will the chair read?
   c. As the car slowed to a halt, would the chair read higher or lower than 20 pounds?

   Questions 7–9 refer to the following situation:

   Two people, one of mass 75 kg and the other of mass 150 kg, are skating. There is essentially no friction. They stand at rest, and then the 75 kg skater puts her hands on the other skater and pushes.
5. Both skaters will begin to move. Why?
6. What direction will the 150 kg skater move in, compared to the 75 kg skater?
7. Compare the acceleration of the 150 kg skater to that of the 75 kg skater.
8. A football kicker kicks a ball in the air. Identify the action/reaction forces when the ball is kicked. Once in the air, identify the action/reaction forces, if there are any. Ignore air resistance, air pressure, and wind.
9. Find the tensions in each string if the ball in the figure has a mass of 100.0 kg.

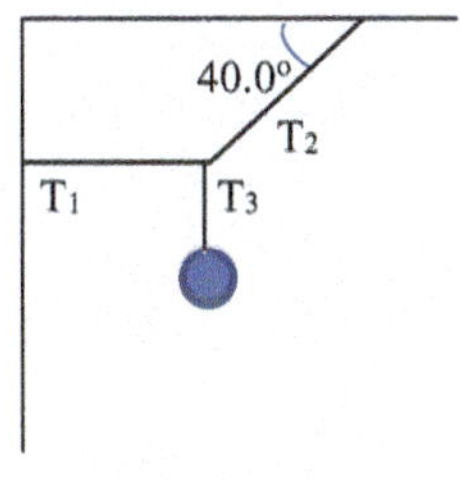

10. Consider the variation on Atwood's machine shown to the right. If $m_1 = 50.0\,\text{kg}$, $m_2 = 30.0\,\text{kg}$, and $m_3 = 30.0\,\text{kg}$, calculate the acceleration of the system and determine the tension in each string.

11. Consider the same Atwood's machine shown to the right. If someone were to grasp m_1 and pull down, what force would be required to keep m_1 moving downwards at a constant speed?

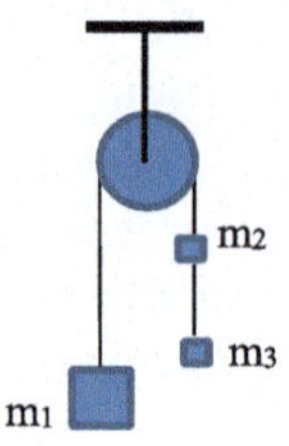

12. Consider the following drawing:

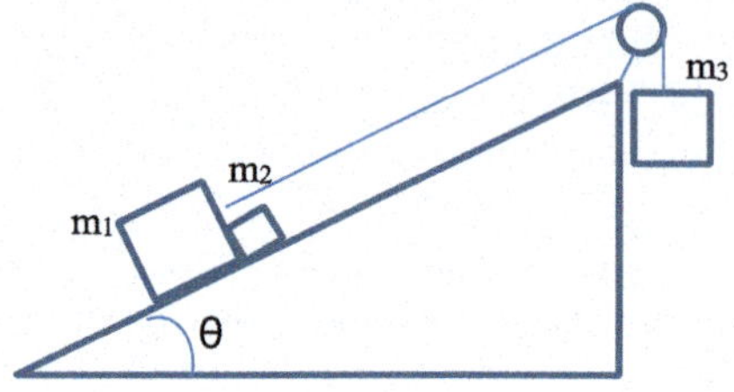

If $m_1 = 50.0\,\text{kg}$, $m_2 = 15.0\,\text{kg}$, and $\theta = 35.0°$ what must m_3 be to make m_1 and m_2 accelerate up the incline at $1.00\ \text{m/s}^2$? Assume that the pulley and string have no mass and experience no friction. However, take the friction between the blocks and the incline into account, assuming $\mu_k = 0.200$.

13. For the situation in the problem above, draw and label all of the forces that work on m_1. Include the magnitudes of the forces in your drawing.

14. A bosun's chair (developed for use on ships) consists of a chair attached to a rope that is slung over a pulley. As illustrated on the right, the person sitting on the chair then pulls down on the rope, and the chair lifts up. Suppose the chair and the person have a mass of 125 kg. What force must the person pull down with in order to lift the chair at a constant speed?

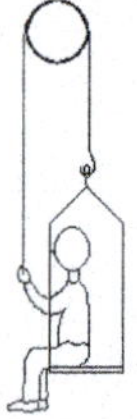

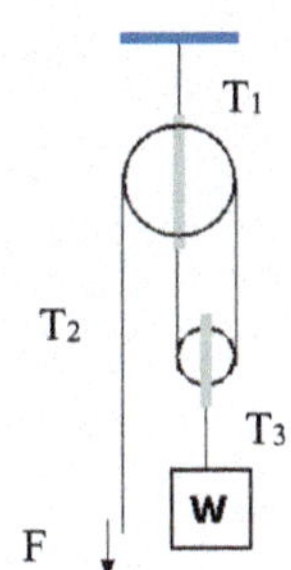

15. A block-and-tackle system is set up as shown to the right. There are three separate ropes. The first is the one which attaches the top pulley to the ceiling. The second is the one that goes through both pulleys and is attached at the bottom of the upper pulley. The last one attaches the weight (W) to the lower pulley. When you pull down on the rope, the weight is lifted up. Calculate the force needed to lift the weight at a constant speed. Leave your answer in terms of W.
16. In the situation shown to the right, what is T_1 in terms of W?
17. Suppose the W in the block-and-tackle system shown was a bosun's chair, with the F being supplied by the occupant of the chair. What force would be required to lift the chair at a constant speed?
18. In the drawing below, a 10.0 kg block is attached to the wall by a rope and then placed on top of a 50.0 kg block. The 50.0 kg block is then pulled with a force of 300.0 N. If μ_k between the two blocks is 0.250 and μ_k between the 50.0 kg block and the floor is 0.350, what is the acceleration of the 50.0 kg block and the tension in the rope?

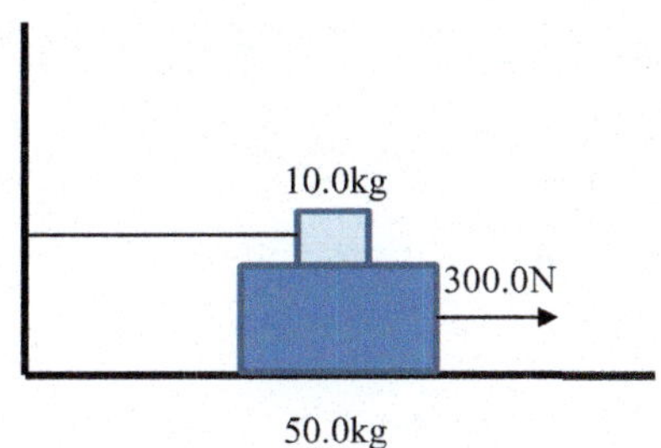

Next Level

19. What is the maximum speed of a 1.5 mm diameter bubble of air moving to the top of a bottle of water? ($C_D = 0.48$ for the drop; $\rho_{water} = 997\,kg/m^3$; $\rho_{air} = 1.225\,kg/m^3$)
20. What is the fastest speed of a brass ball bearing 2.0 mm in diameter falling to the bottom of a can of oil? ($C_D = 0.25$ for the ball bearing; $\rho_{oil} = 800\,kg/m^3$; $\rho_{brass} = 8587\,kg/m^3$)

Work, Energy, and Momentum

3

Chapter Summary

In physics, the concept of energy is not well defined. It is generally considered to be the capacity to do work. There are many different types of energy we deal with in physics. Energy cannot be created or destroyed but it can be converted from one type of energy to another. We also know that energy and mass are related. This chapter focuses on the properties of energy. Some of the themes of the chapter are represented in Figs. 3.1, 3.2 and 3.3.

Main Concepts in This Chapter

- Work and Energy
- Momentum, Impulse, and Recoil

3.1 Introduction

So far we have shown how to describe the motion of an object using the equations of motion plus Newton's force laws. Similar information can be obtained by analyzing the energy and momentum of the situation. Note that both of these approaches are correct. One is not right and the other wrong for a specific problem. However, depending on the knowns and unknowns, one approach might be a much easier way to solve the problem than the other approach. In this chapter we will review the basic concepts of energy and momentum and then apply them to different physical situations.

R. C. Powell, *Physics*, Undergraduate Lecture Notes in Physics,
https://doi.org/10.1007/978-3-032-09361-5_3

3.2 Work and Energy

The concept of energy is abstract and somewhat difficult to grasp. It is related to work which is easier to understand, so let's begin by discussing work. The concept of *work* has a specific definition in physics, given by the dot product of the force and the distance a force displaces an object,

$$W = \mathbf{F} \cdot \mathbf{X} = F \cos\theta \Delta x. \tag{3.1}$$

Here Δx is the distance an object moves and $F \cos \theta$ is the component of the force that caused the motion parallel to the direction of motion. θ is the angle between the applied force and the direction of motion. Thus, you can push on an object as hard as you can, but you do no work unless the object moves. Work is a scalar quantity. The unit of work in the SI system is a Joule which has dimensions of kg m^2 s^{-2}.

A force acting over a distance is called a conservative force if the amount of work done depends only on the initial and final positions and not on the path taken. Gravity is an example of a conservative force. The amount of work done by a dissipative force does depend on the path taken. Friction is an example of a dissipative force.

If we understand work, we can then consider the concept of energy. Energy is defined as **the ability to do work**. Although energy exists in many forms, they can all be classified as one of two types. The first is energy associated with motion. This is called kinetic energy and is expressed mathematically as

$$\mathrm{KE} = mv^2/2. \tag{3.2}$$

Note that the v in the equation refers to the scalar speed not the vector velocity.

The second type of energy is associated with the capability to do work that is stored in an object. This is called potential energy. This stored energy can be due to many different types of sources such as gravitational energy or chemical energy. Since by now we know something about gravity, we can use gravitational potential energy as an example. The expression for this type of potential energy is,

$$\mathrm{PE} = mgh. \tag{3.3}$$

This is the object's weight multiplied by its height above the surface of the earth. A rock sitting at the top of a hill, for example, has gravitational potential energy. It is not in motion, but if it were to roll down the hill, it would start moving faster and faster, thus gaining kinetic energy.

Where does the kinetic energy come from? Well, as the ball rolls down the hill, its potential energy is converted into kinetic energy. Note that the amount of potential energy an object has depends on where we have chosen to define the zero of potential energy in our coordinate system. The value of h in Eq. 3.3 is the height above this point of PE = 0 J. The expressions for chemical potential energy and other types of potential energy will be quite different. For example, in Chap. 7 we will define the potential energy of a charged particle in an electric field. Both kinetic and potential energy are scalar quantities, and the units of energy are Joules.

One of the most important laws of physics is conservation of energy.

This law states that for a closed system (no transfer of mass in or out), energy cannot be created or destroyed but can only be converted from one type of energy to another type.

This can be written as,

$$\mathrm{TE} = \mathrm{KE} + \mathrm{PE}, \tag{3.4}$$

$$\Delta\mathrm{TE} = 0. \tag{3.5}$$

In this expression TE is equal to the total energy of the system. This law holds for a closed system, which means there is no exchange of matter between the system and its surroundings and in addition there are no external forces acting on the system including important forces such as friction and air resistance.

The law of conservation of energy can be used to work problems as long as we know the expression to use for potential energy. Mechanical energy is the total energy of a system when the potential energy is due to gravity. In this case Eq. 3.4 becomes,

$$\mathrm{TE} = \mathrm{KE} + \mathrm{PE} = mv^2/2 + mgh \tag{3.6}$$

and it is this quantity that must be conserved.

If there are external forces acting on the system, the work done by the sum of these forces is equal to the change in the kinetic energy of the system,

$$\Delta\mathrm{KE} = W. \tag{3.7}$$

This equation is called the work-energy theorem.

We have just introduced a lot of equations. To show you how to work with problems from an energy perspective, you should perform the following experiment that utilizes Eq. 3.3.

Experiment 3.1
Ping Pong Pendulums

Supplies:

- Two ping pong balls
- Thread
- Scissors

- Cellophane tape
- Double-stick cellophane tape
- Ruler
- Pencil
- Sheet of lined paper (If you don't have lined paper, you can make the lines yourself.)
- Two nails (Small finishing nails are ideal.)
- Hammer
- A board that is at least a few inches wide and a few inches long. It must be thick enough to hammer the nails into the edge of the board.

Introduction: Colliding pendulums provide an excellent opportunity to study the energy concepts you learned in your first-year physics course.

Procedure:

1. Take the board and hammer the nails into one end of the board so that they are roughly centered and about 4 cm away from each other. The board should look like the following sketch if viewed from above:

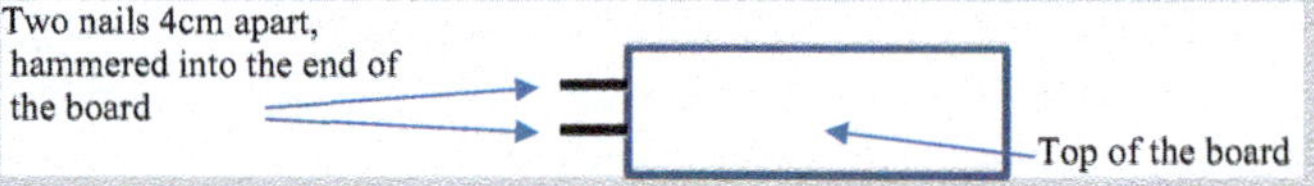

2. Cut two lengths of thread approximately 25 cm long. Use the regular (not the double-stick) cellophane tape to attach one end of each thread to each of the ping pong balls. Now you have two ping pong balls, each of which hangs from a thread.
3. Place the board on a desk or some other surface with a solid wall beneath it. For example, I used the top of a filing cabinet. You are going to hang the ping pong balls from the nails, but there must be a surface behind those ping pong balls to which you can tape the paper.
4. Tie the free end of each thread to each of the nails in the board. You now have two ping pong balls hanging from threads which are attached to the nails in the board. Adjust the board so that the nails hang far enough over the edge so that the balls are very close to but not touching the surface behind them.
5. The ping pong balls should be just barely touching each other. Adjust them so that they are at the same height. The best way to adjust them is to take the ball that is hanging lower and start to wrap its thread around the nail from which it hangs. That will raise the ball up until it is even with the other ball.
6. Now you are going to tape the lined paper to the surface which is behind the ping pong balls. If your paper does not have lines, make lines which

are 1.0 cm apart. The paper should be taped so that the lines run horizontally, level with the floor, and one of the lines should run through the center of the two ping pong balls. Your setup should look something like this:

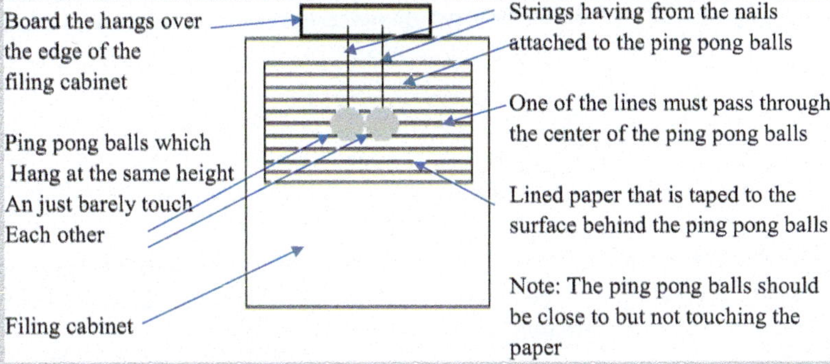

7. If you have paper with lines, measure the distance between the lines.
8. Now you are ready to begin the experiment. Pull one ball up and to the side so that the string stays taut. In essence, you want to turn this ball into a pendulum. Release the ball and let it swing.
9. Note what happens. When the swinging ball hits the ball that is still hanging straight down, the swinging ball stops and the ball that was just hanging there begins to swing upwards. It eventually reaches a maximum height and falls back down, hitting the other ball which is now hanging stationary. Once again, the ball that was moving stops, and the ball that was just hanging there rises again, starting the whole process all over again.
10. Why does this happen? When you pull the one ball up and back, you are giving it potential energy. How do you do that? You *work* on it. Work can either add energy to a system or take energy away, depending on the nature of the work. In this case, your work (lifting the ball) adds potential energy to the system. When you release the ball, that potential energy is converted into kinetic energy, and the ball begins to move. When it collides with the hanging ball, almost all of its energy is transferred to the other ball via the collision. Thus, the ball that was originally moving stops, because it has no more energy. However, the ball that was hanging now moves, because it has all sorts of kinetic energy. The ball then rises until all of that kinetic energy is into potential energy. Once that happens, the ball stops rising and starts falling, converting its potential energy back into kinetic energy.
11. Okay, now it is time to get a bit more detailed. Stop the balls so that they both hang down again, and note the line that runs through the center of the balls. We will call that line the "zero line," because it represents the

converted lowest position the balls can reach. Thus, it corresponds to a height of zero.

12. Now, lift one ball up and back again so that the string stays taut. Lift it high enough so that the center of the ball is even with the line which is 5 lines above the zero line. This corresponds to a height of 5 times the distance between the lines. If the distance between the lines is 0.85 cm, for example, this corresponds to a height of 4.25 cm.
13. Release the ball, and use the lines to read the maximum height that the other ball reaches after the collision. If the situation were ideal, this height would be identical to the first one. However, it will not work out that way. The height reached by the second ball will be lower than the height at which the first ball was released.
14. Now think about the energetics of this situation. Why doesn't the second ball reach the height from which the first ball was released? Well, in addition to the potential and kinetic energy considerations I have discussed so far, there is one more consideration: the work done by friction. As the balls move and collide, friction (mostly air resistance) works against the motion. This removes energy from the system. You can actually use your data to determine how much work friction performed.
15. The mass of an official ping pong ball is 2.2 g. Use Eq. 3.3 to calculate the potential energy of the first ball before it was released. Remember, mass must be in kg, and height must be in meters (standard units) for the units to work out. If the lines on the paper were 0.85 cm, apart, for example, the PE of the first ball would be 9.2×10^{-4} J.
16. Use the maximum height of the second ball to calculate the maximum potential energy of the second ball. Ideally, this should be the same as the potential energy of the first ball. After all, as the first ball fell, it converted all of its potential energy into kinetic energy. When it hit the second ball, the second ball received that kinetic energy and then began to rise, converting the kinetic energy back into potential energy. If that were the end of the story, then, the maximum potential energy of the second ball would be the same as that of the first. However, it will be lower, because friction worked on the balls, removing energy from the system.
17. Calculate how much energy friction removed from the system by taking the difference between the two potential energies. That number is the work done by friction.
18. Now I want you to modify your experiment a bit. Cover the two balls with double-stick tape so that when they collide, they will stick to one another.
19. Repeat the experiment. Notice that in this situation, the two balls stick together and rise to about half of the height that the second ball rose to in the first part of the experiment. Why? Well, when the balls stick

together, the object rising has twice as much mass. Thus, by Eq. 3.3, it needs to reach only half of the height to have the same potential energy.

20. Clean everything up.

This experiment should have helped to demonstrate the concepts of energy. Remember, in the absence of an outside force, the kinetic energy plus the potential energy of a system must stay constant. However, if an outside force is working on the system, that work can either add energy or take energy away from the system.

Fig. 3.1 Sledding down hill. *Credit* Michal Frackawiak, Creative Commons Share Alike 2.0

The work done on an object is equal to the change in its kinetic energy. If the work done is positive, it adds kinetic energy to the object, increasing its speed. If the work done is negative, it lowers the kinetic energy of the object, decreasing its speed. In the experiment, friction removed energy from the balls by working on them as they fell and as they collided. In this case, then, the work done was negative. Since work and energy have the same units, any work done either directly adds to or subtracts from the kinetic energy of the system.

Consider the following example.

Example 3.1

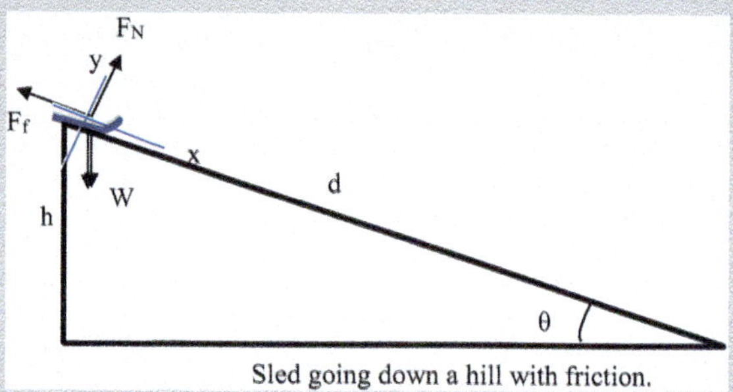

Sled going down a hill with friction.

As an example, consider the problem of a sled with a mass of 1 kg sliding for 20 m on a hill as shown in the figure above. If the height of the incline is 10 m and its angle is 30°, what is the sled's speed at the bottom of the hill if it starts from rest and the coefficient of friction between the sled and the hill is 0.5?

The sketch of the problem is shown above along with the coordinate system and force diagram. The knowns and unknowns are:

Knowns: $m = 1\,\text{kg}$; $h = 10\,\text{m}$; $\theta = 30°$; $\mu_f = 0.5$; $v_i = 0$; $d = 20\,\text{m}$.

Unknowns: v_f.

We have solved problems like this using Newton's Laws and the equations of motion. Now we will solve it using energy analysis. The change in potential energy for the sled from the top of the hill to the bottom is:

$$\Delta\text{PE} = mgh - 0 = (1\,\text{kg})\left(9.81\,\text{m/s}^2\right)(10\,\text{m}) = 98\,\text{J}.$$

If no friction is present, conservation of energy tells us this must equal the increase in kinetic energy of the sled,

$$\Delta KE_{\text{no friction}} = 98\,\text{J}.$$

However, we lose some of this kinetic energy to the work needed to overcome friction. The force of friction is

$$F_f = \mu_f F_N = (0.5)mg\cos 30^\circ$$
$$= (0.5)(1\,\text{kg})(9.81\,\text{m/s}^2)(0.866) = 4.25\,\text{N}.$$

The work done by the sled to overcome the force of friction is

$$W_f = F_f d = (4.25\,\text{N})(20\,\text{m}) = 85\,\text{J}.$$

Since this force acts opposite to the direction of motion, the angle between v and d is 180° so the work is negative. Subtracting this from the final KE with no friction gives,

$$KE_f = KE_{\text{no friction}} - W_f = 98\,\text{J} - 85\,\text{J} = 13\,\text{J}.$$

So,

$$mv_f^2/2 = 13\,\text{J} \text{ or } v_f = \left[(13\,\text{J})(2)/(1\,\text{kg})\right]^{1/2} = 5.1\,\text{m/s}.$$

This solution is much less complicated than the one we encountered using the equations of motion.

Now try the following Student problems to make sure you remember these energy concepts.

Student

3.1 Consider a block that is sitting at the top of a hill (height = 1.5 m). It starts from rest and slides to the bottom. Ignoring friction, what will its speed be at the bottom of the hill? Note: you do not need the mass of the block to answer this question!

3.2 Now let's consider friction in the problem above. Suppose the mass of the block is 150.0 g. In addition, suppose its speed at the bottom of the hill is 4.0 m/s. How much work did friction do as the block slid down the hill?

3.3 A block slides down a ramp ($h = 1.00\,\mathrm{m}$, $\theta = 45.0°$). It then slides across a level floor (length $= 0.200$ m) and up another ramp ($h = 0.250\,\mathrm{m}$, $\theta = 30.0°$) to another flat surface. If the coefficient of kinetic friction between all of the surfaces and the block is 0.150, what will the speed of the block be when it reaches the top of the second ramp?

3.4 An Atwood's machine (see Chap. 2) contains a 40.0 kg mass and a 20.0 kg mass. Neglecting friction, the mass of the string, and the mass of the pulley, what is the speed of the masses once they have each traveled 0.500 m?

Another concept related to work and energy is power. Power is defined as the rate of doing work. Mathematically this is expressed as

$$P = \Delta W/\Delta t. \tag{3.8}$$

Its units are J/s which is defined as a Watt. Because of the relationship between work and energy, power can also be considered as the rate at which energy is being used.

In the example problem we just worked of a sled sliding down a hill doing work to overcome friction, the power exerted can be calculated by first determining the time of travel. If this turns out to be 10 sec, then the power used to overcome friction is

$$P = \Delta W/\Delta t = 85\,\mathrm{J}/10\,\mathrm{s} = 8.5\,\mathrm{W}$$

Now, according to Eq. 3.1, $W = F \cdot x \cdot \cos\theta$. If force and displacement are parallel ($\theta = 0°$), then Eq. 3.1 becomes $W = F \cdot x$. If we substitute that into Eq. 3.8, we get:

$$P = F \cdot v. \tag{3.9}$$

This equation is relatively useful for calculating the power required in certain situations. However, be sure you understand its limits. Equation 3.9 is only applicable when force and displacement are parallel. Also, it is only applicable when velocity is constant. If velocity changes, or if the force and displacement are not parallel, then you must use Eq. 3.8. The following example shows you how this works.

Example 3.2

You are designing an elevator that has a mass of 1200.0 kg and must be capable of lifting at least 1000.0 kg of load. You estimate that an average frictional force of 5000.0 N fights its motion. What average power must the elevator motor provide in order to be able to accelerate the elevator and its full load at 2.00 m/s² upwards? Assume that the motor must accelerate the elevator until it reaches its "cruising speed" of 3.50 m/s. Once it reaches that speed, how much power will the motor need to supply to keep the elevator moving upwards?

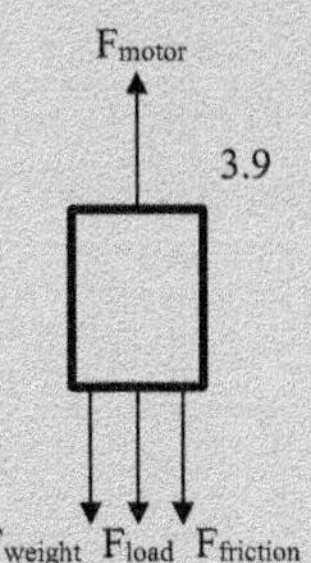

$$\text{Knowns: } m = 1200.0\,\text{kg}; \ \text{load} = 1000.0\,\text{kg};$$

$$F_{\text{friction}} = 5000.0\,\text{N}; \ a = 2.00\,\text{m/s}^2; \ v = 3.5\,\text{m/s}$$

$$\text{Unknowns: } P; \ P_{\text{v}}$$

You have to be very careful in doing a problem like this, because you are given a speed and some forces, and your first inclination might be to use Eq. 3.9. However, the first part of the problem discusses the situation when the elevator is *accelerating*. Thus, velocity is not constant, and as a result, we must use Eq. 3.8. Our first task, then, is to determine the work that the motor must do. To determine that, we need to know the force exerted by the motor.

As you see in the drawing above, the force exerted by the motor (F_{motor}) is working against three other forces. The weight of the elevator (F_{weight}) pulls down; the weight of the load that is being carried also pulls down (F_{load}); and the force of friction (F_{friction}) fights the motion, so it pulls down as well. To get the force required for acceleration, we sum up all of the forces and set them equal to mass times acceleration. We will call upward motion positive, leading to this equation:

$$F_{\text{motor}} - F_{\text{weight}} - F_{\text{load}} - F_{\text{friction}} = m \cdot a$$

What is the mass in the equation? Well, both the elevator and the load are being accelerated, so the mass is the sum of the two (2200.0 kg). Now let's plug in all of the numbers we know:

$$F_{\text{motor}} - (1200.0\,\text{kg})(9.81\,\text{m/s}^2) - (1000.0\,\text{kg})(9.81\,\text{m/s}^2) - 5000.0\,\text{N} = (2200.0\,\text{kg})(2.00\,\text{m/s}^2)$$

$$F_{\text{motor}} = 3.10 \times 10^4\,\text{N}$$

Now, what is the distance? For that, we need to remember our one-dimensional motion equations, because we know the acceleration, and we know the final cruising speed. Thus:

$$\begin{aligned} v^2 &= v_o^2 + 2ax \\ (3.50\,\text{m/s})^2 &= 0 + 2(2.00\,\text{m/s}^2)x \\ x &= 3.06\,\text{m}. \end{aligned}$$

We can now calculate work.

$$W = Fx\cos\theta = \left(3.10 \times 10^4\,\text{N}\right)(3.06\,\text{m})\cos 0^\circ = 9.49 \times 10^4\,\text{J}$$

Next we can determine time from our one-dimensional motion equations:

$$\begin{aligned} v &= v_o + at \\ 350\,\text{m/s} &= 0 + \left(2.00\,\text{m/s}^2\right)t \\ t &= 1.75\,\text{s}. \end{aligned}$$

Finally, at this point we can calculate the power exerted by the motor,

$$P = W/t = \left(9.49 \times 10^4\,\text{J}\right)/(1.75\,\text{s}) = 5.42 \times 10^4\,\text{W}$$

For the second part of the problem remember that to keep the elevator going at a constant velocity, the sum of the forces is zero,

$$\begin{aligned} F_{\text{motor}} - F_{\text{weight}} - F_{\text{load}} - F_{\text{friction}} &= 0 \\ F_{\text{motor}} &= 2.66 \times 10^4\,\text{N} \end{aligned}$$

Also, since the velocity is constant and the force and displacement are parallel, we can use Eq. 3.9

$$P = Fv = \left(2.66 \times 10^4\,\text{N}\right)(3.50\,\text{m/s}) = 9.31 \times 10^4\,\text{W}.$$

Notice that the power delivered once the elevator stops accelerating is *greater* than the power delivered while it is accelerating. That might seem strange, since the force required is lower once the elevator is no longer accelerating. However, remember that power is work divided by time. As the elevator accelerates, the motor applies a constant force, but the faster the elevator moves, the longer the distance traveled (and hence the greater the work) each second. Thus, the power consumption *increases* with increasing speed. So the power given in the first part of the problem is the *average* power. More power is used at the end of the acceleration period than is used at the beginning of the acceleration period. Once the speed is constant, the work per unit time is constant, so the power is constant as well.

Now try the following Student problems.

Fig. 3.2 Bat hitting a baseball. Creative Commons Share Alike 2.0

Fig. 3.3 Recoil of billiard balls. Creative Commons Share Alike 4.0

Student

3.5 A worker pushes a 150.0 kg crate horizontally along the floor at a constant rate of 1.55 m/s. If the coefficient of kinetic friction between the crate and the floor is 0.251, what is the power he exerts?

3.6 A machine is rated at a power of 500 W. If it exerts 10 N of force to push a 20 kg create at a constant speed, how long will it take to move the create 25 m? What is the coefficient of kinetic friction?

3.3 Momentum, Impulse, and Recoil

There is one final type of motion problem we need to consider. These are problems involving collisions with moving objects. In order to deal with collisions, we need to introduce the concept of momentum.

Anything that is moving has momentum. Momentum can be thought of as the quantity of motion that an object has. In physics, the official definition of momentum is the product of an object's mass and its velocity,

$$\mathbf{p} = m\mathbf{v}. \quad (3.10)$$

Note that momentum is a vector in the same direction as the velocity vector. The units of momentum are kg m/s.

Newton's second law can be rewritten in terms of momentum by using the definition of acceleration as the change in velocity with respect to time,

$$\mathbf{F} = m\mathbf{a} = m\Delta\mathbf{v}/\Delta t = (m\mathbf{v}_{\mathrm{f}} - m\mathbf{v}_{\mathrm{i}})/\Delta t = (\mathbf{p}_{\mathrm{f}} - \mathbf{p}_{\mathrm{i}})/\Delta t = \Delta\mathbf{p}/\Delta t \quad (3.11)$$

This is a different way to express Newton's second law. The important consequence from this expression is that if the net external force acting on a system is zero, the change in the momentum of the system is zero. This is called the law of conservation of momentum. Conservation of momentum along with the conservation of energy are two of the most important concepts in physics.

Sometimes it is useful to rewrite Eq. 3.11 as

$$\mathbf{J} = \Delta\mathbf{p} = \mathbf{F}\Delta t \quad (3.12)$$

where **J** is referred to as the impulse the force **F** gives to the system over time Δt. It is a vector in the direction of the external force and has the same units as momentum. This is sometimes called the impulse-momentum theorem. The force applied to the system may change during the time it is active in the collision process. If so, the average value of the force is used in Eq. 3.12.

Note that according to Eq. 3.12, there are two things that can make the impulse (change in momentum) larger, a greater magnitude of the force and a greater contact time for the force. So to get maximum range when hitting a ball, you need to swing hard and have a long follow through.

Special Topic

When you are running, the impact of your foot on the ground causes a jarring to your leg that can be harmful. This is caused by the change in momentum of your foot coming down on the ground and pushing off again. This is proportional to two factors: the force of the ground on your foot and the time during which this impact occurs. Since the momentum change is constant, the force must decrease if the time of impact increases. This can be accomplished by wearing running shoes that are heavily padded to cushion the impact.

These concepts of momentum are best illustrated by examples.

Example 3.3

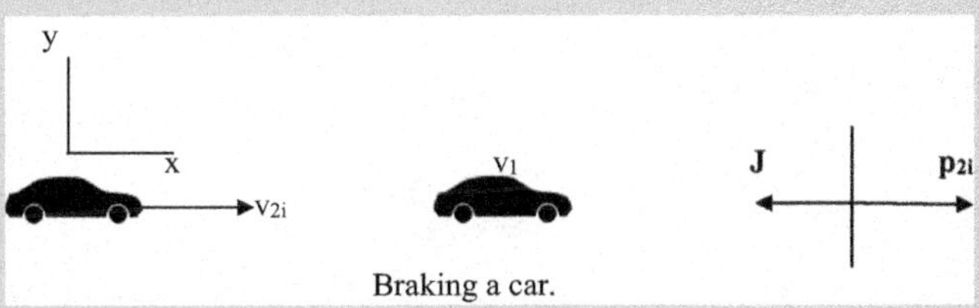

Braking a car.

Consider the motion of two cars as shown in figure above. Car 1 is stopped in the middle of traffic. Car 2 is traveling at 25 m/s in the same direction in the same lane of traffic. It has a mass of 1500 kg. When the driver of car 2 sees the stalled car in front of him, he breaks for 5 s to bring his car to a full stop. What impulse does this breaking have on car 2 and what is the average breaking force?

As always, we begin by sketching the problem as shown in the figure along with a vector diagram. Then we list the knowns and unknowns:

Knowns: $m_2 = 1500\,\text{kg}$; $v_{2i} = 25\,\text{m/s}$; $v_{2f} = 0\,\text{m/s}$; $\Delta t = 5\,\text{s}$; $v_1 = 0\,\text{m/s}$.
Unknowns: J; F_b.

Now we can find the initial and final momentum of car 2,

$$p_{2i} = m_2 v_{2i} = (1500\,\text{kg})(25\,\text{m/s}) = 37{,}500\,\text{kg}\,\text{m/s}^2.$$

$$p_{2f} = m_2 v_{2f} = 0\,\text{kg m/s}^2.$$

Now using the impulse-momentum theorem gives,

$$J = p_f - p_i = 0 - 37{,}500\,\text{kg m/s}^2 = -37{,}500\,\text{kg m/s}^2.$$

Also, $J = F_b \Delta t$ or

$$F_b = J/\Delta t = -37{,}500\,\text{kg m/s}^2/5\,\text{s} = -7500\,\text{N}.$$

Thus, the impulse on the car's motion due to breaking is 37,500 kg m/s^2 in the direction opposite to its initial velocity. The breaking force is 7500 N in the opposite direction of the motion.

Example 3.4

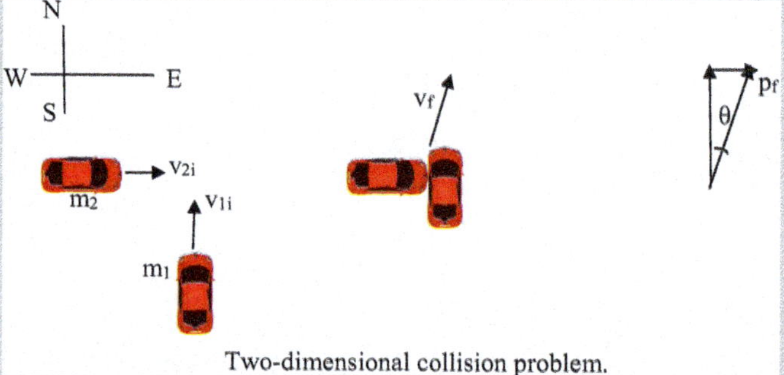

Two-dimensional collision problem.

As another example of this type of problem, consider the collision between these two cars shown in the figure above. In this case the two cars are not traveling in the same direction, so it becomes a two-dimensional problem and we must treat velocity and momentum as vectors. Car 1 is traveling north while car 2 is traveling east and they collide at an intersection. After the collision, the cars move together. Assuming no outside forces act on the two-car system, find the velocity of the two cars after the collision.

The sketch and vector diagram of the problem is shown in the figure above. The knowns and unknowns are:

Knowns: $m_1 = 1000\,\text{kg}$; $v_{1i} = 20\,\text{m/s}$; $m_2 = 1500\,\text{kg}$; $v_{2i} = 25$ m/s.

Unknowns: $\mathbf{v}_f$.

The numbers for this problem are the same as the previous problem but now we have to account for the directions of the velocity and momentum vectors. To do this, we break the vectors into their x- and y-components and apply the law of conservation of momentum to each set of components.

$$p_{fx} = p_{2i} = m_2 v_{2i} = (1500\,\text{kg})(25\,\text{m/s}) = 37{,}500\,\text{kg m/s}.$$
$$p_{fy} = p_{1i} = m_1 v_{1i} = (1000\,\text{kg})(250\,\text{m/s}) = 20{,}000\,\text{kg m/s}.$$

Then from the right triangle in the right of the figure,

$$p_f = \left(p_{fx}^2 + p_{fy}^2\right)^{1/2} = 4.24 \times 10^4\,\text{kg m/s}.$$

Thus, the final velocity is,

$$v_f = p_f/m_f = \left(4.24 \times 10^4\,\text{kg m/s}\right)/(2500\,\text{kg}) = 16.96\,\text{m/s}.$$

Now that we know the magnitude of the final velocity, we need to know its direction. Using the vector triangle shown in the figure, the angle θ is given by

$$\theta = \sin^{-1}(p_1/p_f) = \sin^{-1}(20{,}000\,\text{kg m/s}/42{,}400\,\text{kg m/s}) = 28.14°.$$

So, after the collision, the two cars travel together at a speed of 16.96 m/s in a direction 28.4° east of north.

To demonstrate that you understand these examples, work the following Student problem.

Student

3.7 Consider the situation of two cars shown in the figure below. In this case the lead car (car 1 m_1 = 1000 kg) is moving at 20 m/s and the trailing car (car 2 m_2 = 1500 kg) at 25 m/s and they are on ice so breaks do not work. This causes a collision in which car 2 rear-ends car 1, and they continue moving in the same direction locked together. What is the final velocity of the two cars after the collision?

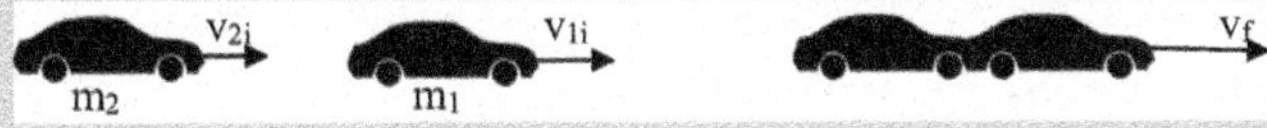

Collisions can be classified as elastic and inelastic. In elastic collisions kinetic energy is conserved. In an inelastic collision kinetic energy is not conserved. For example, in the collision between two automobiles, a significant amount of kinetic energy may be converted to the work done in denting the metal of the car. This changes the car's kinetic energy into heat energy of the molecules of the bending metal. Momentum is conserved in both types of collisions.

Let's look at some other examples of collisions.

Example 3.5

Two billiard balls of equal mass hit each other in a glancing (not head-on) collision. One ball is at rest, and the other ball is moving at 5.00 m/s. After the collision, the ball that was initially moving has a velocity of 2.89 m/s at 54.7°. What is the velocity of the ball that was initially at rest?

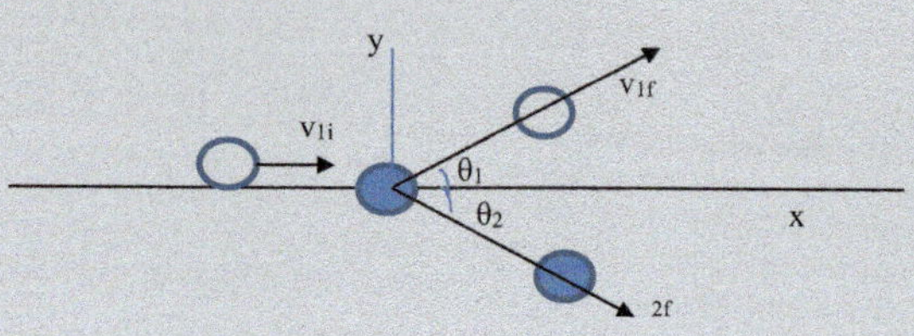

Knowns: $v_{i1} = 5.00\,\text{m/s};\ v_{if} = 2.89\,\text{m/s};\ \theta_1 = 54.7°$

Unknowns: v_{2f}

To solve this kind of problem, we split it up into two, one-dimensional problems. It turns out that the y-dimension is a bit easier in this case, since initially, there is no momentum in the y-dimension. Thus,

$$0 = v_{1f}\sin\theta_{1f} + v_{2f}\sin\theta_{2f}$$

Solving for the final velocity of ball 2 gives

$$v_{2f} = -v_{1f}\sin\theta_{1f}/\sin\theta_{2f}$$
$$v_{2f} = (-2.89\,\text{m/s})(\sin 54.7°)/\sin\theta_{2f} = -2.36\,\text{m/s}/\sin\theta_{2f}$$

That gives us two unknowns, but that's okay, because we have another dimension to investigate.

Recognizing that the masses will cancel in the x-dimension as well, we get:

$$v_{1i} + 0 = v_{1f}\cos\theta_{1f} + v_{2f}\cos\theta_{2f}$$
$$(5.00\,\mathrm{m/s}) = (2.89\,\mathrm{m/s})\cos 54.7^\circ + v_{2f}\cos\theta_{2f}$$
$$v_{2f}\cos\theta_{2f} = 3.33\,\mathrm{m/s}$$

Let's take the equation we got for v_{2f} from the y-dimension and stick it into this equation

$$3.33\,\mathrm{m/s} = (-2.36\,\mathrm{m/s})(\cos\theta_{2f})/\sin\theta_{2f}$$
$$\tan\theta_{2f} = (-2.36\,\mathrm{m/s})/(3.33\,\mathrm{m/s})$$
$$\theta_{2f} = -35.3^\circ$$

Notice that we used $\cos\theta/\sin\theta = 1/\tan\theta$ to solve this. You already should have this ingrained in your head from trigonometry. Now based on the way the *drawing* is made, that is θ_{2f}. However, the angle is not really defined properly. To give the properly defined angle, you must subtract it from 360.0° (it is in quadrant IV), so the properly defined angle is 324.7°.

Now we can use the relationship between v_{2f} and $\sin\theta_{2f}$ to get the speed:

$$v_{2f} = (-2.36\,\mathrm{m/s})/\sin\theta_{2f} = (-2.36\,\mathrm{m/s})/\sin 324.7^\circ = 4.08\,\mathrm{m/s}$$

So the velocity of the second ball is 4.08 m/s at 324.7°.

This example is an important principle in collision physics. Let's calculate the energy of the system *before* the collision:

$$\mathrm{TE} = \mathrm{PE} + \mathrm{KE} = 0 + 1/2 \cdot m \cdot v^2 = 1/2 \cdot m \cdot \left(25.0\,\mathrm{m^2/s^2}\right)$$

We don't know the mass of the billiard balls, but we can just leave it in terms of "m." What is the energy *after* the collision? At that point, there are two balls moving and still no potential energy. Thus:

$$\mathrm{TE} = \mathrm{PE} + \mathrm{KE} = 0 + 1/2 \cdot m \cdot v_{1f}^2 + 1/2 \cdot m \cdot v_{2f}^2$$
$$= 1/2 \cdot m \cdot \left(8.35\,\mathrm{m^2/s^2}\right) + 1/2 \cdot m \cdot \left(16.6\,\mathrm{m^2/s^2}\right)$$
$$\mathrm{TE} = 1/2 \cdot m \cdot \left(25.0\,\mathrm{m^2/s^2}\right)$$

The total energy is the same before and after the collision. That tells you that *this collision is elastic*. You were not told this to begin with, because you did not need that fact to solve the problem. Nevertheless, the calculations show that the collision is elastic.

This has an important result. Look at the two angles. Use their values to calculate the *angle between the velocities of the two balls.* The relative angle between the velocities of the two balls is $54.7° + 35.3° = 90.0°$. This is no accident. When two objects of equal mass collide elastically in a glancing collision and one mass is initially at rest, the relative angle between the objects after the collision will always be 90.0°. This is a useful bit of information you should know.

When two equal masses have a glancing, elastic collision and one of them is initially at rest, they will move away at a right angle relative to each other.

Now try one more Student problem.

Student

3.8 Two balls of equal mass collide with one another in a glancing, elastic collision. One ball is at rest, and the other moves with a velocity of 3.5 m/s towards the ball at rest. If the ball that was initially moving travels at an angle of 40.0° relative to its initial velocity, what is its speed?

Another important type of force we need to consider is a recoil force. Newton's third law tells us that for every force there is an equal and opposite force. This describes a recoil. One place that recoil plays an important role is in rocket engines. The expulsion of the burned particles of fuel creates a recoil force that propels the rocket. Similarly, firing a bullet out of a gun produces a recoil force on the gun. To demonstrate this, consider the following example.

Example 3.6

Consider a rocket with a mass of 500 kg that is launched by expelling 50 kg of burned fuel at a velocity of 1000 m/s. What is the launch speed of the rocket?

The sketch of the problem is given to the right. The knowns and unknowns are:

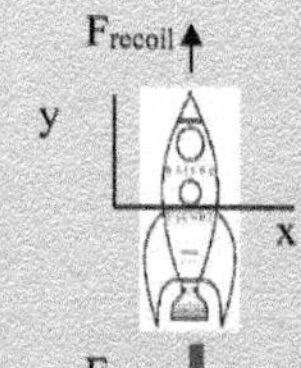

Knowns: $v_i = 0\,\text{m/s}$; $m_r = 500\,\text{kg}$; $m_{fuel} = 50\,\text{kg}$; $v_{fuel} = 1\,\text{km/s}$.

Unknowns: v_r.

The initial momentum of the rocket plus fuel is zero since they are at rest. Since momentum must be conserved,

$$0 = P_{rf} + P_{ff} = m_r v_r + m_{fuel} v_{fuel}.$$

So,

$$v_r = -m_{fuel}v_{fuel}/m_r = -(50\,kg)(-1000\,m/s)/(500\,kg) = 100\,m/s.$$

The launch speed of the rocket is 100 m/s in the upward direction. This is a recoil force to the force of the fuel ejected in the downward direction.

Try the one last Student problem for this chapter.

Student

3.9 Find the recoil velocity of a 5 kg rifle firing a 15 g bullet at a speed of 1000 m/s.

Next Level

Not all collisions are elastic. For inelastic collisions, we still have conservation of momentum and a energy but kinetic energy is not conserved since some off it. As an example, it is used to do work or create heat. Consider the collision shown in the figure. Conservation of momentum allows us to solve for the velocity after the collision,

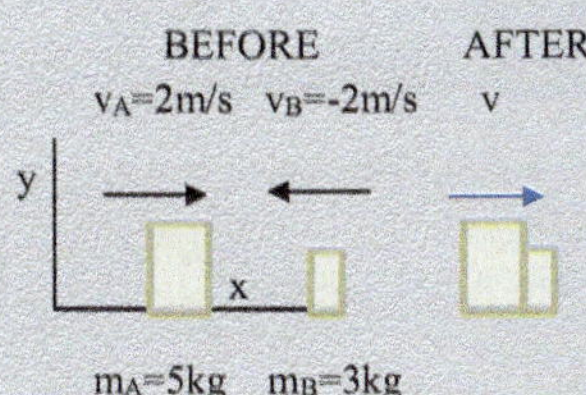

$$m_A v_A + m_B v_B = (m_A + m_B)v \quad \text{so}$$
$$v = (m_A v_A + m_B v_B)/(m_A + m_B) = 0.5\,m/s$$

Before the collision, the total kinetic energy is

$$KE = m_A v_A^2/2 + m_B v_B^2/2 = 16\,J.$$

After the collision, the total kinetic energy is

$$KE = (m_A + m_B)v^2/2 = 1\,J$$

In this collisions 15 J of energy is lost to deformation of the colliding object and heat.

Thus total energy is conserved but kinetic energy is not.

Summing Up

In this chapter you learned about the concepts of energy and momentum and how important they are in describing collisions between moving objects. It is important to remember that each of these concepts is conserved in any physical event, that is, neither energy nor momentum are created in the event, and they only change from one type of energy or momentum to another type. We will use these concepts and their conservation laws when we study different areas of physics in future chapters.

Answers to the Student Problems

3.1

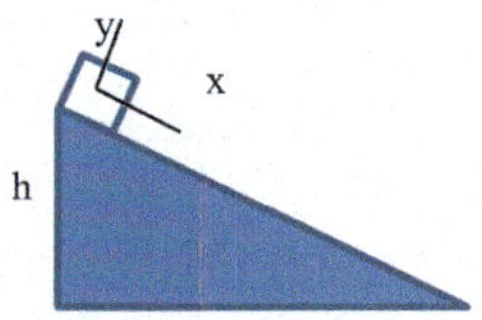

Knowns: $h = 1.5$ m; $v_i = 0$

Unknowns: v

When sitting on top of the hill, the block has only potential energy. Thus, its total energy is:

$$\text{TE} = \text{PE} + \text{KE} = mgh + 0 = m(9.81\,\text{m/s}^2)(1.5\,\text{m})$$

We can't get a number for TE, but that's okay, because we can just continue on and hope that we can either solve for mass or that it will cancel out. At the bottom of the hill, the total energy will be all kinetic.

$$\text{TE} = \text{PE} + \text{KE} = 0 + mv^2/2$$

We don't know either m or v, so that's the best we can do. Now, energy conservation says that (ignoring outside forces like friction which work on the system) the total energy must stay the same. Thus:

$$m(9.81\,\text{m/s}^2)(1.5\,\text{m}) = mv^2/2$$

Now we see that mass cancels, leaving only v as the unknown:

$$v = \sqrt{2(9.81\,\text{m/s}^2)(1.5\,\text{m})} = 5.4\,\text{m/s}$$

3.2

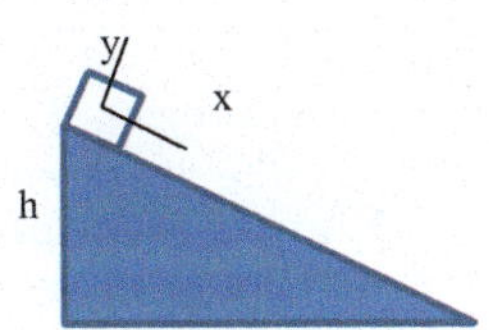

Knowns: $h = 1.5$ m; $v_i = 0$; $m = 150.0$ g; $v = 4.0$ m/s

Unknowns: W

Had friction done no work in the problem above, the block would have a speed of 5.4 m/s.

Friction, however, worked against motion, lowering the kinetic energy. How much energy does it have at the end?

$$\mathrm{TE} = \mathrm{PE} + \mathrm{KE} = 0 + mv^2/2 = (0.1500\,\mathrm{kg})\left(4.0\,\mathrm{m/s^2}\right) = 1.2\,\mathrm{J}$$

Notice that we had to convert from grams to kg to keep things in the standard units. That tells us how much energy the block had at the end. How much energy did it have at the beginning?

$$\mathrm{TE} = \mathrm{PE} + \mathrm{KE} = mgh + 0 = (0.1500\,\mathrm{kg})\left(9.81\,\mathrm{m/s^2}\right)(1.5\,\mathrm{m}) = 2.2\,\mathrm{J}$$

Why was there more energy at the beginning than at the end? Friction *worked* on the block, removing kinetic energy. Thus, the difference between the energy before and after (2.2 J − 1.2 J) represents the work done by friction. Thus, friction did 1.0 J of work.

3.3

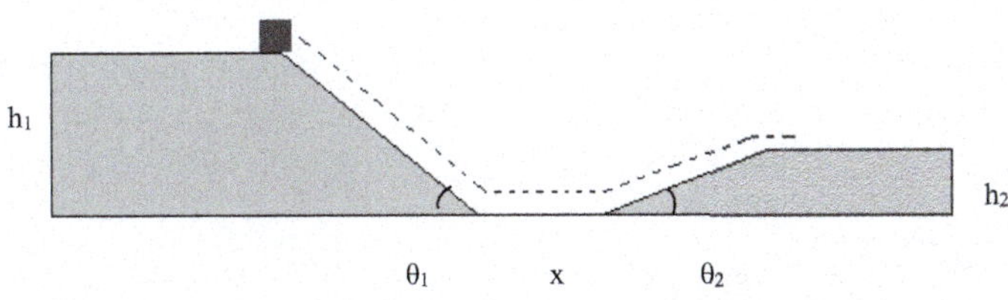

Knowns: $h_1 = 1.00$ m; $h_2 = 0.250$ m; $\theta_1 = 45.0°$; $\theta_2 = 30.0°$; $x = 0.200$ m; $\mu_k = 0.150$

Unknowns: v_2

At first, the block has only potential energy. Thus, the total energy is:

$$\mathrm{TE} = \mathrm{KE} + \mathrm{PE} = 0 + mgh = m \cdot \cdot = m\left(9.81\,\mathrm{m/s^2}\right)(1.00\,\mathrm{m}) = 9.81\,\mathrm{mJ}$$

Notice that we have to leave "m" in as a variable. Hopefully, it will cancel out later. If there were no friction, then the block would have 9.81 m J at the end of the journey. However, there is friction. Thus, we must calculate the work that friction does and remove that energy from the total energy of the object. This is where we have to be a bit careful. On the inclines, the normal force is *not* $m{\cdot}g$. It is $m{\cdot}g{\cdot}\cos\theta$, because only a portion of the weight is perpendicular to the surface. Thus, the frictional force on the first ramp is:

$$\begin{aligned} f &= \mu_k F_N = (0.150)mg\cos\theta_1 \\ &= (0.150)m(9.81\,\mathrm{m/s^2})\cos(45.0°) = 1.04\,\mathrm{mN} \end{aligned}$$

Okay, that's the frictional force while the block is on the first incline. To get the work done by friction, we multiply the force by the distance and by the cosine of the angle between force and displacement. What is that angle? Well, the frictional force works *along* the surface of the incline. We can calculate the distance down the incline using trigonometry:

$$\begin{aligned} &\sin\theta = \text{opposite/hypotenuse} \\ &\text{length of incline} = 1.00\,\mathrm{m}/\sin(45.0°) = 1.41\,\mathrm{m} \end{aligned}$$

Of course, the block slides *down* the incline, so that's the direction of the displacement. The frictional force works *up* the incline, so the angle between force and displacement is 180.0°.

The work done by friction while sliding down the first ramp, then, is:

$$W_{\mathrm{ramp1}} = (1.04\,\mathrm{mN})(1.41\,\mathrm{m})\cos(180.0°) = -1.47\,\mathrm{mJ}$$

Work is negative because it is *removing* energy from the system. That's just one part of the trip, however. When the block slides between the ramps, the normal force is $m{\cdot}g$, because the surface is flat. Friction works along the surface, so the displacement is the length of the flat surface, 0.200 m. However, friction and displacement are opposed, so the angle is 180.0° again. Thus, the work done by friction during that segment is:

$$\begin{aligned} W_{\mathrm{flat}} &= fx\cos(180.0°) = \mu_k mgx\cos(180.0°) \\ &= -(0.150)m(9.81\,\mathrm{m/s^2})(0.200\,\mathrm{m}) = -0.294\,\mathrm{mJ} \end{aligned}$$

There is still one more segment. As the block slides up the next ramp, friction works on it again, but the normal force is less now that it is once again on the ramp. To get the work, we will need to know the displacement again, which is 180.0° relative to friction:

$$\begin{aligned} &\sin\theta = \text{opposite/hypotenuse} \\ &\text{length of incline} = 0.250\,\mathrm{m}/\sin(30.0°) = 0.500\,\mathrm{m} \end{aligned}$$

We can now calculate the work done by friction in the last segment:

$$\begin{aligned} f &= \mu_k F_N = (0.150) mg \cos\theta_2 \\ &= (0.150) m \left(9.81\,\mathrm{m/s^2}\right) \cos\left(30.0^\circ\right) = 1.27\,\mathrm{mN} \\ W_{\mathrm{ramp2}} &= (1.27\,\mathrm{mN}) \cdot (0.500\,\mathrm{m}) \cdot \cos\left(180.0^\circ\right) = -0.635\,\mathrm{mJ} \end{aligned}$$

The *total* work done by friction is just the sum of the work done in each segment, or − 2.40 mJ. This adds to the original energy, to give us the total energy after the block reaches the final flat surface:

$$\mathrm{TE}_{\mathrm{after}} = 9.81\,\mathrm{mJ} - 2.40\,\mathrm{mJ} = 7.41\,\mathrm{mJ}$$

The block has potential energy (it is higher than the floor) and kinetic energy. Thus:

$$\begin{aligned} &\mathrm{TE} = \mathrm{PE} + \mathrm{KE} = mgh + mv^2/2 \\ &7.41\,\mathrm{mJ} = m\left(9.81\,\mathrm{m/s^2}\right)(0.250\,\mathrm{m}) + mv^2/2 \\ &v = 3.15\,\mathrm{m/s} \end{aligned}$$

So the mass did, eventually, cancel.

3.4 The sketch is given below. The knowns and unknowns are:

Knowns: $m_1 = 20.0$ kg; $m_2 = 50.0$ kg; $x = 0.500$ m

Unknowns: v

In Atwood's machine, the heavier block travels down, and the lighter block travels up. Thus, the lighter block gains potential energy, but since it is accelerating, it gains kinetic energy as well. The heavier block loses potential energy and gains kinetic energy. To determine how all of this "pans out," we first need to define potential energy *relative* to something (Fig. 3.4).

Remember, potential energy has to be defined relative to a point, so let's choose the most convenient point. We know that, in the end, the heavy block will move at least 0.500 m down, since we are asked to calculate speed at that point. Thus, let's make that the definition of zero potential energy. If anything is 0.500 m lower than the point the blocks started, we will call that zero potential energy.

Okay, then, what is the total energy before the masses are released?

There is no kinetic energy, but both masses have potential energy:

$$\begin{aligned} \mathrm{TE} &= \mathrm{PE} + \mathrm{KE} = m_1 g h_1 + m_2 g h_2 \\ \mathrm{TE} &= (20.0\,\mathrm{kg})\left(9.81\,\mathrm{m/s^2}\right)(0.500\,\mathrm{m}) + (40.0\,\mathrm{kg})\left(9.81\,\mathrm{m/s^2}\right)(0.500\,\mathrm{m}) \\ &= 294\,\mathrm{J} \end{aligned}$$

Fig. 3.4 Atwood's machine

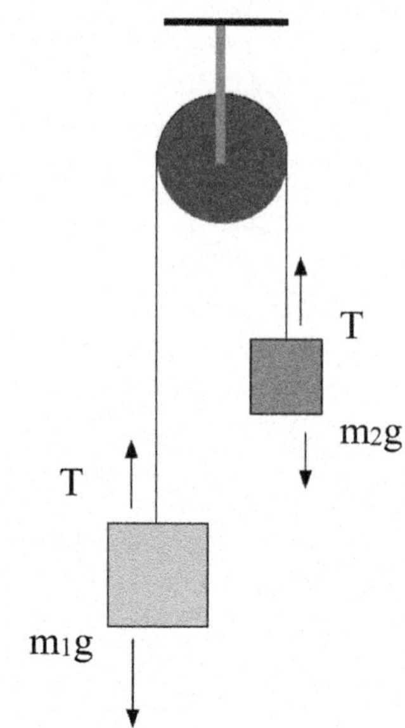

After the masses have traveled 0.500 m, both will have kinetic energy, but the heavy one (m_2) will be at the point we have defined as zero potential energy, so it will have *no* potential energy.

The lighter mass (m_1), however, will now be 1.000 m *above* the zero potential energy point, so it will definitely have potential energy.

$$
\begin{aligned}
\text{TE} &= \text{PE} + \text{KE} = m_1 g h_1 + m_1 v_1^2/2 + m_2 v_2^2/2 \\
294\,\text{J} &= (20.0\,\text{kg})\left(9.81\,\text{m/s}^2\right)(1.000\,\text{m}) + (20.0\,\text{kg})v_1^2/2 + (40.0\,\text{kg})v_2^2/2 \\
98\,\text{J} &= (10.0\,\text{kg})v_1^2 + (20.0\,\text{kg})v_2^2
\end{aligned}
$$

You might think that we are stuck here because we have one equation and two unknowns.

However, remember that these masses start at rest, and, since they are attached by a string that doesn't "bunch up," they accelerate at the same rate. Thus, they will always have the same speed since they start off with the same velocity and experience acceleration of the same magnitude. Thus,

$$
\begin{aligned}
v^2 &= 98\,\text{J}/3.0\,\text{kg} \\
v &= 1.8\,\text{m/s}
\end{aligned}
$$

Note that you could have used Newton's Laws to solve this, as discussed in the previous chapter.

3.5

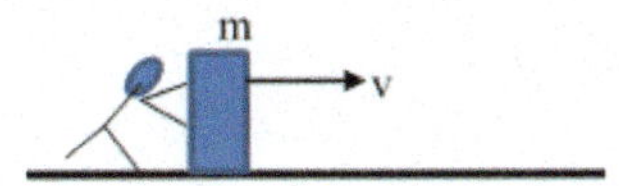

Knowns: $m = 150.0$ kg; $v = 155$ m/s; $\mu_k = 0.251$

Unknowns: P

Force and displacement are parallel, and velocity is constant. Thus, we can use Eq. 3.9. To use that equation, we need to know the force that is doing the work and the velocity.

We know the velocity. What about the force? To be moving the crate at a constant velocity, Newton's Second Law says that the sum of the forces on the crate must be zero. Thus, the worker must supply just enough force to counteract friction.

$$f = \mu_k F_N = (0.251)mg = (0.251)(150.0\,\text{kg})\left(9.81\,\text{m/s}^2\right) = 369\,\text{N}$$

Now we can use Eq. 3.9

$$P = Fv = (369\,\text{N})(1.55\,\text{m/s}) = 572\,\text{W}$$

The units work out here because a N m is a Joule; thus, the units multiply to give J/s, which is a Watt.

3.6

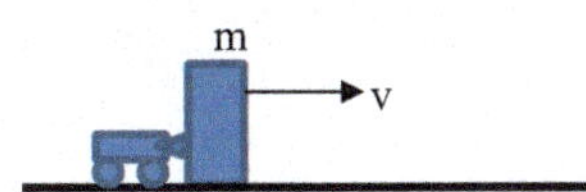

Knowns: $P = 500$ W; $m = 20.0$ kg; $x = 25$ m; $F =$ 10 N

Unknowns: t; μ_k

A machine is rated at a power of 500 W. If it exerts 10 N of force to push a 20 kg crate at a constant speed, how long will it take to move the create 25 m? What is the coefficient of kinetic friction is?

This is a simple application of Eq. 3.9.

$$P = Fv = (10\,\text{N})v = 500\,\text{W}$$
$$v = P/F = (10\,\text{N})/(500\,\text{W}) = 2.0 \times 10^{-2}\,\text{m/s}$$

Now we can use our kinetic equation of motion to get the time.

$$x = vt = 25\,\text{m} = \left(2.0 \times 10^{-2}\,\text{m/s}\right)t$$
$$t = 25\,\text{m}/\left(2.0 \times 10^{-2}\,\text{m/s}\right) = 1250\,\text{s}$$

Since the crate is moving at a constant speed, the force pushing on it is equal to the force of friction working against it.

$$F = f = \mu_k mg = 10\,\mathrm{N} = \mu_k(20.0\,\mathrm{kg})\left(9.81\,\mathrm{m/s^2}\right)$$
$$\mu_k = 10\,\mathrm{N}/(196.2\,\mathrm{N}) = 0.05$$

3.7

Knowns: $m_1 = 1000$ kg; $v_{1i} = 20$ m/s; $m_2 = 1500$ kg; $v_{2i} = 25$ m/s
Unknowns: $\mathbf{v}_\mathrm{f}$

Since they are on ice, we can assume there is no friction in this problem, and therefore there are no external forces acting on the two-car system. In this case, the law of conservation of momentum holds so,

$$p_\mathrm{f} = p_\mathrm{i} \text{ or } (m_1 + m_2)v_\mathrm{f} = m_1v_{1i} + m_2v_{2i}.$$

Since all the velocity and momentum vectors are in the same direction, we can work with only the magnitudes of these quantities. Substituting in the numbers gives,

$$\begin{aligned} v_\mathrm{f} &= (m_1v_{1i} + m_2v_{2i})/(m_1 + m_2) \\ &= \left[(1000\,\mathrm{kg})(20\,\mathrm{m/s}) + (1500\,\mathrm{kg})(25\,\mathrm{m/s})\right]/(2500\,\mathrm{kg}) \\ &= 23\,\mathrm{m/s}. \end{aligned}$$

So after the collision, the two cars will move together at a speed of 23 m/s in the same direction that they were heading.

3.8

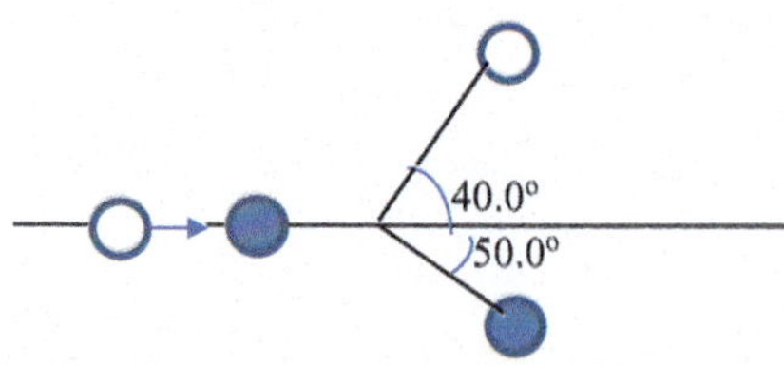

Knowns: $m_1 = m_2$; $v_{1i} = 3.5$ m/s; $v_{2i} = 0$ m/s; $\theta_1 = 40.0°$
Unknowns: v_{1f}

Since the two balls have equal mass, and since the collision is elastic, we know that in a glancing collision, their relative angles will be 90.0°. Thus, if the first ball has a velocity angle of 40.0°, the other ball angles off at 50.0° below the line defined by the collision:

To do the vector math, however, we need to DEFINE the angles properly. The first angle, 40.0° is defined properly already. However the angle we deduced is not 50.0°. It is 310.0°. That's what we need to use in the vector equations.

Now let's conserve momentum in each dimension. Let's start with the y-dimension:

$$m_1 v_{1i} \sin\theta_{1i} + m_2 v_{2i} \sin\theta_{2i} = m_1 v_{1f} \sin\theta_{1f} + m_2 v_{2f} \sin\theta_{2f}$$

Since the masses are equal, they cancel. That leaves:

$$v_{1i} \sin\theta_{1i} + v_{2i} \sin\theta_{2i} = v_{1f} \sin\theta_{1f} + v_{2f} \sin\theta_{2f}$$

Filling in what we know:

$$(3.5\,\mathrm{m/s})(0.00) + (0) \sin\theta_{2i} = v_{1f} \sin 40.0^\circ + v_{2f} \sin 310.0^\circ$$
$$v_{2f} = 0.839 v_{1f}$$

That gives us two unknowns, but that's okay, because we have another dimension to investigate.

Recognizing that the masses will cancel in the x-dimension as well, we get:

$$v_{1i} \cos\theta_{1i} + v_{2i} \cos\theta_{2i} = v_{1f} \cos\theta_{1f} + v_{2f} \cos\theta_{2f}$$
$$(3.5\,\mathrm{m/s}) \cos 0.00^\circ + (0) \cos\theta_{2i} = v_{1f} \cos 40.0^\circ + v_{2f} \cos 310.0^\circ$$
$$(3.5\,\mathrm{m/s}) \cos 0.00^\circ = v_{1f} \cos 40.0^\circ + v_{2f} \cos 310.0^\circ$$
$$= 0.766 v_{1f} + 0.643 v_{2f}$$

Taking the equation from the y-dimension and sticking it into this equation:

$$(3.5\,\mathrm{m/s}) = 0.766 v_{1f} + 0.643(0.839) v_{1f} = 1.31 v_{1f}$$
$$v_{1f} = 2.7\,\mathrm{m/s}$$

That's actually all the problem asked for. You could go back to the equation developed in the y-dimension and find out that $v_{2f} = 2.25\,\mathrm{m/s}$. One way you could see whether or not the answers are correct is to confirm that energy is conserved. Indeed, since the square of the initial speed is equal to the sum of the squares of the final speeds, and since the masses are equal, energy is conserved.

3.9

Knowns: $m_r = 5$ kg; $v = 1000$ m/s; $m = 15$ g

Unknowns: v_r

Initially the bullet and rifle are at rest son the total momentum is zero. It must be the same after the rifle is fired.

$$0 = p_b + p_r = mv + m_r v_r$$

$$v_r = -mv/m_r = \left(15 \times 10^{-3}\,\text{kg}\right)(1000\,\text{m/s})/(5\,\text{kg}) = 75\,\text{m/s}$$

Study Guide for This Chapter

1. A ball hits a wall and bounces straight back along the same path. What is the impulse delivered to the ball by the wall in terms of the ball's mass (m) and the ball's speed (v)?
2. Two objects collide with one another and move away from each other. There are no net forces acting on the objects. Could the collision be elastic? Could it be inelastic? Could it be perfectly inelastic?
3. Consider the situation described in question #3. Suppose the total kinetic energy of the objects after the collision is lower than it was before the collision. What kind of collision was it? Was momentum conserved?
4. Is energy ever conserved in a perfectly inelastic collision? If so, under what conditions?
5. Is momentum ever conserved in a perfectly inelastic collision? If so, under what conditions?
6. A student sees a laboratory notebook in which the following notation is made: "Total Energy is − 2.3 J." The student says that this must be wrong, since energy cannot be negative. Why is the student wrong?
7. In this chapter, we used the equation PE = $m \cdot g \cdot h$ quite often. When is that equation valid?
8. Two machines do *exactly* the same amount of work, but the second requires twice as much power. Assuming both machines are equally efficient, what can you say about the time it takes for each machine to get the job done?

9. Consider the desktop toy made by suspending four balls of equal mass. When ball 1 is pulled to the right and released, it falls, striking the other balls. In response, ball 4 will begin moving at the same speed as that which ball 1 had the instant before the collision. This, of course, is because of momentum conservation. If balls 1 and 2 are both lifted and released together, they will hit balls 3 and 4, and *both* ball 3 and ball 4 will begin moving, each with the same speeds that balls 1 and 2 had the instant before the collision. Once again, momentum is conserved. However, momentum *could* be conserved if ball 4 began moving at twice the speed of balls 1 and 2 and all of the other balls remained motionless. That never happens, however. Why not?

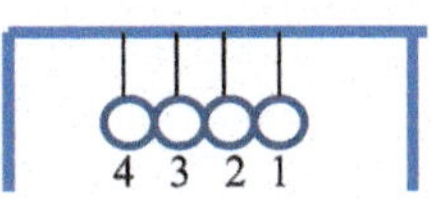

10. A person skis down a hill that is covered with snow. Assume that there is no friction on the snow. If the hill is 150 m high, how fast will the skier be moving at the bottom of the hill? Ignore air resistance and any initial shove that the skier uses to get started.
11. For the skier in problem #1, assume that when he reaches the bottom of the hill, the snow is sparse, and as a result, the coefficient of kinetic friction is 0.550. Assume that the ground at the bottom of the hill is flat. How far will the skier travel before stopping?
12. A student drops a rock off the edge of a cliff. When the rock is halfway down the cliff, what is its speed in terms of its speed right before it hits bottom?

13. A box sits on top of a ramp with a height of 1.00 m and an angle of 45.0°. The μ_k is 0.150. When it reaches the bottom of the hill, it slides on a frictionless surface and goes around a loop (also frictionless) that is 0.300 m high at its tallest point. What is the speed of the box at the top of the loop?

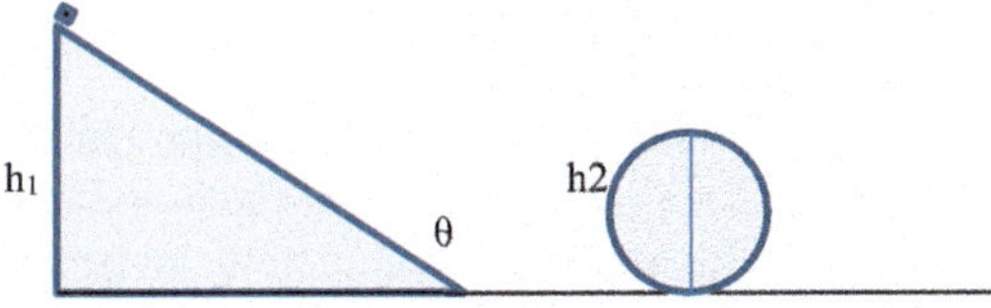

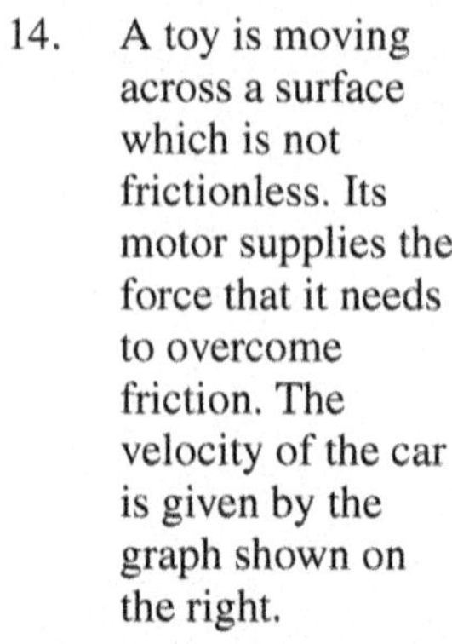

14. A toy is moving across a surface which is not frictionless. Its motor supplies the force that it needs to overcome friction. The velocity of the car is given by the graph shown on the right.

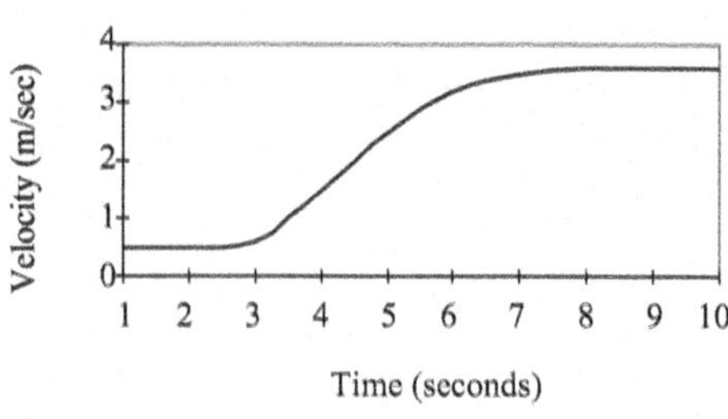

a. Is the motor exerting more force at $t = 8.0$ s or at $t = 5.0$ s?

b. Is the motor exerting more power at $t = 2.0$ s or $t = 9.0$ s?

15. A ball of mass 5.0 kg is kicked with a 55 N force. If the ball begins traveling at 3.5 m/s as a result, how long did the force act on the ball?
16. Two balls of equal mass collide elastically. Ball 1 is initially at rest, and ball 2 moves with an unknown velocity. After they collide, the ball initially at rest moves with a velocity of 2.00 m/s at an angle of 45.0° relative to the initial velocity of the ball that was moving. What was the initial velocity of ball 2?

Next Level

17. Pieces of candy are dropped onto the pan of a spring scale from a height of 1.3 m at a rate of 6 per second. Each one has a mass of 0.028 kg. If the candy collides with the scale completely inelastically, what weight does the scale read after 10 s?
18. Consider Student problem 3.7. How much kinetic energy is lost in this inelastic collision?

Rotational Motion I: Kinematics and Dynamics

4

Chapter Summary

The invention of the wheel in about 3500BC enabled mankind to make use of rotational motion. The wheel is one of the six simple machines that allow us to change the direction or magnitude of forces. It is considered to be one of the most important inventions of all times. We think of it as revolutionizing transportation, but it has many very different uses such as a water wheel or a potter's wheel. Some of the themes of the chapter are represented in Fig. 4.1.

Main Concepts in This Chapter

- Rotational Motion Parameters
- Center of Mass and Moment of Inertia
- Rotational Dynamics

4.1 Introduction

In the previous chapters we have been considering the motion of objects that move from one point to another in space without going through a specific position more than once. This usually referred to as translational motion. There is another type of motion called periodic motion where an object continually returns to the same position over and over again Examples of this are objects that go around in a circle or objects that oscillate on a spring or a swinging pendulum. In this chapter and Chap. 5 we expand our considerations to periodic motion by using circular motion as an example and in Chap. 6 we will discuss pendulum and spring motion.

R. C. Powell, *Physics*, Undergraduate Lecture Notes in Physics,
https://doi.org/10.1007/978-3-032-09361-5_4

Fig. 4.1 Rotational motion example: London Eye Ferris Wheel. *Credit* Martin Falbisoner, Creative Commons Attribution Share Alike 3.0 unported

To treat circular motion, we need to define different parameters and develop different equations than the ones we used for translational motion. We will follow the same pattern of first discussing the kinematics of how to describe circular motion and then the dynamics of what causes circular motion.

4.2 Rotational Motion Parameters

An object undergoing circular motion passes the same point every time it makes a complete rotation of the circle. Because it is continually changing directions, it is always accelerating. We know from Newton's second law that acceleration is caused by a force acting on the object. Figure 4.2 shows a ride at an amusement park that rotates a compartment in which you are standing. It is moving in the counterclockwise direction looking down from the top. The center that it rotates around is called the axis of rotation and counterclockwise rotation is considered to be the positive direction of rotation. One rotation is through an angle of 360° or 2π radians. The force that keeps the body moving in circular motion is called the centripetal force. It is pointed inward toward the center and continually changes the object's direction but not its speed. This results in centripetal acceleration.

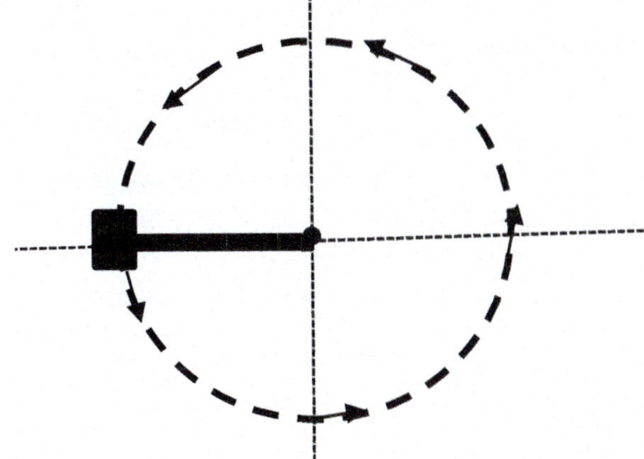

Fig. 4.2 Rotating carnival ride

The first thing we need to do is to define the properties of rotational motion the way we defined the properties of translational motion in Chap. 1. Consider the rotating circle shown in Fig. 4.3. Imagine a wheel that has a stripe along its radius that we can mark as a zero position for time and angle. If the wheel beings to rotate about an axis through its center in a counterclockwise direction the position of the stripe will change. In rotational motion, you can keep track of how much an object has rotated by looking at how the angle has changed with respect to a specific point in time. The more that the wheel rotates, the larger the angle becomes. Thus, angular displacement is a way to keep track of rotational motion like linear displacement is in translational motion.

If an object rotates through an angle of 90.0° ($\frac{1}{2}\pi$ radians), you know that it has completed ¼ of a full rotation. An angular revolution of 180.0° (π radians) means half of a rotation, and if the wheel moves through an angle of 360.0° (2π radians), it makes one full revolution. The angular displacement $\Delta\theta$ can be given in either degrees or radians. (Remember, there are 2π radians or 360° in one revolution of the circle.)

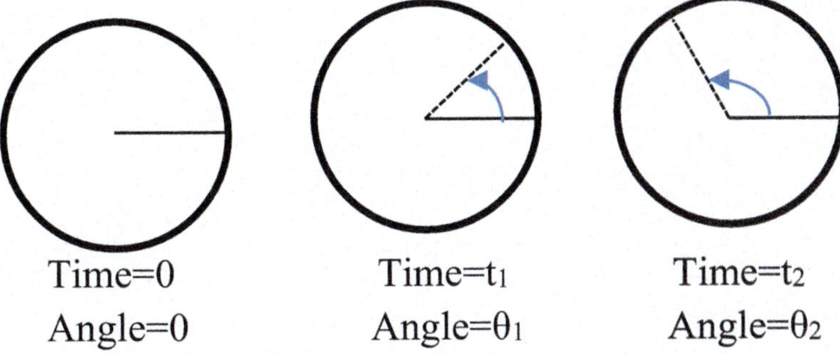

Fig. 4.3 Time and angle of rotational motion

Angular velocity (represented by the lowercase Greek letter omega, ω), is a measure of the rate at which the angle changes. Mathematically:

$$\omega = \Delta\theta/\Delta t \tag{4.1}$$

with units of radians per second. Remember, just as velocity can change over a given time period, angular velocity can change as well. Thus, Eq. 4.1 gives you the ***average* angular velocity** over the time interval. The smaller the time interval, the closer you come to measuring the ***instantaneous* angular velocity**.

We can also describe how fast the object is spinning by how many revolutions it makes in one second. This is called angular frequency, f. This has the units of revolutions per second. It is related to angular velocity by

$$f = \omega/2\pi. \tag{4.2}$$

Angular acceleration (represented by the lowercase Greek letter alpha, α) is the change in angular velocity during a certain time interval:

$$\alpha = \Delta\omega/\Delta t \tag{4.3}$$

with units of rad/s^2. Once again, this is the average angular acceleration, because acceleration could easily change over a given time interval. To get the instantaneous angular acceleration, the time interval would have to be infinitesimally small.

Thus, the *angle* is what we use to keep track of rotational motion. It is the rotational analogue of displacement. The *angular velocity* tells us how the angle is changing. Thus, it is the rotational analogue of velocity. Finally, the *angular acceleration* tells us how the angular velocity is changing, so it is the rotational analogue of acceleration. Since angle, angular velocity, and angular acceleration are simply the rotational versions of displacement, velocity, and acceleration, you might expect that there is a relationship between a translational quantity (like displacement) and its rotational analogue (like angle). These rotational analogues are all related to the corresponding translational quantity by the radius of the object that is rotating, r

$$s = r \cdot \theta \tag{4.4}$$

$$v = r \cdot \omega \tag{4.5}$$

$$a = r \cdot \alpha \tag{4.6}$$

In these equations, "s" is the distance that a given point on the object moves as it is rotating. We use the symbol "s" rather than "x" because "x" is typically considered a linear distance. When an object is rotating, a point on that object will not travel a linear distance. Instead, it will follow an arc. The letter "s" is typically used as an abbreviation for arc length, which is the length of an arc swept out by an object moving in a circle. It is still a term that means "distance," but that distance is around an arc, not along a straight line. The linear velocity of this point v is the change of s with respect to a change in time. Using this fact, Eq. 4.1, the linear and angular velocities of this point are related by Eq. 4.5.

Using Eq. 4.3, the linear acceleration of a point a radial distance r away from the axis of rotation is given by Eq. 4.6. For an object undergoing uniform (constant speed) circular motion with a radius r, the change in its position and velocity vectors in time Δt are proportional

$$\Delta r/(r\Delta t) = \Delta v/(v\Delta t)$$

Since $\Delta v/\Delta t$ is the centripetal acceleration a_c and $\Delta r/\Delta t = v$, this leads to an expression for the centripetal acceleration keeping the object in uniform circular motion

$$v/r = a_c/v$$

$$a_c = v^2/r = 4\pi^2 r/T^2 \tag{4.7}$$

where T is the period of the motion. The period is the time it takes to make one complete revolution. In deriving Eq. 4.7 we used the fact that the velocity is the ratio of the distance of one revolution traveled $2\pi r$ to the time of one revolution T. The period is the inverse of the frequency

$$T = 1/f \tag{4.8}$$

Now let's look at an example of uniform circular motion.

Example 4.1

As an example, consider the carnival ride shown in Fig. 4.2. The compartment is rotating in the counterclockwise direction at a distance of $r = 20$ m from the axis of rotation. It starts from rest and in 30 s its angular frequency is 0.5 revolutions per second. What is its average acceleration and final angular velocity? If you are riding in the compartment, what is your final linear velocity and how far in meters will you travel in the next 10 s?

The figure is given. The knowns and unknowns are:

Knowns: $r = 20\,\text{m}$; $t_1 = 30\,\text{s}$; $f_1 = 0.5\,\text{rps}$; $t_2 = 10\,\text{s}$.

Unknowns : α_1; ω_1; d_2.

Using Eq. 4.2,

$$f_1 = \omega_1/2\pi = 0.5 \quad \text{so} \quad \omega_1 = 2\pi\, 0.5 = 3.14\,\text{rad/s}.$$

Then using Eq. 4.3,

$$\alpha = \Delta\omega_1/\Delta t_1 = (3.14\,\text{rad/s})/(30\,\text{s}) = 0.105\,\text{rad/s}^2.$$

The linear velocity is given by Eq. 4.5 as,

$$v = \omega_1 r = (3.14\,\text{rad/s})(20\,\text{m}) = 62.8\,\text{m/s}.$$

For the second part of the problem, the distance traveled is given by,

$$d = vt_2 = (62.8\,\text{m/s})(10\,\text{s}) = 628\,\text{m}.$$

4.3 Center of Mass and Moment of Inertia

So far when we have described the motion of an object, we have assumed that all the object's mass is concentrated at one specific point and we gave described the motion of that point. This works fairly well for translational motion, but for rotational motion we need to take the shape of the object and its distribution of mass into account. This leads us to two important concepts associated with mass. The first is the center of mass (sometimes called center of gravity). This is essentially the average position of the mass distributed throughout the object. For translational motion, we can apply the kinetic equations of motion to the center of mass to describe the motion of an object. For rotational motion it is important to know the position of the center of mass with respect to the axis of rotation.

The second concept is the object's moment of inertia. For translational motion, an object's mass provides the inertia to resist the change in its motion. For rotational motion, an object's moment of inertia provides the resistance to a change in its motion. This depends on the object's mass, the shape of the object, how the mass is distributed, and the location of the axis of rotation.

These two concepts are discussed below beginning with the center of mass.

The center of mass is a concept that is useful when all of the mass is not concentrated in one place. It gives the effective position where you can assume all the mass to be located and then apply the normal equations of motion to describe how the center of mass moves.

An extended body or a system of individual particles will behave as though all of the mass is concentrated at the center of mass.

Because of this, we need to know how to calculate the position of the center of mass of a physical system. For a system consisting of n objects with mass m_i located at positions x_i, the center of mass coordinate x_{cm} is given by

$$x_{cm} = \frac{\sum_i^n x_i m_i}{\sum_i^n m_i}. \tag{4.9}$$

Remember the Greek sigma Σ indicates summation. In this equation, "x_{cm}" is the position of the center of mass, and "n" is the number of particles in the system. The term "m_1" represents the mass of the first particle; "m_2" represents the mass of the second particle; and so on; "x_1" is the position of the first particle; "x_2" is the position of the second particle; etc. To calculate the center of mass in a system of particles, then, you just multiply the position of each particle by its mass, sum up the products, and then divide by the total mass of the system. That will give you the position of the system's center of mass. If the problem is in two or three dimensions, the center of mass coordinates in each direction can be found using similar equations. This is demonstrated in the following two-dimensional example.

Example 4.2

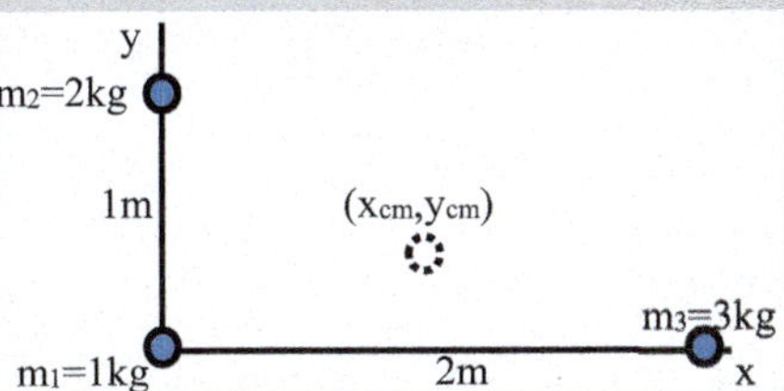

Three mass system.

Consider a system made up of three objects shown at the positions in the figure above. Each object has a different mass as designated in the figure. What are the center of mass coordinates of this system?

Using Eq. 4.9,

$$x_{cm} = \frac{m_1 x_1 + m_2 x_2 + m_3 x_3}{m_1 + m_2 + m_3} = 1\,\text{m}, \quad y_{cm} = \frac{m_1 y_1 + m_2 y_2 + m_3 y_3}{m_1 + m_2 + m_3} = 0.33\,\text{m}.$$

The center of mass is located at $x_{cm} = 1$ m and $y_{cm} = 0.33$ m. If an external force was applied to this system of three objects, it will move as if it were a mass of 6 kg at the position of the center of mass.

So far we have just talked about the center of mass in terms of a system of discrete particles. For an object where the mass is distributed continuously over its shape, it is not possible to use Eq. 4.9 to calculate the center of mass coordinates. The appropriate equation requires the use of integral calculus which is beyond the math required for this course. However, if the mass is uniformly distributed over a simple geometric shape (like a sphere), the center of mass will be at the center of the geometric shape. For this course, we will try to work with this type of object for problems where we need to know the center of mass. Go through the following example to illustrate this.

Example 4.3

A system of three uniform rods is shown to the right. The top rod is twice as massive as the rod on the left, and the rod on the right is three times as massive as the rod on the left. If the top rod has a length of "*L*," and the side rods have lengths of "*K*," what is the center of mass of the system?

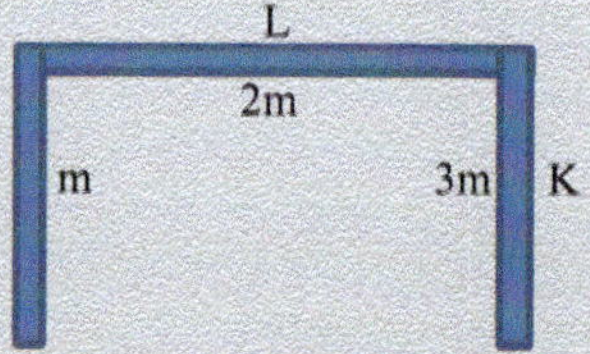

You might think that this problem requires calculus to solve, because each of the rods has a center of mass that must be determined. However, they are uniform rods with a simple geometric shape. Thus, the center of mass of each rod is at the exact center of the rod. Then we really just have a system of three particles. If we define the origin as the upper, left-hand corner of the assembly, the mass labeled "$2m$" is at $x = L/2$, $y = 0$. The mass labeled "m" is at $x = 0$, $y = -K/2$, and the mass labeled "$3m$" is at $x = L$, $y = -K/2$. Now we can use Eq. 4.9:

$$x_{cm} = \frac{m_1x_1 + m_2x_2 + m_3x_3}{m_1 + m_2 + m_3} = \frac{0 + mL + 3mL}{6m} = 2L/3$$

$$y_{cm} = \frac{m_1y_1 + m_2y_2 + m_3y_3}{m_1 + m_2 + m_3} = \frac{-Km/2 + 0 - 3Km/2}{6m} = -K/3.$$

The center of mass, then, is $2/3 \cdot L$ to the right of the upper, left-hand corner, and $1/3 \cdot K$ below the upper, left-hand corner. That makes perfect sense. Since the bar to the right is heavier than the bar to the left, the center of mass is weighted to the right ($2/3 \cdot L$ is to the right of halfway in between the bars). Since the upper bar is heavier than the other two bars, the center of mass is weighted up (– 1/3 · K is only one-third of the way down the bars).

To further illustrate the center of mass of objects, do the following experiment.

Experiment 4.1
The Center of Mass

Supplies:

- A teaspoon (Not a measuring teaspoon—a normal teaspoon you use at the table.)
- A fork roughly as long as the spoon (The fork will probably be a bit longer, but that's okay.)
- Wooden matches
- A glass

Introduction—This experiment demonstrates that a system behaves as if all of its mass is concentrated at its center of mass.

Procedure:

1. Set the glass on the top of a table.
2. Interlock the spoon in the tines of the fork. The middle tines should touch the back of the spoon, while the outer tines should touch the front of the spoon. In the end, the spoon and fork should form a "*V*," as shown in the picture to the right.

3. Take a match and stick one end (not the end you strike—the other end) between the tines of the fork under the spoon so that the entire "*V*" assembly balances on the match.

4. Now balance the entire assembly on the edge of the glass by placing the match on the lip of the glass, as shown in the picture to the right.
5. You might think that the match has something to do with the balance of the assembly on the glass. It does not. The match serves only to provide a surface for the assembly to hang on the glass. To demonstrate this, light a match and use it to light the head of the match that the assembly sits on. The match will burn until it reaches the edge of the glass, and then it will go out.

Carefully use the match you have in your hand to break off the burnt part of the match. The assembly will still stay balanced on the glass. Carefully use the match you have in your hand to break off the burnt part of the match. The assembly will still stay balanced on the glass.

6. Clean everything up.

Why did the fork/spoon assembly balance on the glass even when the match was burnt from one end? It's because the center of mass of the assembly is sitting on the edge of the glass.

Think about the spoon. The majority of the spoon's mass is in the ladle. Thus, the center of mass is on the handle but near the ladle of the spoon. For the fork, the center of mass is near the tines as well, because the mass of the fork is concentrated on the part of the fork that forms the tines. Since the handles both bend in towards the glass, the center of mass of the entire assembly ends up being located in the same plane as the edge of the glass. Since objects behave as if their mass was concentrated at the center of mass, the fork/spoon assembly behaved as if its mass was concentrated on the edge of the glass. Thus, it balanced. That's why the center of mass is so important in physics. An object that is difficult to analyze can simply be replaced by a point that contains all of the object's mass. That point is the center of mass.

One place the concept of center of mass is important is in determining the stability of an object. Consider two vehicles with different centers of mass going around a banked curve as shown in Fig. 4.4. When a car banks, its weight acting at its center of mass, will tend to make the car roll around an axis of rotation into the page where their driver's side wheels contact the road. These axes are shown as red circles in the figure. The car in Fig. 4.4a with a low center of mass is stable because its weight acts within the wheelbase of the tires. This tends to cause a

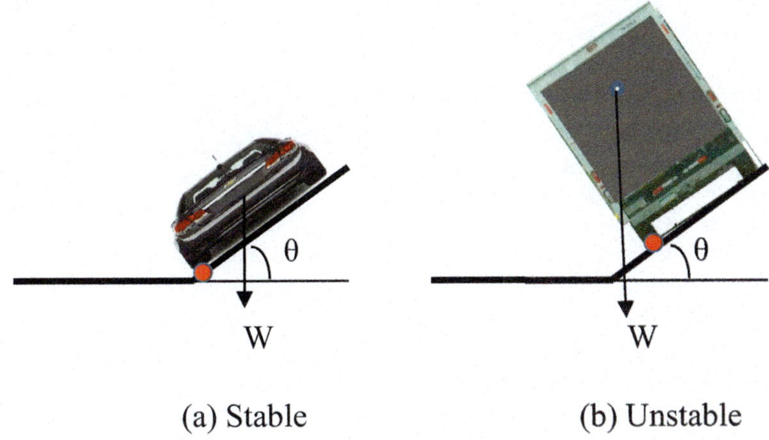

Fig. 4.4 Center of mass on a banked curve

clockwise rotation pushing the car into the road. However, the truck in Fig. 4.4b has such a high center of mass that its weight falls outside its wheelbase. This tends to cause a counterclockwise rotation so it will tip over.

To see how well you understand center of mass, try these Student problems.

Student

4.1 A man ($m = 85.0$ kg) stands in the center of a uniform raft ($m = 150.0$ kg), 2.50 m from one edge. The raft is 15.0 m long and 4.00 m wide.

4.1 A man ($m = 85.0$ kg) stands in the center of a uniform raft ($m = 150.0$ kg), 2.50 m from one edge. The raft is 15.0 m long and 4.00 m wide.

a. What is the center of mass of this system?
b. The raft is floating motionless in the water. If the man walks to the other side of the raft so that he is now 2.50 m from the other edge of the raft, how far will the raft move?

4.2 A shoulder-fired rocket launcher fires a projectile (from ground level) with an initial speed of 125 m/s at an angle of 30.0° relative to the horizontal. The projectile then explodes into ten pieces. Where will the center of mass of those ten pieces be when they hit the ground?

The second mass related concept important in analyzing rotational motion problems is the moment of inertia of an object.

Moment of inertia—A measure of an object's resistance to a change in its state of rotational motion

Thus, the larger the moment of inertia, the harder it is to cause its rotation to change. Once again, then, we find that a physical quantity in rotational motion is an analogue of a physical quantity in translational motion. The *moment of inertia* is the rotational analogue of *mass*.

There are three factors which affect how difficult it is to change the rotation of an object. First, the mass plays a role. The more massive an object, the harder it will be to change the object's rotation. Second, the shape plays a role. Some shapes are just easier to rotate than others. Third, the *axis* about which the rotation occurs is also important. For a system consisting of discrete particles of mass, the moment of inertia I is calculated with the equation

$$I = \sum_{i}^{n} r_i^2 m_i \tag{4.10}$$

where r_i is the radial distance from the center of the object of mass m_i. The units of I are kg m^2. As was the case with Eq. 4.9 for calculating the center of mass, for a continuous distribution of mass over a geometric shape Eq. 4.10 becomes a calculus integral. Since we are not using calculus in this course, we will not pursue this further. Instead we simply list the results of calculating the moments of inertia for several common situations. These are shown in Fig. 4.5. The information it contains will be provided on the test if you need it.

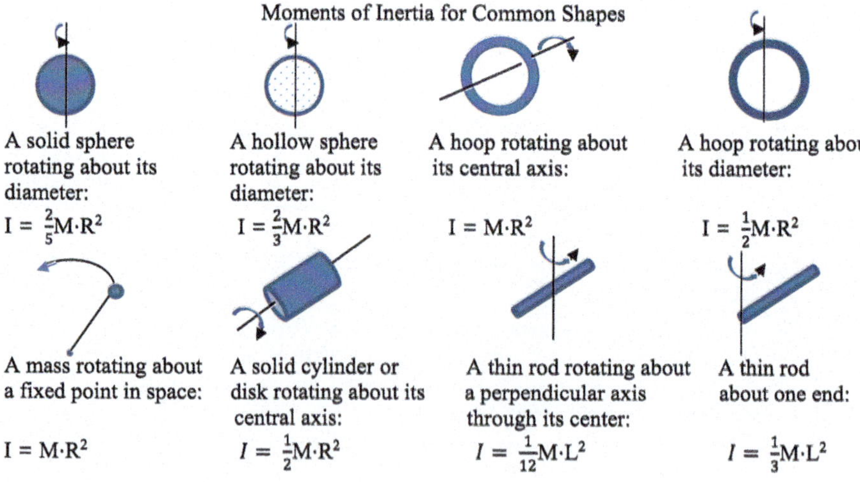

Fig. 4.5 Moments of inertia for common shapes

Remember, the moment of inertia is simply a measure of how an object resists changes to its rotational motion. For example, suppose you had two spheres of equal mass and radius but one was solid.

In the equations, "*I*" represents the moment of inertia; "*M*" represents the mass of the object; "*R*" is the radius; and "*L*" is the length.

Next level

In calculus, the expression for the moment of inertia of an object I is given by an integral, $I = \int_0^M r^2 \mathrm{d}m$ where M is the total mass of the object and $\mathrm{d}m$ is a small increment of mass located at a position r from the axis of rotation, and $\int$ is the integral sign. Consider the rotation of a thin rod about an axis perpendicular to the rod through it center (see Fig. 4.5). The increment of mass can be changed to a small increment of length $\mathrm{d}l$. Then, $I = \int_{-L/2}^{L/2} l^2 \frac{M}{L} \mathrm{d}l$ where L is the total length of the object and l is the distance from the center to the increment $\mathrm{d}l$. The center of the rod is at $l = 0$ and the ends are at $l = \pm \mathrm{L}/2$. The rule for integration of a polynomial in calculus is $\int_A^B x^n \mathrm{d}x = \left(\frac{B^{n+1}}{n+1} - \frac{A^{n+1}}{n+1}\right)$. For our example, the moment of inertia becomes $I = \frac{M}{L}\left(\frac{L^3}{24} + \frac{L^3}{24}\right) = \frac{1}{12}ML^2$. For rotation about an axis at one end instead of in the center, everything is the same except the limits of integration are 0 and L.

$I = \frac{M}{L}\left(\frac{L^3}{3} + 0\right) = \frac{1}{3}ML^2$. This is an example of the *parallel axis theorem*. If you know the moment of inertia about an axis going through the center of mass of an object, the momentum of inertia about any axis parallel to this axis a distance h away is $I = I_{\mathrm{cm}} + Mh^2$. For our example, $I = \frac{1}{12}ML^2 + M(L/2)^2 = \frac{1}{3}ML^2$. If the mass is not distributed uniformly in the object, the math becomes much more difficult.

One was solid while the other was hollow. If both were rotating about their diameters, and you tried to stop them from rotating, which would be harder to stop? Well, if you look at Fig. 4.5, you will see that the moment of inertia of a solid sphere rotating about its diameter is, $I = \frac{2}{5}M \cdot R^2$ while the moment of inertia of a hollow sphere rotating about its diameter is $I = \frac{2}{3}M \cdot R^2$. Since the mass and radius are the same for both spheres, the hollow sphere has the larger moment of inertia. Thus, the hollow sphere would be harder to stop.

4.4 Rotational Dynamics

Now we turn to the dynamics of rotational motion. Remember from your introductory course that analyzing the dynamics of the motion of an object means that we must consider what is causing the motion to occur the way it does. For linear motion an object's velocity was altered by the application of a force. The same is true for rotational velocity except in this case, how we apply the force is critical. This type of entity that causes a change in angular motion is called torque, and it is represented by τ. Torque is a vector quantity with dimensions of Newton‾meters. It is defined by the equation

$$\tau = \mathbf{r} \times \mathbf{F} = Fr \sin\theta. \tag{4.11}$$

F is the applied force, **r** is the vector distance from the axis of rotation to the point of application of the force, and θ is the angle between the **r** and **F**. The $rF \sin\theta$ can be thought of either as the component of the force vector perpendicular to r or the component of **r** perpendicular to the force vector. In the latter case $r \sin\theta$ is called the lever arm. The force will be maximum if it is applied perpendicular to r so $\theta = 90°$. Note that torque is a vector formed by the cross product of the vector from the axis of rotation to the point of force application and the force vector. The direction of the torque vector is given by the right-hand rule. This is clarified in the following example.

Example 4.4

Consider the example shown in the figure.
A nut is rusted in place and will take a torque of 30 Nm to remove it. A force of 120 N is applied at $r = 0.2$ m at an angle of 45° to the end of the wrench as shown. Will this provide enough torque to turn the nut? What if the angle the force is applied is increased to 90°. Will this turn the bolt?

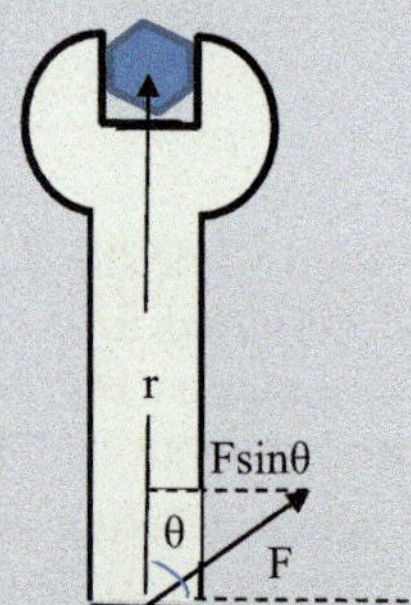

For the first angle of incidence of the force, the resulting torque is $\tau_1 = Fr \sin\theta_1 = (120\,\text{N})(0.2\,\text{m}) \sin 45° = 16.97\,\text{Nm}$ in the counterclockwise direction.

For the second angle of incidence of the force, the resulting torque is,

$\tau_2 = Fr \sin \theta_2 = (120)(0.2) \sin 90° = 24.00$ Nm in the counterclockwise direction.

Both of these numbers are less than the required 30 Nm so this amount of force on this wrench will not loosen the nut. In order to obtain the required amount of torque, either 30 N of additional force must be applied, or a 0.05 m longer wrench must be used. Note that the direction of torque causes rotation in the counterclockwise direction which is the correct way to turn for loosening the nut.

When Newton's second law is applied to rotational motion it is written as,

$$\tau = I\alpha \tag{4.12}$$

where the left side of the equation represents the sum of all the torques acting on the object, I is the object's moment of inertia, and α is the angular acceleration. This is equivalent to Eq. 2.1 for linear motion with torque taking the place of force, moment of inertia taking the place of mass, and angular acceleration taking the place of linear acceleration. As seen in Fig. 4.5, I depends on the magnitude of the body's mass, how this mass is distributed (in other words, its shape), and the location of the axis of revolution.

There is a special situation when the sum of all the torques on a system adds to zero. Then according to Eq. 4.12 the angular acceleration of the system is zero. In this case, the object is either stationary or rotating with a constant angular velocity. The first situation is called static rotational equilibrium, and the second is dynamic rotational equilibrium. You should now do the following experiment to review rotational equilibrium.

Experiment 4.2
Static Rotational Equilibrium

Supplies:

- A wooden pencil
- A wooden ruler
- At least 5 U.S. quarters, preferably more

Introduction—This experiment allows you to use what you learned in the previous section as well as what you learned about static rotational equilibrium in your first-year physics course.

Procedure:

1. Set the pencil on the table.
2. Balance the ruler on the pencil so that neither end of the ruler hits the table. If your ruler is uniform, it will balance when the halfway mark of the ruler is sitting on the middle of the pencil, because that is the center of mass of the ruler. If it is not the center of the ruler, note where it is.
3. Now you are going to weigh your ruler. How are you going to do that without a scale? Put a quarter on the end of the ruler so that one end of the quarter touches the zero mark on the ruler. Read the mark that corresponds to the center of the quarter. If you are using an English ruler, for example, the center of the quarter will be at the ½-inch mark.
4. Now find the point at which the ruler balances again. You will discover that it is not the halfway mark of the ruler. Instead, a shorter portion of the ruler will be on the same side of the pencil as is the quarter.
5. Read the ruler to find the distance from the center of the quarter to the point at which the ruler is balancing on the pencil.
6. Read the ruler to find the distance from the center of mass you determined in step (2) to the point at which the ruler balances.
7. Now think about the rotational motion aspect of this situation. When not balanced, the ruler rotates around the axis defined by the pencil. The rotation is short because it is interrupted by the table. Nevertheless, it *is* rotation. When the ruler balances, there is no rotation, so the sum of the torques is zero. What is supplying the torques? The weight of the quarter supplies one torque. Let's say that rotation in that direction is positive. The weight of the ruler (remember—it is concentrated at the center of mass) provides a torque in the opposite direction, which we will call negative. Thus, we have the following equation to represent the sum of the torques:

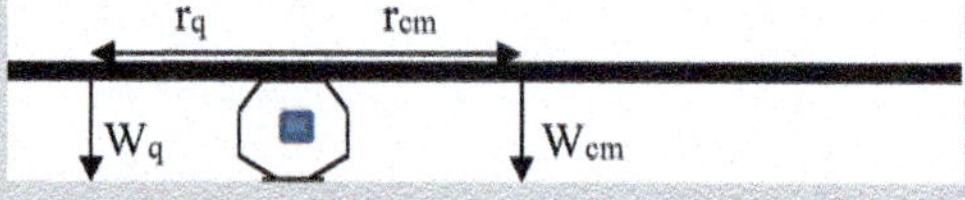

$$W_q \cdot r_q - W_{cm} \cdot r_{cm} = 0$$

In this equation, W_q is the weight of the quarter, r_q is the lever arm distance from the axis of rotation to the center (center of mass) of the quarter, W_{cm} is the weight of the ruler (concentrated at the center of mass), and r_{cm} is the lever arm distance from the ruler's center of mass to the axis of rotation.

8. The mass of a U.S. quarter is 5.75 g. Now you can calculate W_{cm}. You measured r_q and r_{cm} in steps (5) and (6). Thus, the only thing you don't know in the equation above is the weight of the ruler. Thus, you can solve for it. This is actually a very accurate weight for the ruler, providing you did your calculations and measurements correctly.
9. Now that you know the weight of the ruler, I want you to set up another problem. Take the quarter off of the ruler and place the ruler so that the center of mass is even farther away from the pencil than it has been so far. Make sure that the change is significant (7–15 cm).
10. Use the equation in step (7) to calculate the weight of quarters that would be required to balance the ruler with this new setup, assuming that you put the quarters at the same place you had the one quarter before.
11. Divide the weight you calculated by the weight per quarter, and that will tell you the number of quarters you need to balance the ruler now. Most likely, the number will not be an integer.
12. Put the nearest whole number of quarters on the ruler in the same place you had the one quarter in the previous part of the experiment. In other words, if you calculated that you needed 3.2452 quarters to balance the ruler, put 3 quarters on the spot where you had just one before. Note that the ruler still tips to the side without any quarters.
13. Put one more quarter on the pile. Note that the ruler now tips to the side with the quarters. That's because with the addition of another quarter, the quarters are now supplying more torque than is the ruler.
14. You can continue to play with this. In my setup, my ruler had a weight of 0.205 N (a mass of 20.9 g). When I placed the pencil very near the end of the ruler, it could hold up a stack of 10 quarters.
15. Clean everything up.

When an object is not rotating, then, the sum of the torques on the object must equal zero. It turns out that we can combine this concept (static rotational equilibrium) with the concept of static translational equilibrium to be able to analyze some pretty complex situations, as shown in the following example.

Example 4.5

One side of a uniform 50.0 kg rod (length = 1.00 m) is Attached to a wall. The rod is also supported by a rope that is attached to the center of the rod and tied to the wall above the rod at a 40.0° angle. A 150.0 kg block hangs 20.0 cm from the free end of the rod. What is the tension in the rope?

Knowns: l=1.00 m; $m_r = 50.0$ kg; $\theta = 40.0°$; $m_b = 150.0$ kg; $l = 0.800$ m

Unknowns: T 50.0°

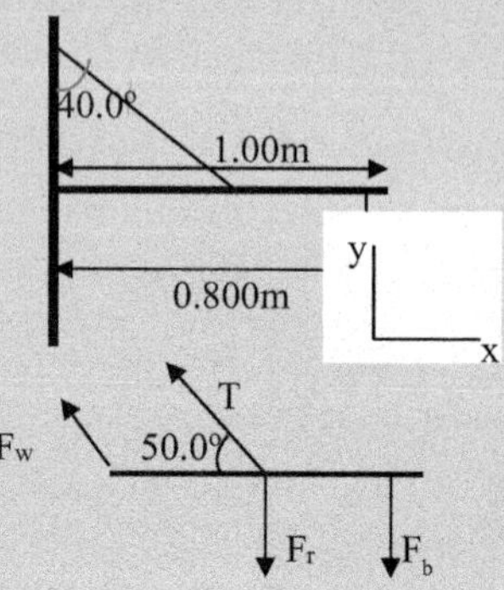

Now we can identify all the forces acting on the rod. The tension in the rope (T) and the weight of the block (F_b) are obvious. The rod has mass as well. Thus, its weight will exert a force downward concentrated at the at the center of mass, F_r. There is still more force. The rod is attached to the wall. Thus, the wall must be exerting a force on the rod (F_w). We do not know the direction of the force so the figure shows it in an arbitrary direction. We can solve for the real direction at the end.

Those are all of the forces. This is clearly a two-dimensional problem, so let's split it up into two, one-dimensional problems using the x,y coordinate directions as shown.

Starting with the y-dimension,

$$T_y + F_{wy} - F_r - F_b = 0$$
$$T\sin(50.0°) + F_{wy} - (50.0\,\text{kg})(9.81\,\text{m/s}^2) - (150.0\,\text{kg})(9.81\,\text{m/s}^2) = 0$$
$$0.766\,T + F_{wy} - 491\,\text{N} - 1470\,\text{N} = 0$$

We know the mass of the rod and the block, so we can multiply by g and turn those into weights: Since the sum of angles in a triangle must be 180.0°, the angle that the rope makes relative to the rod is 50.0°. We are left with two unknowns.

Let's go to the x-dimension and see what we can find out.

$$-T_x - F_{wx} = 0$$

Once again, since we are putting the negatives in ourselves, we use the reference angle to calculate the x-component of the tension:

$$-T \cdot \cos(50.0) - F_{wx} = 0$$
$$-0.643T - F_{wx} = 0$$

This is a problem. We have a total of three unknowns (T, F_{wx}, F_{wy}). If we knew the angle of F_w, we could reduce that list to just two unknowns, but we don't. Thus, we are stuck with two equations and three unknowns.

This is where the concept of static rotational equilibrium comes into play. We also know that the rod is not rotating. Thus, the sum of the torques on the rod must be zero. So, we can just sum up the torques and set them equal to zero. However, to define a torque, we need to determine an axis about which this rod could potentially rotate. There are two. If the rod were to come loose from the wall, it could rotate around its center, where the rope is attached. Alternatively, if the rope were to break, it could rotate around its end where it is attached to the wall. It turns out that *either* axis of rotation can be used to solve the problem. In fact, *any* axis of rotation could be used. However, if we choose the axis *carefully*, it will make the job *a lot* easier. Remember how we define torque. We take the force, multiply by the distance to the axis of rotation (the lever arm), and then multiply by the sine of the angle. Well, if we choose the axis of rotation to be a point at which one of the unknown forces acts, the torque generated by that force will be zero, because the distance between the force and the axis of rotation is zero. Thus, choosing the axis of rotation properly is a great way to get rid of unknown forces in the equation. Since the real mystery force here is F_w, let's choose the axis of rotation to be the point at which the rod attaches to the wall. That way, the torque generated by F_w is zero.

We have defined the axis of rotation. Now we have to define direction. Let's say that clockwise rotation is positive, and counterclockwise rotation is negative. That means the block exerts a positive torque (it causes the rod to rotate clockwise about the end attached to the wall) as does the weight of the rod, and the rope exerts a negative torque (it causes the rod to rotate counterclockwise about the end attached to the wall). Thus:

$$F_b \cdot (r_b) \cdot \sin\theta_b + F_r \cdot (r_{cm}) \cdot \sin\theta_{cm} - T \cdot r_{rope} \cdot \sin\theta_{rope} = 0$$

The distance from the axis of rotation to the block is 80.0 cm, so r_b is 0.800 m. The force exerted by the block is its weight, and the angle between the block and the rod is 90.0°. The force exerted by the rod is its weight, its distance is 50.0 cm (the center of the rod), and the angle is also 90.0°. We don't know the tension, but the distance is 50.0 cm (the rope is attached to the center of the rod), and the angle between the rope and the rod is 50.0°. Thus:

$$(1470\,\text{N}) \cdot (0.800\,\text{m}) \cdot \sin(90.0°) + (491\,\text{N}) \cdot (0.500\,\text{m}) \cdot \sin(90.0°)$$
$$- T \cdot (0.500\,\text{m}) \cdot \sin(50.0°) = 0$$
$$T = 3710\,\text{N}$$

The application of static rotational equilibrium, then, allowed us to solve for the tension in the rope. That is, in fact, all that the problem wanted. Thus, had we started with static rotational equilibrium, we would already be done. For completeness's sake, however, let's figure out F_W.

Using the equation we got from the y-dimension:

$$T \cdot \sin(50.0°) + F_{wy} - 491\,\text{N} - 1470\,\text{N} = 0$$
$$(3710\,\text{N}) \cdot \sin(50.0°) + F_{wy} - 491\,\text{N} - 1470\,\text{N} = 0$$
$$F_{wy} = -881\,\text{N}$$

The negative means that this component is directed *opposite* of the way we defined it. We said it was pointing up, but since we got a negative answer, it must actually point down. Now let's move to the x-dimension:

$$- T \cdot \cos(50.0°) - F_{wx} = 0$$
$$- (3710\,\text{N}) \cdot \cos(50.0°) - F_{wx} = 0$$
$$F_{wx} = -2380\,\text{N}$$

The negative sign once again tells us that the wall force points *opposite* of what we thought. We thought it pointed left, but since we got a negative answer, it must point right. Thus, the arrow we originally drew was not correct. The force of the wall actually pulls slightly down and to the right. If you use the components to calculate the magnitude and direction of the force, you will find that the force is 2.54×10^3 N directed at an angle of 339.7°.

Try these Student problems to show your ability to analyze problems like this.

Student

4.3 A man holds a 15.0-Newton ball in his hand (38.0 cm from the elbow) with his forearm flexed, as shown in the drawing below. The biceps, which flex the forearm, attach to the forearm roughly 3.50 cm from the elbow and pull at a 15.0° angle relative to the vertical. If the man wants to hold the ball stationary as shown, what force must the biceps exert (F_b)? What force must the arm bone (humerus) exert (F_h)?

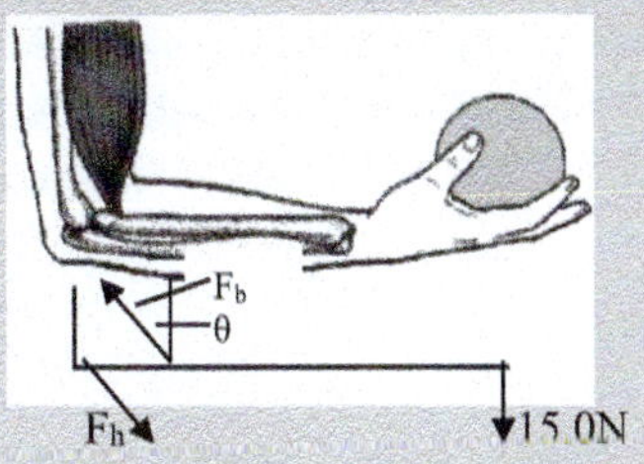

4.4 A hoop is exposed to a torque so that it rotates around its central axis. The same hoop is then stopped and exposed to the same magnitude of torque, but the torque is oriented so that the hoop spins about its diameter. In which of the two situations will the angular acceleration be greater?

One important concept in dealing with this type of physical situation is the lever. The lever (along with the wheel) is one of the six simple machines that help us change the magnitude or direction of a force. The concept of a lever is shown in Fig. 4.6. It consists of a solid object on a pivot point and can be used to increase the mechanical force applied to another object. The applied force is called, the effort and the force that performs the work is called the load. The pivot point is the fulcrum. The increase of an effort force to lift a load by using a lever is called the mechanical advantage. With no friction this is given by

$$\text{MA} = \text{Load}/\text{Effort}. \tag{4.13}$$

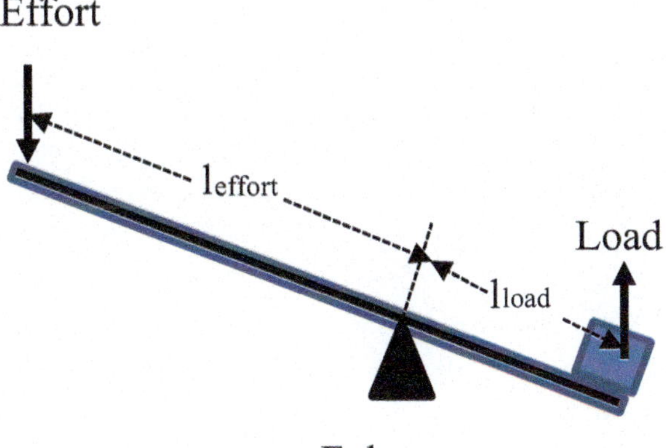

Fig. 4.6 Mechanical advantage of a lever

Example 4.6

As an example, consider a load of 200 kg. Using a lever with $l_{\text{effort}} = 2\,\text{m}$ and $l_{\text{load}} = 1\,\text{m}$, how much effort force is needed to lift the load? What is the mechanical advantage of the lever?

Using Fig. 4.6 as the diagram of the problem, the knowns and unknowns are:

$$\text{Knowns: } l_{\text{effort}} = 2\,\text{m};\;\; l_{\text{load}} = 1\,\text{m};\;\; \text{m}_{\text{load}} = 200\ \text{kg}.$$
$$\text{Unknowns: } F_{\text{effort}};\ \text{MA}.$$

Balancing the torques gives

$$F_{\text{effort}} l_{\text{effort}} = F_{\text{load}} l_{\text{load}} \quad \text{or} \quad F_{\text{effort}} = F_{\text{load}} l_{\text{load}} / l_{\text{effort}}$$
$$= (200)(9.8)(1)/2 = 980\,\text{N}.$$
$$\text{MA} = 200 \times 9.8/980 = 2.$$

So the lever allows the effort force to lift twice as heavy a load as it could lift directly

Note that there is a price you pay for the mechanical advantage. The work done on the load is equal to the work done by the effort. The work done by pushing the lever to a horizontal position in Fig. 4.6 is $(F_{\text{effort}})(l_{\text{effort}}) \sin\theta$. This can be set equal to the work done on the load, $(F_{\text{load}})(l_{\text{load}}) \sin\theta$. The sin θ's cancel out leaving $l_{\text{effort}} = l_{\text{load}}(F_{\text{load}}/F_{\text{effort}}) = (\text{MA}) l_{\text{load}}$. So for the example above, the force is half the load weight but it must be applied for twice the distance the load moves.

Special Topic

The Greek scientist Archimedes is famously quoted as saying "Give me a lever long enough and a fulcrum on which to place it, and I shall move the world." This indicates how important the concept of the lever is. Remember the other simple machines are the wheel, the pulley, the screw, the wedge, and inclined plane. Each of these significantly affects the strength or direction of a force applied to an object.

It is also possible to use a combination of pulleys to obtain a mechanical advantage as discussed below. Now let's look at some other examples of rotational dynamics.

Example 4.7

A mass ($m = 20.0$ kg) hangs on a string that has been wrapped several times around a pulley ($M = 500.0$ g, $r = 35.0$ cm). At first, the pulley is held so that the system is not moving in any way. When the pulley is released, what is the angular acceleration of the pulley and the acceleration of the mass? Neglect friction and the mass of the string, and assume that the string never slips on the pulley.

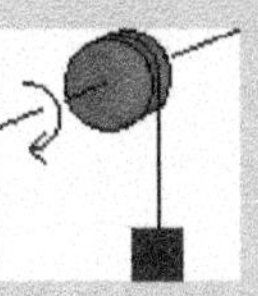

$$\text{Knowns: } m = 20.0\,\text{kg};\ \ M = 35.0\,\text{g};\ \ r = 500.0\,\text{cm};$$
$$\text{Unknowns: } \alpha;\ \ a$$

If we were not dealing with the rotation of the pulley, this would be an easy problem. The only force acting on the block would be gravity, and the acceleration of the mass would be equal to g. However, since the pulley must rotate, its moment of inertia resists a change in its rotational motion. As a result, a tension develops on the string.

This tension works against gravity. The resulting force diagram for the mass is shown on the right. This force diagram leads to the following equation:

$$T - m \cdot g = m \cdot a$$

Note that this equation tells you that we have defined upward motion as positive.

This gives us one equation, but there are two unknowns, because we do not know the value of T or a. However, this is not the only equation. We have another body to consider that of the pulley. The pulley is also exposed to a force, as shown in the diagram on the right. In this diagram, the string pulls down on the pulley at a distance away from its axis of rotation. Thus, the tension on the string provides a torque. Since the string pulls along a line tangent to the disk, and since the radius of any circle or disk is always perpendicular to any line drawn tangent to the circle or disk, the force that the tension exerts is perpendicular to the radius.

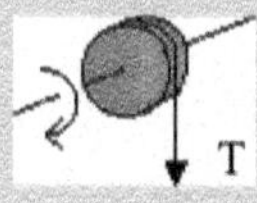

As a result, the angle between **T** and **r** is 90.0°, so the torque is given by

$$\tau = r \cdot F \cdot \sin\theta = (-T) \cdot r$$

where T is the tension of the string and r is the radius of the disk. Why do we have $-T$ instead of T in this? Remember, up is positive. Thus, down is negative. That means the forcesupplies the torque is $-T$, not T: We can take this equation and substitute it into Newton's second law for rotational motion

$$\tau_{\text{net}} = I \cdot \alpha$$
$$(-T) \cdot r = I \cdot \alpha$$

Since we know the shape, mass, and radius of the pulley, we can use Fig. 4.5 to determine I.

Since we know I and also know r, we have two unknowns in this equation—T and α. At this point, we have two equations, but we have three unknowns (T, a, and α). However, α is related to a via Eq. 4.6:

$$a = r \cdot \alpha$$

or

$$\alpha = a/r$$

Putting that into the equation above gives us:

$$(-T) \cdot r = I \cdot a/r$$

Substituting in our first equation for T gives

$$-(ma + mg)r = Ia/r$$

Solving for a gives

$$a = \frac{-mgr^2}{mr^2 + I}$$

The moment of inertia is given by

$$I = MR^2/2 = (0.5000\,\text{kg})(0.350\,\text{m})^2/2 = 0.0306\,\text{kg m}^2$$

Notice the unit on moment of inertia. Although the equation for moment of inertia changes, its standard unit is always kg m^2. Now that we have a number for I, we can calculate a:

$$a = \frac{-mgr^2}{mr^2 + I} = \frac{-(20.0\,\text{kg})(9.81\,\text{m/s}^2(0.350\,\text{m})^2}{(20.0\,\text{kg})(0.035\,\text{m})^2 + 0.0306\,\text{kg m}^2} = -9.69\,\text{m/s}^2$$

Notice two things about the acceleration. First, it is negative. That makes sense, since we defined upward motion as positive. The mass obviously travels downwards, so the acceleration must be negative. Also, notice that it is just slightly smaller than g. This should make sense. After all, the motion of the pulley takes some energy away from the motion of the mass. Thus, the acceleration of the mass is slightly lower than it would be if the mass were to fall freely. The more massive the pulley and the larger the pulley's radius, the smaller the acceleration would be, because it would take more energy to get the pulley spinning.

What about the angular acceleration of the pulley? The acceleration and angular acceleration are related by Eq. 4.6:

$$a = r \cdot \alpha$$

or

$$\alpha = a/r = \left(-9.69\,\text{m/s}^2\right)(0.350\,\text{m}) = -27.7\,\text{rad/s}^2$$

We used our knowledge of Newtons' Second Law and applied it both to translational motion (the motion of the mass) and rotational motion (the motion of the pulley) to give us a more accurate analysis of a system. The more accurate analysis tells us that the mass accelerates at $-9.69\ \mathrm{m/s^2}$ and the pulley has an angular acceleration of $-27.7\ \mathrm{rad/s^2}$.

Now what about that unit for angular acceleration? As you should recall from geometry, a "radian" is a measure of angle. An angle of 2π radians, for example, is the same as 360.0°.

An angle of π radians is 180.0°. When converting from translational motion into rotational motion, the radian unit must be inserted in place of "1" to preserve the integrity of the units. Thus, even though the math makes the units work out to $1/\mathrm{s}^2$, the unit of radian is inserted in place of "1" in order to get $\mathrm{rad/s^2}$. In the same way, when converting from rotational motion into translational motion, the unit radian must be replaced with "1." Why do we have to do this? Well, geometry defines angle as the ratio of the arc length to the radius of a circle. Both of these quantities are measured in meters. Thus, when you divide arc length by radius, the units cancel and you are left with a number that has no units. This is called a **dimensionless quantity**. Physicists cannot stand to see a measurement without units, however. Thus, they insert "radian" where there is no unit. There is no problem with that if you deal purely with rotational motion. However, when we covert from rotational motion to translational motion (or vice versa), we run into a problem with the unit radian. Thus, you remove it when going from rotational motion to translational motion, and you add it when going from translational motion to rotational motion. You will get used to this as you do more problems.

Let's try one more problem to make sure you really understand this. We'll go back to an earlier system we analyzed (Atwood's machine), and see how much the analysis changes when considering the rotational motion of the pulley.

Example 4.8

Let's consider Atwood's machine, but this time, we will include the rotation of the pulley. We ignored it the last time, but that's not really correct. Let's see how much the results change when we analyze the situation more correctly. The machine has two masses (m_1 and m_2). The pulley has mass M and radius r. What is the acceleration of the masses in terms of m_1, m_2, r, M, and g?

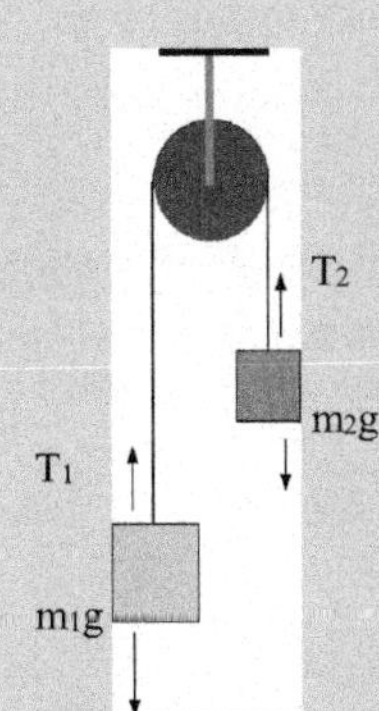

Knowns: m_1; m_2; r; M

Unknowns: a

When we dealt with Atwood's machine before, we determined that the acceleration of the masses was $a = \left(\frac{m_2 - m_1}{m_2 + m_1}\right) g$

This, of course, is not quite right. How does the motion of the pulley affect Atwood's machine. Well, think about the torques on the pulley as shown on the right. There is one torque (T_1) trying to make the pulley rotate counter-clockwise and another torque (T_2) trying to make the pulley rotate clockwise. They each act at the same distance (r) from the axis of rotation. If $T_1 = T_2$, the pulley would not rotate, because there would be two equal torques trying to get the pulley to rotate in opposite directions. Thus, the tension in the string on one side of the pulley (the side with the heavier mass) must be greater than the tension in the string on the other side of the pulley (the side with the lighter mass). That's where the motion of the pulley affects our analysis of Atwood's machine. When we did not consider the motion of the pulley, we just assumed that the tension along the string is constant. That's not true.

The motion of the pulley makes the tension in the string different on one side than it is on the other. Thus, we really have three unknowns in our problem: T_1, T_2, and a. Let's reanalyze Atwood's machine taking this into account.

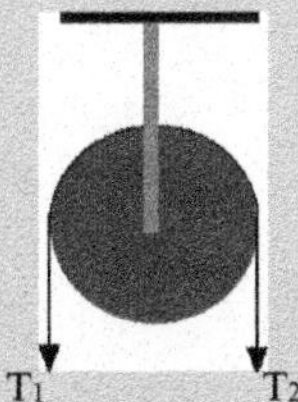

Looking at mass 1 and defining upward motion as positive, we get:

$$T_1 - m_1 g = m_1 a$$

Looking at mass 2, we realize that *downward* motion of m_2 is the same as upward motion of m_1.

Thus, downward motion of m_2 is positive:

$$m_2 \cdot g - T_2 = m_2 \cdot a$$

The last equation comes from looking at the pulley. T_1 and T_2 both exert torques on the pulley. T_1 is negative, but T_2 is positive because downward motion on that side of the pulley is positive. This should make sense, since the torques must have opposite signs because they cause the pulley to rotate in opposite directions. Since the tensions pull perpendicular to the radius, the torques are easy to calculate

$$-r \cdot T_1 + r \cdot T_2 = I \cdot \alpha$$

Since α and acceleration are related, we can substitute for α:

$$-r \cdot T_1 + r \cdot T_2 = I \cdot a/r$$

We can calculate I by looking at Fig. 4.5 again, so now we have three unknowns (T_1, T_2, and a) and three equations. Let's use the equation from m_1 to solve for T_1 in terms of a:

$$T_1 = m_1 \cdot a + m_1 \cdot g$$

Now, let's use the equation from m_2 to get T_2 in terms of a:

$$T_2 = m_2 \cdot g - m_2 \cdot a$$

Finally, we can take these equations for T_1 and T_2 and plug them into the equation from the rotation of the pulley. That will then allow us to solve for a:

$$- r(m_1 a + m_1 g) + r(m_2 g - m_2 a) = Ia/r$$
$$r^2 m_2 g - r^2 m_1 g = a\left(m_1 r_2 + m_2 r^2 + I\right)$$

Solving for acceleration,

$$a = \left(\frac{r^2 m_2 - r^2 m_1}{r^2 m_2 + r^2 m_1 + I}\right) g$$

We have the acceleration in terms of r, m_1, m_2, and g. We just need to replace I with something else. We can do that by looking at Fig. 4.5 and realizing that the pulley is a disk rotating on its central axis. Thus, $I = Mr^2/2$. Substituting this into the equation for acceleration, all the factors of r^2 cancel out and we are left with,

$$a = \left(\frac{m_2 - m_1}{m_2 + m_1 + M/2}\right)g$$

Now notice the difference between this equation and the one that we derived without considering the rotation of the pulley. The only difference is the $M/2$ term in the denominator of the fraction. What does this term do? Well, the more massive the pulley, the lower the acceleration. That should make sense. The more massive the pulley, the more it will resist the motion. Also, notice that if the pulley has no mass ($M = 0$), this equation turns into the original equation we derived. This should also make sense. If the mass of the pulley were zero (impossible, of course), then its moment of inertia ($Mr^2/2$) would be zero, and it would not resist change in its rotational state. Thus, it would not affect the acceleration of the masses.

Now you can test your knowledge of rotational dynamics by working the following Student problems.

Student

4.5 Consider the system shown on the right. You have analyzed it before, but you have ignored the mass of the pulley. Suppose $m_1 = 20.0$ kg, $m_2 = 20.0$ kg, and the pulley has a mass of 1.00 kg and a radius of 10.0 cm. Ignoring friction and the mass of the string, what are the acceleration of the system, the angular acceleration of the pulley, and the tensions in the string?

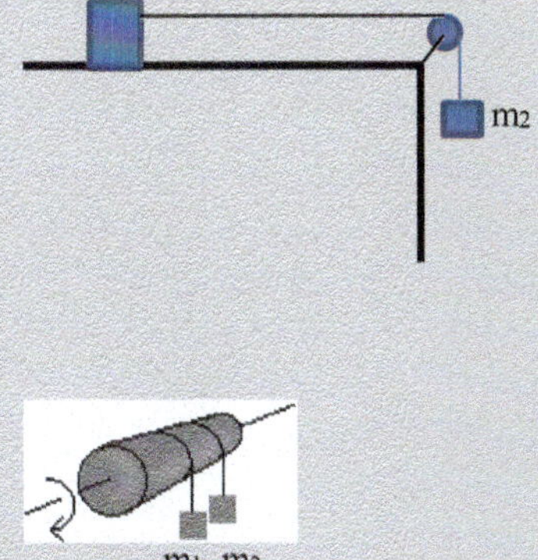

4.6 Two masses are hung from strings wound around a cylinder (mass = M, radius = R) which rotates without friction around its central axis.
Develop an equation in terms of m_1, m_2, M, R, and g for the accelerationof the masses as they descend.

4.7 Given that $m_1 = 1.00$ kg, $m_2 = 750.0$ g, $M = 450.0$ g, and $R = 15.0$ cm, determine the angular acceleration of the cylinder.

In order to obtain a mechanical advantage with pulleys we can set up a multiple pulley system with a movable pulley. This is sometimes called a block and tackle. A two-pulley system is shown in Fig. 4.7. For this case, the weight of the load is distributed equally between the two ropes coming from the lower pulley. This has to be true in equilibrium when the sum of the forces and sum of the torques must both equal zero. Now however, going over the upper pulley the tension translated to the rope on this side is only one half of the load weight. Therefore the force required to lift the load is only one half its weight. That is a mechanical advantage of 2. As was the case with a lever, the work done in lifting the weight a specific distance must equal the work done by the effort force. This means the effort force must pull the rope twice as far as the load is lifted. More elaborate systems with additional pulleys can be constructed with greater mechanical advantages. Every time another pulley is added to the system another rope is involved in supporting the weight of the load, so the MA is equal to the number of pulleys.

In many cases we must work with objects that have both translational motion and rotational motion. In these cases, our analysis must include both translational and rotational dynamics equations. One important aspect of this type of combined motion is the role played by friction to cause the round object to rotate instead of just translate. As shown in Fig. 4.8, the point where a round object touches a surface can experience two torques. The first is the applied torque trying to cause the object to rotate. The second (according to Newton's Third Law) comes from the reaction force which opposes this action. This comes from the friction between the object and the surface at the point of contact. Since this point is on the surface, the lever arm is the radius of the object. If the force of static friction is large enough to off-set the driving force, that point of contact has no instantaneous

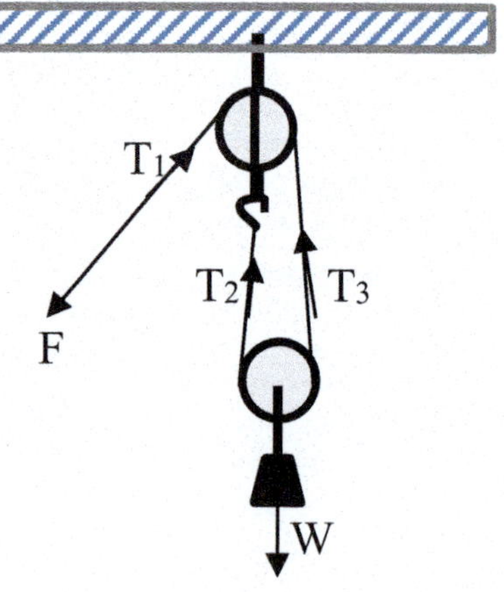

Fig. 4.7 Compound pulley

velocity causing the object to rotate while it moves forward. If the driving torque is stopped by applying brakes, the rotation will stop and friction will cause the object to skid to a stop.

If the driving torque is increased beyond the maximum force of static friction, the round object will spin in place or with reduced translation. This can happen when a car starts too fast or is on snow where the coefficient of friction is very small. Consider the following example.

Example 4.9

A solid disk ($I = mR^2/2$), a thin hollow disk ($I = mR^2$), and a solid sphere ($I = 2mR^2/5$) each of radius R and mass m are rolling without slipping down a 30° inclined plane. Which of these three has the greatest acceleration?

The picture for this problem is shown below.

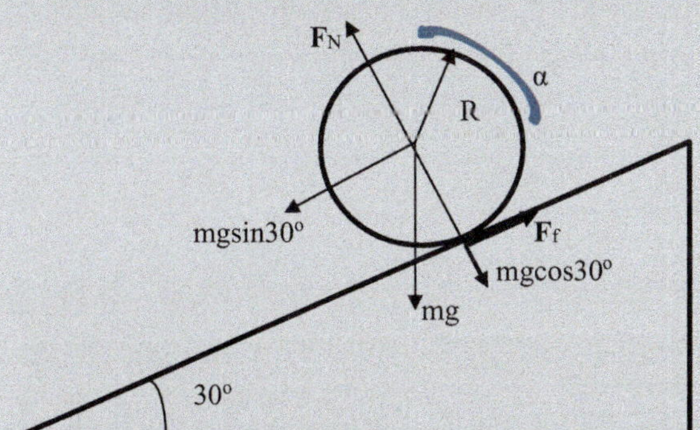

Knowns: R; m; $\theta = 30°$

Unknowns: a

The three forces relevant to translational motion are the component of gravity down the incline $F_g = mg \sin 30°$, friction $F_f = \mu F_N$, and the normal force $F_N = mg \cos 30°$. The torque causing the rotation is the friction force with a moment arm R acting in the counterclockwise direction. Newton's second law for translational and rotational motion gives

$$F = ma = mg \sin 30° - \mu mg \cos 30°$$

Fig. 4.8 Translation with rotation

$$\tau = I\alpha = R\mu mg\cos 30^\circ = Ia/R$$

where the last equation uses the fact that $a = \alpha R$ at the point where the rolling object touches the surface of the plane. Combining these two equations and solving for acceleration gives,

$$a = \frac{mgR^2 \sin 30^\circ}{I + mR^2}.$$

Now inserting the moment of inertia for each object gives

$$\begin{aligned} &\text{Hollow disk}: a = g/4 = 0.25\,\text{g} \\ &\text{Solid disk}: \quad a = g/3 = 0.33\,\text{g} \\ &\text{Sphere}: \qquad a = 5g/14 = 0.357\,\text{g}. \end{aligned}$$

Thus, the acceleration of the sphere is the fastest and the acceleration of the hollow disk the slowest.

Answers to the Student Problems

4.1

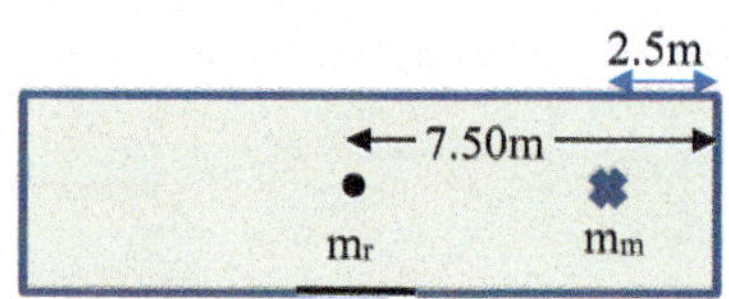

Knowns: $m_{\text{m}} = 85.0$ kg; $m_{\text{r}} = 150.0$ kg; $x_{\text{m}} = 2.5$ m
$x_{\text{r}} = 7.50$ m; $w_{\text{r}} = 4.00$ m; $\Delta x_{\text{m}} = 10$ m
Unknowns: CM; Δx_{r}

a. The raft is uniform, so its center of mass is the very center of the raft. If the raft is 15.0 m long, that means the center of mass is in the middle of the raft, 7.50 m from the edge. The man, on the other hand, is 2.50 m from the edge. In essence, then, we have two masses, 5.00 m from each other, as shown below:

From the right end of the raft, then, the center of mass can be calculated as:

$$x_{\text{cm}} = \frac{m_r x_r + m_m x_m}{m_r + m_m} = \frac{(150.0\,\text{kg})(7.5\,\text{m}) + (85.0\,\text{kg})(2.50\,\text{m})}{(150.0\,\text{kg}) + (85.0\,\text{kg})} = 5.69\,\text{m}$$

The center of mass of the system is the middle of the raft, 5.69 m from the end that the man is standing near.

b. There are no external forces acting on the man/raft system. As a result, momentum is conserved. Initially, the raft and man are motionless, so the momentum is zero. As the man walks, he gains momentum. In order to conserve momentum, the raft will have to move in the other direction. The problem says that the man walks towards the opposite end so that he ends up 2.5 m from that end. That means he walks a total of 10.0 m. He does this in some time interval, which we will call "*t*." Thus, his average velocity is 10.0 m/t, resulting in an average momentum of:

$$p_{\text{man}} = m_{\text{m}}v_{\text{m}} = (85.0\,\text{kg})(10.0\,\text{m})/\text{t}$$

In order to conserve momentum, then, the momentum of the raft must be the opposite. Thus:

$$p_{\text{raft}} = m_{\text{r}}v_{\text{r}} = -(85.0\,\text{kg})(10.0\,\text{m})/\text{t}$$

$$v_{\text{r}} = \frac{-(85.0\,\text{kg})(10.0\,\text{m})/\text{t}}{150.0\,\text{kg}} = -5.67\,\text{m}/\text{t}$$

Since the velocity of the raft must be 5.67 m/t, that means the raft must move 5.67 m in the same time interval. Thus, the raft must move 5.67 m opposite the man.

4.2

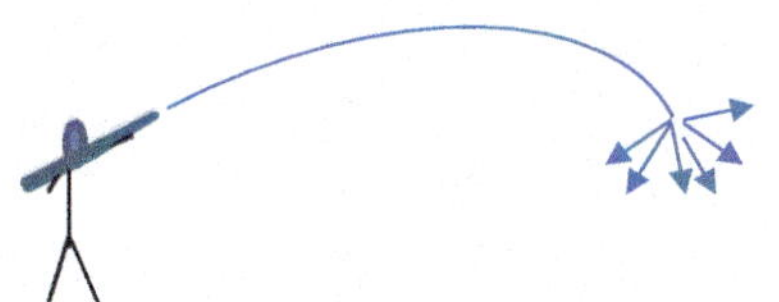

Knowns: v_{o} =125 m/s; $\theta = 30°$

Unknowns: CM

Remember, an entire system can be replaced by its center of mass. Thus, the projectile can be replaced by its center of mass. When the projectile explodes, it still can be replaced by the center of mass of all of the pieces, and that center of mass will continue on its same trajectory. Thus, to determine the final position of the center of mass, just use the range equation, because the center of mass will continue to move as a projectile:

$$R = \left(v_o^2\right)\sin(2\theta)/g = (125\,\text{m/s})^2\sin 60°/9.81\,\text{m/s}^2 = 1380\,\text{m}$$

The center of mass, then, will end up 1380 m from the point at which the projectile was fired.

4.3 The figure is given.

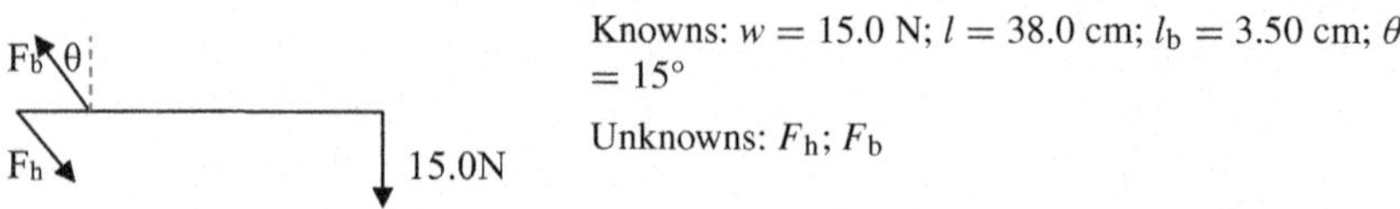

Knowns: $w = 15.0$ N; $l = 38.0$ cm; $l_b = 3.50$ cm; $\theta = 15°$

Unknowns: F_h; F_b

In order for the system to not move, the sum of the torques must be zero, and the sum of the forces must be zero. Let's start with the sum of the torques. We can choose any axis of rotation, but the most reasonable one is the elbow. When we do that, the force exerted by the humerus (F_h) disappears, because the distance from the axis is zero. If we define clockwise motion as positive, the sum of the torques becomes:

$$(15.0\,\text{N}) \cdot (38.0\,\text{cm}) - [F_b \cdot \sin(105.0)] \cdot (3.50\,\text{cm}) = 0$$
$$F_b = 169\,\text{N}$$

Notice that the angle used was 105.0°. That's because the angle is given relative to the vertical. However, that's not the proper definition. The angle must be defined from the horizontal. If the angle relative to the vertical is 90.0°, the angle relative to the horizontal is 90.0° + 15.0°, or 105.0°. Thus, the biceps pull with a force of 169 N at an angle of 105.0° relative to the horizontal.

To get the force that the humerus must exert, we have to sum up the forces in both dimensions. Starting with the horizontal, let's call the horizontal component of the humerus' force F_{hx}. The ball's weight exerts only a vertical force. Thus, we just have to look at the horizontal component of the biceps' force. Now remember, if we use the angles defined properly, then they take care of the signs for us. Thus, we can just add the horizontal component of F_b, because the properly defined angle will give you the proper sign. Alternatively, you can use the reference angle (75.0°—the angle from the nearest horizontal) and put the sign in yourself. Since we already have the properly defined angle, we will use it:

$$F_{hx} + (169\,\text{N}) \cdot \cos\left(105.0°\right) = 0$$
$$F_{hx} = 43.7\,\text{N}$$

That is the x-component of the force. To get the y-component, we sum up the forces in the y-dimension:

$$(169\,\text{N}) \cdot \sin\left(105.0°\right) - F_{hy} - 15.0\,\text{N} = 0$$
$$F_{hy} = 148\,\text{N}$$

These two components then lead us to a magnitude of 154 N and a direction of 286.5° relative to the positive horizontal.

4.4 Remember, the moment of inertia tells you how much the object resists changes in rotational motion. The more the object resists, the less angular acceleration there will be for the same magnitude of torque. According to Fig. 4.5, a hoop that spins about its central axis has a moment of inertia of MR^2. However, when it spins about its diameter, it has a moment of inertia of $MR^2/2$. That means the moment of inertia is *least* when it rotates about its diameter, so the angular acceleration will be greatest when it rotates about its diameter.

4.5 Figure is given. Knowns; $m_1 = 20.0$ kg; $m_2 = 20.0$ kg; $M = 1.00$ kg; $R = 10.0$ cm

Unknowns: a; α; T_1; T_2

Once we take the pulley into account, the tension of the string on each side of the pulley is different. As a result, the tension that pulls to the right on m_1 (T_1) is different than the tension which pulls up on m_2. If we ignore friction, the only force acting on m_1 is T_1. If we define motion to the right as positive, we get:

$$T_1 = m_1 \cdot a$$

On mass two, only gravity and T_2 operate. Now, since we defined m_1 moving to the right as positive, that means m_2 moving down is positive. Thus:

$$m_2 \cdot g - T_2 = m_2 \cdot a$$
$$T_2 = m_2 \cdot g - m_2 \cdot a$$

When we consider the rotational motion of the pulley, we know that the sum of the torques is equal to the moment of inertia ($MR^2/2$) times the angular acceleration (a/R). Thus, we have the equation:

$$T_2R - T_1R = \left(M \cdot R^2/2\right)(a/R).$$

Notice that we use the direction definitions to give us the signs of the torque. Since motion to the right (and down) is positive, T_2 is pulling positively on the pulley and T_1 is pulling negatively on the pulley. If we substitute the equations for T_1 and T_2 that we derived above, we get:

$$(m_2g - m_2a)R = (m_1a)R = \left(M \cdot R^2/2\right)(a/R)$$
$$a = \frac{m_2g}{(m_1 + m_2 + M/2)} = \frac{(20.0\,\text{kg})\left(9.81\,\text{m/s}^2\right)}{(40.0\,\text{kg} + 0.50\,\text{kg})} = 4.84\,\text{m/s}^2$$

Without considering the pulley, the acceleration would have been 4.91 m/s^2. Now that we have the acceleration, we can get the tensions from the equations above. $T_1 = 96.8\,\text{N}$, and $T_2 = 99.4\,\text{N}$. The angular acceleration is given by the translational acceleration divided by R, or 48.4 rad/s^2.

4.6 Figure is given. Knowns: m_1; m_2; M; R

Unknowns: a

Since the angular acceleration of the cylinder will determine the acceleration of the masses, the masses must each have the same acceleration, which we will call "a." This acceleration is the result of two forces working against each other: gravity and the tension in each string. The tension in the string connected to m_1 will be called T_1, and the tension in the string connected to m_2 will be called T_2. Summing up the forces on m_1, calling downward motion negative, gives us:

$$T_1 - m_1 \cdot g = m_1 \cdot a$$
$$T_1 = m_1 \cdot a + m_1 \cdot g$$

Summing up the forces on m_2 gives us:

$$T_2 - m_2 \cdot g = m_2 \cdot a$$
$$T_2 = m_2 \cdot a + m_2 \cdot g$$

That's two equations with three unknowns. However, we can get a third equation from the rotation of the cylinder. The sum of the torques is equal to the moment of inertia ($MR^2/2$) times the angular acceleration (a/R). Thus, we have the equation:

$$-T_1R - T_2R = \left(M \cdot R^2/2\right)(a/R)$$

Since both tensions pull down on the cylinder, they are negative tensions. If we substitute the equations we already have for those tensions, we will reduce this equation to only one unknown:

$$-(m_1a + m_1g)R - (m_2a + m_2g)R = \left(M \cdot R^2/2\right)(a/R)$$
$$a = \frac{-m_1g - m_2g}{m_1 + m_2 + M/2}$$

4.7 Figure given. Knowns: m_1=1.00 kg; m_2 = 750.0 g; M = 450.0 gm; R = 15 cm

Unknowns: α

The best way to get the angular acceleration is to determine the acceleration and then divide by R. The acceleration, according to the equation above, works out to – 8.69 m/s^2. The negative just means that the masses are accelerating downward. When we divide by R (0.150 m), we get an angular acceleration of – 57.9 rad/s^2. Once again, the negative sign gives direction. The cylinder rotates clockwise so that the masses fall down.

Study Guide for This Chapter

1. Imagine two cars approaching one another in opposite directions. The first car is twice as massive as the second car. They move so that the center of mass of the two-car system stays in exactly the same place. What will happen when they collide?
2. Is it possible for a system to be in static rotational equilibrium but not be in static translational equilibrium?
3. Suppose a sphere is rolling without slipping down a ramp under the influence of only gravity and is not losing any energy to friction. Is the sphere in dynamic translational equilibrium?
   Why or why not? Is it in dynamic rotational equilibrium? Why or why not?
4. An object is acted on by only one force. Is it possible for the object to be in any kind of rotational equilibrium? Why or why not?
5. A hollow sphere and a solid sphere each of the same mass are exposed to the same torque.
   Which will experience the greater angular acceleration?
6. Think about a disk spinning about its central axis at a constant angular velocity. Is it possible for the disk to be experiencing a torque?
7. Two identical thin rods are given the same amount of rotational kinetic energy. The first rod spins about a perpendicular axis which passes through its center while the second spins about a perpendicular axis which passes through one of its ends. If the angular velocity of the second rod is ω, what is the angular velocity of the first rod?

8. A piano mover hears that a block-and-tackle system can be used to make lifting heavy objects easier. He constructs the block-and-tackle system shown to the right to lift a grand piano (m = 200.0 kg).

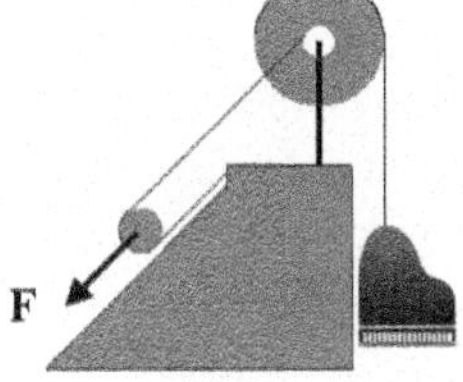

a. If the inner radius of the large pulley is 0.250 m and the outer radius is 1.00 m, what force (F) will the man need to exert in order to pull the grand piano up at a constant rate?

b. How could the man change the setup to actually make his job easier?

9. A child's top ($m = 250.0$ g) is a solid sphere ($r = 6.00$ cm) with a string tied around its center. When the string is pulled, the top spins. Suppose a child pulls in the string with a force of 55.0 N. What would the angular acceleration of the top be?

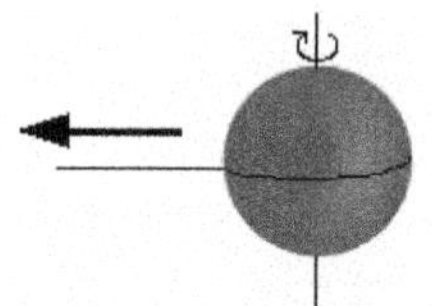

10. Suppose the top in problem #9 was spinning at 5.0 revolutions per second when the string was completely pulled away. What is the angular momentum of the top? Based on the drawing in problem #2, include the direction of the angular momentum.

11. A ball is held on an incline at a height of 75.0 cm. It is then released so that it rolls down the incline without slipping. At the bottom of the incline is a loop which the ball rolls around. If the loop is 50.0 cm high at its highest point, what is the speed of the ball at that point?

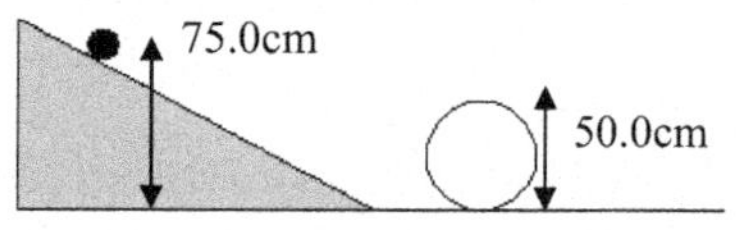

12. A large disk ($m = 25.0$ kg, $r = 1.00$ m) is spinning about its central axis at a rate of 10.0 revolutions per minute. A student drops a ball of clay ($m = 2.50$ kg) on the disk as it spins. The ball sticks on the disk at a radius of 0.750 m from the center. Ignoring friction and assuming that there are no net torques acting on the system, how many revolutions per minute will the disk make after the clay has been dropped?

13. An equilateral triangle is formed by three uniform meter sticks. One of the meter sticks is metal. The other two meter sticks are wooden, and each has a mass that is half that of the metal meter stick. Where is the center of mass of this system relative to the metal meter stick?

14. A uniform, 2.50 m rod ($m = 50.0$ kg) is attached to a wall at one end. The other end is attached to a cable that attaches to the wall. The cable makes an angle of 30.0° relative to the rod, and the rod is perpendicular to the wall. If a 100.0 kg mass hangs from the center of the rod, what is the tension on the cable? What force does the wall apply to the end of the rod?

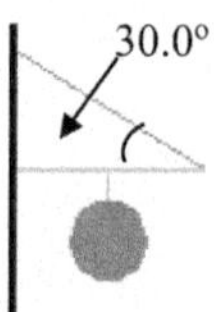

15. A pendulum is constructed of a string (length = 55.0 cm) attached to a bob ($m = 50.0$ g). The bob is pulled so that the string makes a 25.0° angle relative to its hanging position (dashed line).

    a. What is the maximum torque experienced by the bob?
    b. What will be the bob's angular acceleration when it experiences the maximum torque?
    c. What is the minimum torque experienced by the bob?

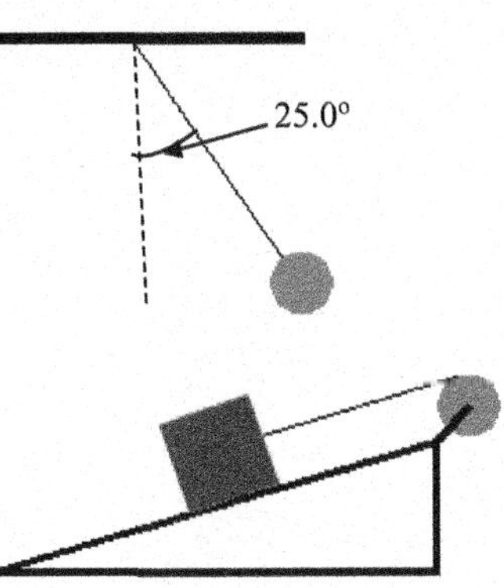

16. A 10.0 kg mass slides down a frictionless ramp as shown to the right. Its acceleration is 3.75 m/s^2. The mass is attached to a string which is wound around a wheel which spins as the mass falls. If the ramp makes a 25.0° angle with the ground, what is the mass of the wheel? Treat the wheel as a disk rotating about its central axis.

17. Consider an Atwood's machine. The heavy mass ($m_2 = 25.0$ kg) is held 2.45 m above the ground, and the lighter mass ($m_1 = 10.0$ kg) sits on the floor. The mass is then released and begins to fall. If the pulley has a mass of 5.00 kg and a radius of 75.0 cm, what will be the speed of the heavy mass the instant before it hits the ground? Ignore the mass of the string, energy loss due to friction, and air resistance.

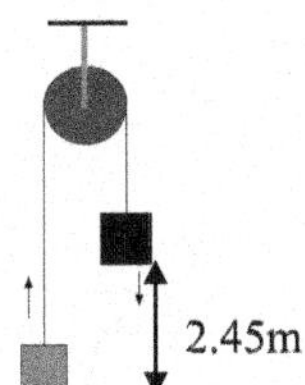

18. Think about a disk spinning about its central axis at a constant angular velocity. Imagine that two pennies are sitting on the disk. Suppose the first penny is close to the center of the disk and the second penny is near the edge of the disk. Compare the angular velocities of the two pennies. Compare the linear velocities of the two pennies.

Next Level

19. Use simple integral calculus to derive the expression for the moment of inertia of a thin hoop rotating about an axis through its center. (Hint: Follow the example in Next Level to convert the increment of mass, dm, to an increment of angle, dθ.)

20. Consider problem 19 again with the axis of rotation going through one edge of the hoop as shown.

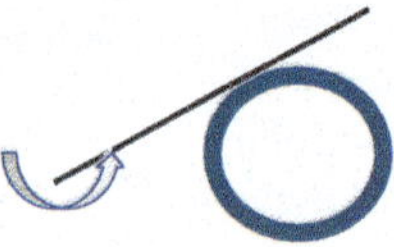

Rotational Motion II: Energy and Gravity

5

Chapter Summary

Rotational motion is a critical part of many aspects of our lives. This includes both the earth's orbit around the sun and the earth spinning on its axis. The details of these motions are what makes our planet inhabitable and our creation so beautiful. Some of the themes of the chapter are represented in Figs. 5.1 and 5.2.

Main Concepts in This Chapter

- Rotational Energy and Momentum
- Circular Motion and Gravity
- Kepler's Model of the Solar System
- Living on a Rotating Earth

5.1 Introduction

Now that we have the tools we need to describe rotational motion, we can consider the concepts of energy and momentum in the context of rotational motion. As we found with our study of translational motion, rotational motion also has conservation concepts associated with energy and momentum. This chapter covers these

R. C. Powell, *Physics*, Undergraduate Lecture Notes in Physics,
https://doi.org/10.1007/978-3-032-09361-5_5

Fig. 5.1 Solar system. *Credit* NASA, public domain

Fig. 5.2 Spinning skater. *Credit* David Carmichael Creative Commons Attribution-Share Alike 3.0

properties of rotational energy and momentum and then uses the force of gravity as an example of a centripetal force causing circular motion.

In addition, we will talk about how the actual motion of the planets in our solar system deviate from this simple, circular motion.

5.2 Rotational Energy and Momentum

In both Example 4.7 and Problem 4.7 dealing with pullies, when the rotation of the pulley is included in the analysis, the resulting acceleration is lower than the results of an analysis which does not consider the rotation of the pulley. This is because it takes energy to rotate the pulley. We can determine how much energy it takes by remembering that translational kinetic energy is given by $mv^2/2$. Then, considering the analogues between translational and rotational motion, the rotational kinetic energy of an object is given by:

$$\mathrm{KE}_{\mathrm{rot}} = I\omega^2/2 \tag{5.1}$$

where the moment of inertia has replaced the mass and the rotational speed has replaced the translational speed. Since the units of I are kg m^2 and the units of ω are rad/s, the units of rotational kinetic energy are kg m^2 rad^2/s^2. However, we drop the radian unit (since it represents a dimensionless quantity), giving us kg m^2/s^2, which is the same thing as a Joule. Thus the unit for rotational kinetic energy is the Joule. Now let's see how to take rotational energy into account when analyzing situations which involve rotational motion.

Example 5.1

A mass (m = 150.0 g) is attached to a string which is wound around a pulley (M = 10.0 g, r = 11.0 cm). The pulley is held still and then released. If the mass started out 35.0 cm above the floor, what speed will it have the instant before it hits the floor? Ignore energy losses due to friction and air resistance.

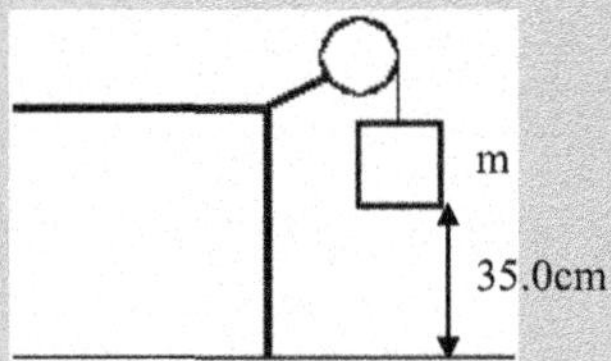

Figure given above. Knowns: m = 150.0 g; M = 10.0 Mg; r = 11.0 cm: h = 35.0 cm
Unknowns: v_f

We already solved a problem similar to this using the rotational equivalent of Newton's Second Law. However, we can solve this problem using energy concepts, since it does not ask about acceleration. It asks only about speed. Think about the total energy of the system before it is released. The pulley is not moving, and neither is the mass. Thus, the only energy is the potential energy of the mass:

$$\mathrm{TE}_1 = \mathrm{mgh}$$

Right before the mass hits the floor, there will be no potential energy, but there will be kinetic energy. The mass will be moving, so it has kinetic energy. However, the pulley will be rotating as well. Thus, the pulley will also have kinetic energy. The total energy will be:

$$\mathrm{TE}_2 = mv^2/2 + I\omega^2/2 = \mathrm{TE}_1$$

so total energy is conserved. The pulley is a disk so we know its moment of inertia is $Mr^2/2$. Also, Eq. 4.5 gives us a relationship between ω and v. We can substitute these into the equation above to give us:

$$\mathrm{mgh} = mv^2/2 + \frac{1}{2}\left(Mr^2/2\right)(v/r)^2$$

Solving for v gives

$$v^2\left(\frac{1}{2}m + \frac{1}{4}M\right) = \mathrm{mgh}$$

$$v = \sqrt{\frac{\mathrm{mgh}}{\frac{1}{2}m + \frac{1}{4}M}} = \sqrt{\frac{(0.1500\,\mathrm{kg})(9.81\,\mathrm{m/s^2})(0.350\,\mathrm{m})}{\frac{1}{2}(0.1500\,\mathrm{kg}) + \frac{1}{4}(0.0100\,\mathrm{kg})}} = 2.58\,\mathrm{m/s}$$

Had we not considered the pulley, the speed would have been 2.62 m/s^2. Notice that the radius of the pulley cancels out of the final equation. Thus, the radius of the pulley does not affect the speed of the system.

As shown above, many of the motion problems you will analyze will involve *both* translational motion and rotational motion. A car, for example, has translational energy as it moves down the street, but its wheels have rotational energy. If you were to analyze the energy supplied by the engine, you would have to take into account the fact that some of the energy causes the rotational motion of the wheels and some of it causes the translational motion of the car. Make sure you understand how to include rotational energy in the analysis of a system by solving the following problem.

Student

5.1 A sphere, a disk, and a hoop all sit at the top of a ramp, as shown to the right. Each of these objects has the same mass.

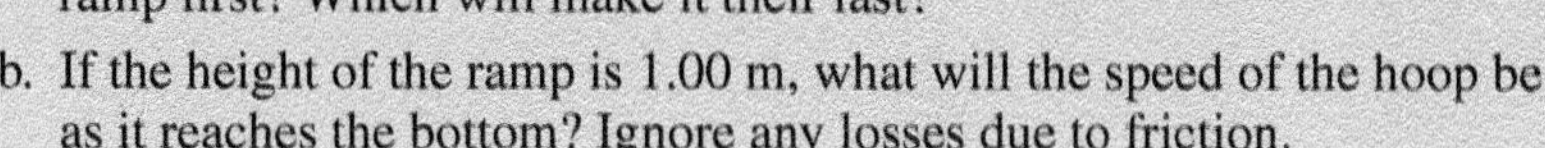

a. Which will make it to the bottom of the ramp first? Which will make it their last?
b. If the height of the ramp is 1.00 m, what will the speed of the hoop be as it reaches the bottom? Ignore any losses due to friction.

c. Even though you were told to ignore the losses due to friction in the problem above, if this is a rotational motion problem, we *must* assume that there is friction. Why?

We have one more rotational analogue to discuss, and that is angular momentum. The formula for angular momentum is:

$$\mathbf{L} = \mathbf{r} \times \mathbf{p} \tag{5.2}$$

In this equation, "**L**" is the angular momentum, "**r**" is the radius of the angular motion, and "**p**" is the translational momentum of a point on the edge of the object which is rotating. Notice that all of the quantities are vectors, because the cross product of two vectors gives us a vector whose direction is given by the right-hand rule, as discussed in the introduction.

In your introductory course, you may have learned a simpler expression for angular momentum that is applicable when the object of interest is moving in a circle. As shown in Fig. 5.3, when an object moves in a circle, its velocity vector is always tangent to the circle. Thus, the angle between the radius of the circle and the velocity is 90°. Then the magnitude of the angular momentum can be calculated from Eq. 5.2 as:

$$L = rp \cdot \sin(90°) = rp \tag{5.3}$$

Since the translational momentum of an object is mv, the magnitude of the angular momentum in this situation is:

$$L = mvr = I\omega \tag{5.4}$$

where we have used the expression for I from Fig. 4.4 appropriate to the situation shown in Fig. 5.3. To get the direction of L point the fingers of your right hand

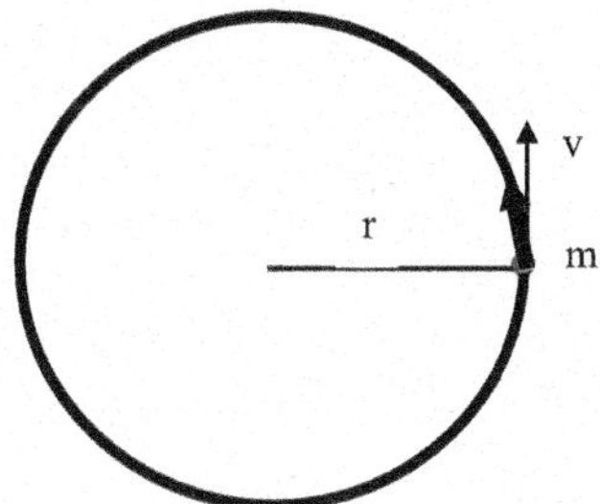

Fig. 5.3 Circular motion

along r and rotate them toward v. Your thumb points out of the page which is the direction of the angular momentum vector.

Newton's second law in terms of circular motion can be written as

$$\tau = I\alpha \tag{5.5}$$

where τ is the net torque on the object, I is the object's moment of inertia and α is the object's angular acceleration. Using the definition for angular acceleration as the change in angular velocity $\Delta\omega$ over a time interval, this expression can be written as,

$$\tau = I\Delta\omega/\Delta t.$$

Similar to what we did for linear momentum, we can write this as,

$$\tau\Delta t = I\Delta\omega = I\omega_{\mathrm{f}} - I\omega_i. = L_{\mathrm{f}} - L_i. \tag{5.6}$$

The left side of Eq. 5.6 is the net torque multiplied by the time interval over which it acts. This is called the angular impulse. It is equal to the change in angular momentum it produces. In general, the magnitude of the angular momentum is given by the product of its moment of inertia and its angular velocity,

$$L = I\omega. \tag{5.7}$$

The units of L are the same as for linear momentum, kg·m^2/s. Equation 5.6 is a statement of the angular impulse-angular momentum theorem. This leads to the important physics law, the called the conservation of angular momentum. This states:

Conservation of angular momentum—if the net external torques acting on a system are zero, its angular momentum will remain constant.

There is one important difference between this law and the law of conservation of linear momentum. Since linear momentum is defined by the product of mass and velocity and since mass is a constant, having constant linear momentum implies that the linear velocity is also constant. This is not true for angular momentum as shown in Eq. 5.7. If L is constant because of no net external torque, the angular velocity ω can still change as long as the momentum of inertia I changes in the opposite direction to compensate for this.

Since angular momentum is a vector, in order for it to be conserved, *neither* the magnitude *nor* the direction can change. To see the consequences of this, perform the following experiment.

Experiment 5.1
The Direction of the Angular Momentum Vector

Supplies

- The front wheel from a bicycle.
- A helper
- A chair or stool that spins around (optional)

Introduction—The direction of the angular momentum vector is hard to visualize, but it is quite easy to feel, given the right conditions.

Procedure:

1. Take the wheel and hold it with both hands. You should be able to hold onto the part that attaches to the bicycle. Grab it with one hand on the part that attaches to one side of the bicycle and the other hand on the part that attaches to the other side. If you are holding it properly, the wheel should be able to spin freely on its axle, and the wheel will be between your arms.
2. Make sure that the wheel is not spinning and turn it so that the wheel is parallel to the floor.
3. Next, turn the wheel so that it is now perpendicular to the floor. Note how easy that was.
4. Next, turn the wheel so that it is once again parallel to the floor and have your helper start spinning the wheel as fast as he can. **Be careful here. If you lose your grip on the wheel, someone could get hurt!**
5. Once the wheel is spinning, have your helper stand back and once again, try to turn the wheel so that it is perpendicular to the floor. Note how much harder it is to turn the wheel.
6. (Steps 6–8 are optional. You should read them even if you do not do them, however.) Sit on a chair that spins, lifting your feet up off of the floor so that the chair will spin easily. Make sure the chair and the wheel are not moving.
7. Once again, hold the wheel so that it is parallel to the floor and have your helper start spinning the wheel as fast as he can. **Be careful here. If you lose your grip on the wheel, someone could get hurt!**
8. With your feet still off of the floor, turn the wheel upside down so that it is once again parallel to the floor but spinning in the opposite direction. What happens?
9. Clean everything up, including putting the wheel back on the bicycle.

Did you notice how much more difficult it was to turn the tire while the wheel was spinning? What causes the difficulty? When the tire is spinning, it has angular momentum. The direction of the angular momentum depends on the way that the tire is spinning. Point the fingers of your right hand from the center of the wheel out and then curl your fingers in the direction that the wheel spins. Your thumb points in the direction of the angular momentum, perpendicular to the wheel. When you turned the wheel, however, the direction of the angular momentum changes to stay perpendicular to the wheel. That goes *against* the conservation of angular momentum. Thus, to turn the wheel, you need to exert a torque. Without the wheel spinning, there is no angular momentum, so turning the wheel is easy. With the wheel spinning, a torque is necessary in order to change the direction of the angular momentum vector. The faster the wheel is spinning, the greater the magnitude of the angular momentum, so the greater the torque required to turn the wheel.

If you did the optional part of the experiment, the chair should have started spinning when you turned the wheel. With your feet off of the ground, you could no longer apply a torque to the system. Thus, when the wheel turned, angular momentum had to be conserved. In order to conserve angular momentum, your chair had to start spinning in order to offset the change induced by turning the wheel. Depending on the friction involved, this might not have been a great effect, because friction applies a torque to the chair. Nevertheless, in a well-oiled chair, you can really start spinning this way.

Study the following example to see how to use the concept of angular momentum conservation in analyzing physical situations.

Example 5.2

A person ($m = 75$ kg) stands at the edge of a merry-go-round (a uniform disk, $M = 150.0$ kg, $r =$ 1.80 m). As the merry-go-round spins freely, the person walks toward the center of the merry-go-round. If the merry-go-round starts out spinning with one revolution every 3.00 s, what will its angular velocity be when the person is only 0.500 m from the center?

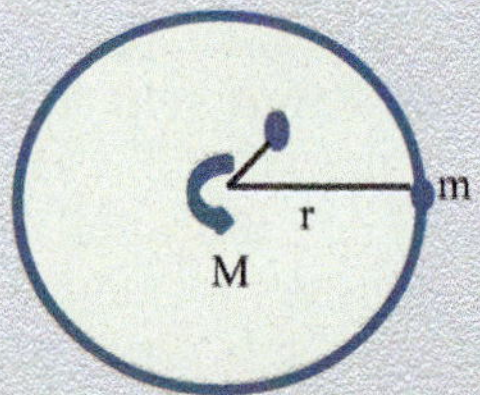

Knowns: $m = 75$ kg; $M = 150$ kg; $r = 1.8$ m; $r_f = 0.5$ m; $\omega_i = 1/3$ rev per second

Unknowns: ω_f

In this problem, there are two things involved in rotational motion: the merry-go-round and the person. The total angular momentum, then, is:

$$L_{tot} = L_{merry\text{-}go\text{-}round} + L_{person} = (I_{merry\text{-}go\text{-}round}) \cdot \omega_{merry\text{-}go\text{-}round} + (I_{person}) \cdot \omega_{person}$$

The merry-go-round is a disk that is spinning around its central axis. According to Fig. 4.4, then, it has a moment of inertia of $\frac{1}{2}\,Mr^2$. The person is an object that moves in a circle. Thus, the person's moment of inertia is mr^2.

To begin with, the person is at the edge of the merry-go-round so both the merry-go-round and the person make one revolution every 3.00 s. In one revolution, the merry-go-round sweeps out an angle of 2π radians. Thus, the angular velocity for both the person and the merry-go-round is:

$$\omega = \Delta\theta/\Delta t = 2\pi\ \mathrm{rad}/3\ \mathrm{s} = 2.09\ \mathrm{rad/s}$$

So at the beginning, the total angular momentum is

$$\begin{aligned} L_{\mathrm{tot}} &= \left(\frac{1}{2}Mr^2\right)(2.09\ \mathrm{rad/s}) + \left(mr^2\right)(2.09\ \mathrm{rad/s}) \\ &= \left(\frac{1}{2}150\ \mathrm{kg}\right)(1.80\ \mathrm{m})^2(2.09\ \mathrm{rad/s}) \\ &\quad + (75.0\ \mathrm{kg})(1.80\ \mathrm{m})^2(2.09\ \mathrm{rad/s}) = 1016\ \mathrm{kg\ m^2/s} \end{aligned}$$

Once again, the radian unit gets dropped.

When the man starts walking, his moment of inertia changes because the radius of his motion changes. Because of conservation of angular momentum, the angular velocity must change to compensate for this. We can calculate how the angular velocity changes by making sure that angular momentum is conserved:

$$L_{\mathrm{tot}} = \left(\frac{1}{2}Mr^2\right)\omega_{\mathrm{f}} + \left(mr^2\right)\omega_{\mathrm{f}}$$

$$1016\ \mathrm{kg\ m^2/s} = \left(\frac{1}{2}150\ \mathrm{kg}\right)(1.80\ \mathrm{m})^2\omega_{\mathrm{f}} + (75.0\ \mathrm{kg})(0.50\ \mathrm{m})^2\omega_{\mathrm{f}}$$

$$\omega_{\mathrm{f}} = \frac{1016\ \mathrm{kg\ m^2/s}}{\left(\frac{1}{2}150\ \mathrm{kg}\right)(1.80\ \mathrm{m})^2 + (75.0\ \mathrm{kg})(0.50\ \mathrm{m})^2} = 3.88\ \mathrm{rad/s}$$

Notice that the merry-go-round speeds up from 2.09 rad/s to 3.88 rad/s. When the man walks to the center, the total moment of inertia decreases, so the angular velocity must increase in order to keep the angular momentum constant.

Make sure you understand these concepts by solving the following problems.

Student

5.2 A figure skater is spinning on the tip of her skates with her arms folded over her chest. She then stretches her arms out until she is holding them parallel to the ice, fully extended.

Ignoring friction, what would happen to the speed at which she is spinning? What would happen if she brought her arms back in again and folded them across her chest?

5.3 Consider the situation discussed in the experiment. A person sits on a chair that is free to spin. The person and the chair form roughly a cylinder which can spin on its central axis. The mass of this "cylinder" is 85.0 kg, and the radius is 15.0 cm. Assume that the person and chair are sitting still. The person holds a wheel as was done in the experiment and has a friend spin the wheel (treat it as a hoop with $M = 5.00$ kg and $r = 13.0$ cm) while it is parallel to the ground. If the wheel is rotating at 10.0 revolutions each second, at what angular velocity will the person and chair start spinning if the wheel is turned upside down so that it is parallel to the ground but spinning in the opposite direction?

5.3 Circular Motion and Gravity

There are many situations in which an object moves in a circle around a central point with a constant angular velocity. To have this type of motion, the object must have a force acting on it pointed inwardly along the radial direction. This type of force delivers zero torque to the object since it is applied parallel to the lever arm. Therefore, there is no angular acceleration. However, there is linear acceleration since the object continues to change the direction of its linear velocity (although not the magnitude of its linear velocity). The direction of this acceleration is radially toward the center in the same direction as the force. This type of force that is responsible for uniform circular motion is called centripetal force. There are several different types of physical situations in nature involving centripetal force. These included simple models of the solar system and of atoms. In this chapter we will discuss the effects of gravitational attraction and in later chapters we will discuss similar effects of electrostatic forces.

Using Eq. 4.7 for centripetal acceleration, the centripetal force required to keep an object of mass m moving with uniform circular motion at a velocity v at a radius r from the center of rotation is,

$$F_c = mv^2/r, \tag{5.8}$$

Newton developed an expression for universal gravitational attraction as,

$$F_g = Gm_1 m_2/r^2. \tag{5.9}$$

This is the attractive force between two objects whose centers are a distance r apart that have masses m_1 and m_2. G is called the universal gravitational constant. It is one of the fundamental constants in physics and has a value of $G = 6.67 \times 10^{-11}\,\mathrm{N\,m^2/kg^2}$. Each of the two objects feels the same force of attraction due to gravity. Sometimes one of the objects is much larger than the other one and we are interested in the gravitational force of the larger body on the smaller body. For instance, consider the force the earth's gravity pull exerts on a small 1 kg object on its surface. Equation 5.9 becomes,

$$\begin{aligned}F_g &= \left(6.67 \times 10^{-11}\,\mathrm{N\,m^2/kg^2}\right)(1\,\mathrm{kg})\left(5.98 \times 10^{24}\,\mathrm{kg}\right)/\left(6.38 \times 10^{6}\,\mathrm{m}\right)^2 \\ &= 9.81\,\mathrm{N}\end{aligned}$$

where 5.98×10^{24} kg is the mass of the earth and 6.38×10^{6} m is the radius of the earth. Note that for an object with a mass other than 1 kg, multiplying the mass times the gravitational acceleration of $9.81\,\mathrm{m/s^2}$ gives the weight of the object which is equivalent to the earth's gravitational force on the object. We used this information in previous chapters.

A large object is surrounded by a gravitational field. The magnitude of the field is defined as the force experienced by a unit test mass at a specific point a distance r away. Thus,

$$g = GM/r^2 \tag{5.10}$$

and the direction is toward the mass M. For an object on the surface of the earth, the earth's gravitational field is $g = 9.81\,\mathrm{N/kg}$ or $9.81\,\mathrm{m/s^2}$ pointed toward the center of the earth.

As an example, consider a model of the moon's orbit around the earth as uniform circular motion driven by the gravitational force of attraction between the moon and the earth.

Example 5.3

What is the force of gravity that the earth ($m_e = 5.98 \times 10^{24}$ kg) exerts on the moon ($m_m = 7.36 \times 10^{22}$ kg) and what is the period of the moon's rotation around the earth? (The distance between the earth and moon is 3.84×10^{8} m.)

The problem is sketched below. The knowns and unknowns are:

Knowns: $m_e = 5.98 \times 10^{24}$ kg; $m_m = 7.36 \times 10^{22}$ kg; $r = 3.84 \times 10^8$ m.

Unknowns: F_g; period of rotation.

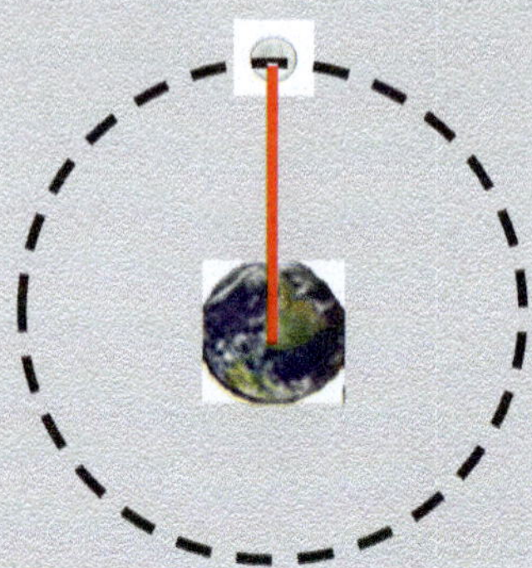

In this case the centripetal force keeping the moon in its orbit around the earth is the force of the earth's gravity. The gravitational force is,

$$\begin{aligned} F_c = F_g &= G m_e m_m / r^2 \\ &= \left(6.67 \times 10^{-11}\,\mathrm{N\,m^2/kg^2}\right)\left(5.98 \times 10^{24}\,\mathrm{kg}\right)\left(7.36 \times 10^{22}\,\mathrm{kg}\right)/\left(3.84 \times 10^8\,\mathrm{m}\right)^2 \\ &= 1.99 \times 10^{20}\,\mathrm{N}. \end{aligned}$$

This is the gravitational force the earth exerts to keep the moon in its orbit.

For uniform circular motion, this force is related to the linear velocity of the moon by Eq. 5.8,

$$F_c = m_m v_m^2 / r$$

or solving for the velocity,

$$v_m = (F_c r / m_m)^{1/2}.$$

Substituting the numbers gives

$$v_m = \left(1.99 \times 10^{20}\,\mathrm{N} \times 3.84 \times 10^8\,\mathrm{m}/7.36 \times 10^{22}\,\mathrm{kg}\right)^{1/2} = 1.02 \times 10^3\,\mathrm{m/s}.$$

The moon will travel at this velocity through the distance of one orbit which is the circumference of the orbit,

$$C = 2\pi r = 2\pi\left(3.84 \times 10^8\,\mathrm{m}\right).$$

Then for one orbit, $C = v_m t$, so

$$t = C/v_m = 2\pi\left(3.84 \times 10^8\,\mathrm{m}\right)/\left(1.02 \times 10^3\,\mathrm{m/s}\right) = 23.65 \times 10^5\,\mathrm{s}.$$

Converting this to days gives

$$t = 23.65 \times 10^5\,\mathrm{s}\ (1\,\mathrm{min}/60\,\mathrm{s})(1\,\mathrm{h}/60\,\mathrm{min})(1\,\mathrm{day}/24\,\mathrm{h}) = 27.34\,\mathrm{days}.$$

This result for the orbit time of the moon is consistent with our observations of the motion of the moon.

Newton's law of gravity gives a simple model for our solar system, eight planets orbiting the sun. (Scientists decided that Pluto didn't meet the criteria for being called a planet.) In the simplest model, each planet is in a circular orbit with the sun at the center. Gravitational attraction between the sun and the planet provides the centripetal force holding the planet in its orbit. The radius of the orbit depends on the velocity of the planet. Using Eq. 5.8 for centripetal force, and setting it equal to the gravitational force given by Eq. 5.9, $F_c = F_g$ so

$$m_{\mathrm{planet}} v^2/r = G m_{\mathrm{planet}} m_{\mathrm{sun}}/r^2$$

so

$$r v^2 = G m_{\mathrm{sun}}.$$

Since the mass of the sun is 2×10^{30} kg and we know the universal gravitation constant G, the product of the radius of a planet's orbit times its velocity squared is given by

$$r v^2 = 13.3 \times 10^{19}\,\mathrm{m^3/s^2}.$$

Mars is observed to have a velocity of 2.4×10^4 m/s so this equation predicts the radius of its orbit to be 2.3×10^{11} m. This is consistent with the fact that it takes mars 1.8 earth years to orbit the sun.

When dealing with problems involving gravity near the surface of the earth, we used a simple expression for gravitational potential energy given in Chap. 3 as

$$\mathrm{PE} = \mathrm{mgh} \tag{5.11}$$

This assumes a constant value of g for the gravitational acceleration throughout the problem.

Sometimes we want to analyze the motion of satellites, planets, comets, etc., which are far from the earth. To do this we need to have an equation that calculates gravitational potential energy regardless of how far you are from the earth. It would also be nice to have an equation that tells us the gravitational potential energy associated with *any* two objects. Equation 5.12 does both. For any two objects whose masses are m_1 and m_2 that are separated by a distance of r, the gravitational potential energy is:

$$PE_{grav} = -Gm_1m_2/r \tag{5.12}$$

Notice that the potential energy due to gravity is negative. That's important. Remember what a negative potential energy means. It means that the object in question is *bound.* When an object is near enough to the earth to feel its gravity, it is bound to the earth in some way. The moon, for example, is bound to the earth because it is caught in earth's gravitational field. You are bound to the earth because you are caught in its field. Thus, the potential energy due to gravity is *negative.*

This may seem really odd to you, since the equation you have been using for the potential energy due to gravity up to this point is not negative. That's because the equation assumes we are dealing with things near the surface of the earth that will never leave the surface of the earth. Thus, the fact that they are bound is irrelevant. Also, it assumes that the force due to gravity is constant, which is a good assumption as long as your height relative to the surface of the earth is very small compared to the radius of the earth. However, if you get rid of those assumptions, you must define the gravitational potential energy as negative in order to be consistent with all of the other rules we have drawn up so far.

Notice that there is one similarity between Eqs. 5.11 and 5.12. They both state that the higher you go, the larger the potential energy. As "h" increases, the potential energy in Eq. 5.11 gets larger. Thus, the higher you are, the more gravitational potential energy you have. Equation 5.12 agrees. The larger the distance between the two objects, the smaller the negative number given by the equation. Well, the smaller the negative number, the larger the value of that number.

Equation 5.12 allows us to consider many interesting situations, such as the one given in the example below.

Example 5.4

In order for a rocket to leave the earth and head to the moon or another planet, it must essentially "escape" earth's gravitational field. With what minimum speed must a rocket be launched in order to do this? (mass of the earth = 5.98×10^{24} kg, radius of the earth = 6.37×10^{6} m)

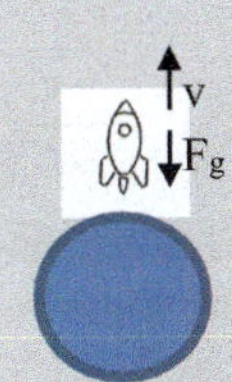

Knowns: $m_e = 5.98 \times 10^{24}$ kg; $r_e = 6.37 \times 10^6$ m
Unknowns: v_e

When a rocket is at the surface of the earth, it has a potential energy given by Eq. 5.12. It also has a kinetic energy given by ½mv^2. That's its total energy. As the rocket rises, its potential energy increases, because as Eq. 5.12 indicates, the farther the rocket gets from earth, the less negative the potential energy becomes. As a result, the kinetic energy decreases. If a rocket were to just lose all of its kinetic energy when it finally has no more potential energy from the earth's gravity, it would just barely escape the earth's gravitational attraction.

If the rocket has neither potential nor kinetic energy, its total energy is zero. The Law of Conservation of Energy states that the total energy must always be the same. Thus, to get a rocket to leave earth's orbit, the kinetic energy plus the potential energy must equal zero.

$$\mathrm{TE_{before}} = \mathrm{TE_{after}}$$

$$m_{\mathrm{rocket}} v^2/2 - G m_{\mathrm{rocket}} m_{\mathrm{earth}}/r_{\mathrm{earth}} = 0$$

$$v_e = \sqrt{\frac{2Gm_{\mathrm{earth}}}{r_{\mathrm{earth}}}} = \sqrt{\frac{2(6.67 \times 10^{-11}\,\mathrm{Nm^2/kg^2})(5.98 \times 10^{24}\,\mathrm{kg})}{6.37 \times 10^6\,\mathrm{m}}} = 1.12 \times 10^4\,\mathrm{m/s} \tag{5.13}$$

The rocket, then, must attain a speed of at least 1.12×10^4 m/s. This number, of course, neglects such inconvenient things as air resistance, so in reality, a rocket must have a higher speed in order to leave the earth's gravitational field. One thing to note here is that when the rocket escapes earth's gravitational field, it has a gravitational potential energy of zero. At that point, the potential is not negative, indicating that the rocket is no longer bound to earth. That's another way of saying that the rocket has escaped earth's gravitational field: its gravitational potential energy with respect to the earth is no longer negative.

Notice that the speed which the rocket needs to escape earth's gravity is not dependent on the mass of the rocket. It is only dependent on the radius and mass of the earth. As a result, *any* object that wants to leave earth's gravitational field must achieve the same minimum speed. This is called the escape velocity for earth.

Escape velocity—The minimum speed necessary for an object to escape the gravitational field of another object.

If we have the mass and radius of any planet (or any other object), we can calculate the escape velocity for that planet.

Next Level

Gravity Inside the Earth

If an object is buried deep inside the earth, the gravitational force it experiences from the earth is different from the one it experiences when it is outside the earth. As shown in the figure, an object of mass m a distance $(R - r)$ from the earth's surface will experience a gravitational force toward the center of the earth due to the spherical mass between the object and the earth's center, $F = Gm\frac{M'}{r^2}$ where M' is the mass of the inner sphere. Using the usual assumption for uniform mass distribution, the mass density is $\rho = M/V = M'/V'$ so $M/(\frac{4}{3}\pi R^3) = M'/(\frac{4}{3}\pi r^3)$ and $M' = M\frac{r^3}{R^3}$. The gravitational attraction is then given by $F = Gm\frac{Mr}{R^3}$. Note that on the surface of the earth, $r = R$ and the force is the usual Newtonian force of universal gravitational attraction.

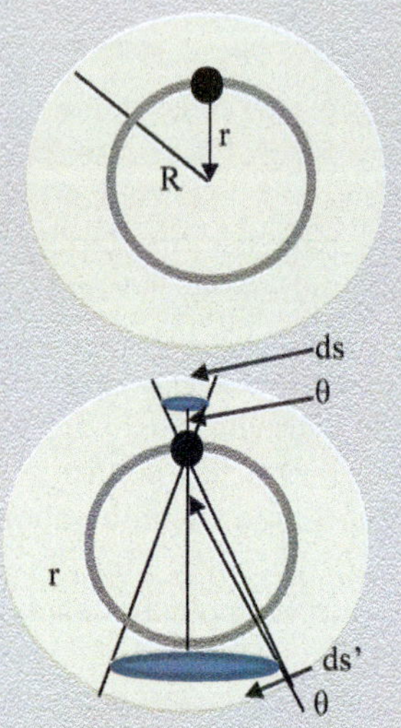

Why does the rest of the mass of the earth not provide a contribution to the gravitational force on the object? This can be seen in the similar triangles of the second part of the figure.

The mass in the volume of the disk above the object creates a force in the upward direction that depends on the mass of the disk, $m = \rho\pi(r_1\theta)^2 x$, and the distance from the object to the disk's center of mass r_1^{-2}. x is the thickness of the disk. The radial distance cancels out. The disk below the object creates exactly the same magnitude of force but in the downward direction. Again the distance from the object to the center of mass of the disk cancels out. Since these two forces have equal magnitudes but are in opposite directions they cancel out. Similar arguments can be made for the rest of the mass outside of the object's position.

Now try your hand at the following problem.

Student

5.4 Suppose you are adjusting the orbit of a satellite ($m = 1576\,\text{kg}$), which is orbiting the earth ($m = 5.98 \times 10^{24}\,\text{kg}$, $r = 6.37 \times 10^{6}\,\text{m}$) 1015 km above its surface. You want to use the satellite's engines to move the satellite into an orbit that is 1505 km above the surface of the earth.
   a. What is the total energy of the satellite before you make the adjustment to the orbit?
   b. What is the total energy of the satellite after you make the adjustment?
   c. How much work did the engines have to do in order to make the adjustment?

5.4 Kepler's Model of the Solar System

The actual planetary orbits are not exactly circular. They are slightly elliptical. The motion of the planets is more accurately explained by Kepler's Laws of planetary motion. Newton was able to write his Universal Law of Gravitation by examining the work of **Johannes Kepler**. Kepler was the student of a man named **Tycho Brahe**, an astronomer who collected *volumes* of data regarding the known planets and their positions in the sky. Kepler studied this data intensely, and he eventually was able to distill this data into three basic laws which are called **Kepler's Laws**. These laws are best understood in the context of a figure, such as the one given below.

The first thing to notice about the figure is that it has the earth orbiting the sun. This went against the established view of the heavens at the time, which held that the earth sat at the center of the solar system and the sun (as well as all of the planets) orbited the earth. Even Tycho Brahe, Kepler's teacher, held to a modified version of this view. Kepler, however, was convinced by the data that the sun was at the center of the solar system and that the planets orbited around it. We know today, of course, that Kepler was right.

Also, Fig. 5.4 shows **Kepler's First Law**.

Kepler's First Law—A planet moves in an elliptical orbit, with the sun at one focus.

Remember from geometry that an ellipse is simply an oval, and it is defined by two foci. An ellipse contains the set of all points in which the sum of the distances between the foci and the point is equal to a given number. When discussing ellipses, mathematicians typically say that the ellipse has a **semimajor axis** (as illustrated in the figure) and a **semiminor axis** (as illustrated in the figure). Mathematicians can also define an ellipse by its **eccentricity**.

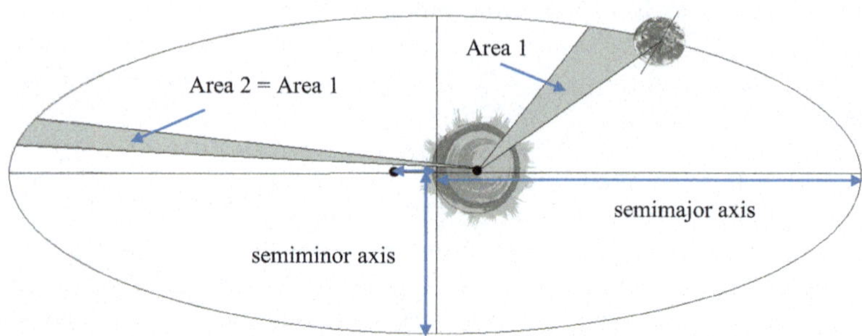

Fig. 5.4 Motion of the Earth around the Sun

Eccentricity is defined as the distance between the center of an ellipse and either focus divided by the length of the semimajor axis. Mathematically, if the distance from the center to a focus is "*c*" and the length of the semimajor axis is "*a*," we can say that the eccentricity (*e*) is given by:

$$e = c/a \tag{5.14}$$

The eccentricity is a way of discussing how "oval" an ellipse is. The higher the eccentricity, the more oval the ellipse.

The ellipse drawn in Fig. 5.4 is exaggerated quite a bit compared to real planetary orbits. The ellipses in which the planets move are nearly circular. For example, the earth orbits the sun in an ellipse whose semimajor axis only about one hundredth of a percent larger than its semiminor axis. Another way to say this is that the eccentricity of a planet's orbit is very close to zero. Remember, from the definition of eccentricity, the eccentricity of a circle is zero. The eccentricity of the earth's orbit, for example, is about 0.017.

Kepler's Second Law explicitly deals with the time it takes for planets to make one complete orbit.

Kepler's Second Law—As a planet orbits the sun, a line drawn from the sun to the planet sweeps out the same area in a given time interval.

As shown in Fig. 5.4, if you draw a line from the sun to the planet and then calculate the area swept out in a given time interval, the area is the same regardless of where the planet is on the ellipse. This is essentially a statement that the speed of the planet increases as the planet approaches the sun. When a planet is closer to the sun, it must cover a greater distance around the ellipse in order to sweep out an area that is equal to the area swept out when the planet is far from the sun. Now once again, since the elliptical orbits followed by the planets are very close to circles, this difference is minor. However, Kepler's Laws also apply to comets whose orbits are ellipses with large eccentricities. The difference in speed for a comet when it is close to the sun compared to when it is far from the sun is significant.

Kepler's Third Law relates the orbits of all of the planets.

Kepler's Third Law—The ratio of the period of the planet's orbit squared to the length of the semimajor axis cubed is the same for all planets.

If we call the planet's orbital period "*T*" and the length of the semimajor axis "*a*," Kepler's third law says T^2/a^3 is constant for all planets. If we assume that the semimajor axis of the ellipse is the average orbital distance from the planet to the sun $\overline{r}$,

$$(T_1/T_2)^2 = (\overline{r_1}/\overline{r_2})^3 \tag{5.15}$$

where the 1 and 2 subscripts refer to two different planets.

For example, the average distance of the earth from the sun is 1.5×10^{11} m and the average distance of mars from the sun is 2.28×10^{11} m. Thus, Kepler's third law predicts

$$(T_{\text{mars}}/T_{\text{earth}})^2 = \left(2.28 \times 10^{11}/1.5 \times 10^{11}\right)^3 = 3.51.$$

So

$$T_{\text{mars}} = 1.87\,T_{\text{earth}}.$$

In other words, mars orbits the sun in 1.87 earth years.

Combining Newton's second law of motion with the expression for centripetal force and the universal law of gravitation for the force given by Eqs. 5.8 and 5.9 gives

$$Gm_{\text{sun}}m_{\text{planet}}/r_{\text{p}}^2 = m_{\text{p}}4\pi^2 r_{\text{p}}/T^2.$$

Then the period of the orbit is given by

$$T = 2\pi\sqrt{r_{\text{p}}^3/Gm_{\text{sun}}}. \tag{5.16}$$

Note that by substituting the mass of the earth for the mass of the sun, this expression can be used to calculate the rotation period of a satellite orbiting the earth.

Newton suggested that *all bodies* that have mass experience a mutual gravitational force with *all other bodies* that have mass, so he called it a "universal law" of gravity.

Newton's Universal Law of Gravitation is expressed in Eq. 5.9 as:

$$F = Gm_1m_2/r^2$$

where "m_1" and "m_2" are the masses of the objects in question, "r" is the distance between the objects, and "G" is the gravitational constant which is $6.67 \times 10^{-11}\ \mathrm{Nm^2\ kg^2}$. Remember, the gravitational force is *mutual.* In other words, the earth exerts a gravitational force on the moon and the moon exerts an *equal but opposite* force on the earth. The same is true for any two objects. A ball falls, for example, because the earth exerts a gravitational force on the ball. In response, the ball exerts an equal but opposite force on the earth. Thus, both the ball and the earth are attracted to one another with an equal force. As the ball falls to the earth, then, the earth also rises to meet the ball. Of course, since the earth is so massive relative to the ball, the earth's acceleration is negligible compared to that of the ball.

Special Topic
Kepler had no idea *why* the planets obeyed his laws, he just knew that all of the observations he and his teacher made were consistent with these laws. Many religious leaders opposed Kepler's work. Kepler himself was a devout Christian. In fact, he once wrote, "I wanted to become a theologian. For a long time, I was restless. Now, however, behold how through my effort God is being celebrated in astronomy." (As quoted in Gerald Holton, *American Journal of Physics* 24 (May1956): 340–351). You do not have to be in a full-time church ministry position to serve God.

Example 5.5
Show that Newton's Law of Universal Gravitation actually gives Kepler's Third Law.

Let's assume that the orbit of a planet around the sun is circular. (Since the eccentricity of a planet's orbit is nearly zero, we can go ahead and make the assumption.) For circular motion to occur, there must be a centripetal force, the strength of which is given by Eq. 5.8. In this case, gravity supplies the centripetal force, and the strength of the gravitational force is given by Eq. 5.9. Equating these two forces gives

$$m_p v_p^2/r = Gm_p m_s/r^2$$

where "m_p" is the mass of the planet; "m_s" is the mass of the sun; "v_p" is the speed of the planet in its orbit; and "r" is the distance from the center of the planet to the center of the sun.

Notice that the mass of the planet cancels because it is on both sides of the equation.

Now, let's think about how to calculate "v_p." The planet makes a full orbit once each period (denoted as "T"). In that time, it travels a total distance of $2\pi r$. Thus, the average speed of the planet is:

$$v_p = 2\pi r/T$$

Substituting the expression for v_p into the first expression gives

$$\frac{4\pi^2 r}{T^2} = \frac{Gm_s}{r^2}$$

So,

$$\frac{T^2}{r^3} = \frac{4\pi^2}{Gm_s}$$

This equation is a statement of Kepler's Third Law. Everything on the right side of the equation is the same, regardless of the planet in the solar system. Thus, it tells us that the period squared divided by the radius cubed is the same for all planets in the solar system.

Now try some problems.

Student

5.5 Calculate the average speed of the earth in its orbit. Is this speed perfectly constant? If not, when does the earth move at its fastest speed? When does it move at its slowest speed? (mass of the sun = 1.99×10^{30} kg, average radius of earth's orbit = 1.50×10^{11} m)

5.6 A **geosynchronous** orbit is an orbit in which the position of the orbiting body relative to any point on the planet's surface remains the same. At what distance must a satellite orbit the earth so as to maintain a circular, geosynchronous orbit? (mass of the earth = 5.98×10^{24} kg).

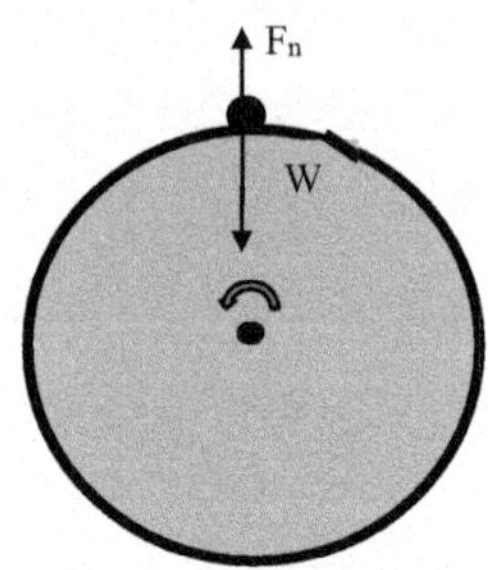

Fig. 5.5 Forces on an object at the earth's surface

5.5 Living on a Rotating Earth

The fact that our planet is rotating about its axis gives us the night and day cycle that is so important to life on earth. It also means that we are constantly experiencing centripetal acceleration. What are the consequences of this?

Remember in Chap. 2 we discussed how your weight is the force the earth pushes up on you in response to gravitational attraction. However, as long as the earth is spinning on its axis, the weight that a scale measures is not exactly the strength at which you are being pulled to the center of the earth. There is a centripetal force keeping you in circular motion. That force interferes with the measurement of your true weight. Figure 5.5 shows the forces acting on an object on the surface of the earth.

There are two forces acting on the object: the force due to gravity (W for "weight") and the normal force (F_n). When you weigh something, you are actually measuring F_n, not W, because a scale measures the force with which *it must push* to keep the object from falling. Thus, when you measure the weight of something, you are really measuring F_n. If the earth were not rotating on its axis, these two forces would cancel out exactly so that $F_n = W$. In that situation, the fact that you are measuring F_n instead of W is irrelevant, since they are both equal. However, since the earth is spinning, we know that these two forces *cannot* be equal. If they were equal, there would be no net force acting on the object and, as a result, it could not move in a circle, and it would fly off of the earth. Since the object does not fly off of the earth, there must be a net force which supplies the centripetal force necessary for the object to move in a circle. Since the only two forces at play are W and F_n, we therefore know that F_n must be slightly weaker than W, so that the net force is pointed toward the center of the earth, which must be the direction of the centripetal force.

Thus:

$$F_c = W - F_n$$

Plugging in the equation for centripetal force in terms of velocity gives

$$mv^2/r = W - F_n$$

To get the velocity of the object, let's assume that the object is sitting on the equator. If that is the case, the object moves a distance of $2\pi r$ in a total time of 24 h (one day), which we will call T, the period of the earth's rotation.

$$\frac{4\pi^2 mr}{T^2} = W - F_n$$

Now remember, F_n is the measured weight, and W is the actual weight. Therefore:

$$\text{measured weight} = F_n = W - \frac{4\pi^2 mr}{T^2}$$

How large an effect is this? We can easily calculate it. Let's suppose a "typical" adult male (mass = 75.0 kg) weighs himself. Since the radius of the earth is 6.37×10^6 m, and the period of the earth's rotation (T) is 24 h (86,400 s), we can actually determine the amount that the rotation of the earth takes away from the man's true weight:

$$\text{measured weight} = (75.0\,\text{kg})\left(9.81\,\text{m/s}^2\right) - \frac{4\pi^2(75.0\,\text{kg})\left(6.37 \times 10^6\,\text{m}\right)}{(86{,}400\,\text{s})^2}$$

$$= 736\,\text{N} - 2.53\,\text{N} = 733\,\text{N}$$

Thus the effect of the earth's rotation on the weight of an object is quite small, but it does exist.

Note that the spinning earth provides a centripetal force just like gravity. The friction between an object on the rotating surface and the earth keeps it from slipping but the centripetal force holds it on the surface like gravity. This can be useful in designing space stations.

Another interesting effect associated with the earth's rotation is the Coriolis effect. This is associated with the fact that the linear velocity you experience due to the earth's rotation depends on where you are on the earth's surface. This is shown in Fig. 5.6. We know from Eq. 4.5 that the linear velocity for an object undergoing rotational motion is equal to the rotational velocity multiplied by the radius of the circular motion. Objects on the surface all rotate around the same axis with the same angular velocity but the radius of their motion is different as shown in Fig. 5.6. Near the equator the radius of the motion is larger than it is for an object located further into the northern (or southern) hemisphere. Thus the translational velocity for an object on the surface of the earth is greatest at the equator and gets less as you move toward the poles.

If you stand on the equator and launch a projectile directly north, it will travel on a circular path to the east. This is because the earth is rotating from west to east and imparts an eastward component to the initial velocity of the projectile when it is launched. We just showed that the radial velocity of an object on the surface of the earth is greatest at the equator and becomes less as you move toward the

Fig. 5.6 Velocity on the surface of a rotating earth

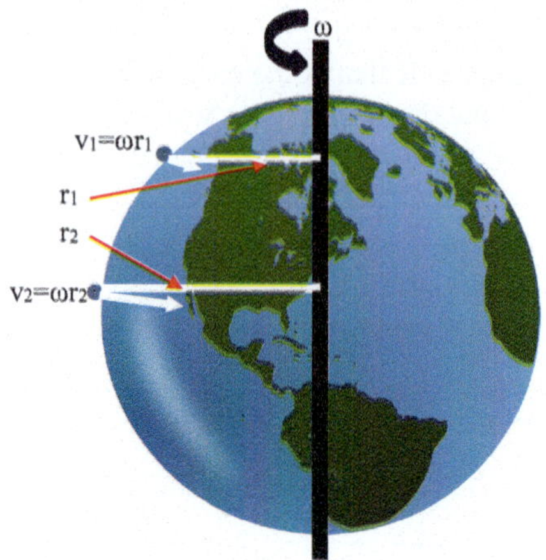

poles where the radius from the axis of rotation to the surface is less. Thus as the projectile moves north, the earth rotating under it has a smaller radial velocity than it did at the equator. Similarly, if you launch a projectile directly south from somewhere in the northern hemisphere, it will travel in an arc toward the west because as it gets further south the earth's radial velocity to the east is greater than the eastward component of the initial velocity of the projectile. This Coriolis effect is responsible for the direction of wind rotation.

Special Topic

From studying the weather, meteorologists know that wind is directed out of an area of high barometric pressure and into an area of low pressure. In the northern hemisphere, the Coriolis effect causes the wind going north out of a high pressure area to curve to the east while wind going south out of the same area will curve to the west. This sets up a clockwise rotation of the wind around a high pressure area. For a low pressure area the wind going into it from the south will curve to the east while the wind going into it from the north will curve to the west. This sets up a counterclockwise rotation of the wind around a low pressure area. These are in the opposite directions in the southern hemisphere.

Now try one final problem for this chapter.

5.7 What are the fastest linear velocities you can have due to the earth's rotation?

We end our discussion of circular motion with the following comment. Some forces are apparent forces instead of real forces. The opposite of centripetal force is called centrifugal force. Centrifugal force tries to throw an object out away from the center of its rotation. This effect can be used in a device such as a centrifuge that spins a sample containing particles of different weights and the differences in their centrifugal forces can be used to separate the sample into different types of particles. However, centrifugal force is an apparent force, not a real force. It results from the inertia of an object not wanting to change its direction of motion without any real force pushing or pulling on it.

Summing Up

You now know that rotational motion is one type of periodic motion. In the next chapter you will learn about two other types of periodic motion. Later in Chap. 16 we will use our knowledge of rotational motion to describe an electron orbiting the nucleus of an atom. Another important concept you learned about in this chapter is Newton's theory of gravity and how important it is to the solar system we live in. In Chap. 15 we will return to the topic of gravity as described by Einstein's theory.

Answers to the Problems

5.1 Figure given.

a. Think about the energy involved. All three objects start with no kinetic energy and the same potential energy (Mgh). When they get to the bottom, they will each have translational kinetic energy ($Mv^2/2$) because they are moving from one point to another. However, they will also have rotational kinetic energy ($I\omega^2/2$). Since they all start with the same total energy, they will all end with the same total energy. However, since the moments of inertia are different, some will have *more* of that energy as rotational kinetic energy, leaving *less* for translational kinetic energy. The one with the least translational kinetic energy will travel slowest. A sphere rotating on its central axis has a moment of inertia of $(2/5)MR^2$; a disk rotating that way has a moment of inertia of $MR^2/2$; and a hoop rotating that way has a moment of inertia of MR^2.

Thus, the hoop has the highest moment of inertia. This means more energy is needed for the rotational motion, so it has the least translational kinetic energy. In the same way, the sphere has the lowest moment of inertia, so less energy is needed to get it rotating. Thus, it will have the greatest translational kinetic energy. As a result, the hoop will reach the bottom last, while the sphere reaches the bottom first.

b. Remember, we start with a potential energy of Mgh. At the bottom of the ramp, all of that energy is converted to kinetic energy of $Mv^2/2 + I\omega^2/2$. We know that $I = MR^2$ for the hoop and $\omega = v/R$.

Thus,

$$\text{Mgh} = Mv^2/2 + \left(MR^2\right)(v/R)^2/2$$
$$v = \sqrt{gh} = \sqrt{\left(9.81\ \text{m/s}^2\right)(1.00\ \text{m})} = 3.13\ \text{m/s}$$

c. In this problem, we assume the sphere, disk, and hoop roll. If there were no friction, they would simply slide, because there would be nothing to provide a torque. After all, gravity works on the center of mass, so it supplies no torque to the objects. Only the friction resulting from the contact of the objects with the ramp generates a torque. Thus, in order for the objects to roll, friction must exist.

5.2

Think about the moments of inertia you have seen so far. They all depend on R^2. Thus, the larger the radius, the larger the moment of inertia. When the skater has her arms tucked into her chest, she has a certain average radius. When she stretches her arms out, that average radius increases. Thus, her moment of inertia increases. If angular momentum ($I\omega$) must be conserved, then as the moment of inertia increases, the angular velocity must decrease. As a result, when she stretches out her arms, she will start spinning more slowly. When she pulls them back in, she will start spinning at her original speed.

5.3

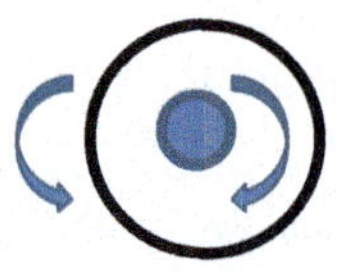

Knowns: $m = 85.0$ kg; $r = 15.0$ cm; $M = 5.00$ kg; $R = 13$ cm; $f_{\text{w}} = 10$ rps

Unknowns: f_{p}

The wheel is a hoop. Thus, we can calculate its angular momentum:

$$L = I \cdot \omega = \left(M \cdot R^2\right) \cdot \omega$$

We know M and R, and we have what we need to calculate ω. The wheel makes 10.0 revolutions per second. Each revolution sweeps out an angle of 2π radians. Thus, the wheel sweeps out 62.8 radians every second. This means the angular velocity is 62.8 rad/s. Now we can determine L:

$$L = (5.00\ \text{kg})(0.130\ \text{m})^2\ (62.8\ \text{rad/s}) = 5.31\ \text{kg m}^2/\text{s}$$

Now think about direction. If the wheel is spinning one way, the right-hand rule will have the angular momentum pointing directly up. If it is spinning the other way, it will be directly down.

When the wheel is flipped over, the direction changes. If it was pointing up before, it points down after. If it pointed down before, it points up after. Suppose it was pointing up before the wheel was flipped. That means $L = +5.31\,\mathrm{kg\,m^2/s}$. When the wheel is flipped, $L = -5.31\,\mathrm{kg\,m^2/s}$ (if up is a positive direction, then down must be negative). To conserve angular momentum, then, the student on the chair will have to start spinning. In order to keep the angular momentum the same, the student must spin so that he or she has an angular momentum of $10.62\,\mathrm{kg\,m^2/s}$ so that the total angular momentum is $10.62\,\mathrm{kg\,m^2/s} + -5.31\,\mathrm{kg\,m^2/s}$, which is $5.31\,\mathrm{kg\,m^2/s^2}$. Now that we know what the student's angular momentum must be, we can determine the student's angular velocity:

$$L = I \cdot \omega = \left(MR^2/2\right)\omega$$

$$\omega = 2L/\left(MR^2\right) = \frac{2(10.62\,\mathrm{km^2/s})}{(85.0\,\mathrm{kg})(0.150\,\mathrm{m})^2} = 11.1\,\mathrm{rad/s}$$

5.4

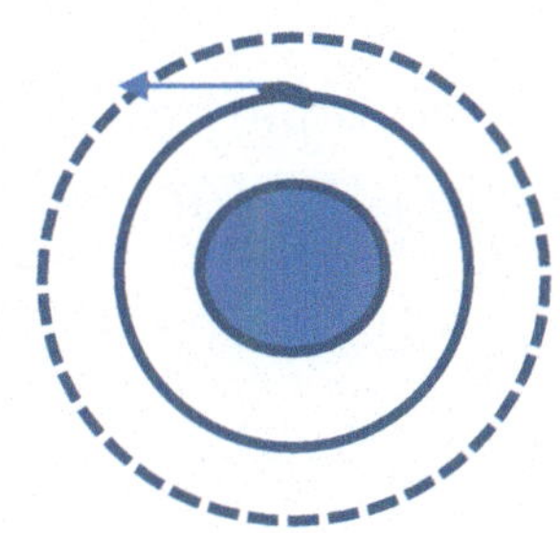

Knowns: $m_s = 1576$ kg; $m_e = 5.98 \times 10^{24}$kg; $r_e = 6.37 \times 10^6$ m; $r_1 = 1015$ km; $r_2 = 1505$ km

Unknowns: TE_1; TE_2; W

a. We can use Eq. 5.12 to calculate the satellite's potential energy. Note that the distance given, however is not r! It is the distance from the surface of the earth. In the equation, r is the distance between the center of masses. Thus, we must add the radius of the earth to the distance given.

$$\begin{aligned}\mathrm{PE}_1 &= Gm_1m_2/r = \left(6.67 \times 10^{-11}\,\mathrm{Nm^2/kg^2}\right)\left(5.98 \times 10^{24}\,\mathrm{kg}\right)\\ &\quad \times (1576\,\mathrm{kg})/\left(6.37 \times 10^6\,\mathrm{m} + 1.015 \times 10^6\,\mathrm{m}\right)\\ &= -8.51 \times 10^{10}\,\mathrm{J}\end{aligned}$$

That's the initial potential energy. What is the initial kinetic energy? To get that, we need to learn the speed. We can determine that by setting the centripetal force equal to the gravitational force:

$$\frac{m_s v_s^2}{r} = \frac{G m_s m_e}{r^2}$$

$$v_{s1} = \sqrt{\frac{G m_s}{r}} = \sqrt{\frac{(6.67 \times 10^{-11}\,\mathrm{Nm^2/kg^2})(5.98 \times 10^{24}\,\mathrm{kg})}{6.37 \times 10^6\,\mathrm{m} + 1.015 \times 10^6\,\mathrm{m}}} = 7.35 \times 10^3\,\mathrm{m/s}$$

Now we can determine the kinetic energy:

$$\mathrm{KE}_1 = m_s v_{s1}^2/2 = (1576\,\mathrm{kg})\left(7.35 \times 10^3\,\mathrm{m/s}\right)^2/2 = 4.26 \times 10^{10}\,\mathrm{J}$$

The total energy, then, is:

$$\mathrm{TE} = \mathrm{PE} + \mathrm{KE} = -8.51 \times 10^{10}\mathrm{J} + 4.26 \times 10^{10}\mathrm{J} = -4.25 \times 10^{10}\,\mathrm{J}$$

Why is the total energy negative? Remember, the satellite is bound to the earth, and negative energies generally represent a bound system.

b. We have to do all of this calculation again for the new orbit:

$$\begin{aligned}\mathrm{PE}_2 = G m_1 m_2/r &= \left(6.67 \times 10^{-11}\,\mathrm{Nm^2/kg^2}\right)\left(5.98 \times 10^{24}\,\mathrm{kg}\right)\\ &\times (1576\,\mathrm{kg})/\left(6.37 \times 10^6\,\mathrm{m} + 1.505 \times 10^6\,\mathrm{m}\right)\\ &= -7.98 \times 10^{10}\,\mathrm{J}\end{aligned}$$

Now we can determine the kinetic energy:

$$v_{s2} = \sqrt{\frac{G m_s}{r}} = \sqrt{\frac{(6.67 \times 10^{-11}\,\mathrm{Nm^2/kg^2})(5.98 \times 10^{24}\,\mathrm{kg})}{6.37 \times 10^6\,\mathrm{m} + 1.505 \times 10^6\,\mathrm{m}}} = 7.11 \times 10^3\,\mathrm{m/s}$$

$$\mathrm{KE}_2 = m_s v_{s2}^2/2 = (1576\,\mathrm{kg})(7110\,\mathrm{m/s})^2/2 = 3.98 \times 10^{10}\,\mathrm{J}$$

The total energy, then, is:

$$\mathrm{TE} = \mathrm{PE} + \mathrm{KE} = -7.98 \times 10^{10}\,\mathrm{J} + 3.98 \times 10^{10}\,\mathrm{J} = -4.00 \times 10^{10}\,\mathrm{J}$$

c. Look at what happened. The satellite did have a total energy of $-\ 4.25 \times 10^{10}$ J. After the orbit switch, it has an energy of $-\ 4.00 \times 10^{10}$ J. The total energy *increased* (got less negative) by 2.5×10^9 J. How can that happen? The only way that can happen is if you *work* on the satellite with that much energy. Thus, the engines did 2.5×10^9 J of work on the satellite.

5.5

Knowns: $r_{avg} = 1.50 \times 10^{11}$ m; $m_s = 1.99 \times 10^{30}$ kg

Unknowns: v_{avg}

Calculating the details of a planet's orbit usually starts by setting the centripetal force required to keep the planet in circular motion equal to the gravitational force, which is actually supplying the centripetal force:

$$\frac{m_e v_{avg}^2}{r_{avg}} = \frac{G m_e m_s}{r_{avg}^2}$$

$$v_{avg} = \sqrt{\frac{G m_s}{r_{avg}}} \sqrt{\frac{(6.67 \times 10^{-11}\,\text{Nm}^2/\text{kg}^2)(1.99 \times 10^{30}\,\text{kg})}{1.50 \times 10^{11}\,\text{m}}} = 2.97 \times 10^4\,\text{m/s}$$

Thus, the average speed of the earth in its orbit is 2.97×10^4 m/s.

This is the average speed, but it is not constant. The earth moves faster when it is close to the sun and slower when it is far from the sun. You can figure that out either with Kepler's Second Law or from your knowledge of centripetal force. In Kepler's Second law, the planet sweeps out equal areas in equal times. When the earth is far from the sun, it must move more slowly to sweep out an equal area than when it is close to the sun (see Fig. 4.8). From a centripetal force point of view, the higher the centripetal force, the higher the speed. When the earth is closest to the sun, the gravitational force (which supplies the centripetal force) is higher, so the speed is higher as well.

5.6

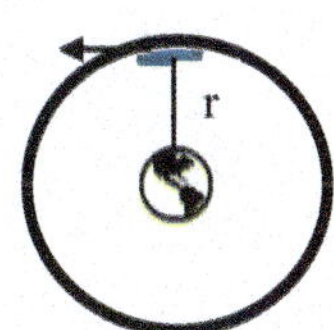

Knowns: $m_e = 5.98 \times 10^{24}$ kg; $T = 24$ h

Unknowns: r

Since the orbit is geosynchronous, we know the speed. How do we know the speed? Well, the satellite must stay in the same spot relative to the earth at all times. The only way it can do this is to have an orbital period of 24 h. That way, it will orbit at the same speed that the earth is rotating. In a single orbit, the satellite moves a distance equal to the circumference of the orbit, or $2\pi r$. It does so in 24 h, so the speed is $2\pi r$ divided by 24 h (which is 86,400 s)

$$\frac{m_s v_s^2}{r} = \frac{G m_e m_s}{r^2}$$

$$\frac{\left(\frac{2\pi r}{86{,}400\,\mathrm{s}}\right)^2}{r} = \frac{G m_e}{r^2}$$

$$r^3 = \frac{G m_e (86{,}400\,\mathrm{s})^2}{4\pi^2} = \frac{(6.67 \times 10^{-11}\,\mathrm{Nm^2/kg^2})(5.98 \times 10^{24}\,\mathrm{kg})(86{,}400\,\mathrm{s})^2}{4\pi^2}$$

$$r = 4.23 \times 10^7\ \mathrm{m}$$

Please note that this is the distance from the center of the earth, not the surface of the earth, because "r" in the universal law of gravitation refers to the distance between the center of mass of each object.

5.7 At the equator, the distance from the rotation axis to an object on the surface is the radius of the earth, $r = 6.371 \times 10^6$ m. At the poles the distance from the axis of rotation is 0.0 m. The angular velocity is $\omega = 2\pi\,\mathrm{radians/24\,h} \times 1\,\mathrm{h/60\,min} \times 1\,\mathrm{min/60\,s} = 7.27 \times 10^{-5}\,\mathrm{s^{-1}}$. Thus

$$v_{\mathrm{equator}} = \omega r_{\mathrm{equator}} = \left(7.27 \times 10^{-5}\,\mathrm{s^{-1}}\right)\left(6.371 \times 10^6\,\mathrm{m}\right) = 463.1\,\mathrm{m/s}$$

$$v_{\mathrm{pole}} = \omega r_{\mathrm{pole}} = \left(7.27 \times 10^{-5}\,\mathrm{s^{-1}}\right)(0\,\mathrm{m}) = 0.0\,\mathrm{m/s}$$

Study Guide for This Chapter

$$\left(G = 6.67 \times 10^{-11}\,\mathrm{Nm^2/kg^2}\right)$$

1. An object of mass m has a weight of w at the surface of the earth. If it is raised to an altitude that is equal to the radius of the earth, what is its mass and weight?
2. Suppose a new planet was discovered whose mass is 1/10 that of earth and whose radius is 1/5 that of earth. What is the gravitational acceleration at the surface of that planet?
3. The distance from Mercury to the center of the sun is 0.387 times that of earth. How long does it take for Mercury to orbit the sun?
4. Suppose you had a way of measuring the speed at which the earth is traveling in its orbit. How could you use it to tell when the earth was farthest from and closest to the sun?
5. If you analyze the motion of a block supported by a pulley system, why is the answer different if you ignore the motion of the pulley or take it into account?

6. If you launch a missile straight north from the Panama Canal, will it land closer to Florida or Baja California? Why?
7. A satellite is orbiting the earth at a radius of R and has a kinetic energy of KE. If the satellite moves to an orbit whose radius is $3R$, what is the new kinetic energy in terms of KE?
8. A satellite is orbiting Neptune with a radius of R and an orbital period of T. What is the mass of Neptune in terms of R, T, and G? This is one way to measure the mass of a planet, since R and T are observable from earth and G is a constant.
9. Mars has a mass of 6.42×10^{23} kg and a radius of 3.37×10^{6} m. If a rock is dropped from a height of 5.51×10^{6} m from the surface of the planet, at what speed will the rock be traveling when it hits the planet's surface? Ignore any resistive forces.
10. Two masses (m_1 and m_2) are in deep space, at rest a distance R apart. They are not being acted on by any other forces, and they start to move under their mutual gravitational attraction.

    In terms of G, m_1, and m_2, and R, what is the acceleration of each mass when they are a Distance $R/2$ from one another?
11. Suppose you had a spherical space station whose radius is 3500.0 m. To pro vide the same gravitational effects as the surface of the earth, at what speed (in m/s) must the space station rotate?
12. A satellite of earth ($m_e = 5.98 \times 10^{24}$ kg) orbits the earth once every 12.0 h. What is the radius of the satellite's orbit?
13. If you wanted to increase the radius of the satellite ($m = 1123$ kg) in problem #12 by 25.0%, how much energy would be required?

Next Level

14. What is the force of gravity on an object with a mass of 200 kg buried 5.0×10^{5} m below the surface of the earth?
15. What is the force of gravity of the same object in problem 19 buried the same distance under the surface of Mars? (See problem 9 for the parameters of Mars.)

Simple Harmonic Motion

6

Chapter Summary

Oscillations are another type of periodic motion like circular motion where an object returns to the same point in space many times. If it is controlled and orderly, it is called simple harmonic motion. If it is wild and uncontrolled it is called chaotic. Pendulums and masses on springs are both examples of simple harmonic motion that have important applications. Some of the themes of the chapter are represented in Figs. 6.1, 6.2 and 6.3.

Main Concepts in This Chapter

- Harmonic Oscillators
- Motion of a Pendulum
- Mass on a Spring

6.1 Introduction

In Chap. 5 we used circular motion as an example of a type of periodic motion where an object continually returns to the same position. Two other examples of this are objects that oscillate on a spring or a swinging pendulum. These are the types of motion discussed in this final chapter covering the classical mechanics area of physics. They are examples of simple harmonic motion and can be analyzed by using either the equation of motion approach or the conservation of energy and momentum approach. Simple harmonic motion is the periodic motion exhibited

R. C. Powell, *Physics*, Undergraduate Lecture Notes in Physics,
https://doi.org/10.1007/978-3-032-09361-5_6

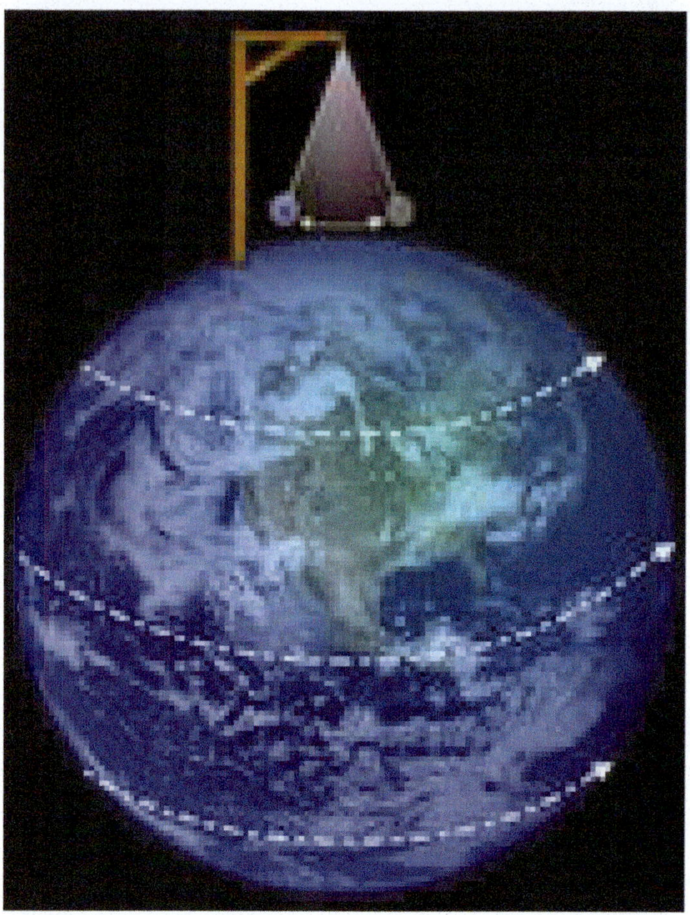

Fig. 6.1 Foucault pendulum at north pole. *Credit* Kralljа, Creative Commons Attribution-Share Alike 3.0

by a system that has a force whose magnitude is proportional to its displacement from equilibrium trying to restore it to its equilibrium position.

6.2 Harmonic Oscillators

The relationship between rotational motion and oscillatory motion can be demonstrated by the following example. Consider an object undergoing circular motion as shown in Fig. 6.4.

The two-dimensional circular motion is projected on the wall as one-dimensional linear motion.

Fig. 6.2 Grandfather Clock. *Credit* AlejandroLinaresGarcia, Creative Commons Attribution Share Alike 3.0

If the object is moving along its circular path at constant speed (obviously *not* constant velocity), it will always take the object the same amount of time to make a full circle. If you project that motion into one dimension, when the object starts out at the top of the circle, it will be at the top of the one-dimensional projection. When it reaches the top of the circle again, it will reach the top of the one-dimensional projection again. Thus, it will travel down to the bottom of the one-dimensional projection and then back up to the top. It will always take the same amount of time to make this trip, because the object always takes the same

Fig. 6.3 Atomic clock. *Credit* NIST, public domain

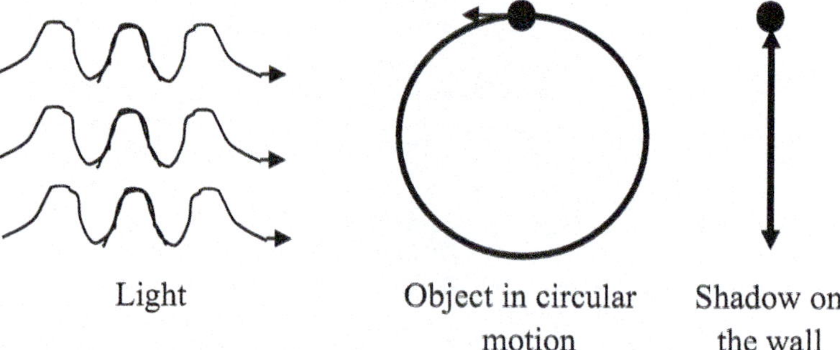

Fig. 6.4 Circular motion projected to linear motion

amount of time to make one trip around the circle. The kind of motion that occurs in one dimension is simple harmonic motion.

The motion of an object undergoing simple harmonic motion is described by the equations giving its position, speed, and acceleration as

$$\begin{aligned} x &= x_{\max} \cos(\omega t + \delta) \\ v &= -\omega x_{\max} \sin(\omega t + \delta) \\ a &= -\omega^2 x_{\max} \cos(\omega t + \delta) \end{aligned} \tag{6.1}$$

where the amplitude of the motion A is the maximum displacement, x_{max}, the angular frequency of the oscillation is $\omega = 2\pi/T$ where T is the period of the oscillation. The units of ω are radians/s. Finally, the term "δ" is called the phase angle. The phase angle depends on the initial conditions. It is a constant that helps us "line up" the cosine function with the motion of the mass. Look at what happens to Eq. 6.1 when $t = 0 : x = x_{max} \cos(\delta)$. Now, suppose the motion of the system was started at the maximum amplitude of the motion. If that is the case, the displacement is usually considered positive. Since the distance the system is originally pulled from equilibrium defines the amplitude, the equation above will be correct if $\delta = 0$ so $\cos(0) = 1$. Thus, $x = A$, which is correct. However, suppose the motion was started by *compressing* the system rather than stretching it. If that were the case, the initial displacement of the mass would be – A, not A. In this case we have to make $\delta = \pi$. Since $\cos(\pi) = -1$, so the equation will give us $x = -A$ when $t = 0$. Note that we used π for the angle instead of 180° because ω has the units radians/second. That means the angles must be in radians in this situation. The phase angle, then, allows us to offset the cosine function so that the initial conditions are described by the equation. Once we do that, the equation will be valid for all times.

The time $t = 0$ can be defined at any time. The most reasonable definition of $t = 0$ is the instant that the motion is started. If that is the case, δ will be 0 or π depending on the direction the system is initially pushed or pulled. However, $t = 0$ need not be defined in the most reasonable way. For example, we could start the system in motion, and then define $t = 0$ as the time at which it passes some arbitrary mark on the floor. Thus, the phase angle can be any value between 0 and 2π, depending on how $t = 0$ is defined.

To determine the phase angle we need to line up the cosine function so that it is consistent with our initial condition. In other words, we have to determine the initial condition and solve for δ. Thus:

$$\delta = \cos^{-1}(x_0/A) \tag{6.2}$$

where x_0 is the displacement from equilibrium at $t = 0$. As stated before, for most cases $t = 0$ will be defined as the moment in which the motion begins. If that is the case, then $x_0 = A$ and δ is 0. Thus, the cosine function is better to use than the sine function in Eq. 6.1.

Figure 6.5a shows how the position, velocity, and acceleration change in one-half period of the periodic motion. The displacement goes from a maximum in the positive direction through equilibrium to a maximum in the negative direction. The velocity is zero at the end points of the motion where the object stops to turn around. In this first half of the period it reaches a maximum negative value at the equilibrium point. The acceleration is maximum at the end points where the velocity is changing directions. It goes to zero at the equilibrium point.

Simple harmonic motion can also be analyzed from an energy perspective. As long as there is no external force doing work on the system, the total energy will always be constant. As shown in Fig. 6.5b, the energy changes back and forth

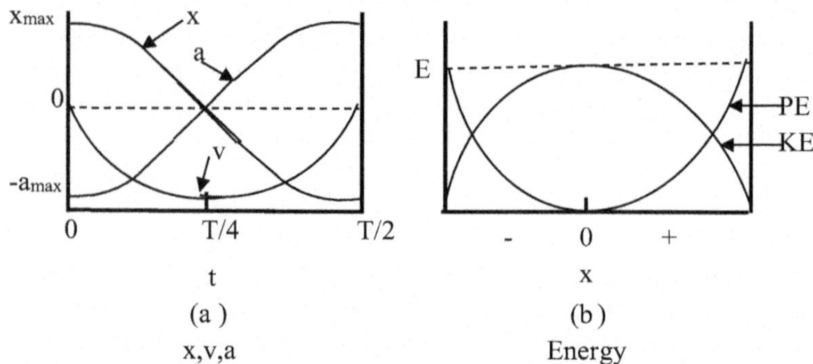

Fig. 6.5 Simple harmonic motion

between kinetic energy and potential energy. At the end points of the motion, the kinetic energy is zero since the object stops and turns around. Maximum KE occurs as the object passes through the equilibrium point where v is maximum. Potential energy is maximum at the end points of the motion and zero at the equilibrium point.

The following two sections discuss two different examples of simple harmonic motion.

As shown in Fig. 6.6, a pen attached to a swinging pendulum or an oscillating mass on a spring will trace out a sine or cosine type of function on a paper moving by at a constant speed. Remember that the shapes of the sine and cosine functions are the same and are simply offset by 90° (or $\pi/2$).

This is consistent with the first of the three expressions in Eq. 6.1 that describes the position of an object undergoing simple harmonic motion.

6.3 Motion of a Pendulum

Consider the case of a pendulum (mass on a string) shown in Fig. 6.7. The external force acting on this system is the force of gravity acting on the mass. When the pendulum is hanging straight down ($\theta = 0°$) the tension in the string is equal and opposite to the weight of the mass. In this position the total force on the system is zero and it is in equilibrium. If the mass is displaced an angle θ away from its equilibrium, the force of gravity must be broken up into its vector components along the string ($mg \cos \theta$) and perpendicular to the string ($mg \sin \theta$). Since the component which pulls perpendicular to the string is pointed back towards the equilibrium position, it is a *restoring force*.

$$F_{\text{restoring}} = -mg \sin \theta$$

The negative sign simply indicates that the force is opposite the direction of the displacement.

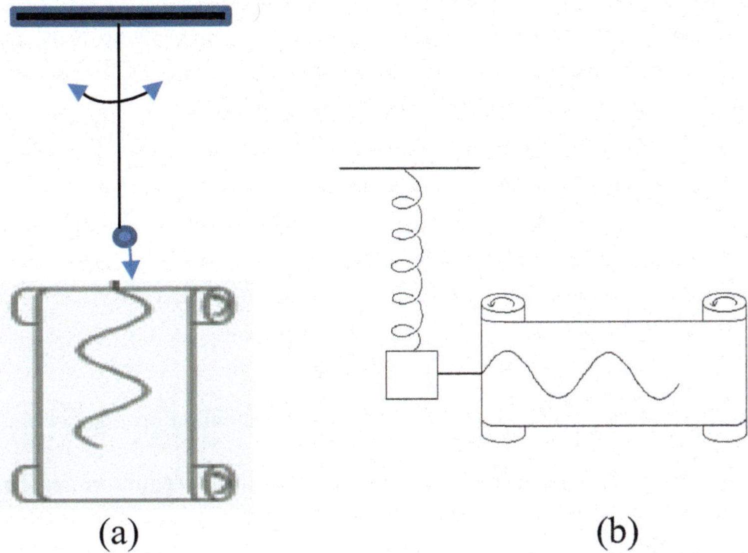

Fig. 6.6 **a** Pendulum and **b** mass on a spring

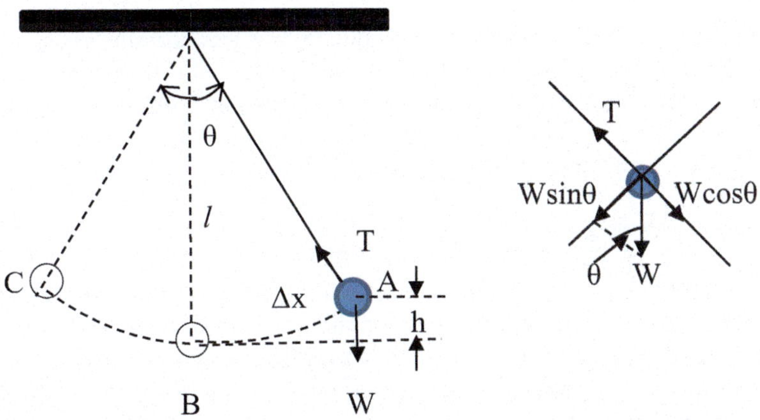

Fig. 6.7 Pendulum motion

The force diagram for this is shown in Fig. 6.7. As shown in the diagram, the tension in the string is equal and opposite to the downward component of the weight along the string. The restoring force in the diagram is the component of the weight perpendicular to the string. This is the force that moves the pendulum back toward its equilibrium position.

If the displacement from equilibrium is very large, the motion of the pendulum will be unpredictably chaotic. However, if there are no forces like air resistance or friction acting on the pendulum and you slowly displace it by a relatively small angle before letting it go, the pendulum will continue to swing back and forth undergoing simple harmonic motion. In the real world, friction and air resistance will cause the motion of the pendulum to slow and finally stop.

When an angle is small, the sine of that angle is approximately equal to the value of the angle (in radians). Mathematically, then,

For small θ, $\sin\theta \approx \theta$.

Let us now assume that the angle of displacement in a pendulum is small enough for the approximation above to be valid. If that is the case, the restoring force becomes:

$$F_{\text{restoring}} = -(\text{mg})\theta$$

Notice the structure of this equation. For a given pendulum on a given planet, mg is a constant.

The angle, θ, is a measure of the displacement from equilibrium. Thus, the restoring force in this case is directly proportional to the displacement from equilibrium. That tells us that:

When the displacement in a pendulum is small, the pendulum exhibits simple harmonic motion.

In general, when the angle is $\pi/10$ (about 18°) or less, the sine of the angle is within 2% of the angle itself. Thus, as long as a pendulum is displaced 18° or less from equilibrium, it essentially exhibits simple harmonic motion.

At the two end points of the pendulum's motion it comes to a complete stop and reverses its direction of motion. At these two points the pendulum's velocity goes to zero so its kinetic energy is also zero. Its speed and kinetic energy are both maximum when the pendulum is at its equilibrium point at the bottom of its arc of motion. Conversely, its potential energy due to gravity is maximum at the end points of its motion and zero at its equilibrium position. Since the pendulum is moving with simple harmonic motion, we should be able to determine its period. The rigorous derivation of the formula requires calculus, but with some simplifying assumptions we can obtain a good approximation. In the real world, friction and air resistance will cause the motion of the pendulum to slow and finally stop.

Conservation of energy requires that the maximum potential energy and maximum kinetic energy are equal. We can use this to estimate the period of the pendulum, that is the time for it to make one complete cycle. First we equate the maximum KE and PE and solve for the maximum velocity which occurs at point B

$$\text{KE}_{\text{max}} = \frac{1}{2}mv_{\text{max}}^2 = \text{PE}_{\text{max}} = mgh$$

or

$$V_{\text{max}} = \sqrt{2gh}. \tag{6.3}$$

Next, we use the equation of motion relating time and distance traveled, Eq. 1.4,

$$\Delta x = v_i t + \frac{1}{2} a t^2.$$

For our situation the initial velocity at point A is zero and the time is the time to go from A to B which is one-fourth of a complete period T. The most important point to note about this equation is that its derivation assumed a constant acceleration while we are dealing with an acceleration that changes between points A and B. For small angles, we can use the average acceleration during this time period as a reasonable approximation of the situation. However, we must keep in mind that this is just an approximation and to get an exact expression requires calculus which is beyond the math used in this book. Then the distance traveled from A to B assuming the start time at A is taken to be time zero is approximated as

$$\Delta x = 0 + \frac{1}{2} \frac{v_{\mathrm{max}}}{t_{\mathrm{AB}}} t_{\mathrm{AB}}^2$$

Solving for time gives,

$$t_{\mathrm{AB}} = 2\Delta x / v_{\mathrm{max}}$$

For small angles, the arc traveled can be approximated by $\Delta x = l\theta$ and the height at position A is $h = \Delta x \sin\theta$ so $h = l\theta^2$. In the final step we have used the fact that for small angles the sine of an angle is equal to the angle itself. (To use this approximation, the angle must be measured in radians.) Substituting the expressions for Δx and v_{max} (Eq. 6.3) into the expression for t_{AB} gives,

$$t_{\mathrm{AB}} = (2l\theta)/\sqrt{2gl\theta^2} = \sqrt{2l/g}.$$

For one complete cycle the period of the pendulum is

$$T = 4t_{\mathrm{AB}} = 4\sqrt{2}\sqrt{l/g} = 5.7\sqrt{l/g}.$$

The exact expression for the period of a pendulum is given by,

$$T = 2\pi\sqrt{l/g} = 6.3\sqrt{l/g} \qquad (6.4)$$

Note that our estimate is very close to the exact value, so our use of the small angle approximation is legitimate. Also note that the period only depends on the length of the pendulum, not on its mass.

Example 6.1

As an example, consider a pendulum with a mass of 5 gm, a maximum velocity of 0.2 m/s, and a period of 2 s. What is the length of this pendulum and how high will it rise above its equilibrium point?

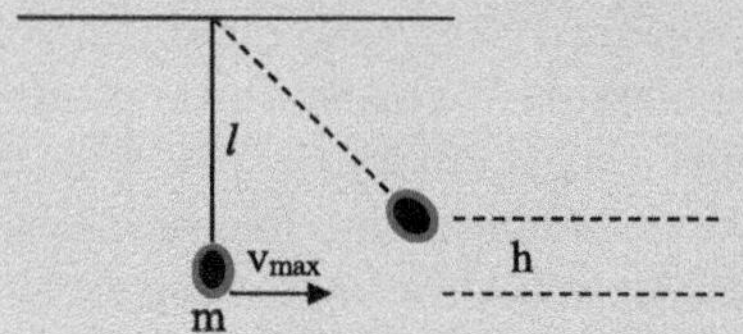

Knowns: $m = 5\,\text{gm}$; $v_{\max} = 0.2\,\text{m/s}$; $T = 2\,\text{s}$

Unknowns: l; h

Once you sketch the problem and list the knowns and unknowns, you can use Eqs. 6.3 and 6.4,

$$T = 2\pi\sqrt{l/g} = 2\,\text{s},$$
$$v_{\max} = \sqrt{2gh} = 0.2\,\text{m/s}.$$

The first of these can be solved for the pendulum's length and the second for its height,

$$l = \frac{T^2 g}{(2\pi)^2} = \frac{(2s)^2(9.81\,\text{m/s}^2)}{(2\pi)^2} = 0.99\,\text{m},$$
$$h = \frac{v_{\max}^2}{2g} = \frac{(0.2\,\text{m/s})^2}{2(9.81\,\text{m/s}^2)} = 0.002\,\text{m}.$$

Note that the mass of the pendulum does not enter the problem.

Instead of a simple mass on a string, a physical pendulum is a mass distributed over an arbitrary shape that swings back and forth from one end point. It acts like a pendulum with the mass located at its center of mass (see Chap. 4) and its motion is determined by the moment of inertia associated with its shape. This is shown in Fig. 6.8.

At small displacement angles a physical pendulum undergoes simple harmonic motion.

Equation 6.4 for the period of a pendulum is generalized to

$$T = 2\pi\sqrt{\frac{I}{mgd}} \tag{6.5}$$

where d is the distance from the fixed point to the center of mass. Note that $I = ml^2$ for a mass on a string and $d = l$ so Eq. 6.5 simplifies to Eq. 6.4 for a simple pendulum. For a rod with a uniform mass oscillating about a fixed end

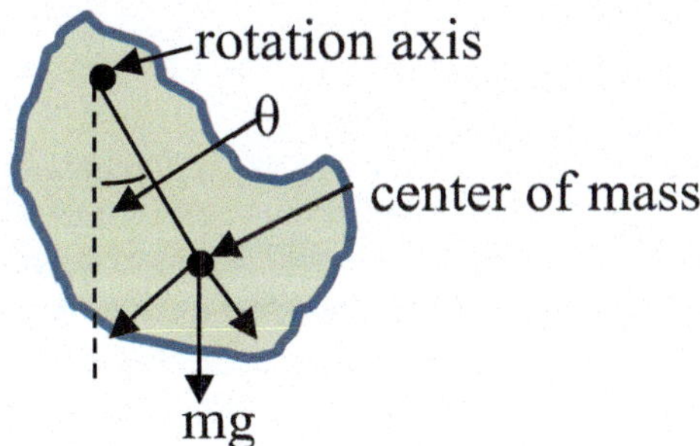

Fig. 6.8 Physical pendulum

point, $I = ml^2/3$ and the center of mass distance is $d = l/2$. So in this case the period of the physical pendulum found from Eq. 6.5 is

$$T_{\text{rod}} = 2\pi\sqrt{\frac{ml^2/3}{mgl/2}} = 2\pi\sqrt{\frac{2l}{3g}} = (0.82)T_{\text{simple pendulum}}.$$

This shows that the period of this physical pendulum is less than the period predicted for a simple pendulum.

Now do the following experiment to demonstrate the difference between a simple pendulum and a physical pendulum.

Experiment 6.1
The Simple Pendulum and the Physical Pendulum

Supplies:

- Thread
- A piece of wood like you used in Experiment 4.1. If you still have that piece of wood, it will work well.
- A thin nail that is at least ½ of an inch long
- A drill or a nail that is much thicker than the nail listed above
- Hammer
- A few washers or nuts
- A wooden ruler that is about 30 cm (1 foot) long. Make sure that it's okay to put a hole in this ruler. That's part of the experiment.
- A few heavy books
- Stopwatch

Introduction—This experiment demonstrates that the amount of mass on the end of a simple pendulum does not affect the period of the pendulum. It also introduces a physical pendulum.

Procedure:

1. Take the piece of wood and hammer a nail into the edge of the board.
2. Place the piece of wood on the edge of a table or cabinet and lay the books on top of the wood to anchor it. The nail should be sticking way out past the edge of the table or cabinet.
3. Cut a piece of string so that it is about twice as long as the ruler. Tie a single washer to one end, but leave plenty of extra string on that end so that you can tie more washers to it later.
4. Tie the other end of the string to the nail. You need to tie it so that the distance from the nail to the center of the washer is the full length of the ruler.
5. Let the washer hang straight down. You now have a simple pendulum whose length is the same as that of the ruler.
6. Displace the pendulum a small angle (less than 18°) away from equilibrium and then release. The moment you release, start the watch.
7. You want to count ten periods and then stop the watch. Thus, the washer must come back to the same point from which you released it 10 times.
8. Take the time it took for the pendulum to complete 10 periods and divide by 10. That is the period of the pendulum.
9. Since you know the length of the pendulum (the length of the ruler), you can use Eq. 5.3 and the acceleration due to gravity to determine the period of the pendulum. Make sure your units are consistent!
10. Compare the period you measured in Step #8 to the period predicted in Step #9. If they are different (to the significant figures allowed), the measured period is probably a bit larger.

    That's because air resistance slows the pendulum down with each swing.
11. Now add another washer to the pendulum and repeat Steps #7 and #8.
12. Add another washer and repeat Steps #7 and #8 again.
13. Add one more washer and repeat Steps #7 and #8 again.
14. Compare the periods measured in Step #8 with those measured in Steps #11–13.
15. Take the string off of the nail.
16. Now you want to make a pendulum out of your ruler. To do that, drill a hole in the ruler.

    The hole needs to be in the middle of the ruler as close to one end as possible without splitting the ruler. If you do not have (or are not allowed to use) a drill, you can use a thicker nail to make the hole. Just make sure the hole is large enough so that the thin nail which is on the board will easily fit inside.
17. Slip the nail on the board through the hole in the ruler so that the ruler hangs from the nail.

You now have a physical pendulum.

18. Displace the physical pendulum a small angle from equilibrium and let it swing. Once again, measure the time it takes to complete 10 periods. Divide your measurement by ten, and you will get the period of the physical pendulum.
19. Compare the period you measured in Step #18 to the one you measured in Step #8.
20. Clean everything up

What happened in the experiment? You should have found that the periods measured in Steps #8, and Steps #11–13 were essentially the same. They will not be exactly the same because of experimental error. However, they should be close (within 5% or so). You might have noticed a slight decrease in the period as you tied on more washers. If so, that's because the greater mass was not as strongly affected by air resistance as was the lighter mass. If you think your periods have a lot of variation to them, think about how you changed the mass.

Between Step #8 and Step #13, you quadrupled the mass. However, the period changed very little compared to that. This demonstrates that the period of a simple pendulum is independent of the mass used.

Now let's move on to the second part of the experiment. How did the period of the physical pendulum compared to that of the simple pendulum? It should have been significantly less (about 15–20%). Why? Well, think of the type of motion the pendulum is exhibiting. Not only is it essentially moving as a simple harmonic oscillator, but it is also exhibiting *rotational motion*. After all, the ruler experiences a torque (from gravity) which causes it to rotate about a fixed point (the nail). The ruler does not make complete rotations around the nail, but it is still rotating with the nail as its axis of rotation.

In the second part of the experiment, you had a physical pendulum. If you used a standard, 12.00-inch ruler, the moment of inertia is given by $I = mL^2/3$ where $L = 30.48$ cm. Now, it turns out that this is a bit of an approximation. This equation is for a *rod* rotating about one end. The ruler is more of a plank. Nevertheless, the approximation is actually quite good. The other parameter we need to determine is d, the distance from the fixed point to the center of mass. The center of mass of the ruler is the center of the ruler, which is ½L. Thus,

$$T = 2\pi\sqrt{\frac{I}{mgd}} = 2\pi\sqrt{\frac{mL^2/3}{mgL/2}} = 6.28\sqrt{\frac{2(0.3048\,\text{m})}{3(9.81\,\text{m/s}^2)}} = 0.904\,\text{s}.$$

What was the period of your physical pendulum? It should have been something around 0.90 s, if you used a standard, 12-inch ruler. A physical pendulum, then, behaves much like a simple pendulum. However, you just have to use a more general equation to calculate its period.

At this point, try the following problem.

Student

6.1 A nail is driven through the edge of a uniform hoop and into a wall. The hoop is then displaced a small angle from equilibrium and released so that it swings back and forth. Its period is 0.754 s. What is the radius of the hoop? (The moment of inertia of a hoop rotating about a point on its edge is $I = 2MR^2$.)

Special Topic

A Foucault Pendulum is a very long, heavy pendulum that was used to measure the rotation of the earth. It is easiest to envision if you imagine setting up such a large pendulum at the north or south pole. The pendulum swings back and forth in a plane that cuts across the pole. As it is moving in its arc, the plane is rotating with respect to the earth around the pole. Thus a marker attached to the pendulum bob will trace out a line in the snow that rotates with the earth's rotation. At other latitudes on the surface of the earth, the speed of rotation of the plane of oscillation will be different and more difficult to calculate. However, it still shows the rotational motion of the earth.

6.4 Mass on a Spring

Another type of periodic motion is exhibited by a mass on a spring such as that shown in Fig. 6.9. Again, if the displacement is small and we can ignore dissipative forces such as friction and air resistance, the mass will undergo simple harmonic motion about its equilibrium position. This is because the force of a spring is a restoring force; that is it always acts to push or pull the block back toward its equilibrium position.

Empirically, it was found that this force is equal to the displacement of the block from equilibrium multiplied by a constant that describes the strength of the spring,

$$F = -k\Delta x. \tag{6.6}$$

Here Δx is the distance the block is displaced from equilibrium and k is the force constant for the spring. The units of k are Newtons per meter. The minus sign is necessary to give the direction of the force as opposite to the direction of the displacement. This expression is sometimes called Hooke's Law.

In terms of analyzing the energy of a block on a spring, at the equilibrium position the potential energy is zero and the kinetic energy is maximum, $K =$

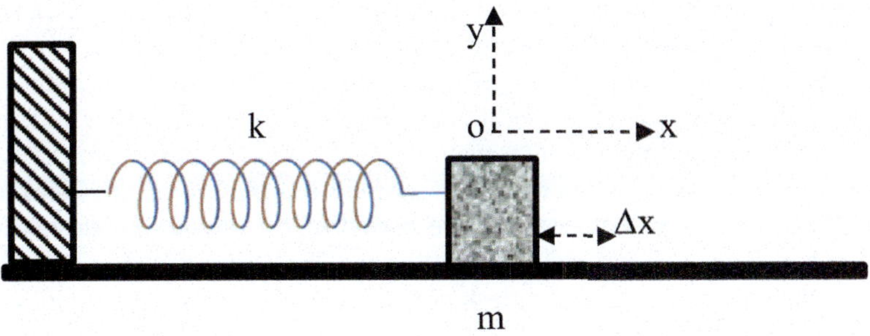

Fig. 6.9 Mass on a spring

$\frac{1}{2}mv_{\text{max}}^2$. At the extreme end points of the motion, the kinetic energy is zero and the potential energy is maximum. We can derive an expression for this potential energy using conservation of energy and the work-energy theorem we discussed previously in Chap. 3. For this case, Eqs. 3.1 and 3.7 tell us that the potential energy will be equal to the work done against the restoring force by displacing the block an amount x from equilibrium,

$$\Delta\text{KE} = \Delta\text{PE} = W_{\text{r}} = F_{\text{avg}}X = \left(\frac{1}{2}kx\right)x$$

So, for a spring

$$\text{PE} = kx^2/2. \tag{6.7}$$

This is the elastic potential energy. Again, we have made an assumption of small displacements. Since the actual restoring force changes continually with displacement, a rigorous derivation of this expression requires calculus which is beyond our math. Instead we have used the average force taken as one-half the difference between the initial and final force. This should be close to the correct amount for small displacements.

Think about what this equation means. When $x = 0$, PE $= 0$. That means the spring has no potential energy when it is at its equilibrium position. This makes complete sense, since the spring exerts no force at that point. However, as the spring is stretched or compressed, it gains potential energy. That potential energy is stored in the spring until the mass is released, allowing the spring to do work.

What happens when the mass is released? At the instant it is released, the mass has no kinetic energy, because it is not moving. However, there is all sorts of potential energy stored in the spring. Thus, the spring begins converting potential energy into kinetic energy. This gets the mass moving. As the mass moves, more potential energy is converted to kinetic energy, and the mass begins to move faster. However, when it reaches $x = 0$, there is no more potential energy in the spring. However, there is a lot of kinetic energy in the mass. Thus, the mass begins to compress the spring. This, however, reduces the speed of the mass, decreasing its kinetic energy. Of course, the potential energy of the spring increases as a result. Thus, when the mass eventually stops, the spring will have a lot of potential energy stored up. That potential energy will start being converted to kinetic energy, and the mass will start sliding the other way.

Two other quantities of interest in describing the motion of the block on the spring are its maximum velocity and the period of its motion. An expression for the first of these can be derived by equating the expression for maximum kinetic energy at the equilibrium point with the expression for maximum potential energy at one of the end points,

$$\frac{1}{2}mv_{\max}^2 = kx_{\max}^2/2$$

or

$$v_{\max} = \sqrt{k/m}x_{\max}. \tag{6.8}$$

It is more difficult to derive the expression for the period of the motion since the restoring force varies with displacement so calculus in necessary. However, we can find an expression for T through drawing an analogy between this type of simple harmonic motion and uniform circular motion we discussed in Chap. 5. The period for circular motion is given by the circumference divided by the velocity of motion,

$$T = 2\pi r/v.$$

If we assume that the radius of the circular is equivalent to the maximum displacement of the mass on a spring, we can substitute the velocity in Eq. 6.8 into this expression to give,

$$T = 2\pi x_{\max}/\sqrt{k/m}x_{\max} = 2\pi\sqrt{\frac{m}{k}}. \tag{6.9}$$

This should be accurate as long as we are working with simple harmonic motion. (Note that if you use half the maximum velocity for the average velocity and use $4x_{\max}$ as the distance traveled, the result is 8 $\sqrt[2]{k/m}$ as opposed to the 6.3 $\sqrt[2]{k/m}$ in Eq. 6.9.)

All of this should be review for you from your first-year physics course. Let's now think of this situation in a little more detail with the following examples.

Example 6.2

As an example, consider a mass of 2 kg on a spring shown in Fig. 6.9. It has taken a force of 2 N to displace it 0.2 m. What is the maximum velocity the block will achieve when it is released and what is the period of its motion?

Knowns: $m = 2\,\text{kg}$; $F = 2\,\text{N}$; $x = 0.2\,\text{m}$.

Unknowns: v_{max}; T

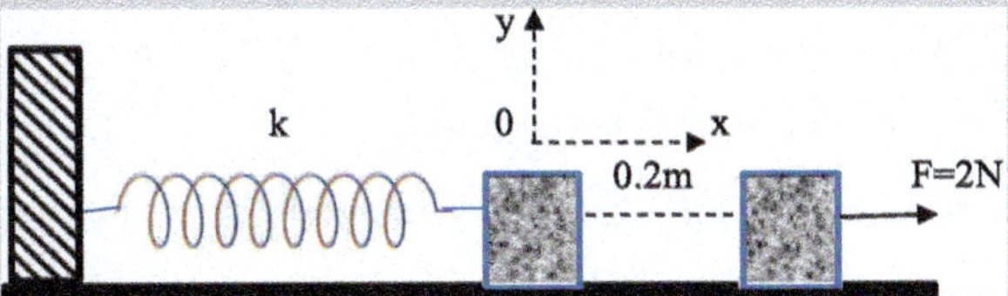

The restoring force when the block is released will be equal and opposite to the displacement force. For displacement in the positive x-direction, this is given by Eq. 6.6 as

$$F = k\Delta x.$$

Since we know the displacement force and the amount of displacement, we can solve for the spring constant,

$$k = F/\Delta x = 2\,\mathrm{N}/0.2\,\mathrm{m} = 10\,\mathrm{N/m}.$$

Now that we know the spring constant, we can use Eq. 6.8 to calculate the maximum velocity the block will have when it travels by its equilibrium point,

$$v_{max} = \sqrt[2]{k/m}x_{max} = \sqrt{\frac{10\,\mathrm{N/m}}{2\,\mathrm{kg}}}(0.2\,\mathrm{m}) = 0.45\,\mathrm{m/s}.$$

Also, Eq. 6.9 can be used to calculate the period of motion,

$$T = 2\pi\sqrt{\frac{m}{k}} = 2\pi\sqrt{\frac{2\,\mathrm{kg}}{10\,\mathrm{N/m}}} = 2.8\,\mathrm{s}.$$

Example 6.3

As another example, consider the block attached to two springs as shown in Fig. 6.10. The two spring constants are $k_1 = 100\,\mathrm{N/m}$ and $k_2 = 200\,\mathrm{N/m}$ and the block has a mass of 10 kg. If the block is displaced by 0.2 m to the right and then allowed to oscillate back and forth, what are its maximum velocity and its period of oscillation?

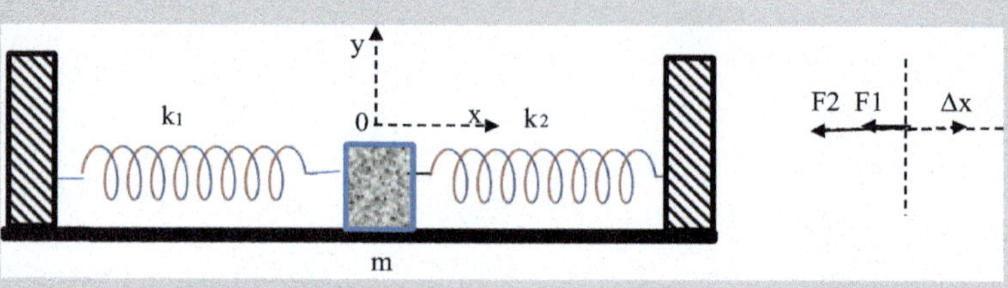

Fig. 6.10 Mass with two springs

The knowns and unknowns are:

Knowns: $m = 10\,\text{kg}$; $k_1 = 100\,\text{N/m}$; $k_2 = 200\,\text{N/m}$; $x = 0.2\,\text{m}$.

Unknowns: v_{max}; T.

Each spring exerts a restoring force to try to move the block back to its equilibrium position. If the displacement is to the right, both the restoring forces will be to the left. Since they are in the same direction, they simply add together.

$$F_1 = k_1 x = 100x\,\text{N},$$
$$F_2 = k_2 x = 200x\,\text{N},$$

so the total restoring force on the block is,

$$F_T = 300x\,\text{N}.$$

This shows that the system acts like a block on a spring with a spring constant of 300 N/m. Thus, we can again use Eqs. 6.8 and 6.9 to find the unknowns. The maximum velocity of the block is

$$v_{\text{max}} = \sqrt[2]{k/m}x_{\text{max}} = \sqrt{\frac{300\,\text{N/m}}{10\,\text{kg}}}(0.2\,\text{m}) = 1.1\,\text{m/s}.$$

$$T = 2\pi\sqrt{\frac{m}{k}} = 2\pi\sqrt{\frac{10\,\text{kg}}{300\,\text{N/m}}} = 1.15\,\text{s}.$$

Example 6.4

In the mass/spring system illustrated in Fig. 6.9, assume that the block has a mass of 50.0 g, the spring constant is 1.50 Newtons/meter, the block is pulled back 15.0 cm, and the mass of the spring can be ignored.

a. **If we ignore friction, what is the maximum speed of the mass and where will it occur?**
b. **Now take friction into account. If the coefficient of kinetic friction is 0.100, what will the maximum speed of the block be?**

Knowns: $m = 50.0\,\text{g}$: $k = 1.50\,\text{N/m}$; $x_{\text{max}} = 15.0\,\text{cm}$; $\mu_{\text{k}} = 0.100$.

Unknowns: v_{max}.

To solve part (a), we just need to think about the energy. When the block is pulled back 15.0 cm from the equilibrium position, the spring stores up a certain amount of potential energy:

$$\text{PE} = kx^2/2 = (1.50\,\text{N/m})(0.150\,\text{m})^2 = 0.0169\,\text{J}$$

While pulled back, the block has no kinetic energy, because it is at rest. Thus, the total energy is also 0.0169 J. When will the speed be greatest? It will be greatest when all of the potential energy is converted to kinetic energy. Thus, when potential energy is zero, the kinetic energy will be 0.0169, and we can solve for the speed using Eq. 6.8,

$$v_{\text{max}} = \sqrt[2]{k/m}x_{\text{max}} = \sqrt[2]{(1.50\,\text{N/m})/0.0500\,\text{kg}}(0.150\,\text{m}) = 0.822\,\text{m/s}$$

This maximum velocity will occur when the block is at equilibrium position.

Now let's take friction into account. Since we have the mass and the coefficient of kinetic friction, we can calculate the frictional force.

$$F_f = \mu k \cdot F_n = (0.100) \cdot (0.0500\,\text{kg}) \cdot \left(9.81\,\text{m/s}^2\right) = 0.0491\,\text{N}$$

Friction works against the motion. How much work does it do? Well, the block travels a distance of 15.0 cm to get back to the equilibrium position. That displacement is 180° to the direction of the frictional force (the frictional force opposes the motion), so calculating the work done is easy:

$$W = F \cdot x \cdot \cos\theta = (0.0491\,\text{N}) \cdot (0.150\,\text{m}) \cdot \cos(180) = -0.00737\,\text{J}$$

The work is negative because it removes energy from the system. Thus, once friction is taken into account, the kinetic energy of the mass at the equilibrium position will be 0.0169 J minus 0.00737 J, or 0.0095 J. Thus, the speed of the mass after friction is taken into account is:

$$\mathrm{KE} = mv^2/2$$

$$0.0095\,\mathrm{J} = (0.0500\,\mathrm{kg})v^2/2$$

$$v = \sqrt[2]{\frac{2(0.0095\,\mathrm{J})}{0.0500\,\mathrm{kg}}} = 0.62\,\mathrm{m/s}$$

Try analyzing situations like this by solving the following problem.

Student

6.2 In the mass/spring System Given in Example 6.4 (Part B), What is the Total Distance that the Block Will Travel Before Coming to Rest?

To summarize our analysis of the motion a mass on a spring, for low displacements it undergoes simple harmonic motion with an amplitude of x_{max} and a period given by Eq. 6.9. The frequency of the motion is given by

$$f = 1/T = \frac{1}{2\pi}\sqrt{\frac{k}{m}} \tag{6.10}$$

in units of s^{-1} which are called Hertz, abbreviated as Hz. The unit is named in honor of Heinrich Rudolf Hertz, a German physicist who demonstrated that electricity can be transmitted in electromagnetic waves. As you will see in future chapters, electromagnetic waves are really just light waves. Hertz's research led to the development of the wireless telegraph, radio, and (eventually) television. The energy of the vibrating mass on a spring goes between totally potential energy at the amplitude positions to totally kinetic energy at the equilibrium position. Note that the angular frequency is related to f by

$$\omega = 2\pi f \tag{6.11}$$

with units of radians per second.

Let's use these concepts to analyze the following example.

Example 6.5

A 15.0 kg mass is attached to two springs as shown below. The system is arranged so that neither spring is stretched when the mass is at rest in the position shown in the figure. The spring constant of the first spring is 157 N/m, and the spring constant of the second spring is 234 N/m. If the mass is displaced 15.0 cm from the position shown below, what is the amplitude and frequency of its motion? Ignore friction.

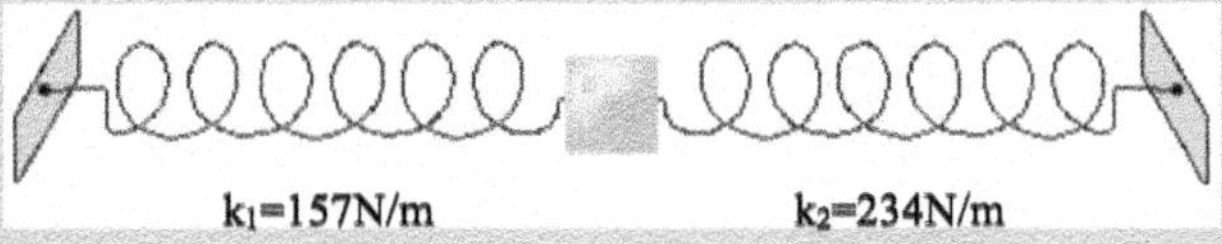

Knowns: $m = 15.0\,\text{kg}$: $x = 15.0\,\text{cm}$; $k_1 = 157\,\text{N/m}$; $k_2 = 234\,\text{N/m}$.
Unknowns: x_{max}; f

The key to understanding any mass/spring system is knowing the spring constant. In this case, we have two springs that are both pulling on the mass. How does that affect things? The basis of simple harmonic motion is Eq. 6.6. Any system that has a force which behaves as Eq. 6.6 dictates that the system will exhibit simple harmonic motion. Thus, let's see whether or not this system really does conform to Eq. 6.6. Consider the forces on the mass.

When the mass is sitting as shown in the figure, neither spring is stretched. Thus, there are no forces acting on the object (at least not in the horizontal dimension), and the object is in its equilibrium position. Once the mass is displaced, however, there will be *two* forces acting on the mass. Let's suppose we displace the mass to the right of its equilibrium position. If that happens, the spring on the left will be stretched and thus will pull back on the mass. As a result, it will exert a force that attempts to move the mass to the left. The spring on the right, however, will be compressed. Thus, it will push the mass away, which is also a force to the left. The sum of the forces on the mass, then, are shown in the figure below:

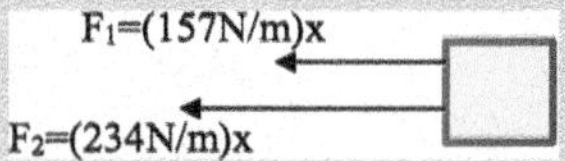

This is equivalent to a total force to the left of

$$F_{\text{tot}} = (391\,\text{N/m})x.$$

In the end, then, the two springs simply combine as if they were one stronger spring. Thus, this mass/spring system is really the same as a mass/spring system in which the mass is connected to only one spring whose spring constant is 391 N/m. As a result, the frequency is given by Eq. 6.10:

$$f = \frac{1}{2\pi}\sqrt{\frac{k}{m}} = \frac{1}{2\pi}\sqrt{\frac{391\,\text{N/m}}{15.0\,\text{kg}}} = 0.813\,\text{Hz}.$$

The amplitude of the motion is equal to the original displacement or 10.0cm.

The mass/spring system in the example, then, will simply oscillate back and forth. It will travel 15.0 cm to the left of its equilibrium position, then it will reverse course and travel to the right until it reaches 15.0 cm right of its equilibrium position, which is the point from which it was released. It will make 0.813 of these trips each second.

Now work the following problems.

Student

6.3 A mass/spring system is constructed with three springs and a 20.0 kg mass as shown to the right. When it is sitting as shown, none of the springs are stretched or compressed. It is then displaced 10.0 cm to the right and released. What is the period of its motion? What is the amplitude of its motion? What is its maximum speed and where does it occur? Ignore friction.

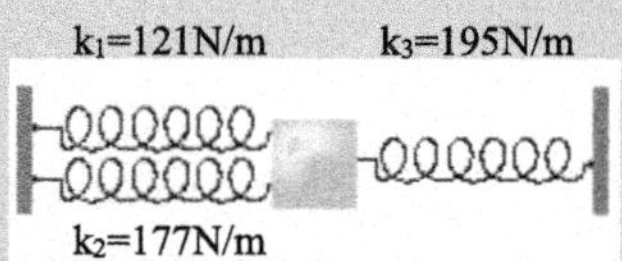

6.4 A mass/spring system is constructed with a 15.0 kg mass attached to a single spring. It is set up horizontally as shown in section A of the figure to the right. When displaced from equilibrium, it oscillates with a frequency of 1.25 Hz. The system is then hung vertically as shown in section B of the figure. When the system is hung vertically, how far will the spring be stretched when the mass is at its new equilibrium position?

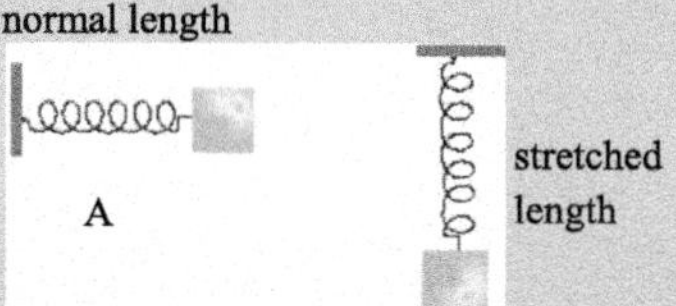

Let's do one last example treating a mass on a spring as simple harmonic motion.

Example 6.6

A mass/spring system is composed of a 10.0 kg mass and a spring whose springconstant is 151 N/m. The mass is displaced − 50.0 cm from equilibrium and then released.

The experimenter examines the motion for a moment and then defines $t = 0$ as the instant that the mass passes through its equilibrium position. What is the displacement of the mass at $t = 15.0$ s?

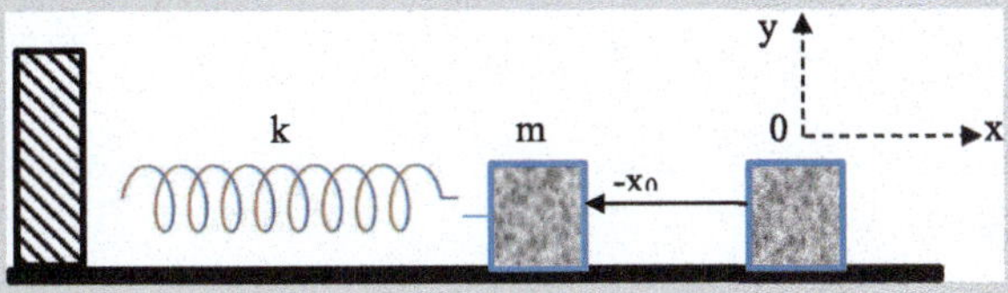

Knowns: $m = 10.0$ kg.

$k = 151$ N/m.

$x_0 = -50.0$ cm.

$t = 0$ s @ $x = 0$ m.

Unknowns: x @ $t = 15.0$ s.

Before we were given Eq. 6.1, the only positions we knew in a mass/springsystem were the points at which the velocity of the mass was 0 (that occurs at the maximum distance from equilibrium), and the point at which the speed of the mass was greatest (which occurs at the equilibrium position). With Eq. 6.1, we can determine the position of the mass at *any* time. First, however, we must set up the equation. To do that, we need to know what A, ω, and δ are.

Figuring out the amplitude (A) of the motion is easy. The mass is displaced − 50.0 cm away from equilibrium in order to get the motion started. As you already know, the energy of the situation forbids the mass from traveling any farther than 50.0 cm from equilibrium. That means the amplitude of the motion is 50.0 cm. What about the angular frequency (ω)? Well, ω is given by Eq. 6.11. This equation, however, contains the period of the motion. The period is given by Eq. 6.9. Thus, the angular frequency is:

$$\omega = \sqrt{\frac{k}{m}} = \sqrt{\frac{151\,\mathrm{N/m}}{10\,\mathrm{kg}}} = 3.89\,\mathrm{rad/s}.$$

We now know A and ω. There is only one thing more that we need: δ. To figure that out, we use Eq. 6.2. Time zero is defined as the time at which the mass reaches its equilibrium position. Thus, $\mathbf{x}_0$ is zero, because $\mathbf{x}$ is the displacement from equilibrium:

$$\delta = \cos^{-1}(x_0/A) = \cos^{-1}(0/A) = \pi/2$$

Substituting these three values into Eq. 6.1 for position gives

$$x = A\cos(\omega t + \delta) = (50.0\,\text{cm})\cos[(3.89\,\text{rad/s})t + \pi/2]$$

That equation gives us the displacement from equilibrium at any time, t. Thus, to answer the question, we just plug 15.0 s in for t:

$$x = (50.0\,\text{cm})\cos[(3.89\,\text{rad/s})(15.0\,\text{s}) + \pi/2] = -47.6\,\text{cm}$$

The mass, then, is − 47.6 cm from its equilibrium position. Usually, a negative displacement means the spring is compressed. However, the sign convention was not mentioned in the problem. Before you move on, make sure you can solve the equation above on your calculator.

Remember, the angles are in *radians*, not degrees. Thus, you must change the mode on your calculator so that it knows the angles are in radians.

To complete the analysis of the problem in Example 6.6, we can write the equations for the velocity and the acceleration of the mass according to the second and third parts of Eq. 6.1

$$v = -\omega A\sin(\omega t + \delta) = -(50.0\,\text{cm})(3.89\,\text{rad/s})\sin[(3.89\,\text{rad/s})t + \pi/2]$$
$$a = -\omega^2 A\cos(\omega t + \delta) = -(50.0\,\text{cm})(3.89\,\text{rad/s})^2\cos[(3.89\,\text{rad/s})t + \pi/2]$$

The plot of position, velocity, and acceleration versus time then looks like Fig. 6.5a but with the time axis shifted so at t = 0 s the position is at the equilibrium point and expanding in the positive direction.

As you look over the curves, in Fig. 6.5, notice that acceleration and displacement basically behave as opposites: when displacement is large and positive, acceleration is large and negative. When displacement is small and positive, acceleration is small and negative. That should make sense, given Eq. 6.1. Notice also that velocity behaves like acceleration that is shifted by 90°. That should make sense both physically and mathematically. Physically, velocity is 90° behind acceleration because when acceleration is at its maximum (the maximum displacement), velocity is at its minimum. By the time the force of the spring has worked on the mass enough to increase the speed, the mass is closer to equilibrium, so the acceleration has decreased. Mathematically, velocity is a sine function, but acceleration is a cosine function so they are simply shifted 90° relative to each other.

Try the following problem to make sure you understand these equations and how to use them.

Student

6.5 A mass/spring system is composed of a 150.0 kg mass attached to a spring whose spring constant is 225 N/m. The mass is displaced − 15.0 cm from equilibrium and released at $t = 0$. What are the position, velocity, and acceleration of the mass at $t = 3.00$ s? What are the maximum distance from equilibrium, maximum speed, and maximum acceleration of the mass?

6.6 You have a mass on a spring system ($m = 1.0$ kg; $k = 2.0$ N/m) that oscillates with a period of T_p. If you want to make a pendulum with the same period, how long does the suspension bar have to be?

Next Level

Resonance

One oscillating system is said to be in resonance with another oscillating system if they are vibrating in phase with the same vibration frequency. The frequency of a system that can be modeled as a pendulum or a mass on a spring can be found using the relationship between frequency and period given by Eq. 6.9 plus the expressions for periods of oscillation given by Eqs. 6.4 or 6.8. The natural frequency of vibration for a system is the frequency at which a system tends to oscillate in the absence of any driving or damping force. The motion pattern of a system oscillating at its natural frequency is called the normal mode of the system.

One important type of resonance occurs when one of the oscillating systems exerts a periodic driving force on the other system. If they are in resonance, the force will occur when the second system is at the maximum displacement of its oscillation and thus act to increase its amplitude during each oscillation cycle. This will cause the system to oscillate at a larger amplitude then it would if the force were applied at a non-resonant frequency. If the amplitude increases too much the system may no longer exhibit simple harmonic motion. The motion can become chaotic and produce unwanted vibrations. These can cause catastrophic failure of the system. Resonance failures have been responsible for many disasters with machinery, bridges, and buildings. One interesting demonstration is a high-frequency sound wave hitting the resonant vibrational frequency of a wine glass and shattering it shown in Fig. 6.11. The resonant frequency of the wine glass depends on its shape and the type of glass it is made of. Engineers have learned to design things to minimize the detrimental effects of resonance vibrations.

Fig. 6.11 Breaking a wine glass with a resonant sound wave

Resonance phenomena occur with all types of vibrations. These include mechanical resonance, acoustic resonance, electromagnetic resonance, and nuclear magnetic resonance (NMR). Musical instruments are designed to produce specific resonant frequencies as discussed in Chap. 12.

Special Topic

Oscillating systems have many important applications in the modern world. One important example is keeping time. Old time grandfather clocks used pendulums. Then there were vibrating quartz oscillators. Now the standard for time is kept by atomic clocks at the National Institutes of Standards and Technology. This is based on the resonant frequency of atomic oscillations that will be discussed more in Chap. 16. The Cesium atomic clock has a resonance frequency of 9,192,631,770 cycles per second. This is used to define one second of time to very high precision.

Summing Up

In this chapter we learned about simple harmonic motion. What we learned will be important when we treat wave motion in Chaps. 11 and 12. Here we applied our knowledge of harmonic motion to both regular and physical pendulums and to the vibration of a mass on a spring. These both have important applications.

Answers to the Problems

6.1

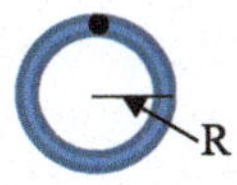

Knowns: $T = 0.754$ s; $I = 2mR^2$

Unknowns: R

According to Eq. 6.5, the period of a physical pendulum depends on the moment of inertia. Since we have an equation for the moment of inertia, we can stick that into Eq. 6.5 and solve for R:

$$T = 2\pi\sqrt{\frac{I}{mgd}}$$

$$0.754\,\mathrm{s} = 2\pi\sqrt{\frac{2mR^2}{mgd}}$$

The definition of "d" is the distance from the axis of rotation to the center of mass of the hoop. The nail is stuck through the edge of the hoop, so the distance from the nail to the center of the hoop (which is the center of mass) is the radius.

$$R = \frac{(0.754\,\mathrm{s})^2(9.81\,\mathrm{m/s^2})}{2(2\pi)^2} = 0.0706\,\mathrm{m}$$

6.2

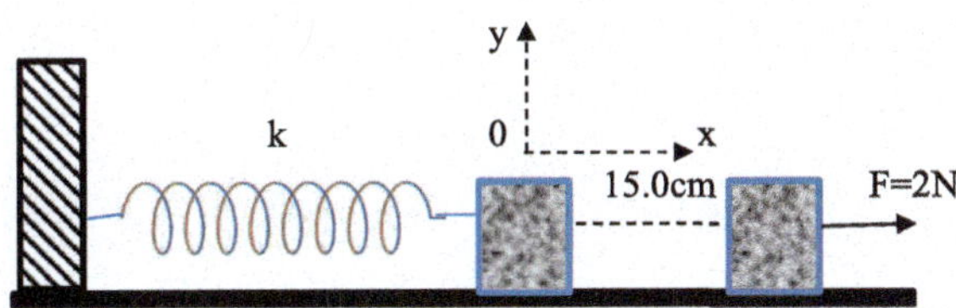

Knowns: $m = 50.0$ g: $k = 1.50$ N/m; $x_{\mathrm{max}} = 15.0$ cm; $\mu_{\mathrm{k}} = 0.100$
Unknowns: x_{f}

The spring constant and initial displacement of the block give us the potential energy with which the system begins. There is no initial kinetic energy, as the block is being held.

$$\mathrm{TE} = \mathrm{KE} + \mathrm{PE} = 0 + kx^2/2 = (1.50\,\mathrm{N/m})(0.150\,\mathrm{m})^2 = 0.0169\,\mathrm{J}$$

The work due to friction is easy to calculate, because the block is on a flat surface:

$$W_f = fx\cos(180.0) = \mu_k mgx\cos(180.0)$$
$$= -(0.100)(0.0500\,\text{kg}(9.81\,\text{m/s}^2)x = -(0.0491\,\text{kg m/s}^2)x$$

We can calculate x by realizing that when the total energy plus the work done by friction is zero, the system will stop moving:

$$\text{TE} + W_f = 0$$

$$0.169\,\text{J} - \left(0.0491\,\text{kg m/s}^2\right)x = 0$$
$$x = 0.344\,\text{m}$$

6.3

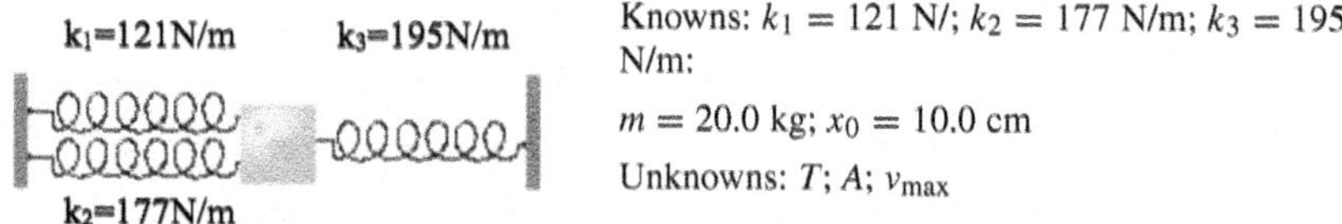

Knowns: $k_1 = 121$ N/; $k_2 = 177$ N/m; $k_3 = 195$ N/m:

$m = 20.0$ kg; $x_0 = 10.0$ cm

Unknowns: T; A; v_{max}

To determine things like the period, we need to know the effective force constant of all of those springs. To do that, we think of the forces involved. If the mass is displaced to the right, the first spring exerts a force of k_1x to the left, because it is trying to pull the mass back to equilibrium position. The second spring exerts a force of k_2x to the left, because it is trying to pull the mass back to its equilibrium position. The third spring exerts a force of k_3x *also to the left*, because it is trying to push the mass back to its equilibrium position. The total force, then, is:

$$F_t = k_1x + k_2x + k_3x = (k_1 + k_2 + k_3)x = (493\,\text{N/m})x$$

and that force is directed opposite of the displacement. Thus, this is simple harmonic motion with an effect spring constant of $k_1 + k_2 + k_3$. Now we can use our simple harmonic motion equations:

$$T = 2\pi\sqrt{\frac{m}{k}} = 2\pi\sqrt{\frac{20.0\,\text{kg}}{493\,\text{N/m}}} = 1.27\,\text{s}$$

The period, then, is 1.27 s. The amplitude is easy. It was displaced 10.0 cm from the equilibrium position, so its amplitude is 10.0 cm. To determine its maximum speed, we have to determine the energy. When displaced, the energy of the mass is potential energy:

$$PE = kx^2/2$$

The mass will be at its maximum speed when all of that energy is converted to kinetic. Thus:

$$kx^2/2 = mv_{max}^2/2$$

$$(493\,N/m)(0.100\,m)^2 = (20.0\,kg)v_{max}^2$$

$$v_{max} = \sqrt{\frac{493\,N/m(0.0100\,m^2)}{20.0\,kg}} = 0.496\,m/s$$

This maximum speed will occur when the mass is at its equilibrium position, because at that point, there will be no potential energy and therefore all of the energy will be kinetic.

6.4

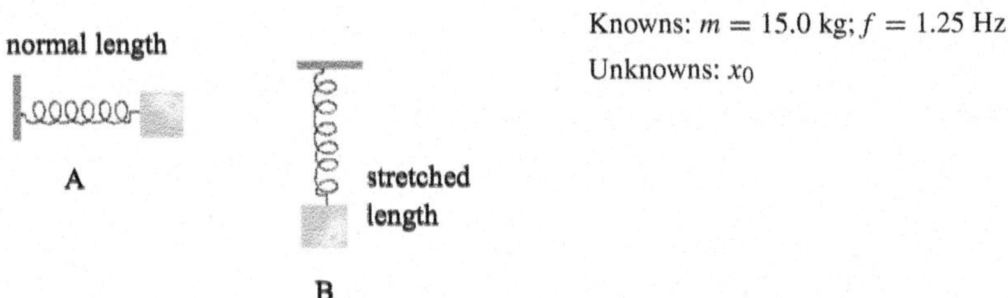

If we know the frequency, we can calculate the spring constant:

$$f = 1/T = \frac{1}{2\pi}\sqrt{\frac{k}{m}} = 1.25\,s^{-1} = \frac{1}{2\pi}\sqrt{\frac{k}{15.0\,kg}}$$

$$k = (2\pi)^2(1.25\,s^{-1})^2(15.0\,kg) = 925\,N/m$$

The units work because a N/m is the same as a kg/s^2. Now that we have the spring constant, all we have to do is think of the force involved when the mass hangs from the spring. When that happens, the mass is pulled by gravity with a force of mg downwards. When the spring pulls back with an equal but opposite force (mg upwards), the mass will come to rest. That will be the final position of the mass. We can calculate that position with Hooke's Law:

$$F = -kx$$

$$(15.0\,\text{kg})\left(9.81\,\text{m/s}^2\right) = -(925\,\text{N/m})x$$

$$x = -0.159\,\text{m}$$

The spring will be stretched 15.9 cm. Make sure you understand the signs in this equation.

Hooke's law calculates the *spring's* force. Thus, the **F** in the equation refers to the force exerted by the spring, which is an upward (positive) force. That's why the force is positive.

The **x** is negative because the mass is displaced downwards.

6.5

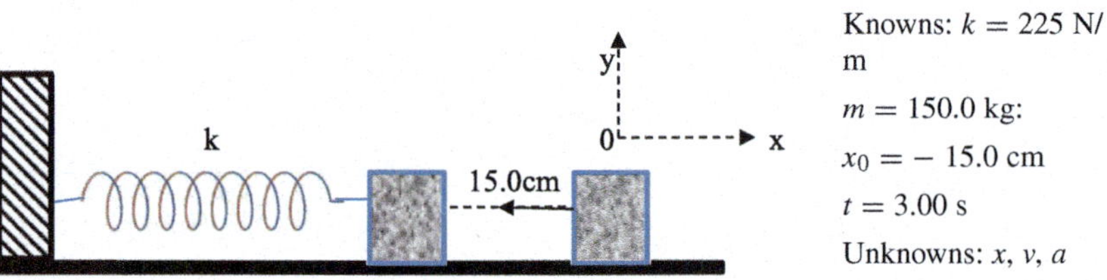

With Eq. 6.1, we can determine the position of the mass at any time. First, however, we must set up the equation. To do that, we need to know what A, ω, and δ are.

Figuring out the amplitude (A) of the motion is easy. The mass is displaced − 15.0 cm away from equilibrium in order to get the motion started. That means the amplitude of the motion is 15.0 cm. Angular frequency is given by $\omega = 2\pi/T$. This equation, however, contains the period of the motion. The period is given by Eq. 5.9. Thus, the angular frequency is:

$$\omega = \sqrt{\frac{k}{m}} = \sqrt{\frac{225\,\text{N/m}}{150.0\,\text{kg}}} = 1.22\,\text{rad/s}$$

where did the radians come from? Since radian is a dimensionless unit, it can be added to or taken from the units as needed. In this case, we need it, because without the radian, we would not know the unit for angle in Eq. 6.1.

We now know A and ω. There is only one thing more that we need: δ. To figure that out, we use Eq. 6.2. Time zero is defined as the time at which the mass is released. Thus, $\mathbf{x}_o$ is – 15.0 cm, because $\mathbf{x}$ is the displacement from equilibrium:

$$\delta = \cos^{-1}(x_0/A) = \cos^{-1}(-15.0\,\text{cm}/15.0\,\text{cm}) = \cos^{-1}(-1) = \pi$$

Now we have the complete equations for position, velocity, and acceleration:

$$x = x_{\max}\cos(\omega t + \delta)$$
$$v = -\omega x_{\max}\sin(\omega t + \delta)$$
$$a = -\omega^2 x_{\max}\cos(\omega t + \delta)$$

At a time of $t = 3.00$ s, the position is given by

$$x = (15.0\,\text{cm})\cos[(1.22\,\text{rad/s})(3.00\,\text{s}) + \pi] = 13.0\,\text{cm}$$

The mass, then, is 13.0 cm from equilibrium, which is the opposite side compared to where the mass started.

The velocity and acceleration at three seconds are given by

$$v = -(1.22\,\text{rad/s})(15.0\,\text{cm})\sin[(1.22\,\text{rad/s})(3.00\,\text{s}) + \pi] = -9.07\,\text{cm/s}$$

$$a = -(1.22\,\text{rad/s})^2(15.0\,\text{cm})\cos[(1.22\,\text{rad/s})(3.00\,\text{s}) + \pi] = -19.4\,\text{cm/s}^2$$

The negative signs simply mean that velocity and acceleration are directed opposite of the displacement.

The maximum distance from equilibrium is easy. It was displaced – 15.0 cm from the equilibrium position, so its maximum distance from equilibrium is 15.0 cm. The maximum acceleration will occur where the maximum force is exerted on the mass. That occurs when the mass is the farthest away from equilibrium. We can calculate the force:

$$F = -kx = -(225\,\text{N/m})(-0.150\,\text{m}) = 33.8\,\text{N}$$

Now we can use Newton's Second Law to determine the acceleration:

$$F = ma$$

$$a = F/m = 33.8\,\text{N}/150.0\,\text{kg} = 0.225\,\text{m/s}^2$$

Since the acceleration is positive and the displacement is negative, the maximum acceleration will be 0.225 m/s^2 opposite of the displacement.

You can actually calculate the maximum acceleration another way. You know the formula for acceleration at any time:

$$\boldsymbol{a} = -\omega^2 A \cdot \cos(\omega t + \delta)$$

When will this be at its maximum? It will be at its maximum when the cosine equals ± 1. After all, cosine varies from $+1$ to -1. Thus, the maximum acceleration will occur when cosine is at its maximum. Thus:

$$\boldsymbol{a} = -\omega^2 A = -(1.22\,\text{rad/s})^2(0.150\,\text{m}) = -0.223\,\text{m/s}^2$$

The numbers are not identical because of the rounding in the calculation of ω.

To determine its maximum speed, we can use the same technique we just used for maximum acceleration. The maximum acceleration will occur when the sine function equals ± 1. Thus

$$\boldsymbol{v} = -\omega \boldsymbol{A} = -(1.22\,\text{rad/s})(0.150\,\text{m}) = -0.183\,\text{m/s}$$

Once again, the negative just means opposite of displacement. Thus, the maximum speed is 0.183 m/s opposite of the displacement.

6.6

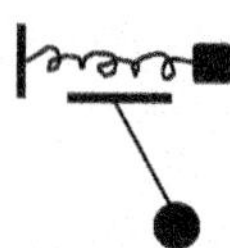

Knowns: $m = 1.0$ kg; $k = 2.0$ N/m; $T_p = T_s$

Unknowns: l

We know the two expressions for the period of a spring and a pendulum so we can equate these two.

$$2\pi\sqrt{\frac{m}{k}} = 2\pi\sqrt{\frac{l}{g}}$$

Solving this for l gives

$$l = gm/k = \left(9.81\,\text{m/s}^2\right)(1.0\,\text{kg})/2.0\,\text{N/m} = 4.9\,\text{m}$$

Study Guide for This Chapter

1. Two mass/spring systems are investigated. The spring constant is the same for each spring, but the mass of the second system is twice that of the first system. They are both displaced the same distance from equilibrium and then released. Compare the following in both systems: total energy, maximum kinetic energy, maximum speed, maximum displacement. Ignore friction.

2. Suppose the two mass/spring systems above were oscillating on a rough surface so that friction was an important consideration. Both systems would eventually stop oscillating. Which would travel the shortest total distance before stopping?
3. A mass/spring system oscillates with an amplitude of A. What is the total distance traveled by the mass in a time equal to the period of the system? A pendulum is displaced by an angle of θ and then released. What is the total angle swept out by the pendulum in a time equal to the period of the system?
4. Potential energy can be negative in many systems. Can the potential energy in a mass/spring system ever be negative? Why or why not?
5. A simple pendulum is hung in an elevator, and its period is measured when the elevator is stationary. Compare that period to the period measured when:
   a. The elevator travels upward at a constant velocity
   b. The elevator accelerates upward
   c. The elevator cable is cut, and the elevator is in free fall
6. The motion of two pendulums is studied. They have identical lengths, and gravitational acceleration is identical for both. Nevertheless, the periods are different. How can you explain this?
7. A mass/spring system ($m = 15.0$ kg) is constructed as shown to the right. The first spring has a force constant of 112 N/m while the second spring has a force constant of 235 N/m. When the mass is displaced from equilibrium and released, what is the period of the motion? (HINT: The springs will stretch different distances when a force is applied to the mass.)

8. A mass/spring system is set in simple harmonic motion. What fraction of the system's energy is kinetic when the mass is one-half of its maximum distance from equilibrium?
9. The displacement of a mass/spring system from its equilibrium position follows this equation:

$$x = (10.0\,\text{cm})\cos[(11.5\,\text{rad/s})t + \pi/2]$$

10. For the mass/spring system in problem #3, what are the maximum speed and acceleration of the block? Where do they occur?
11. A mass/spring system (mass = M, k = K) is set in simple harmonic motion on a frictionless surface with an amplitude of A. When the block is halfway between equilibrium and its maximum distance from equilibrium, a block (mass = ¼ M) is added on top of the first mass.

Please note that this is a perfectly inelastic collision.

a. What is the ratio of the frequency of oscillation before the block was added to the frequency after the block was added?
b. What is the ratio of the amplitude before the block was added to the amplitude after the block was added?
c. Suppose the extra block was added when the mass was at A rather than ½ A. Would that change either of the previous two answers? Why or why not?

12. What is the ratio of the period of a simple pendulum of length L to a pendulum made of a uniform rod of length L that pivots on one end? The moment of inertia of a rod pivoting on one end is $I = ML^2/3$.

Next Level

13. A child swinging on a swing set can be modeled as a simple pendulum. For a swing on a chain that is 4 m long, will you get the swing to go higher if you push it every 4 s or every 3 s?
14. A damped, driven harmonic oscillator is shown in the figure. It consists of a normal mass m on a spring of spring constant k driven up and down by an external force **F** at a frequency f_D. The motion is damped by friction represented by the damping constant b. The motion is described by Eq. 6.1

$x = x_{max} \cos(2\pi f_D t + \delta)$
where the amplitude is frequency-dependent

$$x_{max} = \left(\frac{F_o}{2\pi m}\right)\left[\frac{1}{\sqrt{\left(f_D^2 - f_R^2\right)^2 + \left(\frac{b f_D}{m}\right)^2}}\right].$$

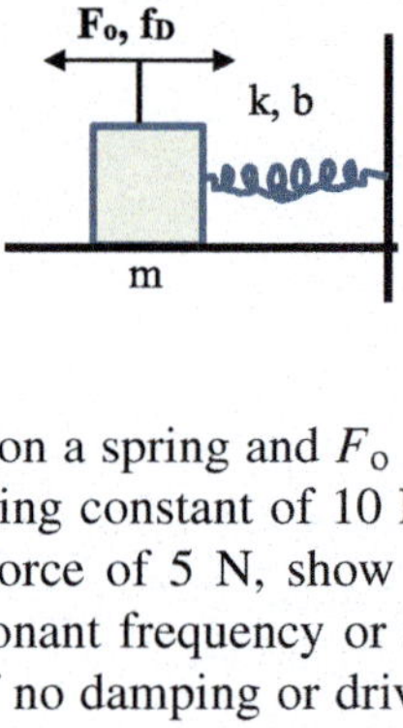

Fo, fD
k, b
m

Here f_R is the resonant frequency of the mass on a spring and F_o is the amplitude of the driving force. For a mass of 2 kg, a spring constant of 10 N/m, a damping constant of 5 kg/s, and a maximum driving force of 5 N, show how the amplitude changes if the system is driven at its resonant frequency or at $f_D = 2.0\,s^{-1}$. (Note, the resonant frequency is the same as if no damping or driving forces were present.)

Electrostatics

7

Chapter Summary

At this point we switch from classical mechanics to electromagnetic theory. Learning to generate and use electricity was one of the greatest advances in the history of mankind. Not only did it provide light to eliminate the problems with darkness, it provided power to run all of the devices and major pieces of equipment we use. We have become a society based on electricity. The next four chapters are dedicated to studying the properties of this important phenomenon. Some of the themes of the chapter are represented in Figs. 7.1, 7.2, 7.3, 7.4, 7.5 and 7.6. Figure 7.1 shows electricity traveling through the air. Figure 7.2 shows wires for conducting electricity. Figure 7.3 shows transfer of electricity through touch. Figure 7.4 shows a place where electrical particles are accelerated. Figure 7.5 shows different shapes of capacitors for storing electrical charge. Figure 7.6 shows electrical discharge in nature.

Main Concepts in This Chapter

- Coulomb's Law
- Electric Field
- Potential Energy and Electrical Potential
- Capacitors

R. C. Powell, *Physics*, Undergraduate Lecture Notes in Physics,
https://doi.org/10.1007/978-3-032-09361-5_7

7.1 Introduction

The second general area of physics we will study is electricity and magnetism. This sounds like two different areas of study, but in physics they are closely related. We begin by learning about static electricity and then go on to studying the flow of electrical current before we discuss magnetism and the important connection between electricity and magnetism. In later chapters, we will learn that all matter is made up of elementary particles each of which has specific properties such as size, mass, and electrical charge. Although we don't completely understand the origin of the electric charge of a particle, we do know some of its important properties.

There are two different types of electrical charges that we call positive and negative. Any piece of matter will have many electrical charges in it, but if it has an equal number of positive and negative charges, it is electrically neutral. If it has more positive charges than negative charges it is positively charged while if it has an excess of negative charges, it is negatively charged. The unit of charge is the coulomb (C). In Chap. 16, we will learn than an atom is made up of electrons orbiting around a nucleus that contains protons. The charge on an electron is negative and exactly the same magnitude as the charge on a proton which is positive. This amount is 1.60×10^{-19} C which is called the elementary charge.

An important law of physics is that charge is conserved. Charge cannot be created or destroyed. However, charge can flow from one part of a physical system to another. If two pieces of matter with different conditions of electrical charge come into contact with each other, excess charge from one piece can move to the

Fig. 7.1 Spark from a Tesla coil. *Credit* Arne Groh, Creative Commons Attribution-Share Alike 3.0

Fig. 7.2 Conducting wire surrounded by insulating shield. *Credit* Scott Ehardt, public domain

Fig. 7.3 Child with van de Graaff generator. *Credit* U.S. Department of Energy, public domain

Fig. 7.4 Fermi Lab particle accelerator. *Credit* Reidn Hahn, U.S. Department of Energy, public domain

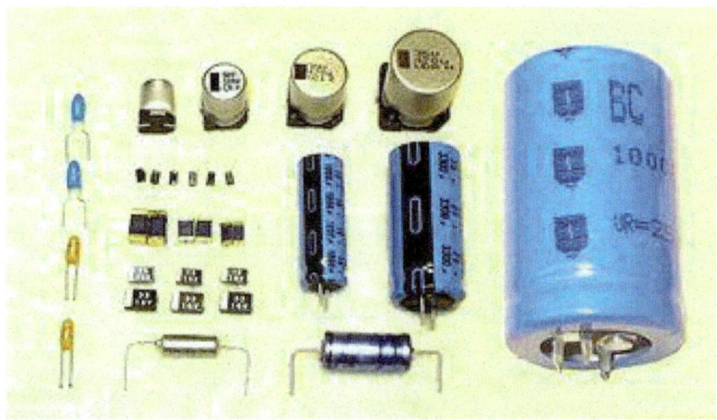

Fig. 7.5 Different capacitors. *Credit* Elcap, Creative Commons zero

other piece but the total charge of both pieces remains the same. The earth is so large it can accept almost any amount of excess charge. Attaching an object with an excess of electrical charge to the earth is called grounding.

Another important concept is that like charges repel each other while unlike charges attract each other. The electrostatic forces between particles act at a distance without touching each other. This is similar to the gravitational force between objects having mass. So a positively charged particle will be pulled toward a negatively charged particle but pushed away from another positively charged particle.

Fig. 7.6 Lightning strikes. *Credit* Hansueli Krapf, Creative Commons Attribution-Share Alike 3.0

Matter can be classified as one of three types depending on the mobility of the electric charges that are in the material. Conductors are materials in which the charges can easily move from one place to another. These are useful for electric circuits that are discussed in Chap. 8. Insulators are materials in which the charges of the material are fixed in place. The third type of materials are called semiconductors. These are materials where the amount charge mobility can be controlled. Semiconductors are useful in microelectronic devices discussed in Chap. 14. This chapter is focused on electrostatics which is the study of charged particles at rest.

The properties discussed above can be used to create a charge on a piece of matter in one of two ways. If you have a piece of matter with excess charge and bring it in contact with a piece of matter that is electrically neutral, the second piece of matter will be "charged" by the first piece due to the excess charge flowing from one to the other. The sign of the charge on the second piece of matter will be the same as the sign of the charge on the first piece of matter. This is charging by conduction. The second type of charging is by induction. In this case the first piece of charged matter is brought very close to the second piece of neutral matter but without making contact. The force due to the charges on the first piece of matter attracts charges of opposite sign in the second piece of matter causing the charges to move to that area of the matter leaving it charged.

7.2 Coulomb's Law

The fact that charged particles attract or repel each other indicates that there is an electrostatic force acting on them. From empirical observations of experimental measurements, Coulomb's Law was developed to describe the electrostatic force between two charged entities,

$$F = Kq_1q_2/r^2. \tag{7.1}$$

In this expression, q_1 and q_2 are the electrostatic charges on the two entities given in Coulombs while r is the distance between the centers of the two. The K is a physical constant called the Coulomb constant. Its magnitude is

$$K = 9.0 \times 10^9\,\mathrm{N\,m^2/C^2}. \tag{7.2}$$

Note that with this definition of K in Eq. 7.1, the units of force come out to be Newtons as they must. The Coulomb constant can be expressed in terms of the permittivity of free space, ε_0,

$$\varepsilon_0 = 8.85 \times 10^{-12}\,\mathrm{C^2/N\,m^2}.$$

This reflects how well an electric field is transmitted in free space. K and ε_0 are related through the expression

$$K = (4\pi\varepsilon_0)^{-1}. \tag{7.3}$$

Equation 7.1 gives the magnitude of the electrostatic force which is the same on entities one and two but in opposite directions. The positive or negative signs of the charges are not used in this equation. The direction of the force is along the line connecting their two centers pointing either toward or away from the other entity depending on whether they are unlike or like charges. The strength of the force decreases as the square of their distance apart. In some cases, a charged object fixed in space creates a force on a movable charged object. In this case, it is common to designate the fixed charge with a capital Q and the movable charge with a small q.

The problem-solving technique for electrostatics is the same as we used for classical mechanics. Begin by sketching the problem, including a coordinate system and reference point. Then list the knowns and unknowns and apply the appropriate physics equations. Let's use this technique on Example 7.1.

Example 7.1

Find the electrostatic force on particle B in Fig. 7.7. The amount of charge on each particle and the distances separating the particles are given in the "Knowns" list below.

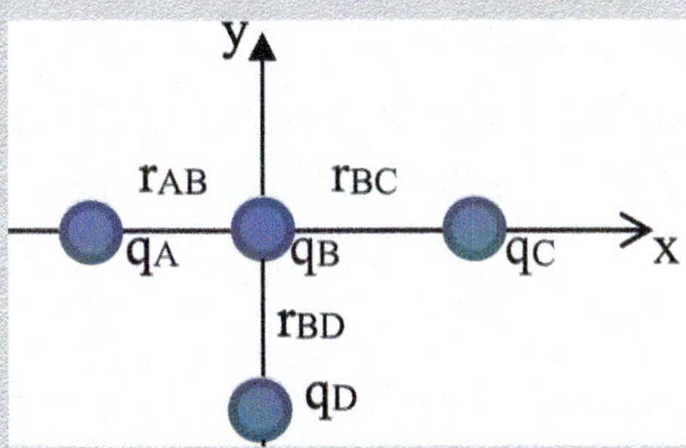

Fig. 7.7 Electrostatic force on four particles

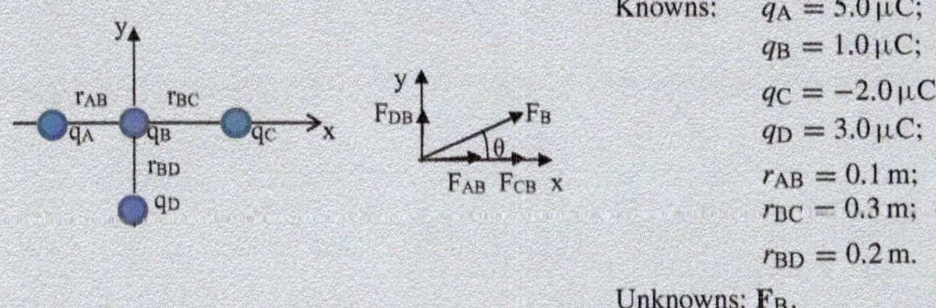

Knowns: $q_A = 5.0\,\mu\text{C}$;
$q_B = 1.0\,\mu\text{C}$;
$q_C = -2.0\,\mu\text{C}$;
$q_D = 3.0\,\mu\text{C}$;
$r_{AB} = 0.1\,\text{m}$;
$r_{BC} = 0.3\,\text{m}$;
$r_{BD} = 0.2\,\text{m}$.

Unknowns: $\mathbf{F}_B$.

Note that the charges are given in microcoulombs. See the table in Appendix B for the correct power of ten. The force diagram on the right shows the directions of the forces that the three charged particles exert on particle B.

The three electrostatic forces on particle B are given by Coulomb's Law from Eq. 7.1:

$$F_{AB} = Kq_Aq_B/r_{AB}^2 = \left(9 \times 10^9\,\text{N}\,\text{m}^2/\text{C}^2\right)\left(5 \times 10^{-6}\,\text{C}\right)\left(1 \times 10^{-6}\,\text{C}\right)/(0.1\,\text{m})^2$$
$$= 4.5\,\text{N};$$
$$F_{CB} = Kq_Cq_B/r_{BC}^2 = \left(9 \times 10^9\,\text{N}\,\text{m}^2/\text{C}^2\right)\left(2 \times 10^{-6}\,\text{C}\right)\left(1 \times 10^{-6}\,\text{C}\right)/(0.3\,\text{m})^2$$
$$= 0.20\,\text{N};$$
$$F_{DB} = Kq_Dq_B/r_{DB}^2 = \left(9 \times 10^9\,\text{N}\,\text{m}^2/\text{C}^2\right)\left(3 \times 10^{-6}\,\text{C}\right)\left(1 \times 10^{-6}\,\text{C}\right)/(0.2\,\text{m})^2$$
$$= 0.675\,\text{N}.$$

Then the x- and y-components of the total force on B are given by:

$$F_x = F_{AB} + F_{CB} = 4.5\,\text{N} + 0.2\,\text{N} = 4.7\,\text{N}$$
$$F_y = F_{DB} = 0.675\,\text{N}$$

where we have accounted for the fact that the AB force is repulsive while the CB force is attractive.

Now we can use the right triangle in the force diagram with the Pythagorean theorem to obtain the magnitude and direction of the force on B.

$$F_B = \sqrt{F_x^2 + F_y^2} = \sqrt{(4.7\,\text{N})^2 + (0.675\,\text{N})^2} = 4.75\,\text{N}$$
$$\theta = \sin^{-1}(F_y/F_B) = \sin^{-1}(0.675\,\text{N}/4.75\,\text{N}) = 8.17^\circ.$$

So the electrostatic force on particle B in Fig. 7.7 is 4.75 N in a direction 8.16° toward the y-axis from the x-axis.

To make sure you remember how to use Coulomb's Law, work the following problems.

Student

7.1 Four charges are arranged at the vertices of a square, as shown to the right. The two gray spheres are charged to $-1.0\,\text{mC}$. The two white spheres are equally charged. The net force on both gray spheres is zero.

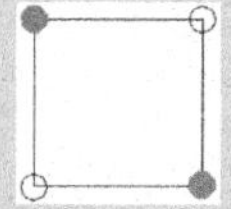

a. What is the charge on each of the white spheres?

b. If a particle with a charge of $+2.5\,\text{mC}$ were placed at the center of the square, what would be the magnitude and direction of the force acting on it?

7.2 A ping pong ball ($m = 2.2\,\text{g}$) is charged at $+0.50\,\mu\text{C}$ and placed at the bottom of a tube. Another ping pong ball is charged to $+0.25\,\mu\text{C}$ and dropped into the tube. What is the distance between the centers of the balls when the system comes to rest?

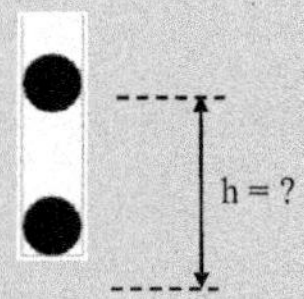

Next perform the following experiment.

Experiment 7.1
A Repulsive Application of Coulomb's Law

Supplies:

- Aluminum foil
- Thread
- Scissors
- Tape
- A blown-up balloon
- A plastic or wooden ruler (It cannot have conductive material in it.)
- An open door inside the house

Introduction—In this experiment, you get to pit gravity and the electrostatic force against one another. This will allow you to calculate the charge that you are able to give two little balls of foil.

Procedure:

1. Cut two lengths of thread so that they are both about 80 cm long.
2. Pull foil out of the box so that you have about 30 cm of foil and then tear it off. You now have a rectangle of foil. Measure the length and width of the rectangle.
3. Cut the rectangle precisely in half and wad each half into a tiny ball. As you wad the foil up, put one end of each thread into each ball so that the foil balls each become attached to one end of each thread.
4. Use tape to attach the strings to the same point at the top of the open door's frame. Adjust the length from the point of attachment to the balls so that the balls hang right next to one another. That way, the setup looks something like this:

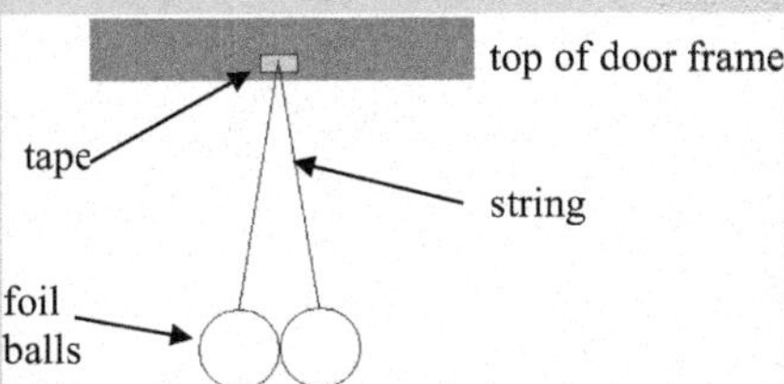

5. Take the balloon and rub it vigorously in your hair. Please note that for this experiment to work well, your hair must be clean and dry. Rubbing the balloon up against a cat would be ideal, because cat hair is meticulously clean. However, I doubt that you will get the cat to consent!

This will induce a negative charge on the balloon as it picks up stray electrons from your hair.

6. In one hand, hold the thread right above the foil balls. Then, touch the balls with the balloon. You are trying to transfer charge from the balloon to the balls. Make sure you don't touch the balls with your hands, or the charge will be conducted through you and away from the balls.
7. Pull the balloon away and gently let the thread go. Because both balls should now be negatively charged, and they should repel one another. Thus, they should move away from one another. If they do not, you did not charge the balls. Try again. It may take a little practice for you to get the balls charged up.
8. At first, the balls will probably move around a lot. You can stop the motion by holding the threads and then gently releasing them again. As long as you don't touch the balls themselves, you should not lose much charge.
9. Eventually, the balls will reach an equilibrium position where they hang in air because their mutual electrostatic repulsion equals the gravitational force pulling them together. Use the ruler to measure the distance between the centers of the balls as they hang there.
10. Measure the distance from the point of attachment on the door frame to the center of each ball. The two distances should be the same.
11. Regular Reynolds wrap is 1.65×10^{-5} m thick, while heavy-duty Reynolds wrap is 2.39×10^{-5} m thick, and the density of aluminum is 2700 kg/m^3. Use this information and the length and width of the foil rectangle you measured in step #2 to calculate the mass of aluminum used. Cut that mass in half so that you have the mass of each ball.
12. The information you got in steps 9, 10, and 11 should be all you need to calculate the charge that you were able to give to the balls, assuming that they ended up equally charged. Try to do the calculation yourself, by balancing the gravitational force, the tension in the strings, and the electrostatic repulsion. If you cannot get it, look at the end of the answers to the "on your own" problems. The solution is there.
13. Clean everything up. Please note: you will use the foil balls again, so don't throw them away. Keep them on their strings as well.

Remember, you can charge an object in two ways. You can charge by conduction (which you did in this experiment) or induction when you bring a charged object close to another object without touching it. These techniques were discussed in your introductory course.

7.3 Electric Field

There is another way of looking at the case of two physical objects producing forces on each other without touching. This is especially useful if the interest is focused on the effects that one of the objects has on several other objects near to it. For example, the sun produces gravitational forces on each of the different planets in the solar system and object B in Fig. 7.7 produces electrical forces on the other three objects near it. Somehow the main object of interest has produced a change in the space surrounding it that allows it to transmit a force to any object entering that space. This change in space is called a field, and our two examples are of a gravitational field and an electrostatic field.

Any point in space can be tested to see if there is an electrostatic field there by placing a small amount of positive charge at that point and detecting any force on it. Any detectable force would be described by Coulomb's Law, Eq. 7.1. The force divided by the test charge gives a constant number that is the magnitude of the field E at that point,

$$E = F_r/q = KQ/r^2. \tag{7.4}$$

The force per unit charge gives the magnitude of the electric field in the units of N/C. Here we have used the convention that Q is the charge that creates the field and q is the test charge that experiences the field. The sign of the charge Q is not used in Eq. 7.4 since this gives only the magnitude of the field and the direction is found from the force on a positive charge at that point in space.

The presence of an electric field can be depicted by field lines as shown in Fig. 7.8. These lines are represented by arrows in the direction of the force that the field will exert on a positive test charge. Figure 7.8a shows that the field lines for a positive charge are directed radially outward from the charge while (b) shows that the field lines for a negative charge are directed radially inward toward the charge. Note that the lines diverge as you get farther away from the charge which is consistent with the $1/r^2$ factor in Eq. 7.4.

If more than one charged object is producing an electric field at the same point in space, the fields from each of the objects can be added vectorially to determine the total field at that point. Examples are shown in Fig. 7.9. On the left-hand side of the figure, we have the electric field generated by two equal but opposite charges. This is often called an electric dipole. An electric dipole consists of two opposite electric charges separated by a certain distance in space. Notice the electric field produced by this electric dipole. Since electric field lines point in the direction that a positive charge would move if placed in the field, the lines point from the positive charge to the negative charge. In other words, the lines "leave" the positive charge and "enter" the negative charge. Since the charges are equal in magnitude, the number of lines for each is the same. Thus, every line that "leaves" the positive charge "enters" the negative charge. The lines are curved because of the results of the vector addition of the two fields.

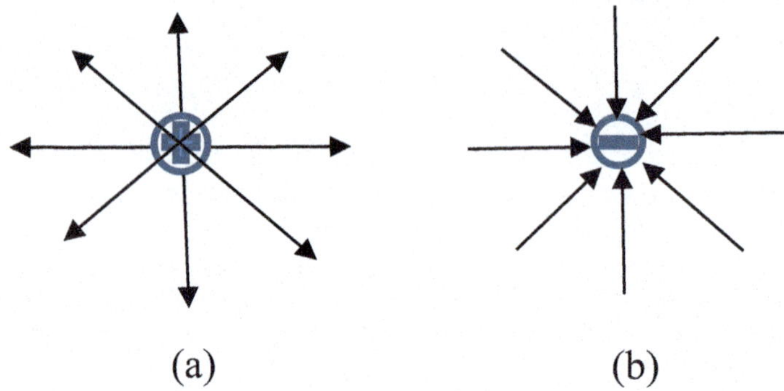

Fig. 7.8 Field lines for positive and negative charges

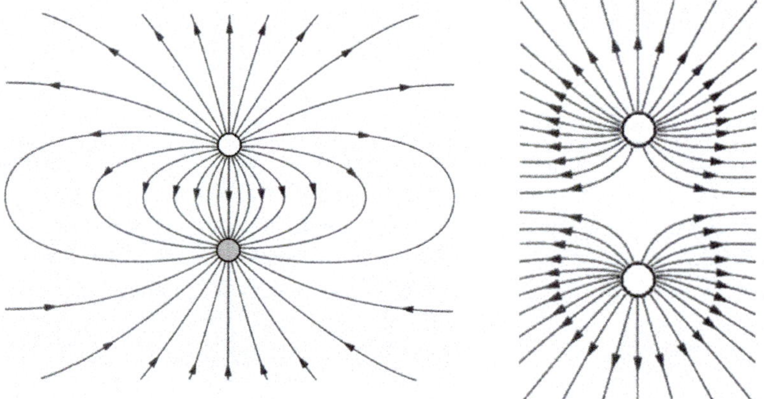

Fig. 7.9 Electric fields generated by two charges

On the right-hand side of the figure, the electric field produced by two equal positive charges is drawn. Note that there are the same number of lines "leaving" each charge, but they do not travel from one charge to another, because a positive charged placed in the vicinity of these two charges would not travel from one positive charge to the other. Instead, it would be repelled by *both* charges. Note that there are no electric field lines midway between the two charges. That's because, if a positive charge is placed midway between the two charges in the figure, it would not move. It would be repelled by each charge, and the two repulsive forces would cancel out, resulting in no net force on the charge. If there is no net force on the charge, there is no electric field.

You can use the two drawings in Fig. 7.8 and your own reasoning to determine the electric field lines produced by several charges placed near one another. Just remember that electric field lines "leave" positive charges and "enter" negative charges; that the field lines will point in the direction that a positive charge would move if placed in the field; and that the more force experienced by a charged particle placed in the field, the denser the field lines. Try your hand at drawing some complex electric fields given in the situations below.

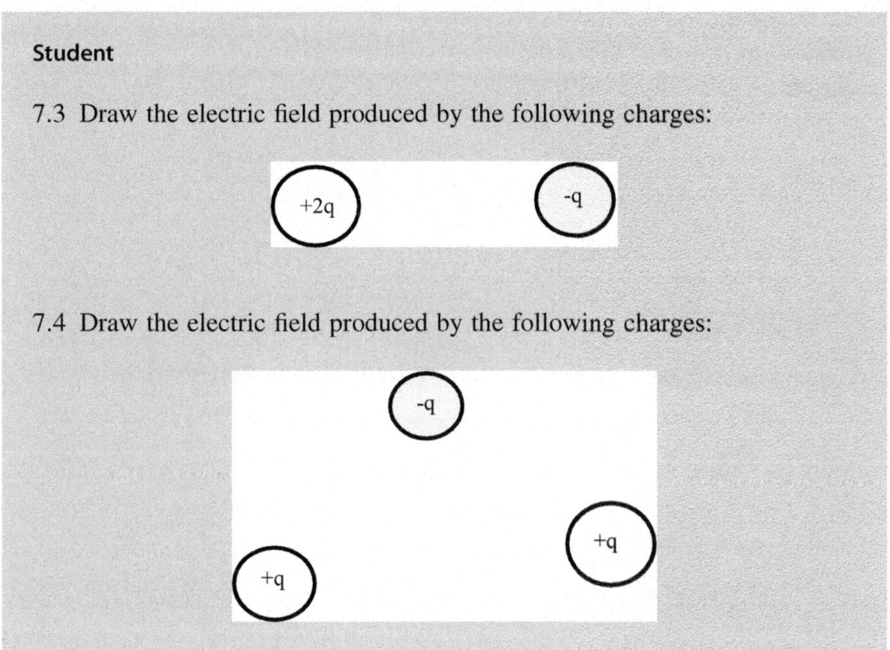

Student

7.3 Draw the electric field produced by the following charges:

7.4 Draw the electric field produced by the following charges:

If you want to calculate the force that a charged particle experiences when it is placed in an electric field, you simply multiply the electric field strength by the charge of that particle. From Eq. 7.4

$$\mathbf{F} = q \cdot \mathbf{E} \tag{7.5}$$

Notice that force and electric field are both vectors in this equation. That should make sense, since force is a vector and charge is not. "E" is not expressed as a vector in Eq. 7.4 because this equation calculates only the strength of the electric field. Like the electrostatic force, you have to reason out the direction yourself. However, once you have **E**, you can use the sign of the charge in Eq. 7.5 to determine direction. Thus, unlike Eq. 7.4, you do use the sign of the charge in Eq. 7.5.

These concepts are used in the following examples.

Example 7.2

Consider the charge distribution shown in Fig. 7.10. What is the electric field at the center of the coordinate system due to charges q_1 and q_2? Also, if charge q_3 is placed at the center of the coordinate system, what is the electrostatic force exerted on it by this field?

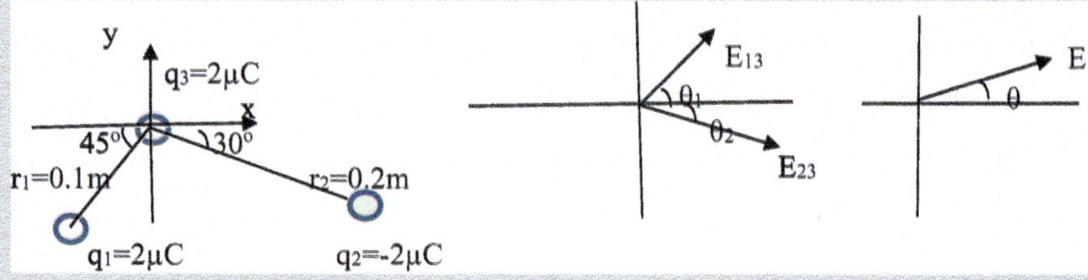

Fig. 7.10 Electric field

The first thing to do is to sketch the problem as shown in Fig. 7.10 as well as the vector diagram for the forces. The knowns and unknowns are:

Knowns: $q_1 = 2\,\mu C$; $q_2 = -2\,\mu C$; $q_3 = 2\,\mu C$; $r_1 = 0.1\,m$; $r_2 = 0.2\,m$; $\theta_1 = 45°$; $\theta_2 = 30°$.

Unknowns: $\mathbf{E}_0$; $\mathbf{F}_3$.

The electric fields at the origin of the coordinate system due to charges q_1 and q_2 are given by Eq. 7.4,

$$E_1 = Kq_1/r_1^2 = (9 \times 10^9\,\mathrm{Nm^2/C^2})(2 \times 10^{-6}\,\mathrm{C})/(0.1\,\mathrm{m})^2$$
$$= 1.8 \times 10^6\,\mathrm{N/C}\ \text{at}\ 45°\ \text{above the}\ x\text{-axis.}$$

$$E_2 = Kq_2/r_2^2 = (9 \times 10^9\,\mathrm{Nm^2/C^2})(2 \times 10^{-6}\,\mathrm{C})/(0.2\,\mathrm{m})^2$$
$$= 4.5 \times 10^5\,\mathrm{N/C}\ \text{at}\ 30°\ \text{below the}\ x\text{-axis.}$$

The directions of the fields are determined by like charges repelling and unlike charges attracting. To find the total field, E_1 and E_2 are added vectorially.

$$E_{1x} = E_1 \cos 45° = (1.8 \times 10^6\,\mathrm{N/C})(0.707) = 1.27 \times 10^6\,\mathrm{N/C}$$
$$E_{2x} = \mathrm{E}_2 \cos 30° = \left(4.5 \times 10^5\,\mathrm{N/C}\right)(0.866) = 3.9 \times 10^5\,\mathrm{N/C}$$
$$\text{So}\ E_x = E_{1x} + E_{2x} = 16.610^5\,\mathrm{N/C}.$$

$$E_{1y} = E_1 \sin 45^\circ = \left(1.8 \times 10^6\,\mathrm{N/C}\right)(0.707) = 1.27 \times 10^6\,\mathrm{N/C}$$

$$E_{2y} = E_2 \sin 30^\circ = \left(4.5 \times 10^5\,\mathrm{N/C}\right)(0.5) = -2.25 \times 10^5\,\mathrm{N/C}$$

$$\text{So}\, E_y = E_{1y} + E_{2y} = 10.4510^5\,\mathrm{N/C}.$$

Then the Pythagorean theorem gives the magnitude of the field as

$$\begin{aligned} E &= \sqrt{E_x^2 + E_y^2} = \sqrt{\left(1.66 \times 10^6\,\mathrm{N/C}\right)^2 + \left(1.045 \times 10^6\,\mathrm{N/C}\right)^2} \\ &= 1.96 \times 10^6\,\mathrm{N/C} \\ \theta &= \tan^{-1}\left(E_y/E_x\right) = \tan^{-1}(1.045\,\mathrm{N/C}/1.66\,\mathrm{N/C}) = \tan^{-1} 0.6295 \\ &= 32.2^\circ \text{ above the } x\text{-axis.} \end{aligned}$$

The force on charge q_3 is given by Eq. 7.4 as
$F_3 = q_3 E = \left(2 \times 10^{-6}\,\mathrm{C}\right)\left(1.96 \times 10^6\,\mathrm{N/C}\right) = 3.92\,\mathrm{N}$ in the same direction as the field since it is a positive charge.

Example 7.3

A + 5.00 mC charge placed 25.0 cm away from a + 15.0 mC charge, as shown in the diagram to the right. At what point is the electric field equal to zero?

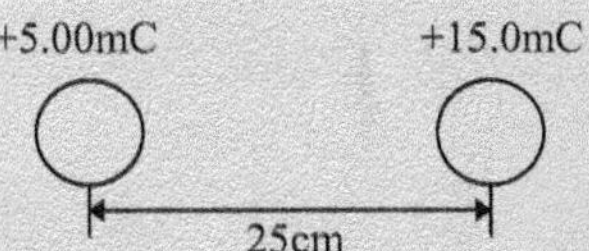

Knowns: $q_1 = +5.00\,\mathrm{mC}$; $q_2 = \mathrm{C}$; $15.0\,\mathrm{mC}$; $r = 25\,\mathrm{cm}$.
Unknowns: x for $E = 0\,\mathrm{N/C}$.
The electric field will equal zero when the vector addition of the two particles' electric fields equals zero. We know instinctively that the electric field must be zero somewhere in between the two particles. If the particles had equal charge, it would be in the middle. Since one particle has more charge than the other, it won't be in the middle; however, it will be on a line connecting the two charges. Thus it turns this into a one-dimensional problem. Let the point of zero electric field be a distance of "x" from the + 5.00 mC charge and a distance of (25.0 cm − x) from the + 15.0 mC charge. We can now use Eq. 7.3 to determine the electric field from each particle and then add them together as vectors. Since this is now a one-dimensional problem, we just use signs to designate the direction. However,

since Eq. 7.3 does not use signs, we have to figure that out on our own. The electric field points in the direction that a positive charge would accelerate if placed in the field. Thus, the electric field from the + 5.00 mC charge points to the right, and the electric field of the + 15.0 mC charge points to the left.

If we define motion to the right as positive, then we get:

$$\mathbf{E}_{\text{tot}} = \mathbf{E}_{5.00} + \mathbf{E}_{15.0}$$

$$\mathbf{E}_{\text{tot}} = \frac{k(0.00500\,\text{C})}{x^2} - \frac{k(0.0150\,\text{C})}{(0.250\,\text{m} - x)^2} = 0$$

$$-2.00x^2 - (0.500\,\text{m})x + 0.0625\,\text{m}^2 = 0$$

$$x = \frac{0.500\,\text{m} \pm \sqrt{(0.500\,\text{m})^2 - 4(-2.00)(0.0625\,\text{m}^2)}}{2(-2.00)}$$
$$= -0.342\,\text{m or } 0.0915\,\text{m}$$

Since the negative solution makes no sense (that would place it left of the + 5.00 mC charge), the other solution must be the correct one. Thus, the point of zero electric field is 9.15 cm to the right of the 5.00 mC charge.

Example 7.4

Three charges are arranged at the vertices of a right triangle, as shown in the figure to the right. What is the value of the electric field at the center of the hypotenuse? What force would a – 5.00 μC charge experience if placed there?

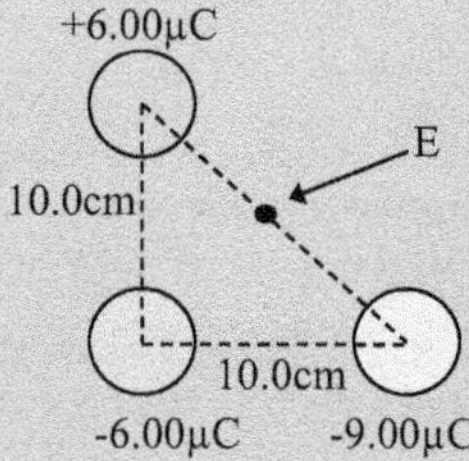

Knowns: $q_1 = 6\,\mu C$; $q_2 = -6\,\mu C$; $q_3 = -9\,\mu C$; $r_{12} = 10.0\,m$; $r_{23} = 10.0\,m$
Unknowns: E at the middle of r_{13}; F on $-5.00\,\mu C$

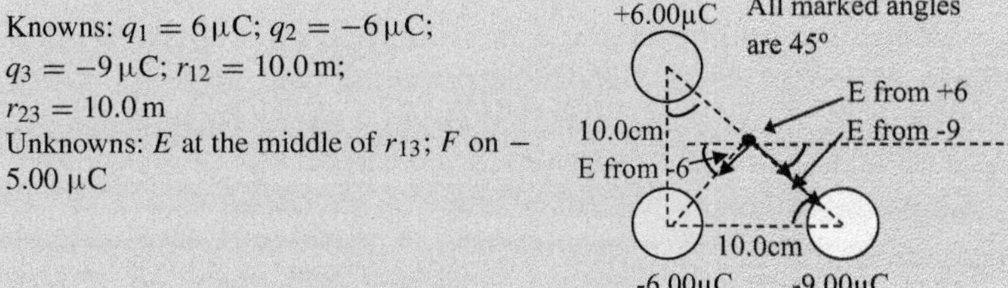

In order to calculate the electric field at the point shown in the figure, we will have to add all of the individual electric fields together as vectors. Thus, we need to figure out some geometry. The hypotenuse of the triangle (via the Pythagorean theorem) is 14.1 cm. That means the point is 7.05 cm from both the + 6.00 μC charge and the − 9.00 μC charge. The dashed line drawn from the − 6.00 μC charge to the point is one leg of a right triangle whose hypotenuse is 10.0 cm and whose other leg is 7.05 cm. Thus, the point (via the Pythagorean theorem) is 7.09 cm away from that charge.

At the point in which we are interested, the + 6.00 μC charge will push a positive charge directly away from it, so the vector for that electric field is oriented along the hypotenuse, towards the − 9.00 μC charge. The − 9.00 μC charge will pull a positive charge towards it, so its electric field is oriented in the same direction as that of the + 6.00 μC charge, but it is stronger (same distance, greater charge), so that vector is a bit longer. Finally, the − 6.00 μC charge pulls a positive charge towards it, so its electric field is pointed towards the right angle of the triangle.

Since the two legs of the big right triangle are equal in length (10.0 cm), this is a 45-45-90 triangle. Thus, the other two angles in the big triangle are 45°. When two parallel lines are cut by a transversal, alternating interior angles are congruent. Thus, the angle between the horizontal and the vectors from the + 6.00 μC and − 9.00 μC charges is 45° as well. Of course, to do vector math properly, we really need to define our angle counterclockwise from the horizontal, so the angle for both of those vectors will be 315°.

A line drawn from the center of the − 6.00 μC charge to the point in which we are interested bisects the right angle of the triangle, since it goes from the vertex to the center of the hypotenuse. That means the angle between that line and each leg of the triangle is 45°. Once again, when two parallel lines are cut by a transversal, alternating interior angles are congruent.

Thus, the angle between the horizontal and the vector from the − 6.00 μC charge is 45°. Defined properly for vector math, however, that is 225°. Now we know all of the angles involved.

Please note that since all of the angles came from strict geometry rules, they are infinitely precise. Thus, they will not enter into the significant figure calculations.

Well, now that we have worked out the geometry, all we have to do is chug through the vector math. Let's call the electric field from the + 6.00 μC charge vector A, the electric field from the − 9.00 μC charge vector B, and the electric field from the − 6.00 μC charge vector C.

First, we get the components:

$$A_x = \frac{kQ}{r^2}\cos\theta = \frac{(8.99\times10^9\,\mathrm{Nm^2/C^2})(6.00\times10^{-6}\,\mathrm{C})}{(0.0707\,\mathrm{m})^2}\cos 315^\circ$$
$$= 7.67\times10^6\,\mathrm{N/C}$$

$$A_y = \frac{kQ}{r^2}\sin\theta = \frac{(8.99\times10^9\,\mathrm{Nm^2/C^2})(6.00\times10^{-6}\,\mathrm{C})}{(0.0707\,\mathrm{m})^2}\sin 315^\circ$$
$$= -7.67\times10^6\,\mathrm{N/C}$$

$$B_x = \frac{kQ}{r^2}\cos\theta = \frac{(8.99\times10^9\,\mathrm{Nm^2/C^2})(9.00\times10^{-6}\,\mathrm{C})}{(0.0707\,\mathrm{m})^2}\cos 315^\circ$$
$$= 1.15\times10^7\,\mathrm{N/C}$$

$$B_y = \frac{kQ}{r^2}\sin\theta = \frac{(8.99\times10^9\,\mathrm{Nm^2/C^2})(9.00\times10^{-6}\,\mathrm{C})}{(0.0707\,\mathrm{m})^2}\sin 315^\circ$$
$$= -1.15\times10^7\,\mathrm{N/C}$$

$$C_x = \frac{kQ}{r^2}\cos\theta = \frac{(8.99\times10^9\,\mathrm{Nm^2/C^2})(6.00\times10^{-6}\,\mathrm{C})}{(0.0707\,\mathrm{m})^2}\cos 225^\circ$$
$$= -7.59\times10^6\,\mathrm{N/C}$$

$$C_y = \frac{kQ}{r^2}\sin\theta = \frac{(8.99\times10^9\,\mathrm{Nm^2/C^2})(6.00\times10^{-6}\,\mathrm{C})}{(0.0707\,\mathrm{m})^2}\sin 225^\circ$$
$$= -7.59\times10^6\,\mathrm{N/C}$$

Now that we have the components, we just need to add them together to get the total electricfield:

$$E_x = 7.67\times10^6\,\mathrm{N/C} + 1.15\times10^7\,\mathrm{N/C} - 7.59\times10^6\,\mathrm{N/C}$$
$$= 1.16\times10^7\,\mathrm{N/C}$$

$$E_y = -7.67\times10^6\,\mathrm{N/C} - 1.15\times10^7\,\mathrm{N/C} - 7.59\times10^6\,\mathrm{N/C}$$
$$= 2.68\times10^7\,\mathrm{N/C}$$

Thus,

$$E = \sqrt{(1.16 \times 10^7\,\mathrm{N/C})^2 + (-2.68 \times 10^7\,\mathrm{N/C})^2} = 2.92 \times 10^7\,\mathrm{N/C}$$

$$\theta = \tan^{-1}\left[\frac{-2.68 \times 10^7\,\mathrm{N/C}}{1.16 \times 10^7\,\mathrm{N/C}}\right] = -66.6^\circ$$

Since this vector has a positive x-component and a negative y-component, it is in the fourth quadrant of the Cartesian coordinate plane. This means to define the angle properly, we must add 360.0° to it. Thus, the electric field is 2.92×10^7 N/C directed at an angle of 293.4°.

The problem also wants us to calculate the force experienced by a − 5.00 μC charge placed at that point. That's a quick application of Eq. 7.5. Now remember, in this equation, we must use the sign of the charge, as that helps determine direction:

$$\mathbf{F} = q \cdot \mathbf{E} = -(5.00 \times 10^6\,\mathrm{C})(2.92 \times 10^7\,\mathrm{N/C}) = -146\,\mathrm{N}$$

Since the result is negative, it means the force is directed *opposite* the electric field. Thus, we could say that the force is 146 N opposite of the electric field, or we could determine what opposite means. If the electric field is oriented at 239.4°, a vector pointed in the opposite direction would point at 59.4°. Thus, we could say the force is 146 N at an angle of 59.4°.

Now you should be ready to work the following problems.

Student

7.5 Consider the midpoint of a dipole. The two charges which make up the dipole are + 5.00 mC and − 5.00 mC, and they are fixed in space 50.0 cm from each other. What is the strength of the electric field at the midpoint? What is the force experienced by a + 14.0 mC charge placed there?

7.6 Three charges (− 12.1 mC, − 13.5 mC, and 11.0 mC) are placed at the vertices of an equilateral triangle whose sides are 35.0 cm long, as illustrated to the right. What is the strength of the electric field at the center of the triangle?

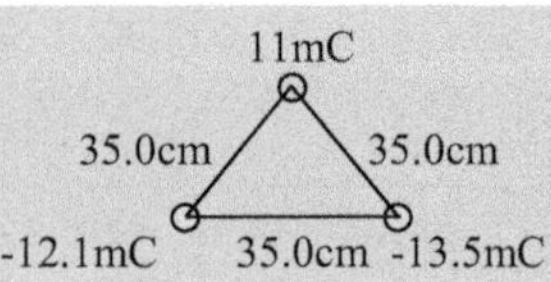

7.4 Potential Energy and Electrical Potential

Imagine a charged particle that is being held in place in an electric field. As soon as that particle is released, it will begin to accelerate. If it is positively charged, it will accelerate in the direction of the electric field. If it is negatively charged, it will accelerate in the opposite direction of the electric field. Either way, it *will* accelerate. That means it will gain kinetic energy because the electric field *will do work* on the particle. That work adds directly to the kinetic energy. If the electric field can do work on the particle, the charged particle must have *potential energy* while it is in the electric field. This is related to but not equal to the concept of electrical potential.

Because of the force that is present, work is done when a charged object is moved from one place to another in an electric field. Consider two unlike charged particles that attract each other. To pull one of the charges further away from the other one requires doing work against the electrostatic force pulling them together. This work is stored as potential energy in the charged particle at its new position, and the greater its charge the greater its increase in potential energy. From the work-energy theorem we learned previously, the magnitude of the change in potential energy is equal to the work done, $\Delta\text{PE} = W$. Since the work is equal to the force in the direction of motion times the distance moved, $W = F_{//}d$, and the electrostatic force is the charge of the moving particle times the electric field it is in, $F = qE$, the change in potential energy in a uniform electric field is given by,

$$\Delta\text{PE} = qEd = q\Delta V. \tag{7.6}$$

The electric *potential difference*, *V*, is defined as the ratio of the work done in moving a positive test charge from one place to another in an electric field to the magnitude of the test charge,

$$V = \text{W}_\text{q}/q = KQ/r. \tag{7.7}$$

where we have used Eq. 7.1. Q is the charge creating the electric field and r is the distance moved by the moving particle. The units of potential difference are

J/C which are called Volts and designated as V. For this equation, the sign of the charge Q must be used so the potential difference can be negative or positive.

When two unlike charges are moved further apart, work must be done on the charges so both the electric potential difference and the potential energy increase (V becomes less negative). When two unlike charges are moved closer together the work done on the charges is negative and both the electric potential difference and the potential energy decreases.

There are two important things to note about electric potential. The first is that this quantity by itself is not very meaningful. It is the difference in V between two points that is important,

$$\Delta V = V_{\mathrm{B}} - V_{\mathrm{A}}. \tag{7.8}$$

This is called the *voltage* between A and B. For example, the strength of a battery is given by the difference in electric potential between its positive and negative posts like 1.5 V. Or a 110 V outlet for a wall plug. Note that the path taken in moving the mobile charge from point A to point B is not important. The difference in the electric potential depends only on the distance moved by the mobile charge in the direction of the electric field. Points with the same value of V are called equipotential points. The zero point for V is arbitrary (as it is for potential energy).

The second important thing to note is that electric potential difference is related to potential energy but is not equal to potential energy. This is obvious from Eq. 7.6 where the mobile charge can be either positive or negative. If it is positive and the charge producing the electric field is negative, when the two charges move closer together both ΔV and $\Delta \mathrm{PE}$ increase in magnitude and are negative. Thus, they both decrease in value. Because of conservation of energy, the decrease in potential energy is converted to kinetic energy. However, if the mobile charge sign is negative, the voltage decreases but the potential energy increases. Thus they are quite different entities as shown in the example below.

Example 7.5

A + 6.5 C charged particle (m = 42.3 kg) is placed 1.4 m from a + 3.4 C stationary charge. If it starts from rest, how fast will the particle be traveling when it is 2.5 m away from the stationary charge?

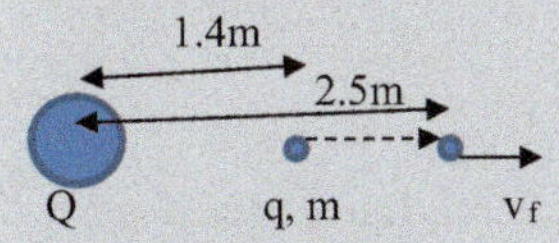

1.4m
2.5m
Q
q, m
v_f

Knowns: $Q = +3.4\,\mathrm{C}$; $q = +6.5\,\mathrm{C}$; $m = 42.3\,\mathrm{kg}$;
$v_i = 0\,\mathrm{m/s}$;
$r_1 = 1.4\,\mathrm{m}$; $r_f = 2.5\,\mathrm{m}$
Unknowns: v_f

To solve a problem like this, we just have to think about the energy involved. As the particle moves, it experiences a change in the electric potential. The potential is originally:

$$V_i = kQ/r = (8.99 \times 10^9\,\mathrm{Nm^2/C^2})(3.4\,\mathrm{C})/(1.4\,\mathrm{m}) = 2.2 \times 10^{10}\,\mathrm{V}$$

The final potential is

$$V_f = kQ/r = (8.99 \times 10^9\,\mathrm{Nm^2/C^2})(3.4\,\mathrm{C})/(2.5\,\mathrm{m}) = 1.2 \times 10^{10}\,\mathrm{V}$$

Thus the potential difference is

$$\Delta V = 1.2 \times 10^{10}\,\mathrm{V} - 2.2 \times 10^{10}\,\mathrm{V} = -1.0 \times 10^{10}\,\mathrm{V}$$

That corresponds to a change in potential energy of:

$$\Delta\mathrm{PE} = \Delta V q$$

$$\Delta\mathrm{PE} = (6.5\,\mathrm{C})(-1.0 \times 10^{10}\,\mathrm{J/C}) = -6.5 \times 10^{10}\,\mathrm{J}$$

This tells us that the particle's potential energy *lowered* by 6.5×10^{10} J. This energy went into kinetic energy. Thus, the kinetic energy of the particle increased by 6.5×10^{10} J. It had *no* kinetic energy to begin with (it was at rest), so the *final* kinetic energy is just 6.5×10^{10} J. Now we can determine the particle's speed from this.

$$\mathrm{KE} = mV_f^2/2$$
$$6.5 \times 10^{10}\,\mathrm{J} = (42.3\,\mathrm{kg})V_f^2/2$$

$$V_f = 5.5 \times 10^4\,\mathrm{m/s}$$

When the particle has traveled to 2.5 m away from the stationary charge, then, its speed is 5.5×10^4 m/s.

Example 7.6

Consider the example problem of two charges $Q = -5 \times 10^{-6}\,\mathrm{C}$ and $q = -2 \times 10^{-6}\,\mathrm{C}$ separated by a distance of 0.1 m. How much work is done in moving q twice as far away? What is the change in electric potential difference and in potential energy?

The first thing to do is to draw a picture of this situation as shown in Fig. 7.11. The next thing to do is to list the knowns and unknowns:

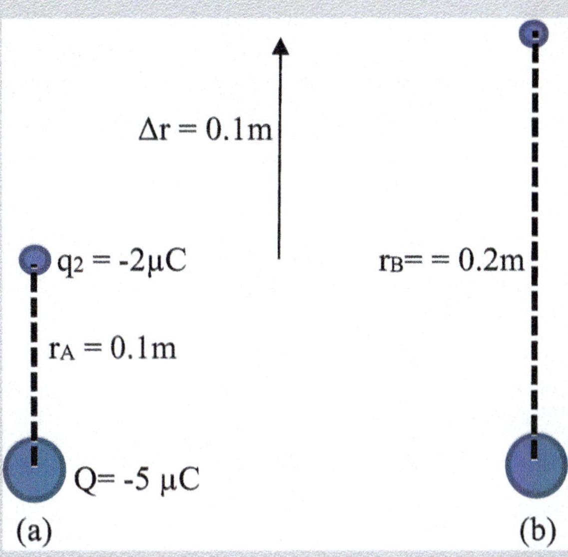

Fig. 7.11 Moving two negative charges

Knowns: $Q = -5 \times 10^{-6} C$; $q = -2 \times 10^{-6} C$; $r_A = 0.1\,\text{m}$; $r_B = 0.2\,\text{m}$.
Unknowns: W; ΔV; ΔP.

Using Eqs. (7.6) and (7.7) we can calculate the electric potential difference between the positions of the charges in part (A) and part (B),

$$\begin{aligned}\Delta V &= V_B - V_A = KQ/r_B - KQ/r_A \\ &= (9 \times 10^9\,\text{Nm}^2/\text{C}^2)(-5 \times 10^{-6}\,\text{C})[(1/0.2\,\text{m}) - (1/0.1\,\text{m})] \\ &= 2.25 \times 10^5\,\text{V}.\end{aligned}$$

The change in potential energy between position A and position B is,

$$\Delta\text{PE} = q\Delta V(-2 \times 10^{-6}\,\text{C})\left(2.25 \times 10^5\,\text{V}\right) = -0.45\,\text{J}.$$

The work done is equal to the change in potential energy,

$$W = \Delta\text{PE} = -0.45\,\text{J}.$$

Thus, separating like negative charges causes a positive change in the electronic potential difference but a decrease in potential energy. The work is done by the charged system of particles and therefore is negative work.

In any kind of energy analysis that we want to do in a situation involving electrical charge, it can be aided by calculating the potential difference. Another important property of electric potentials is that they are additive. Thus, if we have more than one charged particle creating an electric field, the potential at any point in space is simply the sum of the potentials for each individual charge. This is shown in the following example.

Example 7.7

Two stationary charges ($q_1 = -1.5$ mC; $q_2 = -5.6$ mC) are placed 50.0 cm apart. How much work would it take to bring a third charged particle ($q = -3.4$ mC) midway between q_1 and q_2?

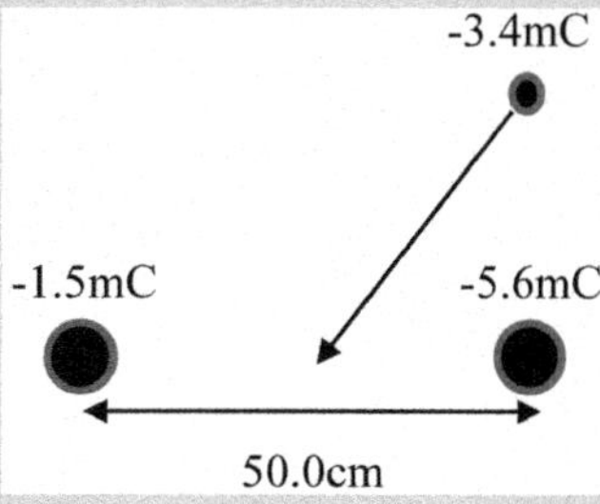

Obviously, it is going to take work to do this, since the negatively charged particle is being repelled by the other two negatively charged particles. All we have to do is consider the electric potentials involved. When the − 3.4 mC particle is far, far away, it experiences no electric potential from the two charges. Thus, $V = 0$. When it is brought midway between the charges, it is 25.0 cm from each of them, and the total electric potential it feels is just the sum of the two individual electric potentials:

$$\begin{aligned} V_{\text{tot}} &= V_1 + V_2 = kQ_1/r_1 + kQ_2/r_2 \\ &= \left(8.99 \times 10^9\,\text{Nm}^2/\text{C}^2\right)(-0.0015\,\text{C})/(0.250\,\text{m}) \\ &\quad + \left(8.99 \times 10^9\,\text{Nm}^2/\text{C}^2\right)(-0.0056\,\text{C})/(0.250\,\text{m}) \\ &= -2.5 \times 10^8\,\text{V} \end{aligned}$$

So, the potential difference is just:

$$\Delta V = -2.5 \times 10^8\,\text{V} - 0\,\text{V} = -2.5 \times 10^8\,\text{V}$$

Now that we know how the electric potential changed, we can figure out how the potentialenergy changed:

$$\Delta\text{PE} = q\Delta V = (-0.0034\,\text{C})\left(-2.5 \times 10^8\,\text{J/C}\right) = 8.5 \times 10^5\,\text{J}$$

From the work-energy theorem in Chap. 3, the change in potential energy is equal to the negative of the work. Thus, the work done is -8.5×10^5 J. The negative means that work was done *against* the electrostatic force. This makes sense since the electrostatic force would push the -3.4 mC charge away from the two stationary charges.

Now try the next three problems.

Student

7.7 An electron has a charge of -1.6×10^{-19} C and a mass of 9.1×10^{-31} kg. Ignoring friction, if an electron starts from rest and travels from the negative side of a 9.0 Volt battery to the positive side, how fast will it be going when it hits the positive side of the battery?

7.8 A charged particle ($m = 1.50$ kg, $q = 12.2$ mC) is fired towards a stationary 15.0 mC charge. If the initial speed of the particle is 2000.0 m/s, how fast will the particle be going when it is 1.00 m from the stationary charge? Assume that the particle was fired very far from the stationary charge.

7.9 Two stationary charges, $q_1 = q_2 = 120.0$ mC, are placed 80.0 cm apart. A third charged particle that is free to move ($m = 1.34$ kg, $q = 88.0$ mC) is placed 25.0 cm above the midpoint of the two charges and released. At what speed will the particle be moving when it is 100.0 cm above the midpoint?

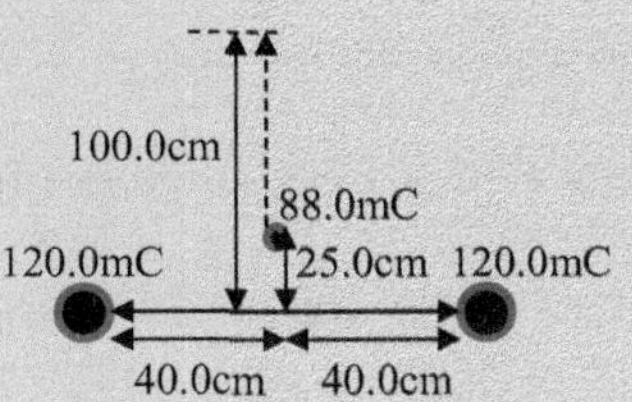

The reason the electrical potential is an important concept is that it tells us something about the electric field itself instead of the effect of the field on a charged particle, which is associated with the concept of potential energy. These are related through Eq. 7.6 which can be rewritten as

$$\text{PE} = qV \tag{7.9}$$

The charge of the individual particle has been factored out, and the electric potential, V, deals only with the properties of the field. It is independent of the charge of whatever particle might be placed in the field. Think about the units involved here. Potential energy has units of Joules. Charge has a standard unit of Coulombs. Thus, in order for Eq. 7.6 to work out to units of Joules, the electric potential, V, must have units of Joules/Coulomb which is a Volt.

When you get a battery, it is rated in terms of Volts. That is a measure of how much the electric field created by the battery can change the potential energy of a charge. The higher the voltage, the more the battery can change a charged particle's potential energy.

However, the voltage does not depend on the charged particle. It depends only on the battery.

That's the purpose of the electric potential. It gives you an idea of how much "push" or "pull" the electric field can provide a charged particle.

Instead of considering the interaction between two particles, let us consider what happens to a charged particle in a uniform electric field. Figure 7.12 shows what happens to a positively charged particle placed in a constant electric field. Why did the particle move from point A to point B? It moved because the electric field *worked* on it. The electric field exerted a force, which Eq. 7.4 says is equal to $q \cdot \mathbf{E}$. That force pushed the particle through the field. Note that based on the drawing and Eq. 7.4, we know that the particle in the picture is a positive particle because the particle moved in the same direction as the electric field. Since the electrostatic force generated by the field is the only force working on the particle, that means the force is pushing in the same direction as the electric field. Thus, the charge must be positive so that Eq. 7.5 gives us a force that is directed the same as the electric field.

That force obviously was exerted over a distance, so work was done. Equation 3.1 tells us that work is the dot product of the force vector ($q \cdot \mathbf{E}$) and the displacement vector ($\mathbf{x}$). Thus, in this case $\mathbf{x}$ is the displacement from point A to point B. From our discussion in Chap. 1, we know that the dot product between two vectors is the same as the magnitude of the vectors times the cosine of the angle between them. In this case,

$$W = q \cdot E \cdot x \cdot \cos\theta \tag{7.10}$$

where θ is the angle between the electric field and the displacement of the charged particle. In Fig. 7.12, the displacement is parallel to the electric field, so $\theta = 0$. Thus, in this case, if we multiply the charge of the particle times the strength of the electric field times the distance from point "A" to point "B," we get the work done on the charged particle by the electric field. Please note that in order for this

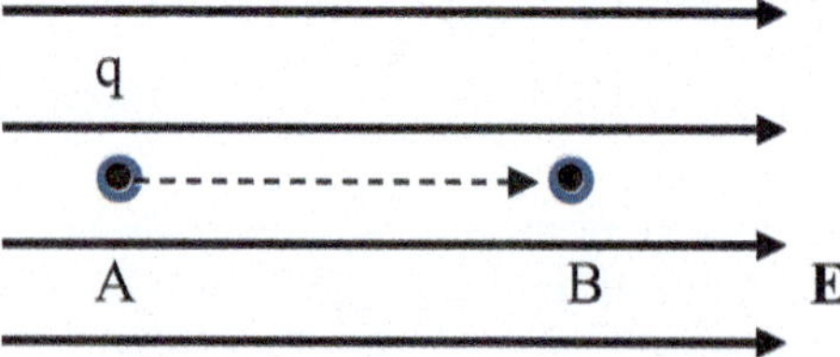

Fig. 7.12 Charged particle in an electric field

equation to work properly, you must include the sign of the charge. If the particle is negatively charged, the negative sign must be included.

Equations 7.9 and 7.10 are only valid for constant electric fields. If the electric field changes while the particle moves more complicated expressions are required to describe the work done. Also note that the electric field provides a conservative force. As defined in Chap. 3, this means that the work done by E depends only on the end points of the movement and not on the path taken. Going from A to B in Fig. 7.12 can be in a straight line or zig-zag and the work done by the field on the charged particle is the same. Remember that for a dissipative force like friction, the work done does depend on the path taken.

Now what does work do to the particle? It changes the kinetic and potential energies of the particle. In Fig. 7.12, the work *increases* the kinetic energy of the particle, making it move faster and faster. If kinetic energy increases, the potential energy must decrease so total energy is unchanged. Thus, any work done alters the kinetic and potential energies according to the equations developed in Chap. 3 as Eqs. 3.4, 3.5, and 3.7,

$$\Delta \mathrm{PE} = -W \tag{7.11}$$

$$\Delta \mathrm{KE} = W \tag{7.12}$$

These formulas really apply to all situations, not just ones involving electric fields. In fact, you actually have used the reasoning behind these equations in previous chapters, realizing that when gravity works on a falling rock, for example, the rock's kinetic energy (KE) increases by the amount of work done while its potential energy (PE) decreases by that same amount.

Consider the following examples.

Example 7.8

Now consider oppositely charged plates as shown in Fig. 7.13. This creates uniform electric field between the plates except near the edges. If the plates are separated by 1.0 m and the field between them is 2000 N/C, what is the potential difference between the plates? What is the work required to move a particle with a charge of 5×10^{-6} C from point A to point B as shown?

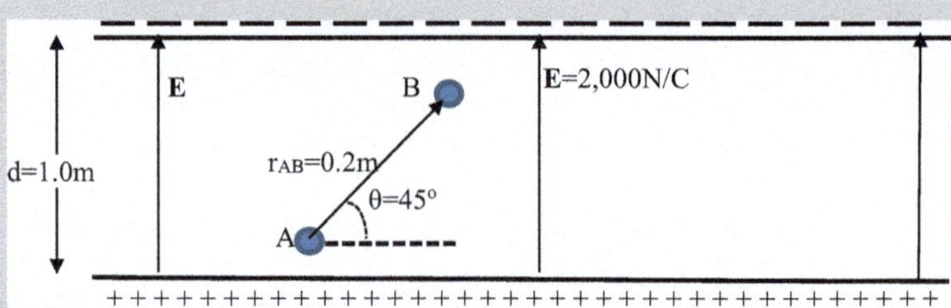

Fig. 7.13 Potential in a uniform electric field

We can use the figure as drawn and list the knowns and unknowns,

Knowns: $d = 1.0\,\text{m}$; $E = 2000\,\text{N/C}$; $q = 5 \times 10^{-6}\,\text{C}$; $r = 0.2 = \text{m}$; $\theta = 45°$.

Unknowns: ΔV between plates; W to move charge.

In a uniform field,

$$\Delta V = Ed = 2000\,\text{N/C} \times 1.0\,\text{m} = 2000\,\text{V}.$$

The work done moving the charge from A to B is given by,

$$W = qEr\sin\theta = \left(5 \times 10^{-6}\,\text{C}\right)\left(2 \times 10^{3}\,\text{N/C}\right)(0.2\,\text{m})\left(\sin 45°\right) = 1.41 \times 10^{-3}\,\text{J}.$$

Since the work is done by the charged system instead of on the charged system, it is negative work.

Example 7.9

A charged particle ($q = -15.0$ mC, $m = 1.0$ kg) enters a constant electric field pointed due west with a magnitude of 25.0 N/C. The particle enters the field with a speed of 2.5 m/s. After a few seconds, it is 5.0 m at an angle of 145 degrees from its initial position. What is its speed?

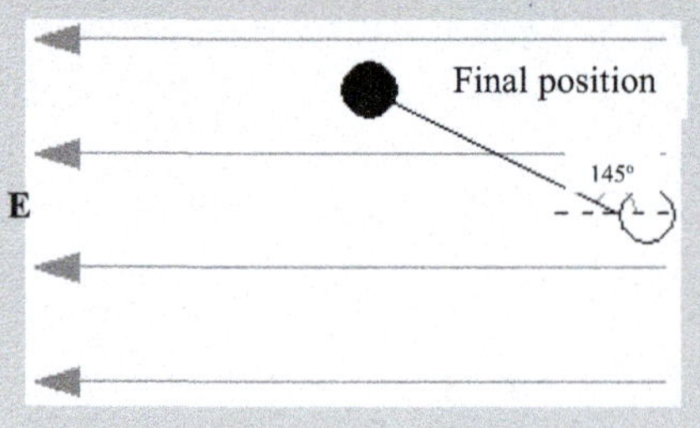

Knowns: $q = -15.0\,\text{mC}$;
$m = 1.0\,\text{kg}$; $E = 25.0\,\text{N/C}$;
$v_i = 2.5\,\text{m/s}$; $\Delta x = 5\,\text{m}$; $\theta = 145°$
Unknowns: v_f

To figure out the new speed, we need to find out the new kinetic energy. We can do that by calculating the work done. Here is where the fact that the electrostatic force is a conservative force comes in handy. Since the electrostatic force is conservative, we can choose a *different path* than the one that the particle actually took. All the particle has to do is end up at the right place. The path doesn't matter. Let's choose the path given by the dashed line below with $\phi = 35°$.

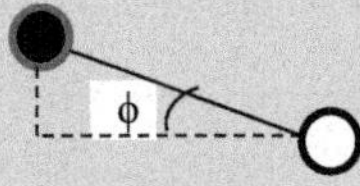

What's nice about this path is that in the second part of the path *no work is done* because the displacement during that part of the journey is perpendicular to the electric field. Thus, the dot product $(q \cdot \mathbf{E}) \cdot \mathbf{x}$ is zero! Thus, work is only done during the first part of the journey which is parallel to **E**. We can calculate that work rather easily. For that side of the right triangle, the length is

$$\text{adjacent side} = (\text{hypotenuse}) \cos\phi = (5.0\,\text{m}) \cos 35° = 4.1\,\text{m}$$

The work done during this leg of the trip is

$$W = qEx \cos 0° = (-0.0150\,\text{C})(25.0\,\text{N/C})(4.1\,\text{m}) = -1.5\,\text{J}$$

Notice that the work is negative. That should make sense. The particle is negative, so the force it experiences is *opposite* the direction of the electric field. Thus, the electric field is slowing this particle down. That means it is removing kinetic energy. We know that $\Delta\text{KE} = -1.5\,\text{J}$. Thus,

$$\begin{aligned} \Delta\text{KE} &= mv_f^2/2 - mv_i^2/2 \\ -1.5\,\text{J} &= (0.5\,\text{kg})\left[v_f^2 - (2.5\,\text{m/s})^2\right] \\ v_f &= 1.8\,\text{m/s} \end{aligned}$$

The final speed, then, is 1.8 m/s, which is slower than the initial speed. This makes sense, since the electric field will push the particle in a direction that is opposed to its initial velocity.

Now try these two Problems.

Student

7.10 Four particles, each of which have the same amount of negative charge, are in an electric field, as shown to the right.
Which has the highest electrical potential energy?

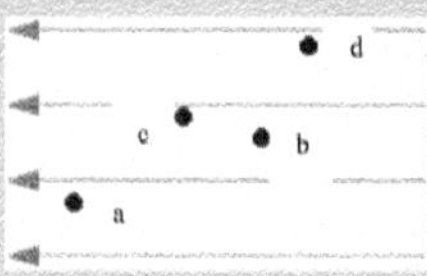

7.11 A physicist takes a particle ($q = 120.0$ mC) and moves it in an electric field (magnitude = 15.0 N/C) as shown in the diagram below. How much work did the physicist do in the process?

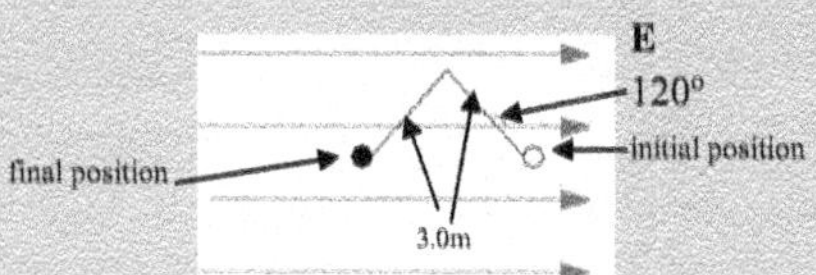

7.5 Capacitors

From Example Problem 7.6, it is clear that it is possible to store electrical energy in a uniform electric field. One device that does this is a capacitor. To understand what a capacitor is, you should perform the following experiment.

Experiment 7.2
Making a Leyden Jar

Supplies

- A plastic or paper cup
- Aluminum foil
- A plastic comb
- Paper clips
- A balloon

Introduction—In 1745, Dutch scientist Pieter van Musschenbroek found that he could store charge in a jar that was lined with metal foil on both the inside and the outside. He demonstrated it at the University of Leyden, and it is therefore called the Leyden jar. The Leyden jar was the forerunner of the modern-day capacitor. In this experiment, you will make a Leyden jar.

Procedure

1. Take the cup and line the lower half of the inside with aluminum foil. Try to make the foil lie as flat as possible against the inside of the cup.

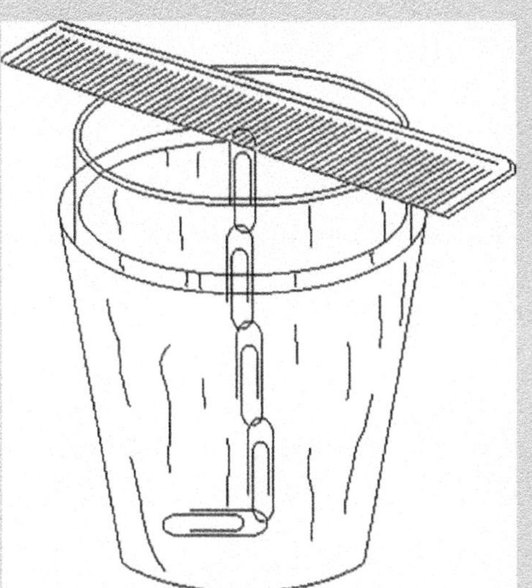

2. Cover the lower half of the outside of the cup with foil as well. In the end, then, you have foil covering the lower half of the cup on both the outside and the inside.
3. Hang a paper clip in the middle of the comb, and then attach more paper clips to it so that a line of paper clips dangle from the comb.
4. Place the comb across the top of the cup. If the paper clip line is not long enough to reach the bottom of the cup where the foil is, add more paper clips. The paper clips must touch each other, and the last one must touch the foil.
5. In the end, your setup should look like the drawing on the right.

6. Blow up the balloon and tie it off.
7. Grab the foil on the outside of the cup with your hand, but be sure you do not touch any of the foil on the inside of the cup.
8. Rub the balloon in your hair to build up charge.
9. Bring the balloon in contact with the tip of the paper clip that is attached to the comb. Make sure the balloon doesn't touch the foil on the outside of the cup at all. It should only touch the paper clip. You should hear a discharge as charge travels from the balloon to the paper clip. That charge will then run down the line of paper clips and into the foil on the inside of the cup.
10. Repeat steps (8) and (9) nine more times.
11. Set the balloon down far from the cup.
12. Slowly bring the index finger of the hand not holding the foil to the paper clip. What happens?
13. If you felt nothing when your finger got near or touched the paper clip, most likely the paper clips are not touching each other well enough to make electrical contact. You need solid electrical contact between each paper clip and the foil.
14. Repeat steps (7)–(11).
15. Now walk away from the setup. You need to tell everyone in the house not to touch the setup.
16. Wait at least an hour, and then grab the outside of the foil with one hand and, once again, bring the index finger of the other hand into contact with the paper clip. What happens?
17. Clean everything up.

What happened in the experiment? You *stored charge*. As you kept touching the charged balloon to the paper clips, negative charges traveled off of the balloon and into the foil.

This built up a negative charge on the inner foil of the cup. That negative charge could not travel to the outside foil, because the cup is an insulator. Thus, the charge just built up on the inner foil. That charge, however, repelled electrons that were in the outer foil. Since your hand was touching the outer foil, those electrons streamed out of the foil, into your hand, and to ground.

This, of course, made the outer foil positive. Thus, as the inner foil became negative, the outer foil became positive.

When you then touched the paper clip with your finger, the excess electrons in the inner foil could travel through your body to the positive outer foil, which you still had in your hand.

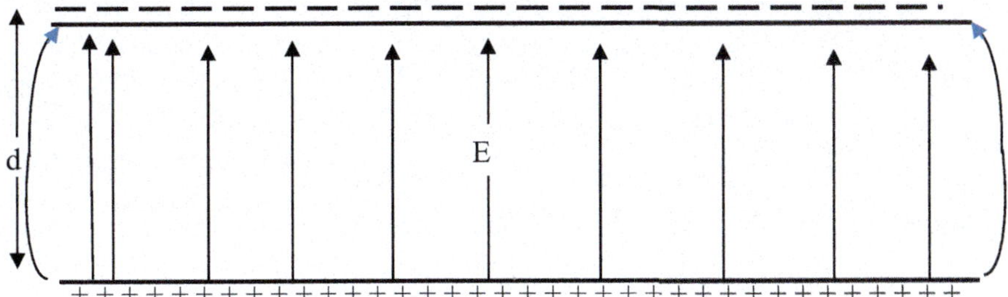

Fig. 7.14 Parallel plate capacitor

As a result, you felt a "shock" from the electrons. When you repeated the experiment but waited at least an hour before touching the paper clip with your finger, you still got the same shock.

Why? The jar was *storing* the charge. If you charge a conductor with an excess charge, that charge will generally dissipate into the air after some time. However, this charge did not. Why?

Because the positive outer foil attracted the negative charges in the inner foil, keeping them there. The Leyden jar, then, is a means of storing charge.

Capacitors are defined as devices which store charge. The Leyden jar was the first capacitor. It stores charge by having a negatively charged conductor separated from a positively charged conductor by an insulator. Today's capacitors do the same thing, but typically with a different geometry. Figure 7.14 illustrates a common type of capacitor, the **parallel plate capacitor**. This consists of a negatively charged plate and a positively charged plate separated by a space filled with either air or an insulator material. The electric field lines shown in the figure are uniform except at the ends where fringing effects occur. Although parallel plate capacitors are quite common, capacitors come in many other configurations such as cylinders.

The strength of an electric field in a capacitor is directly proportional to the amount of charge stored on its plates. The more the charge the greater the field. In the next chapter, we will learn that a capacitor is "charged" by putting it in a circuit where current can flow. A charge in the current trying to flow onto the plates of the capacitor will experience two forces. First the charges of the same sign on the plate the charge is going to will repulse it. Second the voltage driving the current will push it onto the plate. The charging will stop when the repulsive force exceeds the force of the potential difference driving the current. What is the maximum charge that can be stored by a specific capacitor? Obviously the greater the surface area A of the capacitor plates, the greater the amount of charge they can store. However, for many uses of capacitors, space is an issue so they can't just be made arbitrarily large. Another way to increase the charge is to decrease the separation between the plates since a closer distance increases the attractive force from the opposite plate.

The final aspect of a capacitor to consider is the material that is between the two conducting plates. This is some insulating material that does not conduct charge. It can just be air or some type of glass or plastic. The type of material that is used can change the properties of the capacitor. The insulators used for this purpose are called dielectrics. Figure 7.15 shows a picture of the electric field inside a dielectric. Even though the electrons are not free to move in an insulator, the molecules in the substance can orient themselves to become polarized electric dipoles as shown. The electric field associated with these dipoles is in the opposite direction from the field between the two charged plates. Thus the total effective field is less for the same amount of charge on the plates. There is a limit to how large the potential difference between the plates on a capacitor can get. If it gets too large dielectric breakdown will occur and the charges will move across the dielectric to the other plate. This is like a lightning stroke in the atmosphere. The breakdown voltage depends on a property called the dielectric strength of the material, and the separation between the two plates. The breakdown voltage for air is about 24 MV/m and is 100 times greater than this for mica.

These properties of a capacitor can be formulated into several mathematical expressions. First we can define the concept of capacitance C as the ability to hold charge. The larger the capacitance, the more charge that can be stored. From the discussion above for a parallel plate capacitor, capacitance can be defined as

$$C = \varepsilon A/d \tag{7.13}$$

Here ε is the permittivity of the dielectric. It has units of $C^2/N\,m^2$. Remember that Eq. 7.3 relates the permittivity of free space, $\varepsilon_0 = 8.85 \times 10^{-12}\,C^2/N\,m^2$, to the Coulomb constant, $K = 8.99 \times 10^9\,Nm/C^2$. The dielectric constant of the material κ is related to its permittivity through

$$\kappa = \varepsilon/\varepsilon_0 \tag{7.14}$$

It is a dimensionless quantity. The permittivity of a material indicates how well it supports an electric field. The units of capacitance are a Coulomb per volt which is called a farad (F) in honor of British scientist Michael Faraday.

To see how to use this equation, consider the following example.

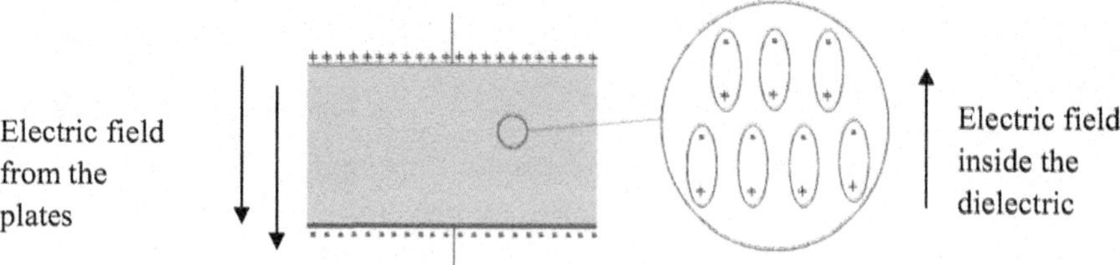

Fig. 7.15 Capacitor with a dielectric

Example 7.9

A physicist would like to make a parallel plate capacitor with a capacitance of 1.0 F using air as the dielectric. If he can place the plates 1.0 mm apart, how big do the plates have to be?

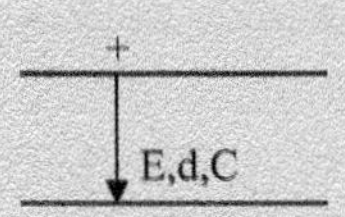

Knowns: $C = 1.0\text{F}$; $d = 1.0$ mm
Unknowns: A

We know the equation for capacitance, and the only unknown in the equation is A, so we

$$C = \varepsilon A/d = 1.0\,\text{F} = \left(8.85 \times 10^{-12}\,\text{C}^2/\text{Nm}^2\right)A/(0.0010\,\text{m})$$

$$A = (1.0\,\text{C/V})(0.0010\,\text{m})/\left(8.85 \times 10^{-12}\,\text{C}^2/\text{Nm}^2\right) = 1.1 \times 10^8\,\text{m}^2$$

Notice how the units work out. When you substitute J/C in for Volt and J for N m, everything cancels except for m^2, which is an area unit. If the plates were square, each plate would have to be about 10,000 m on a side! That's what it would take to make a 1.0 F capacitor out of two plates spaced a millimeter apart. That should give you an idea of how big a farad really is! Using a dielectric with a permittivity two to ten times higher than would help but still require the size of the plates to be extremely large.

Now that we have an expression for capacitance, we can calculate the charge that can be stored in a capacitor. Obviously, the charge will depend on the capacitance. The more capacitance, the more charge that can be stored. Also, it will depend on the potential difference across the capacitor in the circuit charging it, ΔV. Thus, the higher the potential difference between the plates in a parallel plate capacitor, the more charge we can put on the plates. In other words:

$$Q = C\Delta V \tag{7.15}$$

This can be used to calculate the charge stored on a capacitor. Notice that the equation tells us what we already have reasoned out. For a constant potential difference, a higher capacitance will result in a higher Q. For a constant capacitance, a higher potential difference will result in a higher value for Q. However, we can't increase ΔV to an arbitrarily high amount or dielectric breakdown will occur. Remember we said previously breakdown in air occurs at about 24×10^6 V/m.

Next solve the problems below.

Student

7.12 A parallel plate capacitor is made up of two square plates, each of which are 2.21 mm on a side. The distance between the plates is 0.100 mm. How much charge will be stored on the capacitor when it is connected to a 9.00 Volt battery?

7.13 A parallel plate capacitor has a capacitance of 1.20 microfarads. The plates are 1.00 mm apart. The capacitor is charged up so that each plate holds 1.50 mC of charge. Suppose an electron ($q = -1.6 \times 10^{-19}$ C, $m = 9.1 \times 10^{-31}$ kg) is placed right next to the negative plate and then released. When it reaches the positive plate, how fast will it be traveling?

Example 7.8 and problem 7.13 both demonstrated the properties of a charged particle in a uniform electric field such as the field found between the two plates of a capacitor. These examples can be used to discuss the properties of the electric field and electrical energy inside the capacitor. Imagine placing a positive charge next to the positively charged plate in a parallel plate capacitor and letting it go. The change in potential energy of the particle is given by Eq. 7.11 to be

$$\Delta \mathrm{PE} = -W$$

where W is the work done on the particle by the electric field.

Because we are working with a uniform electric field, we can use Eq. 7.10 to calculate the work done by the field on the particle

$$W = q \cdot E \cdot d \cdot \cos\theta$$

So the change in potential energy in moving the charge between the plates is

$$\Delta \mathrm{PE} = -q \cdot E \cdot d \cdot \cos\theta$$

Combining this with Eq. 7.6 gives

$$q\Delta V = -q \cdot E \cdot d \cdot \cos\theta$$

The charge q on each side of the equation cancels and $\theta = 180°$ so $\cos\theta = -1$. This equation then reduces to

$$\Delta V = Ed \tag{7.16}$$

This expression allows us to calculate the electric field in a parallel plate capacitor if we know the potential difference between the plates and the separation between the plates.

Example 7.10
What is the strength of the electric field between two plates of a 3.45 microfarad parallel plate capacitor if it holds 6.54 mC of charge on each plate? The plates are 15.0 cm apart.

+

E,d,C

-

Knowns: $C = 3.45\ \mu\text{f};\ d = 15.0\,\text{cm};\ Q = 6.54\,\text{mC}$

Unknowns: E

To get the electric field, we must know the potential difference. However, since we have C and Q, that is not hard.

$$\Delta V = Q/C = (0.00654\,\text{C})/\left(3.45 \times 10^{-6}\,\text{C/V}\right) = 1.90 \times 10^{3}\,\text{V}$$

The electric field can now be calculated:

$$\Delta V = Ed$$

$$E = \Delta V/d = \left(1.90 \times 10^{3}\,\text{V}\right)/(0.150\,\text{m}) = 1.27 \times 10^{4}\,\text{N/C}$$

Notice that the unit we get in the equation (V/m) can be rearranged to yield the standard unit for electric field (N/C). This is why you will see electric field reported in V/m from time to time.

Although a capacitor primarily stores charge, it also stores energy. After all, it takes work to charge up the plates on a capacitor. Imagine having two parallel plates, each of which is uncharged. If, one at a time, you took electrons off of the bottom plate and put them onto the top plate, you would slowly charge the top plate with a negative charge and the bottom plate with a positive charge. You would be doing *work* to charge the plates.

When you do work in this way, you will be storing up that energy in the capacitor.

Think, for example, about a spring. When you compress the spring, you must work to compress it. However, the energy that you expended in the work gets stored in the spring. Thus, the work becomes *potential energy*. It is the same with a capacitor. The work that you exert in charging the capacitor becomes the potential energy stored in the capacitor. The equation that tells you how much energy is stored in the capacitor is:

$$\text{PE} = Q^2/2C \tag{7.17}$$

Since C can be related to Q and ΔV according to Eq. 7.15, there are a couple of other ways we can express this equation:

$$\text{PE} = C\Delta V^2/2 \tag{7.18}$$

or

$$PE = Q\Delta V/2 \tag{7.19}$$

All three equations are valid, as they are all equivalent to one another.

Example 7.11

A parallel plate capacitor is made from two square plates which are 2.50 cm on each side. The distance between the plates is 1.5 mm. If it takes 4.0 x 10^{-10} J of work to charge the plates up, what is the potential difference between the two plates?

+

E,d,C

-

Knowns: $A = (2.50\,\text{cm})^2$; $d = 1.5$ mm;
$W = 4.0 \times 10^{-10}$
Unknowns: ΔV

The work necessary to charge the capacitor is equal to the energy stored in the capacitor. We are also given enough information to calculate the capacitance:

$$C = \varepsilon_0 A/d = \left(8.85 \times 10^{-12}\,\text{C}^2/\text{Nm}^2\right)(0.0250\,\text{m})(0.0250\,\text{m})/0.0015\,\text{m}$$
$$= 3.7 \times 10^{-12}\,\text{C/V}$$

Since we know the energy and the capacitance, we can use Eq. 7.17 to calculate the potential difference.

$$PE = C\Delta V^2/2$$

$$\Delta V = \sqrt{\frac{2PE}{C}} = \sqrt{\frac{2(4.0 \times 10^{-10}\,\text{J})}{3.7 \times 10^{-12}\,\text{C/V}}} = 15\,\text{V}$$

In the next chapter, we will discuss the use of capacitors in electrical circuits. For now try these last three problems.

Student

7.14 A physicist wants to use a 500.0 microfarad parallel plate capacitor to generate a uniform electric field of 75.0 N/C. If the plates are 15.0 cm apart, what voltage must he use to charge the plates?

7.15 How much energy is stored in that capacitor once the physicist gets it charged?

7.16 As mentioned in the previous section, dielectric breakdown of air occurs at about 24 million Newtons per Coulomb. What is the maximum charge that could be stored in the capacitor in problem 7.14?

Finally, let's consider what happens to excess electrical charges in a conducting material. Since like charges repel each other, an excess of one type of charge inside a conductor where they are mobile will cause them to move as far away from each other as possible. This means the excess charge will move to the surface of the conductor. When they reach equilibrium, they will produce an electric field surrounding the conductor on its outside. The electric field lines will be perpendicular to the surface of the conductor. However, inside the conductor their fields will cancel each other. Thus, there is no electric field inside a conductor. A spherical conductor with excess charge on the surface emits an electric field that acts as if all of the excess charge were concentrated at the center of the sphere.

Next Level
Millikan Oil Drop Experiment

The charge on an electron was first determined by Robert Millikan in 1909 by what is known as the Millikan Oil Drop Experiment. He used a parallel plate capacitor where he could control the electric field between the plates $E = V/d$. Then he sprayed ionized oil drops between the plates of the capacitor and watched them fall. Measuring their terminal velocity in free fall ($E = 0$) and knowing the density of the spherical oil drops, he could determine their mass. Then he turned the field of the capacitor on until the drops stopped falling. At this point the gravitational force equals the electrostatic force $F = qE$ so knowing F and E he could determine the charge on the particle q. What he found was that all of the particles had a charge that was equal to an integer multiple of 1.602×10^{-19} C. Because of this he reasoned that the charge on a single electron must be 1.602×10^{-19} C.

Special Topic
Originally, the screens for televisions, computers, and pieces of scientific equipment were made from cathode ray tubes (CRTs). These operated by having a source shoot a beam of electrons (cathode ray) at a screen containing material that absorbed the electrons and gave off light. The beam passed between the plates of a capacitor in both the x- and y-directions to scan its position so it hit different points on the screen. This allowed it to form an image that would be rapidly refreshed so that it appeared to the eye

to be moving smoothly. CRTs worked pretty well for many years, but their design required that they be over a foot in length to operate at the required definition. This is why old televisions and computers were so bulky. New display technology using liquid crystals or light emitting diodes (see Chap. 14) allows screens to be less than an inch thick. This has led to very thin portable computers and wall mounted televisions.

Summing Up

In this chapter you learned about the forces between particles that have positive and negative electrical charges. You also learned about electric fields and the important differences between electric potential and potential energy. Two important things you learned about electric charge is that charge is conserved, and it can be stored in a capacitor. In the next chapter you will find out how capacitors are used in electric circuits, and in the final chapter you will learn more about the fundamental nature of electrical charge.

Answers to the Student Problems

7.1

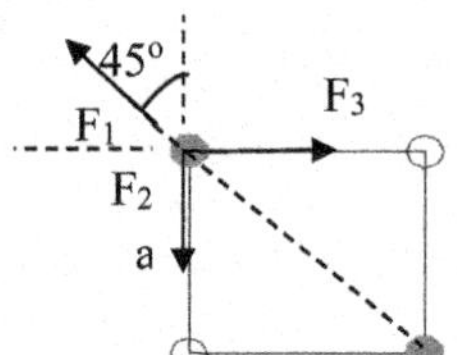

Knowns: $q_g = -1.0\,\text{mC}$; $F_g = 0\,\text{N}$; q_w equal;
$q_t = +2.5\,\text{mC}$ at center
Unknowns: q_w; F_t.

a. Concentrate on the forces affecting one of the gray spheres. The other gray sphere will repel it, and the two white spheres will attract it. The net force must be zero. In order to make the sum of these three vectors equal to zero, the vector pulling straight down ($\mathbf{F}_2$) must counteract the *vertical component* of the repulsive force from the gray sphere ($\mathbf{F}_1$). The distance between the gray sphere and the white ones is a. Since the length of the diagonal is $\sqrt{2}a$, the distance from the gray sphere to the other gray sphere is $\sqrt{2}a$. The diagonal forms a 45-45-90 triangle, so the angle between $\mathbf{F}_1$ and the vertical is 45°. Defined properly for vector addition, however, it is 135°. Since $\mathbf{F}_2$ must counteract the vertical component of $\mathbf{F}_1$:

$$\mathbf{F}_{1y} + \mathbf{F}_2 = \mathbf{0} = \mathbf{k}\mathbf{q}_g\mathbf{q}_g \sin 135°/(\sqrt{2}a)^2 + \mathbf{k}\mathbf{q}_g\mathbf{q}_w/(a)^2$$

$$q_w = -q_g \sin 135°/2 = -\left(-1.0\,\text{s}10^{-3}\,\text{C}\right) \sin 135°/2 = 0.35\,\text{mC}$$

The charge on the bottom white sphere, then, is 0.35 mC. Using the common convention that up is positive, $\mathbf{F}_{1y}$ would be positive and $\mathbf{F}_2$ would be negative. That's where the negative sign comes from in the equation. You can do the same thing for the other white sphere, noting that it must cancel out the horizontal component of $\mathbf{F}_1$. However, since everything is symmetric, the result will be the same. Thus, the white spheres each have a 0.35 mC charge.

b. This one is easy. At the very center of the square, the two positive charges would repel it equally, and the two negative charges would attract it equally. Thus, the net force would be zero.

7.2

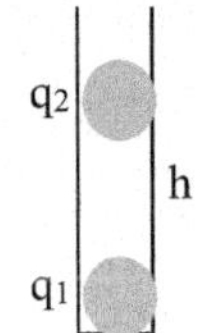

Knowns: $m = 2.2\,\text{gm};\ q_1 = +0.50\,\mu\text{C};\ q_2 = +0.25\,\mu\text{C}$

Unknowns: h

In this situation, the top ping pong ball will come to rest when the force due to gravity ($m \cdot g$) is balanced by the electrical repulsion:

$$F_e + F_g = 0$$

$$kq_1q_2/r^2 = mg$$

$$r = h = \sqrt{\frac{kq_1q_2}{mg}} = \sqrt{\frac{(8.99 \times 10^9\,\text{Nm}^2/\text{C}^2)(5.0 \times 10^{-7}\,\text{C})(2.5 \times 10^{-7}\text{C})}{(0.0022\,\text{kg})(9.81\,\text{m/s}^2)}}$$
$$= 0.23\,\text{m}$$

Since "r" is defined as the distance between the centers of the charges, then $r = h$. Thus, the height is 23 cm.

7.3 Since the positive particle has twice as much charge, it must have twice as many field lines going out of it as the negative charge has going into it. Thus, you get a drawing that looks like the one on the right.

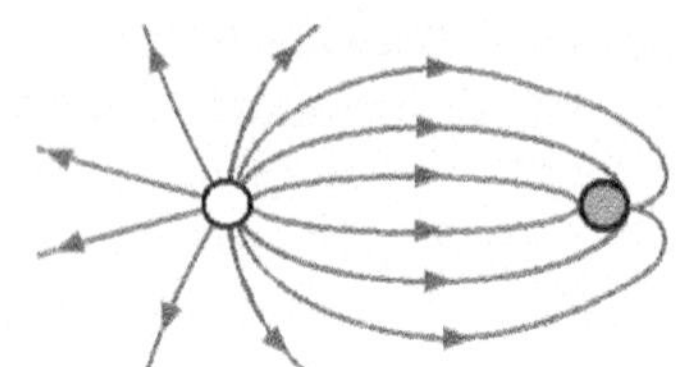

7.4 Since the charges are all the same, the negative charge must have the same number of lines entering it as the positive charges have leaving them. Since there are two positive charges, however, that means only half of the lines leaving each positive charge can end up entering the negative charge.

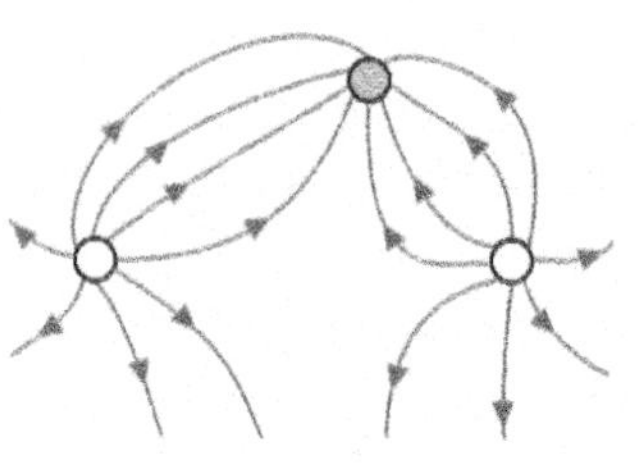

7.5

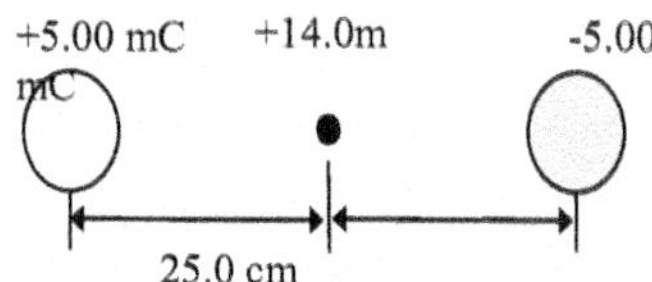

Knowns: $q_{1,2} = \pm 5.00$ mC;
$r = 25.0$ cm; $q_3 = +\ 14.0$ mC
Unknowns: E; F

This problem is not so hard, since the setup makes it one-dimensional.

At the midpoint between the charges, the positive charge creates an electric field pointed to the right (repelling another positive charge), and the negative charge produces an electric field also pointed to the right (attracting a positive charge). Thus, the two electric fields just add together:

$$\mathbf{E_{tot}} = \mathbf{E}_{+5} + \mathbf{E}_{-5}$$

$$\begin{aligned} E_{\text{tot}} &= kq + 5/s^2 + kq_- - r^2 \\ &= \left(8.99 \times 10^9\,\text{Nm}^2/\text{C}^2\right)(0.00500\,\text{C})/(0.25\,\text{m})^2 \\ &\quad + \left(8.99 \times 10^9\,\text{Nm}^2/\text{C}^2\right)(0.00500\,\text{C})/(0.25\,\text{m})^2 = 1.44 \times 10^9\,\text{N/C} \end{aligned}$$

The electric field, then, is 1.44×10^9 N/C pointed to the right.

The problem also wants to know the force felt by a 14.0 mC charge placed there. That's an easy application of Eq. 7.4:

$$\mathbf{F} = q \cdot \mathbf{E} = (0.0140\,\text{C}) \cdot \left(1.44 \times 10^9\,\text{N/C}\right) = 2.02 \times 10^7\,\text{N}$$

Remember, in this equation, direction is taken into account by sign. Thus, the force felt is 2.02×10^7 N to the right.

7.6

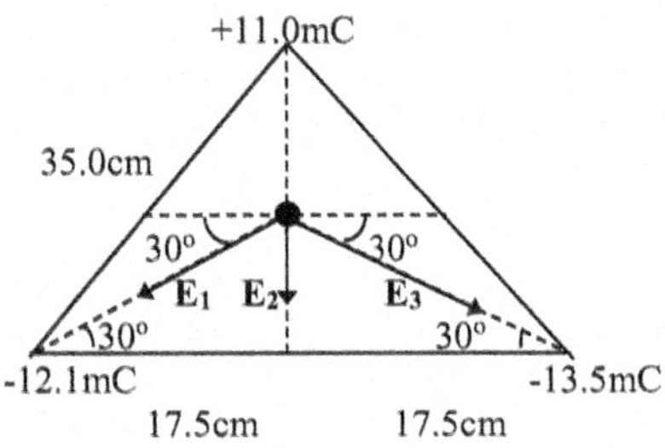

Knowns: 3 charges in an equilateral triangle as shown

Unknowns: E at center

This is more difficult than the previous problem only because we must now add the electricfields in two dimensions. Therefore, we first have to do some geometry. Let's work on distances first. Because the triangle is equilateral, the height of the triangle (the vertical dashed line) bisects the base. Thus, the vertical dashed splits the base into two 17.5 cm segments. The center lies somewhere on that line, but where? If we draw a line from any vertex to the center, it should bisect the 60° angle of that vertex. Thus, the angle from the base of the triangle to the dashed line from the − 12.1 mC charge to the center is 30°. Notice that we now have a right triangle with one leg that is 17.5 cm long, and a 30° angle adjacent to it. We can figure out the hypotenuse of that triangle, which is the distance between the center of the triangle and the − 12.1 mC charge:

$$\text{hyp} = \text{adj}/\cos\theta = 17.5\,\text{cm}/\cos 30^\circ = 20.2\,\text{cm}$$

Since this is an equilateral triangle, all charges must be equidistant from the center. Thus, all distances are 20.2 cm.

What about angles? $\mathbf{E}_2$ is the easiest to figure out. It is pointed straight down, which is an angle of 270°. For $\mathbf{E}_1$ and $\mathbf{E}_3$, you have to remember that the interior angles of an equilateral triangle are all 60°. A line drawn from the center of the triangle to the vertex (the diagonal dashed lines) bisect this angle, so the diagonal dashed lines form 30° angles with the base. When two parallel lines are cut by a transversal, alternating interior angles are congruent. Thus, $\mathbf{E}_1$ and $\mathbf{E}_3$ each make a 30° angle below the horizontal. Converting to the proper definition of angles for vector addition, $\mathbf{E}_1$ has an angle of 210° while $\mathbf{E}_3$ has an angle of 330°. Please note that *all of these angles are exact*, as they all come from geometry definitions. There is no measurement involved. Thus, they do not contribute to the significant figures determination.

Now we are finally ready to do the vector addition.

$$E_{1x} = \left(kQ/r^2\right)\cos\theta = \frac{\left(8.99\times 10^9\,\mathrm{Nm^2\,m^2/C^2}\right)(0.0121\,\mathrm{C})}{(0.202\,\mathrm{m})^2}\cos 210^\circ$$
$$= -2.31\times 10^9\mathrm{N/C}$$

$$E_{1y} = \left(kQ/r^2\right)\sin\theta = \frac{\left(8.99\times 10^9\,\mathrm{Nm^2\,m^2/C^2}\right)(0.0121\,\mathrm{C})}{(0.202\,\mathrm{m})^2}\sin 210^\circ$$
$$= -1.33\times 10^9\,\mathrm{N/C}$$

$$E_{2x} = \left(kQ/r^2\right)\cos\theta = \frac{\left(8.99\times 10^9\,\mathrm{Nm^2\,m^2/C^2}\right)(0.0110\,\mathrm{C})}{(0.202\,\mathrm{m})^2}\cos 270^\circ$$
$$= 0\,\mathrm{N/C}$$

$$E_{2y} = \left(kQ/r^2\right)\sin\theta = \frac{\left(8.99\times 10^9\,\mathrm{Nm^2\,m^2/C^2}\right)(0.0110\,\mathrm{C})}{(0.202\,\mathrm{m})^2}\sin 270^\circ$$
$$= -2.42\times 10^9\,\mathrm{N/C}$$

$$E_{3x} = \left(kQ/r^2\right)\cos\theta = \frac{\left(8.99\times 10^9\,\mathrm{Nm^2\,m^2/C^2}\right)(0.0135\,\mathrm{C})}{(0.202\,\mathrm{m})^2}\cos 330^\circ$$
$$= 2.58\times 10^9\,\mathrm{N/C}$$

$$E_{3y} = \left(kQ/r^2\right)\sin\theta = \frac{\left(8.99\times 10^9\,\mathrm{Nm^2\,m^2/C^2}\right)(0.0135\,\mathrm{C})}{(0.202\,\mathrm{m})^2}\sin 330^\circ$$
$$= -1.49\times 10^9\,\mathrm{N/C}$$

Now we just add the components together:

$$E_x = -2.31\times 10^9\,\mathrm{N/C} + 2.58\times 10^9\,\mathrm{N/C} = 2.7\times 10^8\,\mathrm{N/C}$$
$$E_y = -1.33\times 10^9\,\mathrm{N/C} - 2.42\times 10^9\,\mathrm{N/C} - 1.49\times 10^9\,\mathrm{N/C}$$
$$= -5.24\times 10^9\,\mathrm{N/C}$$

$$E = \sqrt{(2.7\times 108\,\mathrm{N/C})^2 + (5.24\times 109\,\mathrm{N/C})^2} = 5.25\times 10^9\,\mathrm{N/C}$$

$$\theta = \tan^{-1}\left[\frac{-5.24\times 10^9\,\mathrm{N/C}}{2.7\times 10^8\,\mathrm{N/C}}\right] = -87^\circ$$

Since the vector has a positive x-component and a negative y-component, it is in the fourth Cartesian quadrant. That means we must add 360.0° to the answer. So the electric field is 5.25×10^9 N/C directed at 273°.

7.7

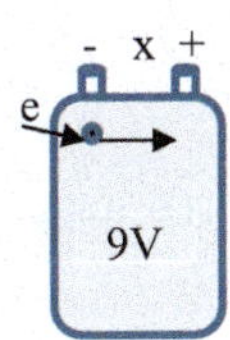

Knowns: $q = -1.6 \times 10^{-19}\,\text{C}$; $m = 9.1 \times 10^{-31}\,\text{kg}$; $V = 9.0\,\text{V}$;
$v_i = 0\,\text{m/s}$

Unknowns: v_f

Since it travels from one side of the battery to the other, it experiences a potential difference of 9.0 V. Thus, it experiences the following change in potential energy:

$$\Delta\text{PE} = q\Delta V = \left(-1.6 \times 10^{-19}\,\text{C}\right) \cdot (9.0\,\text{J/C}) = -1.4 \times 10^{-18}\,\text{J}$$

Since the potential energy decreased, the kinetic energy increased. It had no kinetic energy to start with, so its final kinetic energy is 1.4×10^{-18} J. We can use that to get the speed:

$$\text{KE} = mv_\text{f}^2/2$$

$$v_\text{f} = \sqrt{\frac{2\text{KE}}{m}} = \sqrt{\frac{2\left(1.4 \times 10^{-18}\,\text{J}\right)}{9.1 \times 10^{-31}\,\text{kg}}} = 1.8 \times 10^{6}\,\text{m/s}$$

Notice that's almost 0.5% the speed of light!

7.8

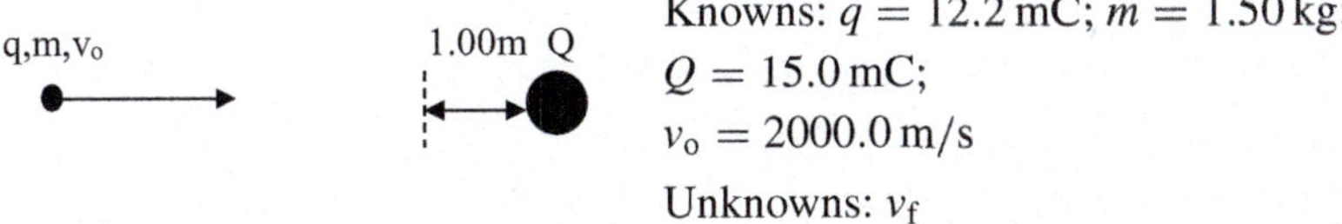

Knowns: $q = 12.2\,\text{mC}$; $m = 1.50\,\text{kg}$;
$Q = 15.0\,\text{mC}$;
$v_\text{o} = 2000.0\,\text{m/s}$

Unknowns: v_f

If the particle starts far from the charge, then the initial potential is zero. When it is 1.00 m from the charge, the potential is:

$$V = kQ/r = \left(8.00 \times 10^{9}\,\text{Nm}^2/\text{C}^2\right)(0.0150\,\text{C})/(1.00\,\text{m}) = 1.35 \times 10^{8}\,\text{V}$$

The potential difference, then, is:

$$\Delta V = 1.35 \times 10^{8}\,\text{Volts} - 0\,\text{Volts} = 1.35 \times 10^{8}\,\text{Volts}$$

Now we can calculate the change in potential energy:

$$\Delta\text{PE} = q\Delta V = (0.0122\,\text{C}) \cdot (1.35 \times 10^8\,\text{V}) = 1.65 \times 10^6\,\text{J}$$

This tells us that the potential energy increased. The only way that can happen is for the kinetic energy to decrease. Thus, the kinetic energy decreased by 1.65×10^6 J. We will have to subtract that from the original kinetic energy:

$$\text{KE}_\text{f} = mv_\text{o}^2/2 - 1.65 \times 10^6\,\text{J} = (1.50\,\text{kg})(2000.0\,\text{m/s})^2/2$$
$$- 1.65 \times 10^6\,\text{J} = 1.35 \times 10^6\,\text{J}$$

The final speed is

$$v_\text{f} = \sqrt{\frac{2\text{KE}_\text{f}}{m}} = \sqrt{\frac{2(1.35 \times 10^6\,\text{J})}{1.50\,\text{kg}}} = 1340\,\text{m/s}$$

The particle slowed down, of course, because the stationary charge repels it. Eventually, the particle would stop and turn around, accelerating away from the stationary charge.

7.9

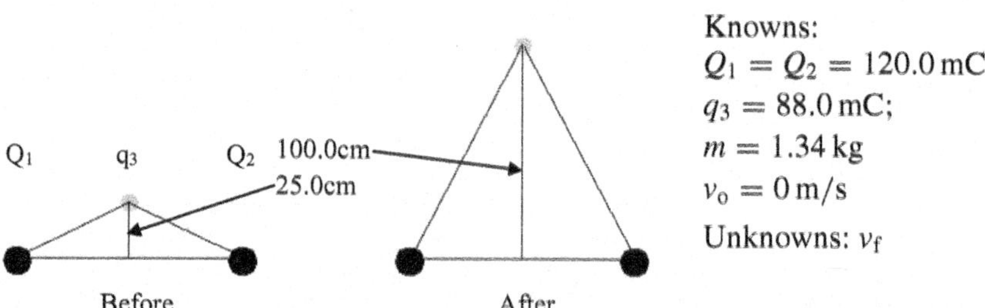

You should recognize that the particle will travel straight up. Both stationary charges repel it, and the horizontal components of that repulsion will cancel out because of the symmetry of the situation. Thus, the particle accelerates straight upward.

To determine the speed, all we need to do is figure out the potential difference that the Particle experiences. To do that, we need the distance from the charges. That's not hard, however, since the charges form right triangles. Before the particle is released, the right triangles each have legs of 40.0 cm and 25.0 cm. Thus, by the Pythagorean theorem, the distance between the stationary charges and the moving charge is 47.2 cm. At the end, the two legs are 40.0 cm and 100.0 cm, making the charges 108 cm apart. Those are the distances we use in calculating the potential.

$$\begin{aligned} V_i &= V_1 + V_2 = kQ_1/r_1 + kQ_2/r_2 \\ &= \left(8.99 \times 10^9\,\mathrm{Nm^2/C^2}\right)(0.1200\,\mathrm{C})/(0.472\,\mathrm{m}) \\ &\quad + \left(8.99 \times 10^9\,\mathrm{Nm^2/C^2}\right)(0.1200\,\mathrm{C})/(0.472\,\mathrm{m}) = 4.57 \times 10^9\,\mathrm{V} \end{aligned}$$

$$\begin{aligned} V_\mathrm{f} &= V_1 + V_2 = kQ_1/r_1 + k_2/r_2 \\ &= \left(8.99 \times 10^9\,\mathrm{Nm^2/C^2}\right)(0.1200\,\mathrm{C})/(1.08\,\mathrm{m}) \\ &\quad + \left(8.99 \times 10^9\,\mathrm{Nm^2/C^2}\right)(0.1200\,\mathrm{C})/(1.08\,\mathrm{m}) \\ &= 2.00 \times 10^9\,\mathrm{V} \end{aligned}$$

The particle, then, experiences a change in potential of:

$$\Delta V = 2.00 \times 10^9\,\mathrm{Volts} - 4.57 \times 10^9\,\mathrm{Volts} = -2.57 \times 10^9\,\mathrm{Volts}$$

Now we can calculate the change in potential energy:

$$\begin{aligned} \Delta \mathrm{PE} &= q\Delta V \\ &= (0.0880\,\mathrm{C}) \cdot \left(-2.57 \times 10^9\,\mathrm{J/C}\right) = -2.26 \times 10^8\,\mathrm{J} \end{aligned}$$

That decrease in potential energy ends up increasing the kinetic energy by the same amount.

Since the particle had no kinetic energy to begin with, that means the final kinetic energy is 2.26×10^8 J.

$$\mathrm{KE_f} = mv_\mathrm{f}^2/2$$

$$v_\mathrm{f} = \sqrt{\frac{2\mathrm{KE_f}}{m}} = \sqrt{\frac{2\left(2.26 \times 10^8\,\mathrm{J}\right)}{1.34\,\mathrm{kg}}} = 1.84 \times 10^4\,\mathrm{m/s}$$

7.10 Looking at the figure, the best way to think about this is to think about how the charges will all accelerate in the electric field. They are all negative, so they will travel opposite the electric field. Thus, they will travel to the right. The ones that can travel farthest will be worked on by the field more than those that do not travel as far. Since the electric field's work will reduce potential energy, the particles that will be worked on more must have more potential energy. Particle a, therefore, has the highest potential energy.

7.11

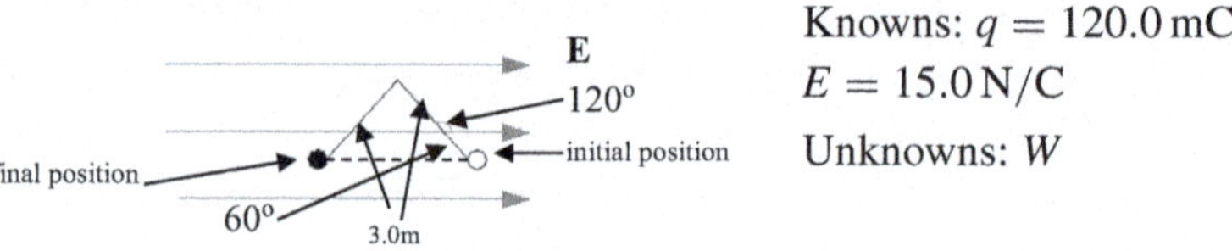

You can do this the hard way or the easy way. In the easy way you connect a line from the particle to its ending location, so you have an equilateral triangle. You know this because the two sides shown in the diagram are the same size, and since the angle shown in the diagram is 120°, the interior angle on that leg is 60°. Isosceles triangles have equal angles on the equal sides, so the other angle is 60° as well. Since all interior angles add to 180, the last angle must also be 60, and thus we have an equilateral triangle. Therefore, the leg drawn from the particle to its ending location is also 3.0 m. Since the electrostatic force is conservative, the path does not matter, so we choose the path where the particle travels straight against the electric field ($\theta = 180°$).

$$W = qEx\cos\theta = (0.1200\,\text{C}) \cdot (15.0\,\text{N/C}) \cdot (3.0\,\text{m}) \cdot \cos\left(180°\right) = -5.4\,\text{J}$$

What does the negative mean? It means that work was done against the field. That makes sense, since a positive particle is accelerated in the opposite direction by the electric field. Thus, the physicist worked against the field.

7.12

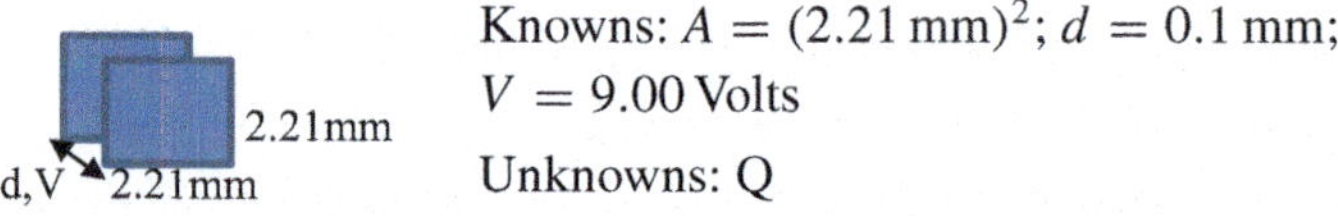

The first part of the problem gives us enough to determine the capacitance:

$$\begin{aligned} C = \varepsilon_0 A/d &= \left(8.85 \times 10^{-12}\,\text{C}^2/\text{Nm}^2\right)(0.00221\,\text{m})^2/(0.000100\,\text{m}) \\ &= 4.32 \times 10^{-13}\,\text{C/V} \end{aligned}$$

If it is hooked up to a 9.00 Volt battery, it has a potential difference of 9.00 Volts. Therefore:

$$\Delta V = Q/C \quad Q = CV = \left(4.32 \times 10^{-13}\,\text{C/V}\right)(9.00\,\text{V}) = 3.89 \times 10^{-12}\,\text{C}$$

That may not sound like a lot of charge, but it means approximately 2×10^7 extra electrons are on the negative plate!

7.13

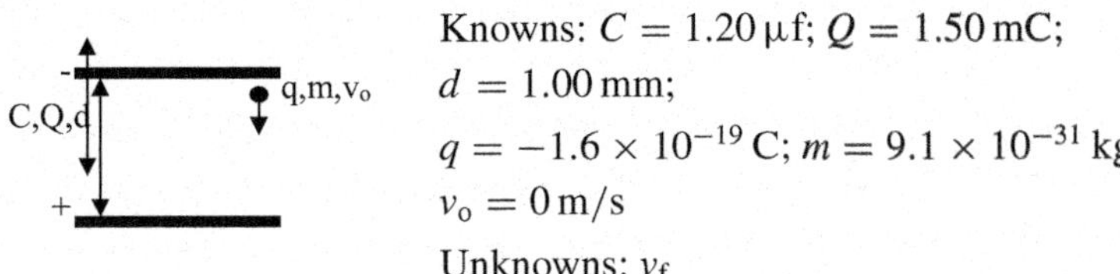

Knowns: $C = 1.20\,\mu\text{f}$; $Q = 1.50\,\text{mC}$;
$d = 1.00\,\text{mm}$;
$q = -1.6 \times 10^{-19}\,\text{C}$; $m = 9.1 \times 10^{-31}\,\text{kg}$;
$v_o = 0\,\text{m/s}$
Unknowns: v_f

This problem is a lot easier than it sounds. The distance between the plates was just put in to fool you. Since we know the capacitance and charge, we can calculate the potential difference between the plates:

$$\Delta V = Q/C$$

$$\Delta V = 0.00150\,\text{C}/1.20 \times 10^{-6} = 1250\,\text{V}$$

Now that we know the potential difference, the change in potential energy can be calculated:

$$\Delta\text{PE} = q\Delta V = \left(-1.6 \times 10^{-19}\,\text{C}\right) \cdot (1250\,\text{J/C}) = -2.0 \times 10^{-16}\,\text{J}$$

That decrease in potential energy ends up increasing the kinetic energy by the same amount.

Since the electron had no kinetic energy to begin with, that means the final kinetic energy is 2.0×10^{-16} J:

$$\text{KE}_f = mv_f^2/2$$

$$v_f = \sqrt{\frac{2\text{KE}_f}{m}} = \sqrt{\frac{2\left(2.0 \times 10^{-16}\,\text{J}\right)}{9.1 \times 10^{-31}\,\text{kg}}} = 2.1 \times 10^7\,\text{m/s}$$

That's almost 10% the speed of light! It turns out that when an object is moving at relativistic speeds, the above equation for kinetic energy is not quite right. Thus, this number is not quite accurate.

7.14

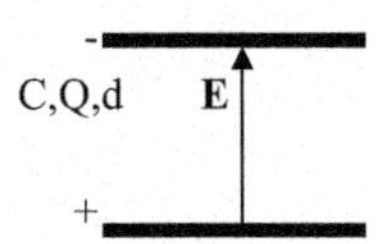

Knowns: $C = 500.0\,\mu\text{f}$; $E = 75.0\,\text{N/C}$; $d = 15.0\,\text{cm}$
Unknowns: V

The capacitance is not important for this problem. To determine the potential difference, we need only know the electric field and the distance between the plates:

$$\Delta V = Ed = (75.0\,\mathrm{N/C}) \cdot (0.150\,\mathrm{m}) = 11.3\,\mathrm{J/C}$$

Since a Volt is a J/C, the potential difference must be 11.3 Volts.

7.15 See problem 7.14. In this case the unknown is the energy stored in the capacitor.

We need to know the capacitance for *this* problem. Since we already know ΔV, and since the capacitance was given in the previous problem, the most reasonable equation to use is Eq. 7.17.

$$\mathrm{PE} = C(\Delta V)^2/2 = \left(5.000 \times 10^{-4}\,\mathrm{C/V}\right)(11.3\,\mathrm{V})^2 = 0.0319\,\mathrm{J}$$

7.16 See problem 7.14. Breakdown voltage for air is 24×10^6 N/C. The unknown is now Q_{max}.

We can't store charge that creates an electric field larger than the dielectric breakdown of the electric field. Thus, the strongest electric field we can have is 24 million N/C. That would lead to the following potential difference.

$$\Delta V = Ed = \left(2.4 \times 10^7\,\mathrm{N/C}\right) \cdot (0.150\,\mathrm{m}) = 3.6 \times 10^6\,\mathrm{V}$$

That potential difference would give the following charge.

$$\Delta V = Q/C$$

$$Q = \Delta VC = \left(3.6 \times 10^6\,\mathrm{V}\right) \cdot \left(5.000 \times 10^{-4}\,\mathrm{C/V}\right) = 1800\,\mathrm{C}$$

Study Guide for This Chapter

1. The electric potential at a given point is zero. Is the electric field necessarily zero? Why or why not?
2. A negative charge is placed in an electric field and released. If it is accelerated by the electric field only, will it move towards a higher or a lower electric potential?
3. Negative work is done on a positive particle as it moves in an electric field. Is the positive particle moving in the direction of the electric field or opposite the electric field?

4. A physicist makes a parallel plate capacitor, but it does not have a high enough capacitance. There are three things he could do to increase the capacitance. What are they?
5. A physicist hooks a capacitor up to a 1.5-Volt battery and measures the charge stored on the capacitor as q. The physicist then hooks the same capacitor up to a 9.0-Volt battery and measures the charge stored as Q. What is Q in terms of q?
6. Suppose that the physicist in question #5 also measured the energy stored in the capacitor. When it was hooked up to the 1.5-Volt battery, the energy was e. When it was hooked up to the 9.0-Volt battery, the energy was E. What is E in terms of e?
7. The electrostatic force between two charged objects is measured. If the distance between the charges is then quadrupled and the charge on each object is doubled, what is the new electrostatic force compared to the old one?
8. A negatively charged particle is fixed so that it cannot move. A positively charged particle is placed in the vicinity and then released. Describe how both the velocity and the acceleration of the positively charged particle changes with time.
9. A negatively charged particle of mass 34.5 g is placed above a large, conductive sheet that has been negatively charged. The electric field at the point over which the particle has been placed is 3.40×10^5 N/C. If the particle hangs at that point without rising or falling, what is the charge on the particle?

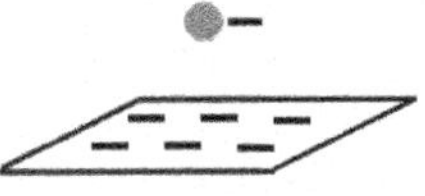

10. Three charges are shown below. The charge labeled "$-q$" is a distance "x" away from the charge "$-2q$," which is a distance "y" from the charge "$-3q$." The charges all lie on a straight line. If the charge labeled "$-2q$" experiences no force acting on it, what is the value of x/y?

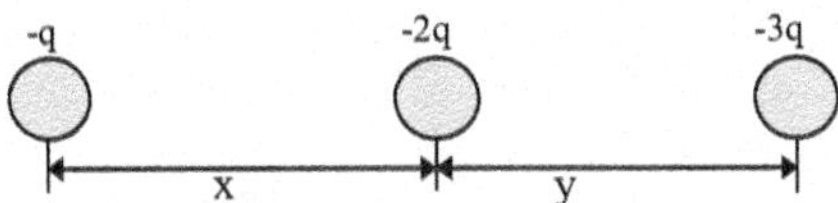

11. Referring once again to problem 10, suppose the charge labeled "$-2q$" was removed and nothing was put in its place. Also, suppose $q = 3.0$ μC. If the distance between the charge labeled "$-q$" and "$-3q$" is 100.0 cm, what is the electric field at the midpoint between the charges?

12. Three charges, each − 4.50 μC, are placed on the vertices of a right triangle whose legs are each 15.0 cm long. What is the magnitude and direction of the electric field at a point that is 4.00 cm from each leg?

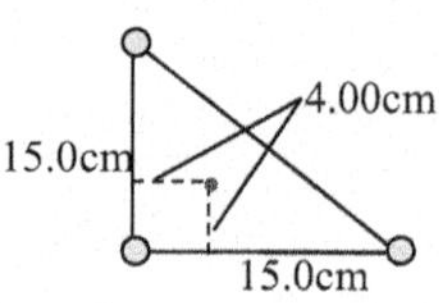

13. If a − 1.50 μC charge were placed at the point shown in problem 12, what electrostatic force would it experience?

14. What is the work required to assemble the system shown to the right?

15. What is the electric potential at the center of the square pictured to the right?

16. What is the electric field at the center of the square pictured to the right?

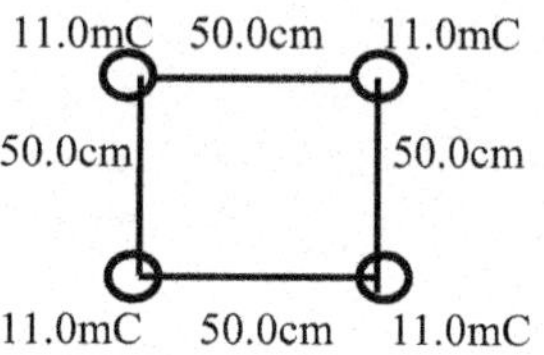

17. A negatively charged particle ($q = -34.5$ mC) is being pushed in the 45.6 N/C electric field shown below. If the particle is forced to travel in a semicircle whose radius of curvature is 15.0 cm, how much work will have been done when it reaches the end of the semicircle?

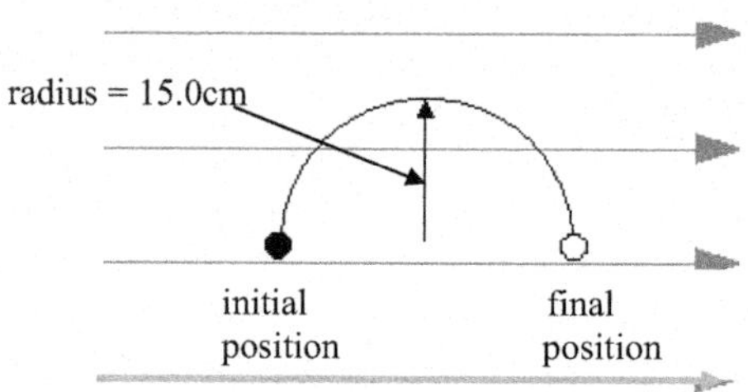

18. A + 3.8 mC charged particle ($m = 5.0$ kg) is shot with an initial velocity of 245 m/s towards a + 1.5 mC stationary charge. If the particle starts out 1.2 m from the stationary charge, how close will it come to the charge before turning around and moving away?

Next Level

19. If you repeat Millikan's oil drop experiment in air with a density of 1.23 kg/m^3 on oil drops with a cross-sectional area of 2×10^{-3} m^2 and a drag coefficient of 0.5, how many electrons are on a drop with a terminal velocity of 2.0 m/s that is brought to a complete stop with an electric field of 7.93×10^{15} N/C.
20. If the oil drop in problem 19 and twice as many electrons on it, what field would it take to stop it and what would be its terminal velocity?

Electrical Circuits 8

Chapter Summary

To put electricity to work for us, we need to generate currents of flowing charges in circuits with components that are designed to convert electric energy into thermal or mechanical energy. These circuits come in many sizes. They can be miniature circuits boards used in our microelectronic gadgets or large generators powering heating and air conditioning systems. This chapter focuses on the analysis of electrical circuits and their basic components. Some of the themes of the chapter are represented in Figs. 8.1, 8.2, 8.3 and 8.4. Figure 8.1 shows a typicle circuit board. Figure 8.2 shows a solonoid. Figure 8.3 shows a set of resistors. Figure 8.4 shows a pacemaker that controls a human heart beat.

Main Concepts in This Chapter

- Electrical Conduction
- DC Circuits with Resistors
- Other Circuit Components
- AC Current

8.1 Introduction

Today's civilization depends on electricity to function. Major power plants are used to generate electricity which is transmitted on power lines to our homes and to industries where it is used for heating and cooling, cooking, and powering

R. C. Powell, *Physics*, Undergraduate Lecture Notes in Physics,
https://doi.org/10.1007/978-3-032-09361-5_8

Fig. 8.1 Printed circuit board. *Credit* Balurbaba, Creative Commons Attribution-Share Alike 3.0

Fig. 8.2 Copper wire solenoid. *Credit* tony_duel, Creative Commons Attribution-Share Alike 2.0

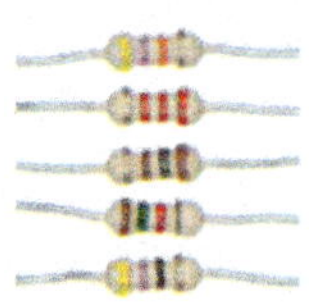

Fig. 8.3 Resistors with color coded resistance. *Credit* Evan-Amos, public domain

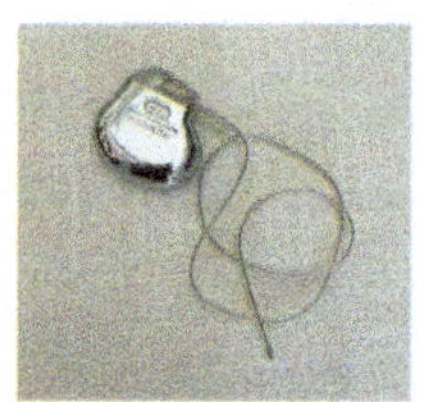

Fig. 8.4 Heart pacemaker. *Credit* Brian Adducci, Creative Commons Attribution-Share Alike 4.0

hundreds of different types of devices. Many small electrical devices depend on batteries for their electrical power. Our lifestyle as we know it today would not exist if we had not learned to generate and control electricity.

Everything that uses electricity depends on electrical currents controlled by electrical circuits. Circuits are made up of currents that are charged particles flowing in conductors. A typical current is made up of electrons flowing in wires made of copper or other metal conductors. The direction of current flow is defined as the direction of the flow of positive charges. Since the charge on an electron is negative, the common convention is to designate the direction of current to be opposite to the direction of the flow of electrons

There are two types of currents: alternating current (AC) and direct current (DC). AC current reverses its direction periodically while DC current continues

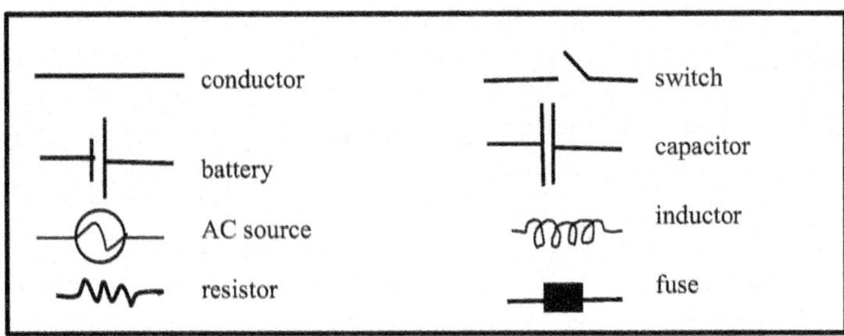

Fig. 8.5 Common circuit elements

in the same direction. It is easier to transmit high power electricity over long distances as AC current, so this what comes into your house and is available in electrical outlets. However, many electrical devices are designed to work with DC currents so they must use an AC to DC converter or batteries that generate DC current directly.

A circuit is made up of a number of circuit elements connected by conductors. Symbols designating some of the most basic circuit elements are shown (Fig. 8.5). The conductors connect the circuit elements and allow current to flow to them. They are shown as a straight line. A switch is an element that interrupts the flow of current, so it turns the circuit on or off. A battery generates a DC current by providing an electric potential in the circuit which causes the electrons to flow.

Special Topic

Throughout the 1880s and 1890s there was a scientific competition now known as the War of the Currents. The focus of this competition was to decide whether the electrical system that was going to be installed in the United States would be based on DC or AC current. Thomas Edison and his backers were pushing for the DC current solution while Nikola Tesla and his backers were trying to convince people we should base our electricity on AC current. One of the major deciding factors in the competition was that AC current can be easily converted to higher or lower voltages using a transformer and DC current cannot. The electricity at the Chicago World's Fair in 1893 was a major demonstration of the use of AC current. This was so successful that it helped convince people to base our electrical distribution system in the United States on AC current. It turns out that distributing electrical power throughout the country's power grid is much easier with AC current than DC current. However, DC current is still the standard for many small devices operated by electricity from batteries.

The internal workings of a battery are typically chemical processes that create positive charged particles at one of its terminals and negative charged particles at the other terminal. The positive terminal is called the anode and the negative terminal the cathode. This separation of charges creates a potential difference between the terminals. When they are connected externally by a conductor, the potential difference forces the electrons (negative charges) to flow out of the cathode, through the conductor, to the anode. This is defined as a current flowing from the anode to the cathode. The potential energy stored in the battery provides it with the ability to force electrons to flow and is called its electromotive force or EMF. A battery goes flat when its chemical energy is used up. The circuit symbol for a battery is two parallel lines of different lengths with the longer one designating the anode and the shorter one the cathode. An AC voltage source is shown as a circle with a sinewave in it.

A resistor is an element made of a low conductivity material that converts some of the energy in the electric current into heat. On a microscopic scale, this happens because the moving electrons carrying the current collide with the atoms of the material causing them to vibrate. Many devices affect a circuit by converting some of its current into heat and thus can be represented as resistors. It should be noted that batteries have an internal resistance to the current flowing in them. Therefore the effective potential difference provided by a battery is its chemical emf minus the voltage loss due to its internal resistance. A resistor is designated by a crooked line.

Capacitors are represented by two parallel vertical lines of the same length. As we learned in Chap. 7, capacitors are used to store charge. Fuses are shown as small black boxes. They limit the flow of current. Inductors store energy and are shown as a coil. Other devices in circuits are things like meters that measure the current or voltage in the circuit, or light bulbs, or motors, etc. These individual components of circuits will be discussed in detail below.

8.2 Electrical Conduction

The electrical properties of solids are discussed in detail in Chap. 14. In this section we are interested specifically in the conduction of electrical current. Current is designated by I and is defined as the amount of charge per unit time flowing past a specific point,

$$I = \Delta q / \Delta t. \tag{8.1}$$

The unit of current is called an Ampere (usually shortened to amp) and designated by A. One amp is equal to the current flow of one coulomb per second.

How much a material impedes the flow of electrical current is called its resistance. The resistance of a material is designated by R and the units of resistance are called Ohms designated by omega, Ω. Some materials have a very high resistance and are called insulators while others have a relatively low resistance and are

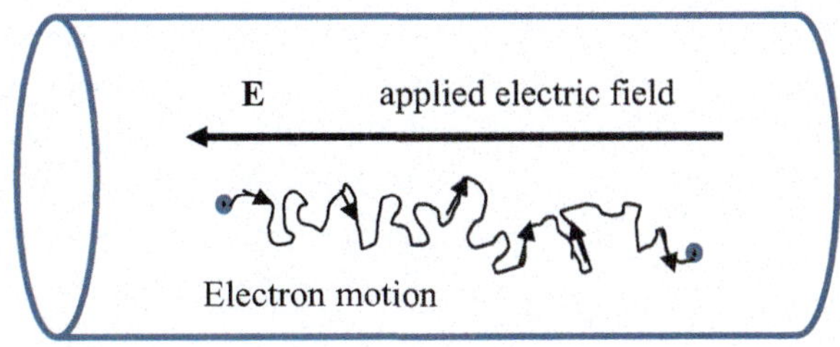

Fig. 8.6 Motion of an electron through a conductor

called conductors. We will learn more about why there is such a large difference in resistance in different materials in Chap. 14. Here we can use Fig. 8.6 to illustrate why resistance occurs. There are a great number of atoms in the conductor, and they all have electrons, which repel the moving electrons. As the moving electrons come near an atom, they will be deflected away. Thus, the electrons must "bump around" a lot in order to get through the conductor.

Notice that the electric field in the diagram is pointing to the left. That means the positive side of the battery is hooked to the right side of the conductor, and the negative side of the battery is hooked to the left side of the conductor. Now, if there were no atoms in the conductor, the mobile electron in the diagram would just move straight to the right (negative charges move opposite the electric field). However, in the conductor, the electron keeps bumping into atoms as it moves. This causes the electron to take a crazy, zigzag path through the conductor. As you might imagine, this slows the electron down quite a bit. In fact, the velocity which an electron moves through a conductor is called the drift velocity of the electron. Typically, the drift velocity of an electron has a magnitude on the order of a few millimeters per second. Thus, even though the electric field has the ability to accelerate free electrons to very high speeds, that ability is severely hampered by the collisions that the electron is constantly making with the atoms in the conductor. As a result, the electrons move relatively slowly through the conductor.

Consider, for example, an electron traveling between the plates of a parallel plate capacitor with air as the dielectric. Suppose the capacitor is charged so that there is a 9 Volt potential difference between the plates. Suppose further that the electron begins its journey from rest at the negative plate. It would be able to travel 10 cm in about 0.36 microseconds. If that same electron were traveling in a copper wire that was hooked up to a 9 Volt battery, it would take the electron several seconds to travel 10 cm! That gives you an idea of how much "interference" exists for an electron traveling through a conductor.

Since the atomic structure of a conductor slows down the speed of an electron as it travels through an electric field, you might imagine this resistance is different for each conductor, because each conductor has its own atomic structure. Thus, physicists have defined a term called resistivity.

Resistivity—A measure of a conductor's inherent resistance to the flow of electricity.

A table of resistivities is given below. Notice that the resistivity of iron is more than 5 times higher than the resistivity of copper. That means iron offers more than 5 times resistance to the flow of electricity than does copper. The resistivity of a material is designated by ρ and has units of Ωm. Resistivity is the inverse of conductivity, $\sigma = 1/\rho$, with units of siemens per meter. Either ρ or σ can be used to designate how well a type of material conducts electricity.

The resistance of a circuit element depends on its material and its geometry. For example, for a wire,

$$R = \rho(L/A) \tag{8.2}$$

where L is the length of the wire, A is its cross-sectional area, and ρ is the resistivity of the material.

The following experiment demonstrates the different factors influencing resistance.

Experiment 8.1

The Factors Which Influence Electrical Resistance

Supplies

- Three #2 pencils (It turns out that *some* pencils will not work in this experiment, as they do not use enough graphite in their leads. Typically, the best pencils to use are pencils that are "natural" or "environmentally friendly," which is also called "green." A few examples of brands that do work are: Papermate© American Natural© pencils, Papermate© Earth Write© pencils, and Papermate© American© pencils.)
- One pencil with a significantly thicker lead than the others. (If you cannot find one of these, or if the one you find does not conduct electricity, go ahead and do the first part of this experiment.)
- Pencil sharpener
- A serrated knife
- Aluminum foil
- Tape
- Three "C" cell batteries
- The cardboard tube from a roll of toilet paper
- Scissors
- A flashlight

Introduction—In this experiment, you will see how geometric factors influence the resistance that carbon offers to the flow of electricity. As you can see from the table, carbon has a large resistivity (2000 times that of copper), but it still does conduct electricity.

Procedure:

1. Take the top off of the flashlight so that you just have the assembly which houses the light bulb.
2. Tear a strip of aluminum foil which is at least 7 inches long and 1.5 inches wide.
3. Look for the point at which the batteries in the flashlight touch the assembly that you have. Typically, this point is directly under the light bulb on the back of the assembly. It is usually a circular metal area surrounded by plastic. This is the conductor which leads to the light bulb. Wad up one end of the aluminum foil strip so that it is the same size as this circular area of metal, and firmly tape it to the metal. Make sure the foil is pressing against the metal.
4. Tear another strip of aluminum foil which is at least 7 inches long and 1.5 inches wide.
5. Look for the return conductor on the assembly. This is typically a ring of metal that surrounds the assembly, and it is typically between the place you just taped the foil and the top of the assembly. When you turn on the switch of the flashlight, the switch usually causes a piece of metal to touch this return conductor. This completes the circuit, giving one continuous loop of conductor from one end of the batteries to the other.
6. Fold one end of the second foil strip so that it is as wide as the return conductor, and firmly tape it to the return conductor. Once again, make sure that the foil is pressing against the metal. Also, make sure that the two foils are not touching, and make sure that they are each attached to their *own* conductor. You cannot have one piece of foil touching both conductors.

7. You now have two conductors which lead to the light bulb in the flashlight. Your assembly should look something like the drawing on the right. To see if everything is hooked up properly, take one of the batteries and put it in between the free ends of each strip of foil. With your fingers, press the end of one foil to one end of the battery and the end of the other foil to the other end of the battery. Don't worry, you won't feel any shock, as the vast majority of the current will run through the circuit, not you. Even if all of the current ran through you, you still would not feel anything, as the current would be very low.

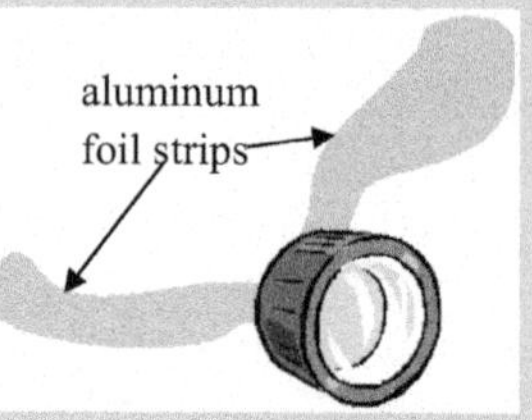

YOU SHOULD NEVER DO SOMETHING LIKE THIS UNLESS A KNOWLEDGEABLE PERSON TELLS YOU TO DO SO! Electricity can be *very* dangerous, and touching a bare conductor could kill you if you don't know what you are doing!

8. If the light did not light when you touched the foil ends to opposite ends of the battery, there is something wrong with the electrical connections at the flashlight assembly. Go back and make sure the foils are firmly touching *only* their conductors and not touching each other.
9. Once you get the flashlight to light in this configuration, you are ready to continue.
10. Take the cardboard tube and use the scissors to cut along its length so that it opens up and can be laid flat.
11. Take one battery and lay it on top of the opened cardboard tube. Position the battery so that half of it is on top of the cardboard and the other half is sticking out over the edge of the cardboard. The flat end of the battery is the one that should be sticking over the edge of the cardboard.
12. Roll the battery up in the cardboard so that the cardboard is tightly-wound around the battery.
13. Tape the cardboard roll so that it does not come unrolled, and also tape the battery to the cardboard at the point where the battery is sticking out of the tube.
14. Drop the other two batteries in the tube. Make sure they are pointing in the same direction as the first battery, so that they all connect up

properly. The third battery should be sticking out of the top. If it does not, cut the tube shorter so that it does.

15. You now should have a roll that contains three batteries stacked on top of one another. It should look something like the drawing on the right.

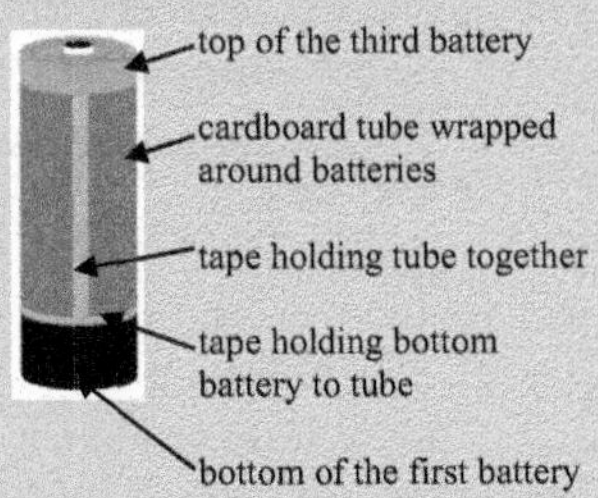

16. You are almost ready to begin the actual experiment. Take one of the #2 pencils and cut it with the serrated knife (use the knife like a saw) so that it is only about one-third of its original length and does not have an eraser on it.
17. Sharpen both ends of the pencil so that the pencil lead is exposed on both ends. However, do not make the ends very sharp. Also, don't expose too much of the lead, as it tends to break easily.
18. Although it is called a pencil "lead," it is actually made of carbon, which has a high resistivity but is still a conductor. You will place this conductor into an electrical circuit with the battery assembly and the flashlight assembly that you made.
19. First, take one more strip of aluminum foil which is at least 7 inches long and 1.5 inches wide. Tape one end to the top battery in the battery assembly.
20. Lay your flashlight assembly on the table, and set the battery assembly on top of the aluminum foil.
21. Finally, wad up the end of the foil coming from the top of the battery assembly as well as the free end of the foil coming from your flashlight assembly.

22. Hold the pencil in between these two wads of foil, pushing the pencil down so that the lead makes a good contact with the bottom foil and pushing the top foil down on the pencil so that good contact is made there. The final assembly should look like the drawing on the right.

23. If you get good contact between the foil and the pencil and the foil, the light should come on. If it does not, you either have a poor electrical connection somewhere, or your pencil does not have a lead that is mostly carbon.
24. Note how bright the light is.
25. Now take a second #2 pencil and use the knife to cut it to half of its original length. Once again, sharpen both ends and put that pencil in your electrical circuit. Once again, note how bright the light is.
26. Finally, take the third #2 pencil and cut just the eraser off. Once again, sharpen both ends and put that pencil in your electrical circuit. Once again, note how bright the light is.
27. Compare the three levels of brightness you observed. What was the pattern?
28. If you could find one, take the pencil with the thicker lead and cut it so that it is the length of one of the three pencils you have already used. Which one does not matter.
29. Sharpen that pencil on both ends and put it in your electrical circuit. Then, take that pencil out and put in the #2 pencil of the same length. What did you notice about the difference in brightness of the light?
30. Clean everything up, but don't throw anything away. You will use these items again.

What happened in the experiment? When you put the #2 pencils into your electrical circuit, you should have noticed that the light was brightest for the *shorter* pencil, somewhat more dim for the medium-length pencil, and dimmest for the long pencil. Why? Think about the electrons flowing through the carbon. Carbon's resistivity is more than 1,200 times that of aluminum. Thus, the longer they have to spend traveling through the carbon, the harder it will be for them to get to the light bulb. As a result, fewer electrons will pass through the light bulb in a given amount of time, so the light bulb will not burn as brightly. That tells you that the resistance of a conductor is proportional to the length. The longer the conductor, the larger the resistance.

What happened when you compared the pencil with the thicker lead to the #2 pencil of the same length? You should have noticed that the light was brighter for the pencil with the thicker lead. Why? Well, in the thicker lead, the electrons have more room to spread out. They will still encounter the same number of carbon atoms, so the collisions between the atoms of the conductor and the electrons will be the same. However, the electrons will *collide with each other* less often the more they can spread out. Thus, the more "room" the electrons have, the lower the resistance. In the end, then, resistance is inversely proportional to the cross-sectional area of the conductor.

These two geometric considerations lead us to the rather simple formula for the resistance of a conducting wire given in Eq. 8.2

Example 8.1

Assume that the pencil lead in your experiment was pure carbon. If the full-length pencil was 6.0 inches (15 cm) long, and the radius of the pencil lead was 0.79 mm, what was the resistance of the pencil?

To calculate resistance, we need resistivity (listed for carbon in Table 8.1), the length (given), and the cross-sectional area. The pencil lead is essentially a cylinder, and the cross sectional area of a cylinder is πr^2.

$$A = \pi \cdot r^2 = (3.14) \cdot (0.00079\ \text{m})^2 = 2.0 \times 10^{-6}\ \text{m}^2$$

Table 8.1 Resistivities of selected materials

Material	Resistivity (Ω m)
Aluminum	2.82×10^{-8}
Carbon	3.50×10^{-5}
Copper	1.72×10^{-8}
Iron	9.71×10^{-8}
Lead	2.06×10^{-7}
Silver	1.59×10^{-8}

Now we can use Eq. 8.2.

$$R = \rho L/A = \left(3.50 \times 10^{-5}\ \Omega\mathrm{m}\right)(0.15\ \mathrm{m})/\left(2.0 \times 10^{-6}\ \mathrm{m}^2\right) = 2.6\ \Omega$$

This example gives you an idea of the resistance that the pencil was giving in your experiment. Although 2.6 Ω is not a lot of resistance compared to many electrical circuits, it was enough to really decrease the brightness of the light. In fact, this resistance is the reason that you had to use three batteries in the experiment, rather than the two that you normally use for a flashlight. With even that small amount of resistance, two batteries (3 V) could not generate enough electrical current to light the light bulb. You had to have a third battery (for a total of 4.5 V) to generate sufficient electrical current. If we needed more voltage, why didn't we just use a 9 V battery? Well, a 9 V battery would have produced too much electrical current, and it probably would have broken the filament in the light bulb as a result. Thus, you needed more voltage than one or two flashlight batteries could provide, but a 9 V battery would have just made too much current!

Now try the following problems.

Student

8.1 You want to make an electrical circuit that carries as much current as possible for the battery you have. If you had the choice of all of the materials listed in Table 8.1, which one would be the best conductor to use in the wires of your circuit?

8.2. A rectangular copper conductor is 1.0 m long and has a resistance of $8.6 \times 10^{-5}\ \Omega$. Assuming that the width and breadth of the conductor are the same, give their value.

8.3 DC Circuits with Resistors

Now that we know something about current flow and resistance, we can review what you learned in your introductory course about simple DC circuits. Figure 8.7 shows schematic diagrams of simple DC circuits involving only resistors as circuit elements. Since many devices are used to turn electricity into heat (heaters, ovens, toasters, etc.) these circuits represent real types of physical situations. On the left side the circuit has a battery to supply the voltage to the circuit, a closed switch to allow current to flow, and a resistor. Note that the direction of current flow is from the plus terminal of the battery to the negative terminal. This is the conventional direction for current even though we know that generally current is carried by negatively charged electrons moving from the negative terminal of the battery to the positive terminal. positive charges.

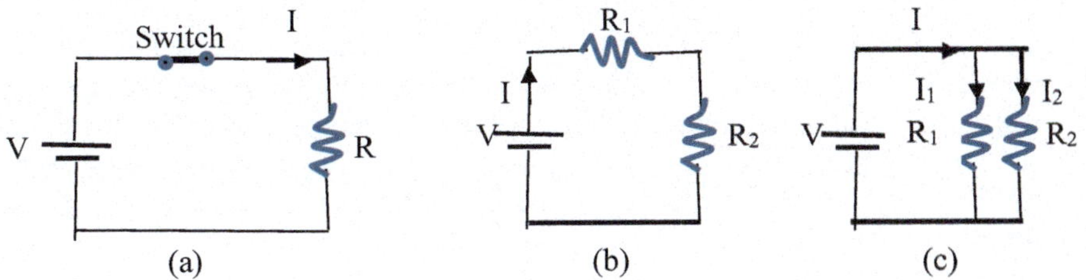

Fig. 8.7 Simple DC circuits

It is really not possible to always draw a light bulb, refrigerator, or whatever is actually connected to your circuit. Since most electrical appliances generate work by offering resistance to the flow of electrons, we can say that appliances are really just resistors. Thus, we can abbreviate them as such. The resistors in Fig. 8.7 represent some kind of electric device.

All conductors have some resistance. Thus, even the wires represented in the circuit offer some resistance to the flow of electricity. However, we generally use a low resistivity metal such as copper for wires, so the resistance is very low—much lower than that of an appliance that uses the electricity to do work. Thus, compared to the resistors in the circuit, the wires have essentially zero resistance.

In the middle of the figure, the resistor in the circuit has been divided into two resistors connected in series. That is, the same current flows through each resistor but the voltage drop across each of them is different. In the diagram on the right, the two resistors are connected in parallel so they have the same voltage drop across each of them but the current flowing through them is different.

One big difference between series and parallel circuits can be seen when a resistor becomes damaged and can no longer conduct electricity. Suppose someone crushed one of the two resistors so that it no longer could conduct electricity. What would happen to the current in in the series circuit? The electricity would stop flowing. There is no longer a complete path from the positive side of the battery to the negative side, so electricity simply will not flow. In the parallel circuit, however, things are different. When one of the resistors is damaged, electricity will no longer flow through that resistor, but there is still a complete path from the positive side of the battery to the negative side through the other resistor. Thus, in a series circuit, when one resistor stops conducting, the circuit is dead. In a parallel circuit, when one resistor stops conducting, *that* resistor is dead, but the other resistor continues to have current pass through it.

Ohm's Law states the relationship between the resistance of a device, the potential difference provided by the battery, and the current flowing through the device is,

$$V = IR. \tag{8.3}$$

Thus, one Ohm is a volt per amp. Any conductor that obeys Ohm's law is called an ohmic conductor. Not all conductors are ohmic.

At this point we are ready to consider simple DC circuits such as those shown in Fig. 8.8. These consist of a battery whose terminals are connected by conductors that go through several resistors. The battery symbol represents the potential difference provided to the circuit from either a regular battery or an electrical outlet on the wall going through a convertor. It causes DC current to flow in the circuit. The resistors represent different kinds of devices that provide resistance to the flow.

Since we generally construct electrical circuits for the purpose of doing some sort of work (light a light bulb, run a toaster, spin a motor, etc.), it is interesting to learn about how much work a circuit can do. We can figure this out by calculating the power dissipated, which as you know is the amount of work per unit time.

$$P = I \cdot V \tag{8.4}$$

In this equation, P stands for power, which and has the unit of Watt, which is a J/S. Now I has units of Amperes, which is also Coulombs per second, and V has units of Volts, which is also Joules per Coulomb. If you multiply Amperes by Volts, then, you get Joules/sec, which is a Watt. There is actually another way to calculate power as well. Since Ohm's Law, Eq. 8.2, gives us an expression for V, we can plug that expression into Eq. 8.3 and get

$$P = I^2 R \tag{8.5}$$

Either equation for power can be used, as they are equivalent.

To make sure you remember how to use all these equations let's look at an example of a simple circuit.

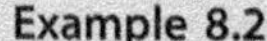

Example 8.2

Given the circuit below, calculate the current that runs through the circuit and the power it generates.

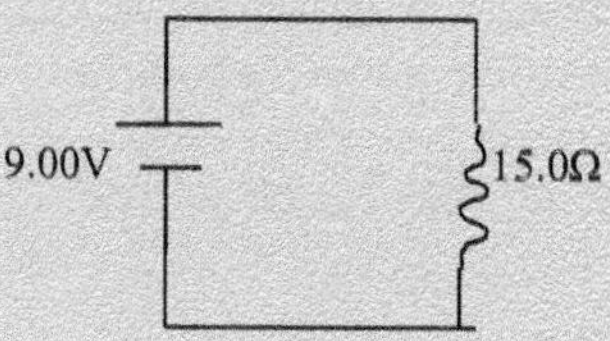

Knowns: $V = 9.00\text{ V}$; $R = 15.0\ \Omega$
Unknowns: I; P

We can get the current from Ohm's Law.

$$V = IR$$

$$I = V/R = (9.00\ \text{V})/(15.0\ \Omega) = 0.600\ \text{A}$$

Now to get the power, we can use either Eq. 8.4 or Eq. 8.5. If we use the former:

$$P = IV = (0.600\ \text{C/s})(9.00\ \text{J/C}) = 5.40\ \text{W}$$

To make sure you understand how to use these equations, work the problem below.

Student

8.3 A physicist wants to design a circuit that has a power output of 14.5 W. She decides to use a 25.0 Ω resistor. What must the voltage be on the battery she wants to use?

Now perform the following experiment to demonstrate the properties of resistors connected in series and parallel.

Experiment 8.2
Resistors in Series and Parallel

Supplies:

- The circuit apparatus you used in the last experiment, including the light, the foil strips, the battery assembly, and the #2 pencils
- More aluminum foil
- Tape
- Serrated knife

Introduction—This experiment will show you the dramatic difference between wiring resistors in series or in parallel.

Procedure:

1. You should have three #2 pencils from the last experiment, sharpened on both ends. You should have a short one, a medium-length one, and a long one. Take the long one and cut it down and sharpen it so that it is the same length as the medium-length pencil.

2. Take a strip of foil and wad it up into a ball.
3. Take the two equal-length pencils (sharpened on both sides) and stick them in opposite ends of the ball so that their leads are fully imbedded in the foil.
4. Stick that assembly in your circuit, as shown on the right.
5. Note the brightness of the light. It is possible that the light will not come on. If so, remove one of the pencils and touch the foil coming from the battery to the foil ball. If the light shines, then you probably had a good connection, but not enough current. That's fine. If the light does not come on, there is probably a bad connection, and you should try to find it and fix it.
6. Take the pencils out of the foil ball and get rid of the foil ball.
7. Put the pencils next to each other and tape them together. Now you have two parallel pencils.
8. Stick the two pencils in the electrical circuit, as shown on the right. Make sure you press the foil against *both* pencil leads.

9. Note the brightness of the light.
10. Clean everything up. You can now throw aw

What happened in the experiment? You should have seen that the light was much, much dimmer in the first part of the experiment than in the second part. Why? In the first part of the experiment, the resistors (the pencils) were in series. All of the current had to travel through both resistors. In the second part of the experiment, the resistors were in parallel. Each electron could travel *either* through one pencil or the other. If you think about it, the second configuration results in *less* resistance than the first. After all, since some electrons go through one resistor and the rest go down the other, the electrons are more spread out. As a result, they interfere with one another less. Thus, wiring resistors in parallel is somewhat like using a resistor with a larger cross-sectional area. Since the electrons can spread out, they do not interfere with each other as much, and resistance goes down.

This can be quantified mathematically. For resistors connected in series, the effective resistance is given by the sum of the individual resistances

$$R_{\text{eff}} = R_1 + R_2 + R_3 + \cdots + R_n = \sum_i R_i. \tag{8.6}$$

where the Greek capitol sigma signifies a sum. The individual resistances can be replaced by the effective resistance when using Ohm's law. For resistors connected in parallel their effective resistance is expressed as the sum of reciprocal resistances

$$1/R_{\text{eff}} = 1/R_1 + 1/R_2 + 1/R_3 + \ldots + 1/R_n = \sum_i 1/R_i \tag{8.7}$$

When resistors are hooked up in series, the effective resistance is the sum of each individual resistance. If two 5 Ω resistors are hooked up in series, for example, the total resistance is simply 10 Ω. However, if those same two 5 Ω resistors are hooked up in parallel, the effective resistance is less than either individual resistor: 2.5 Ω! Wiring resistors in parallel, then, *lowers* the total resistance, while wiring them in series *raises* the total resistance.

Two important things to realize about wiring resistors into circuits. First, resistors wired in parallel all have the same potential difference across each of them. Second, resistors wired in series all have the same amount of current going through each of them. The example below demonstrates these concepts.

Example 8.3

What is the total current running through the following circuit?

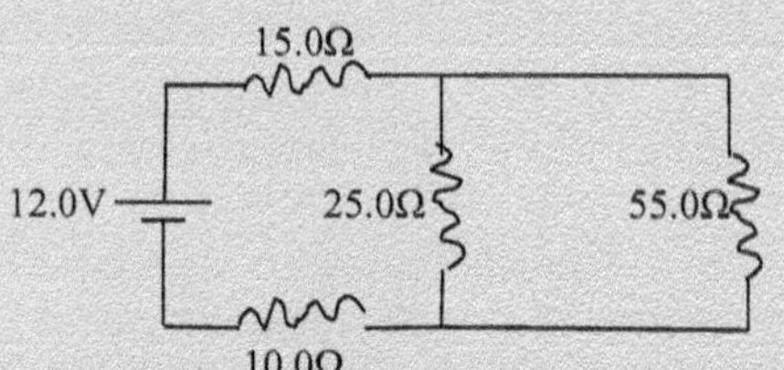

Figure is given.
Knowns: given in figure.
Unknowns: I

One way to analyze a circuit is to reduce it to a very simple circuit. If we could get this circuit down to a battery and a single effective resistor, calculating the total current would be a simple application of Ohm's Law. The trick is reducing the 4 resistors in the circuit to one effective resistor. To do that, we have to know which resistors are hooked in series and which are hooked in parallel.

We imagine the conventional current traveling from the positive end of the battery to the negative end. When the current runs into a "choice" as to which way to go, we have a parallel part of the circuit. When the current has no "choice," we have a series part of the circuit. The first thing to do is to reduce the parallel resistors to one effective resistor, and then we can deal with the series resistors.

Start at the positive side of the battery and begin traveling through the wire. Does the current have a choice as to whether or not to go through the 15.0 Ω resistor? No! All current must go through that resistor. Thus, it is a series resistor, and we can save it for later. Once the current gets through *that* resistor, it *does* have a choice. It can "turn right" and go through the 25.0 Ω resistor, or it can "go straight" and go around the end and travel through the 55.0 Ω resistor. Thus, the 25.0 Ω resistor and 55.0 Ω resistor are hooked in parallel. We can therefore calculate their effective resistance:

$$1/R_{\text{eff}} = 1/R_1 + 1/R_2 = 1/25.0\ \Omega + 1/55.0\ \Omega = 0.0582/\Omega$$
$$R_{\text{eff}} = 17.2\ \Omega$$

Now we can redraw the circuit, replacing the parallel resistors with one effective 17.2 Ω resistor. At this point, all of these resistors are in series. All of the current must go through all three of them. Thus, we can now calculate their effective resistance.

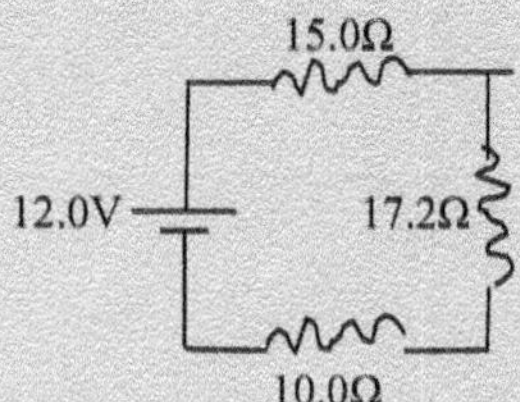

$$R_{\mathrm{eff}} = R_1 + R_2 + R_3 = 15.0\ \Omega + 17.2\ \Omega + 10.0\ \Omega = 42.2\ \Omega$$

All three of these resistors, then, can be replaced by one effective 42.2 Ω resistor. This entire circuit (from a resistance point of view) is therefore equivalent to the following simple circuit.

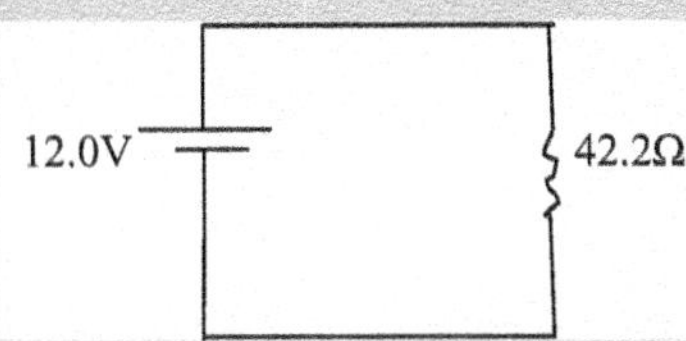

Calculating the current with Ohm's Law is now easy.

$$V = IR$$
$$I = V/R = 12.0\ \mathrm{V}/42.2\ \Omega = 0.284\ \mathrm{A}$$

Now try using this technique on the following problems.

Student

8.4 How much power is drawn by the following circuit?

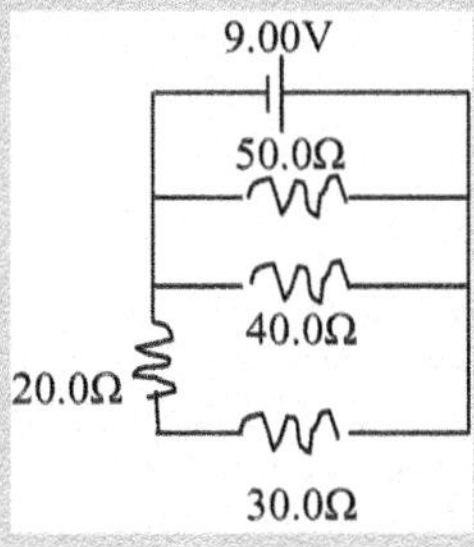

8.5 The following circuit draws 0.500 A. What is the value of the resistor labeled with a question mark?

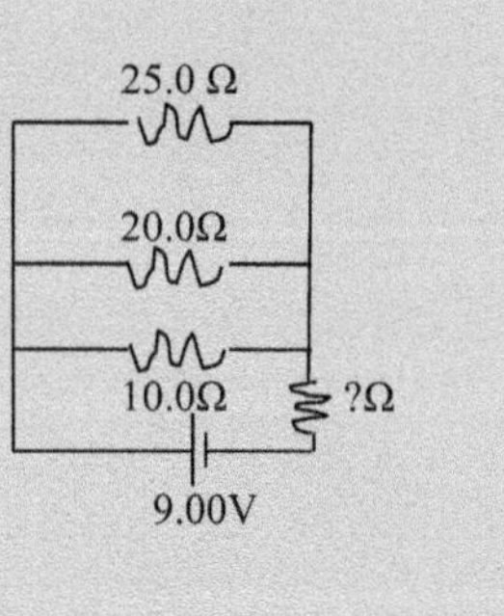

Along with Ohm's Law, we have two important rules to follow in analyzing electrical circuits. These are called Kirchhoff's Laws

Kirchhoff's Loop Law—The change in electric potential around a closed loop is always equal to zero.
Kirchhoff's Junction Law—The sum of the currents entering a junction must equal the sum of currents leaving a junction.

Kirchhoff's Junction Law is the easiest to understand since it is a consequence of conservation of charge. When the current encounters a "choice" as to where it will go, that is called a "junction." If 15A of current enters a junction, when the junction ends, 15A must leave that junction. When the current has a "choice," it can split up. However, that current cannot be lost. When the "choice" ends, the same amount of current must still be flowing.

Kirchhoff's Loop Law requires the use of Ohm's Law. Think about resistors. If current travels through a resistor, the current must be "pushed along," to counter the resistance. This takes work. As a result, the current experiences a *drop* in potential. This results in our first guideline when using Kirchhoff's Laws:

When tracing a loop, if you go across a resistor in the *same direction* as the current, the potential *decreases* by *I R*.

This is just using Ohm's Law in the case of a single resistor. Before passing through a resistor, the current has a higher potential than it does on the other side of the resistor. Thus, if you are following a loop in the same direction as the current, you will record a drop in potential.

When tracing a loop in an electric circuit, it is possible that you will find yourself moving opposite the current. If that happens and you encounter a resistor, you must treat it as an increase in potential:

When tracing a loop, if you go across a resistor *opposing* the direction of current, the potential *increases* by *I R*.

This should make sense as well. If the current experiences a drop in potential, and if you are going opposite the current, you will experience an increase in potential.

Eventually, we will start dealing with circuits that have more than one battery. That makes the circuit more difficult to analyze, but it is not impossible. You just have to follow two guidelines when the loop you are tracing contains a battery:

When tracing a loop, if you encounter the *negative* side of the battery first, the potential *increases* by the battery's voltage.

This should make sense. A battery is a source of potential. If your loop includes a battery and you cross the battery in the direction of the current flow, you get the "push" of the battery, and the potential increases. However, if you go against the current, the opposite happens.

When tracing a loop, if you encounter the *positive* side of the battery first, the potential *decreases* by the battery's voltage.

As an example, let's start with the circuit in Problem 8.5.

Example 8.4

In the circuit diagram below, how much current is running through the 20.0 Ω resistor?

Getting the total current is easy. We just turn all of the resistors into one effective resistor as you learned in the previous section, then we use Ohm's Law. However, the question doesn't ask for the total current. It wants the current going through the 20.0 Ω resistor. To get that we use Kirchhoff's Laws. First, we choose a loop—any loop.

Let's choose the bottom one, as illustrated on the right. We start the loop at the positive side of the battery, and trace the path of conventional current. When we leave the battery, all current is traveling in the loop, because it has not encountered a choice yet. Thus, the current is Itot. However, when the current reaches the top of the loop, it has a choice. It can go through the 10.0 Ω resistor or the 20.0 Ω resistor or the 25.0 Ω resistor. Thus, only a portion of the current (which we will call I_1) goes through the 10.0 Ω resistor.

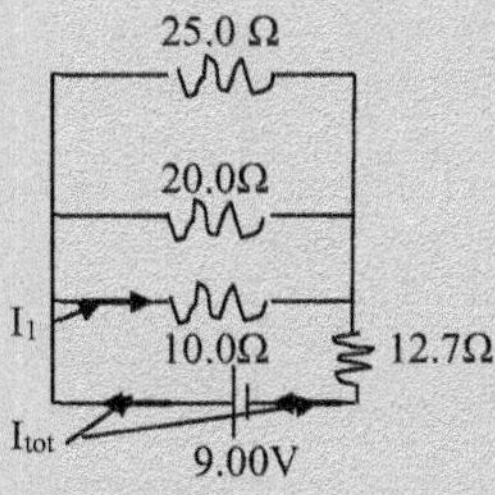

However, once the current reaches the right side of the loop, all of the other branches merge together. Thus, all of the current must come together again. As a result, Itot runs through the 12.7 Ω resistor.

Now we can use Kirchhoff's First Law. The sum of the potentials in this loop must be zero. We are going with the current throughout the loop, so the potential drops across every resistor. By the time we actually *encounter* the battery on our trip, we encounter the negative end first. We started at the positive side of the battery, but we were already past it. We did not actually encounter the battery until the end of the loop, and when we did, we encountered the negative end first. Thus, the potential of the battery is added. This gives us an equation:

$$9.00\,\text{V} - I_1(10.0\,\Omega) - I_{tot}(12.7\,\Omega) = 0$$

That's Kirchhoff's First Law—the change in potential around the loop equals zero. This equation is useful, but we cannot solve it, because it has two unknowns. Thus, we need another equation. For that, we go to another loop.

Let's choose the next logical loop, as illustrated on the right. In this loop, we once again start out in front of the battery. At that point, we are tracing all of the current. Then, we will skip the first branch and head to the second branch. Only a portion of the total current will travel through that branch, so we will call it I_2. Finally, when the branches converge, all of the current flows through the 12.7 Ω resistor, so that is Itot once again. Following the same logic as before, we come up with the following equation for this loop: $9.00\,\text{V} - I_2 \cdot (20.0\,\Omega) - I_{tot} \cdot (12.7\,\Omega) = 0$

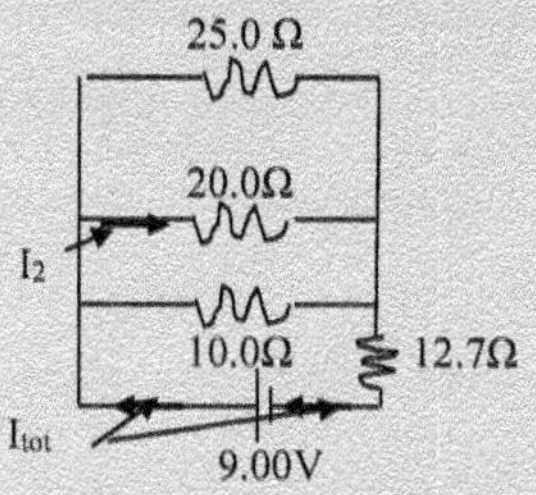

You might not think this helps, because we now have yet another unknown. That's true, but if we keep going, we will eventually get what we need. Let's look at the last loop. Once again, we start in front of the battery and then skip the first two branches but go through the last one. Since only a portion of the current runs through that branch, we will call it I_3. Once again, however, by the time we encounter the 12.7 Ω resistor, all of the branches have merged, so we are dealing with Itot again. This leads to yet another equation:

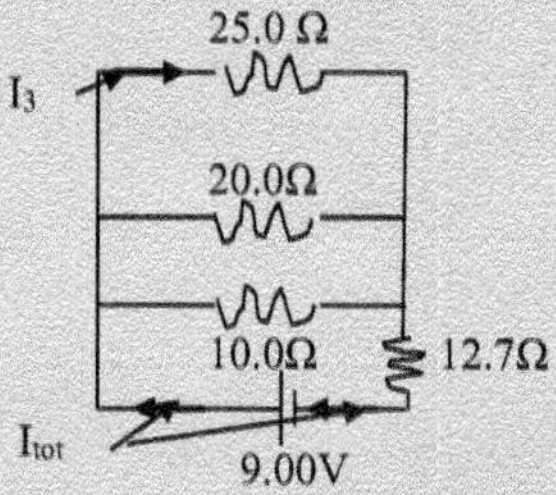

$$9.00\,\text{V} - I_3(25.0\,\Omega) - I_{tot}(12\,\Omega) = 0$$

At this point, we have three equations and four unknowns. However, there is one more equation we can develop. We can use Kirchhoff's Second Law to say that the sum of I_1, I_2, and I_3 must be equal to Itot, because we cannot lose current.

$$I_{\text{tot}} = I_1 + I_2 + I_3$$

Now we have four equations and four unknowns, so we can solve for the four unknowns.

The easiest thing to do is get everything in terms of *I*tot. We can do this using the first three equations. Let's start with the equation from loop 1:

$$9.00\,\text{V} - I_1(10.0\,\Omega) - I_{\text{tot}}(12.7\,\Omega) = 0$$

$$I_1 = \frac{9.00\,\text{V} - I_{\text{tot}}(12.7\,\Omega)}{10.0\,\Omega}$$

Then we can go to the equation from loop 2:

$$9.00\,\text{V} - \text{I}_2(20.0\Omega) - \text{I}_{\text{tot}}(12.7\Omega) = 0$$

$$I_2 = \frac{9.00\,\text{V} - I_{\text{tot}}(12.7\,\Omega)}{20.0\,\Omega}$$

Then we can go to the equation from loop 3:

$$9.00\,\text{V} - I_3(25.0\,\Omega) - I_{\text{tot}}(12.\,\Omega) = 0$$

$$I_3 = \frac{9.00\,\text{V} - I_{\text{tot}}(12.7\,\Omega)}{25.0\,\Omega}$$

Now we can put these three expressions into the last equation and solve for *I*tot.

$$I_{\text{tot}} = I_1 + I_2 + I_3$$

$$I_{\text{tot}} = \frac{9.00\,\text{V} - I_{tot}(12.7\,\Omega)}{10.0\,\Omega} + \frac{9.00\,\text{V} - I_{tot}(12.7\,\Omega)}{20.0\,\Omega} + \frac{9.00\,\text{V} - I_{tot}(12.7\,\Omega)}{25.0\,\Omega}$$

$$I_{\text{tot}} = (42.8\,\text{V})/(85.3\,\Omega) = 0.502\,\text{A}$$

Now that we know *I*tot, we can easily solve for I_2, which is what the question asked for!

$$I_2 = \frac{9.00\,\text{V} - (0.502\,\text{A})(12.7\,\Omega)}{20.0\,\Omega} = 0.131\,\text{A}$$

Think about what we did. We kept choosing loops and developing an equation for each loop until we got to the point where we had as many equations as we had unknowns. Then we solved the equations simultaneously. Kirchhoff's Laws are incredibly useful in circuit analysis. Let's see how they work in more complicated circuits.

Example 8.5

In the circuit below, determine the current that flows through the 15.0 Ω resistor. Also, determine the potential difference between points A and B.

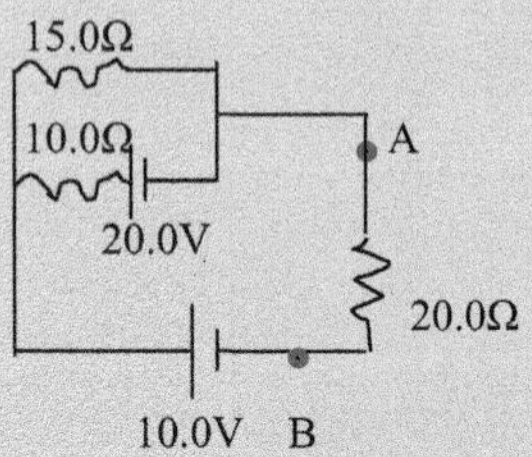

The figure is given.
Knowns: in the figure
Unknowns: I_3; V_{AB}

This is a more complicated circuit because it has two batteries. However, with Kirchhoff's Laws, it is pretty easy to analyze. Let's start with a loop that goes through the 10.0 Ω resistor. Current leaves the 10.0 V battery. We don't know if it is the total current because there is another battery to consider. Thus, we will call it I_1. When it reaches the top of the loop, it can go into the 10.0 Ω resistor or the 15.0 Ω resistor, so this is a branch. Thus, only a part of I_1 will go through the 10.0 Ω resistor, so we will call that I_2. Once the branch merges back, we cannot lose any current, so I_1 will go through the 20.0 Ω resistor. Now let's look at the potentials. When we go through the 10.0 Ω resistor, the potential will drop. The potential will also drop again when we pass through the 20.0 V battery, because we encounter the positive side first. When we go through the 20.0 Ω resistor, the potential will drop, and the potential will increase when we hit the 10.0 V battery. Thus, the equation is:

$$10.0\ \text{V} - I_2(10.0\ \Omega) - 20.0\ \text{V} - I_1(20.0\ \Omega) = 0$$

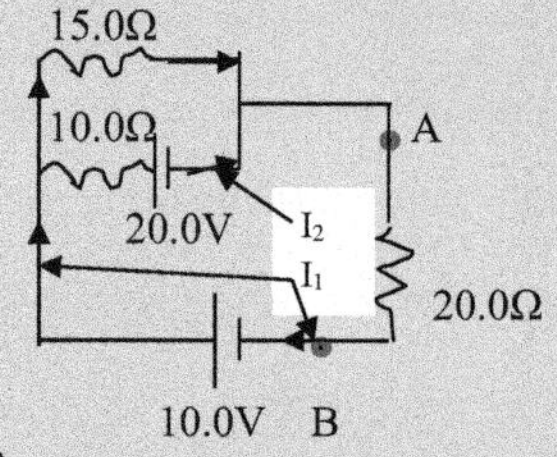

Now, let's choose the other possible loop. In this loop, we are following the other branch. Only a portion of I_1 goes through that branch, so we will call it I_3. It will pass through the 15.0 Ω resistor, and the potential will drop. Then, the branches merge, so all of I_1 will go through the 20.0 Ω resistor, where the potential will once again drop. We then encounter the battery, which raises the potential because we hit the negative side first. In this loop, then:

$$10.0\ \text{V} - I_3(15.0\ \Omega) - I_1 \cdot (20.0\ \Omega) = 0$$

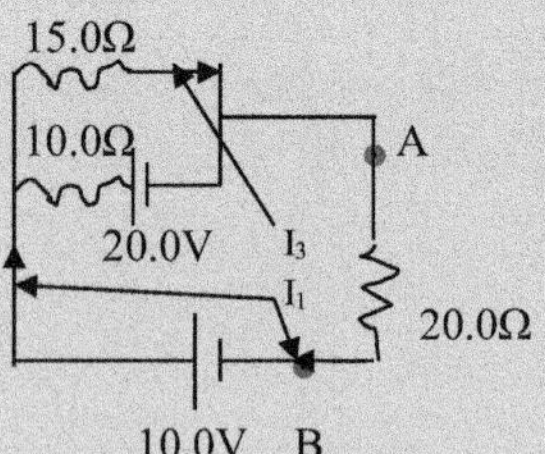

We now have two equations and three unknowns. However, we can use Kirchhoff's Second Rule to get a third equation. Since we cannot lose current:

$$I_1 = I_2 + I_3$$

To solve these equations, let's use the first two to get I_2 and I_3 in terms of I_1:

$$I_2 = \frac{-10.0\ \text{V} - I_1(20.0\ \Omega)}{10.0\ \Omega}$$
$$I_3 = \frac{10.0\ \text{V} - I_1(20.0\ \Omega)}{15.0\ \Omega}$$

We can now use these expressions in the third equation:

$$I_1 = \frac{-10.0\ \text{V} - I_1(20.0\ \Omega)}{10.0\ \Omega} + \frac{10.0\ \text{V} - I_1(20.0\ \Omega)}{15.0\ \Omega} = -0.0769\ \text{A}$$

What does the negative mean? It means we chose the *wrong direction* for current. We had I_1 traveling clockwise. It is really traveling counterclockwise. That's the great thing about Kirchhoff's Laws. You don't have to know how the current flows. Just pick a direction, and if you are wrong, you will find out in the equations!

Now that we have I_1, we can get the other two currents (although we only need one of them).

$$I_2 = \frac{-10.0\ \text{V} - (-0.0769\ \text{A})(20.0\ \Omega)}{10.0\ \Omega} = -0.846\ \text{A}$$
$$I_3 = \frac{10.0\ \text{V} - (-0.0769\ \text{A})(20.0\ \Omega)}{15.0\ \Omega} = 0.769\ \text{A}$$

The question only wanted the current flowing through the 15.0 Ω resistor, which is I_3. Thus, the answer to the first part of the question is 0.769 A.

The last part of the question asks for the potential difference between A and B. If we travel from point A to point B, we go *against* the current I_1. This means the potential *raises.*

$$V = I_1 R = (0.0769\ \text{A})(20.0\ \Omega) = 1.54\ \text{V}$$

This indicates that point A is 1.54 V *lower* in potential than point B, because this is the amount by which the voltage *rises* as we go from point A to point B.

Now try doing the following problems.

Student

8.6 For the circuit that is diagrammed to the right, determine the total current drawn from the battery, the power dissipated by the circuit, and the current that flows through the 15.0 Ω resistor.

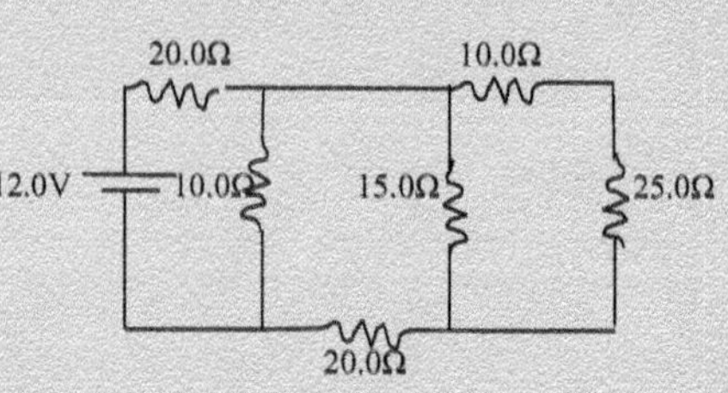

8.7 For the circuit shown on the right, determine the current that runs through *each* resistor. Also, determine the potential difference between points A and B.

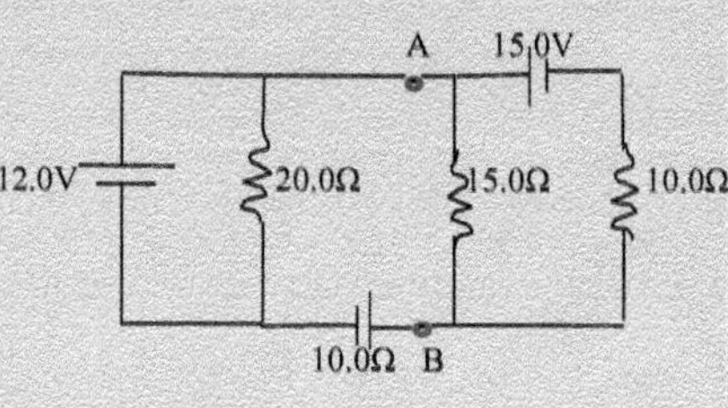

8.8 The circuit diagrammed on the right is called a **Wheatstone Bridge**. It can be used to determine the resistance of an unknown resistor. The bridge is said to be in "balance" when there is no current flowing from point A to point B. If this bridge is in balance, what is the resistance of the resistor labeled with a question mark? (HINT: If no current is flowing from point A to point B, the potential difference between those points is zero.)

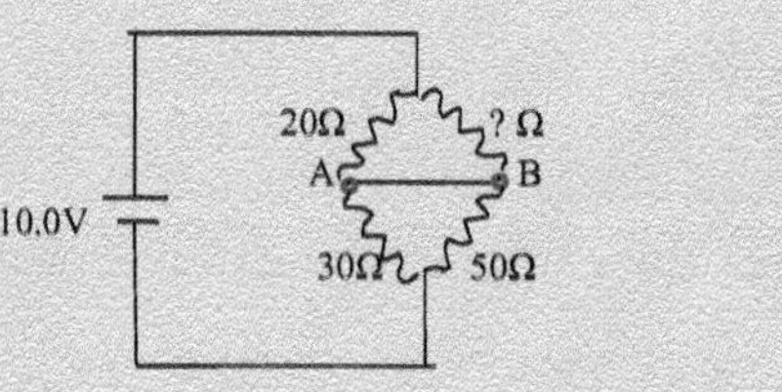

8.4 Other Circuit Components

In these examples, when the resistance in the circuit increases, the current flow decreases. Many circuits include a switch to turn the flow of electricity on and off. This is called a circuit breaker. When the switch is open, no current can flow, and the situation is called an open circuit. When the switch is closed current flows and the situation is called a closed circuit. Sometimes a switch takes the form of a fuse to protect a device from getting too much current. In this case, instead of being controlled manually, the switch is wired in the circuit in the closed position but turns to open automatically if the current gets too high. This can occur if the fuse has a small resistor element that melts when too much heat is generated by the current.

In our circuit diagrams, we have been labeling the battery with a voltage rating, which tells us the electrical potential that the battery is giving to the circuit. However, that voltage rating is not the same as the voltage that is printed on the battery. If we pick up a flashlight battery, for example, we see that it is labeled as a 1.5 V battery. However, if we put that battery in an electric circuit, it will provide less than 1.5 V of potential to the circuit. This is because batteries have internal resistance.

The battery itself will have some resistance to current. Resistance decreases the potential. Remember, as current passes through a battery, the potential drops. Thus, as the battery supplies current to the circuit, its own internal resistance will decrease the potential that the battery can supply. This effect increases with increasing current. Ohm's Law tells us that the more current you pass through a resistor, the more the potential drops. Thus, the higher the current in the circuit, the more discrepancy there will be between the labeled voltage on the battery and the potential it actually supplies. In general, the resistance of the entire circuit is very large compared to the internal resistance of the battery. Thus, in most cases, the internal resistance of the battery can be neglected. However, if you want to be very precise, or if current is very large, you must take the internal resistance of the battery into account.

To make a distinction between the actual potential supplied and the rated voltage of the battery, we make reference to an erroneous concept from the history of physics. Back when physicists didn't understand electricity, they wanted to make analogies between Newton's Laws and electrical phenomena. Thus, when they noticed that a battery caused electrons to move, they assumed that the battery was exerting a force on the electrons. They called this the electromotive force, which is abbreviated as either emf or a fancy-looking E. Today, we know that there really is no such thing as emf. The battery simply supplies a potential difference, and electrons move in response to that potential difference. However, the term emf still remains.

Although we know that emf is really not a correct idea, we will use it to distinguish between a battery's rated voltage and the voltage it can actually supply to a circuit. We will say that the *rated* voltage is the battery's emf, and the actual potential it supplies to the circuit can then be calculated. When no current is flowing, the potential difference of the battery is equal to its emf. The reason the battery doesn't supply all of its emf to the circuit is because its internal resistance causes the potential to drop. The only way resistance can cause potential to drop is when current flows through it. Thus, the emf is equal to the potential difference between the ends of the battery as long as current is not flowing. The following example shows what happens when current starts to flow.

Example 8.6

In the circuit to the right, the battery has an emf of 12.0 V and an internal resistance of 2.00 Ω. Calculate the actual potential that the battery supplies to the circuit if R = 55.0 Ω and if R = 5.00 Ω.

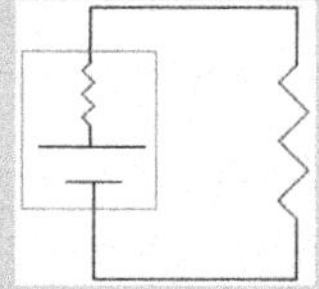

Knowns: emf = 12 V; R_i = 2.00 Ω; R = 55.0 Ω or 5.00 Ω

Unknowns: V

Note that the circuit diagram now shows the battery with its internal resistance along with its emf. When we apply Ohm's Law the internal resistance of the battery is in series with the external resistance.

$$V = I(R_1 + R)$$
$$I = V/(R_i + R) = 12\ \text{V}/(2.00\ \Omega + 55.0\ \Omega) = 0.211\ \text{A}$$

Now that we know the total current, we can calculate the potential drop across the internal resistor of the battery:

$$V = IR_i = (0.211\ \text{A})(2.00\ \Omega) = 0.422\ \text{V}$$

If the internal resistance causes the potential to drop 0.422 V, then the actual potential that the battery provides to the circuit is simply 12.0 V–0.422 V, which equals 11.6 V. In this circuit, then, the battery supplies 11.6 V of potential.

We can now redo the calculation with $R = 5.0\ \Omega$. When that is the case, the total resistance of the circuit (including the internal resistance of the battery) is $7.00\ \Omega$.

$$I = V/(R_i + R) = 12\ \text{V}/(2.00\ \Omega + 5.0\ \Omega) = 1.71\ \text{A}$$

That current causes the following voltage drop over the internal resistor of the battery:

$$V = IR_i = (1.71\ \text{A})(2.00\ \Omega) = 3.42\ \text{V}$$

If the internal resistance causes the potential to drop 3.42 V, then the actual potential that the battery provides to the circuit is simply 12.0 V–3.42 V, which equals 8.6 V. Notice how much lower the actual potential supplied to the circuit is. That's because the larger current in the circuit causes a greater potential drop across the battery's internal resistance.

In most circuits that you will analyze, we will assume that the battery's internal resistance is so small that it does not affect the potential significantly. However, if a problem specifically mentions the internal resistance of a battery, you must take it into account.

Now try the following problem.

Student

8.9 In the circuit drawn to the right, the battery is labeled as a 9.00 V battery. In the circuit, however, it provides only 8.59 V of potential. What is the battery's internal resistance?

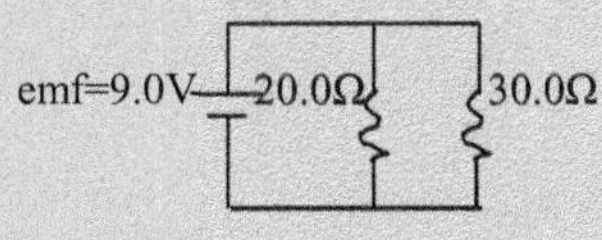

Now we can add another circuit element, a capacitor. As discussed in the previous chapter, a capacitor acts to store a buildup of electrical charges. These can later be discharged as a temporary source of current which is useful for many applications. Some examples include a flashbulb on a camera or smoothing the output of an amplifier. Like other circuit elements multiple capacitors can be hooked up in series or in parallel as shown in Fig. 8.8.

In circuit Fig. 8.8a when the switch is open no current flows so there is no charge on the capacitor. When the switch is closed, the voltage source causes current to flow until the capacitor is completely charged and has a voltage equal to

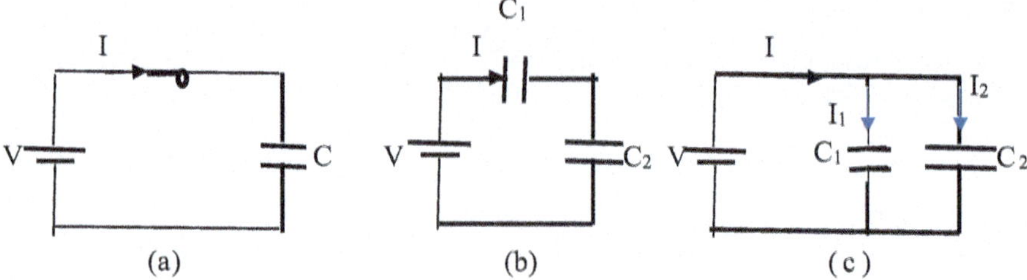

Fig. 8.8 Simple capacitor circuits

the battery voltage. The current then stops flowing until something is changed so the capacitor is discharged. The charge on the capacitor is then drained off causing a temporary current to flow until all the charge is dissipated.

For capacitors connected in series such as the circuit in Fig. 8.8b, the current flowing out of the battery to the first capacitor is the same as the current flowing into the battery from the second capacitor, so the charges built up on the capacitors during charging must be the same. The voltage in the circuit according to Kirkoff's voltage rule is $V = V_1 + V_2$. Using Eq. 7.7 this can be rewritten as $q/C = q_1/C_1 + q_2/C_2$. Since the charges are all the same, the equivalent capacitance for two capacitors connected in series is $1/C = 1/C_1 + 1/C_2$. This can be generalized for any number of capacitors connected in series as

$$1/C = \Sigma_i 1/C_i. \tag{8.8}$$

For multiple capacitors connected in parallel as shown in Fig. 8.8c, the potential difference is the same across each capacitor. The total current flowing from the battery is charging the two capacitors. According to Kirkoff's current rule it is given by $q = q_1 + q_2$. Using Eq. 7.7 this can be expressed in terms of the capacitance, $CV = C_1V_1 + C_2V_2$. Since all the voltages are the same, the equivalent capacitance is given by $C = C_1 + C_2$. This can be generalized for any number of capacitors connected in parallel as

$$C = \Sigma_i C_i. \tag{8.9}$$

To summarize:

When wired in parallel, capacitors have equal potential differences.

When wired in series, capacitors have equal stored charge.

Note that the form of Eqs. 8.8 and 8.9 are the same as Eqs. 8.6 and 8.7 except that series versus parallel connections are reversed for capacitors compared to resistors. This is demonstrated in the following example problem.

Example 8.7

**In the circuit on the right, the battery has a potential difference of 9.00 V. The capacitors have values as follows: $C_1 = 10.0\ \mu F$, $C_2 = 20.0\ \mu F$, $C_3 = 30.0\ \mu F$
Determine the charge stored on each capacitor.**

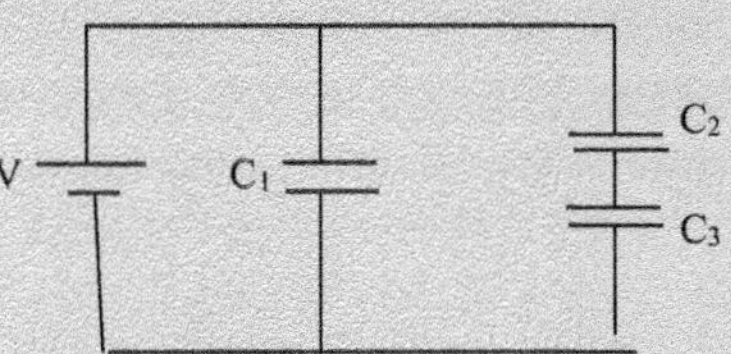

Figure given.

Knowns: $C_1 = 10.0\ \mu F$; $C_2 = 20.0\ \mu F$; $C_3 = 30.0\ \mu F$; $V = 9.00$ V

Unknowns: Q_1; Q_2; Q_3

When faced with several capacitors in series and parallel, we reduce the circuit with effective capacitors. First, let's take $C2$ and $C3$, which are wired in series, and replace them with an effective capacitor:

$$1/C_{\mathrm{eff}} = 1/C_2 + 1/C_3 = 1/20.0\ \mu\mathrm{F} + 1/30.0\ \mu\mathrm{F} = 0.083\ (\mu\mathrm{F})^{-1}$$
$$C_{\mathrm{eff}} = 12.0\ \mu\mathrm{F}$$

Now we just have two capacitors which are hooked in parallel. Thus, we can use Eq. 8.8 to determine their effective capacitance:

$$C_{\mathrm{eff}} = C_1 + C_2 = 10.0\ \mu F + 12.0\ \mu F = 22.0\ \mu F$$

That leaves us with a circuit having just a battery and one capacitor.

The charge stored on the effective capacitor is easy to calculate now, remembering Eq. 7.15 from the previous chapter:

$$\Delta V = Q/C$$
$$Q = \Delta VC = (9.00\ \mathrm{V})\left(2.20 \times 10^{-5}\ \mathrm{C/V}\right) = 1.98 \times 10^{-4}\ \mathrm{C}$$

That's the total charge. We want to know the charge on *each capacitor*. We know that capacitors in parallel must have the same voltage. Thus, in the second step we had the 10.0 μF capacitor wired in parallel with the effective, 12.0 μF capacitor. They each must have a potential of 9.00 V. Since we know V and C for the 10.0 μF capacitor, we can immediately get the charge on it:

$$Q = \Delta VC = (9.00\ \mathrm{V})\left(1.00 \times 10^{-5}\ \mathrm{C/V}\right) = 9.00 \times 10^{-5}\ \mathrm{C}$$

If 9.00×10^{-5} C is stored on the 10.0 μF capacitor, the rest must be stored on the other two:

$$\text{charge on other two} = 1.98 \times 10^{-4}\text{C} - 9.00 \times 10^{-5}\text{C} = 1.08 \times 10^{-4}\text{C}$$

Since capacitors in series must have the same charge, then the other two capacitors each store 1.08×10^{-4} C. You can check that this is correct by using this charge to calculate the potential difference in each of the two capacitors. The total will add to 9.00 V, as is should. Thus, $C1$ stores 9.00×10^{-5} C, while C_2 and C_3 each stores 1.08×10^{-4} C.

Now try the following Problem.

Student

8.10 In the circuit to the right, determine the charge on each capacitor.

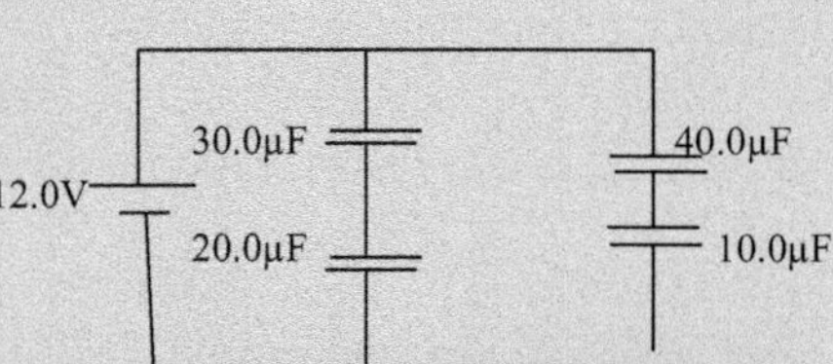

Next Level

RC Circuits

One final important type of circuit involves both resistors and capacitors. This is called an RC circuit. An example is shown in Fig. 8.9. So far we have been considering circuits under steady state conditions. The current and voltage in an RC circuit is transient, that is it depends on time.

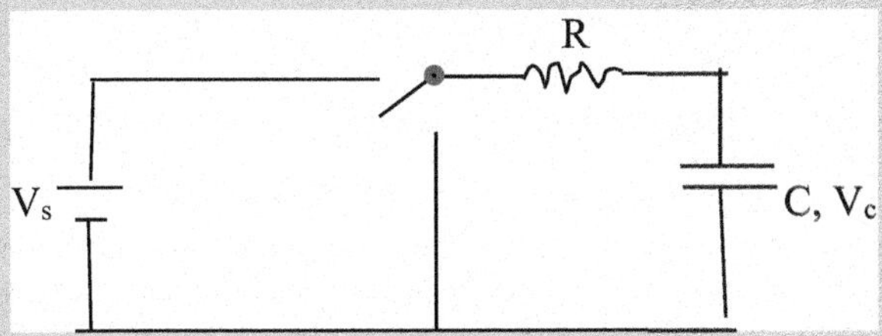

Fig. 8.9 RC circuit

When the switch is open (as shown in the figure), no current flows and there is no charge or voltage on the capacitor. When the switch is closed to the upper branch, current flows from the battery through the resistor and capacitor and back to the negative terminal of the battery. During this time the capacitor is charging. The current and voltage depends on time. The equations describing this time dependence require calculus for their derivation. You will do this in an advanced course, but for now we simple list the equations,

$$I(t) = (V_S/R)e^{-t/\tau} \tag{8.10}$$

$$V_c(t) = V_o\left(1 - e^{-t/\tau}\right) \tag{8.11}$$

with the time constant

$$\tau = RC. \tag{8.12}$$

Here V_s is the potential that the battery provided to the circuit, V_c is the voltage across the capacitor, and V_o is the voltage on the capacitor when it is fully charged. According to these equations, when the switch is closed at time $t = 0$ s, current starts to flow and as charge builds up on the capacitor it opposes the flow of current until it stops when the capacitor is fully charged. The time variation is exponential with a characteristic time constant given by the resistance and capacitance in the circuit. At the same time, the voltage across the capacitor increases exponentially from zero to the maximum value of V_o. This is shown in the left side of Fig. 8.10.

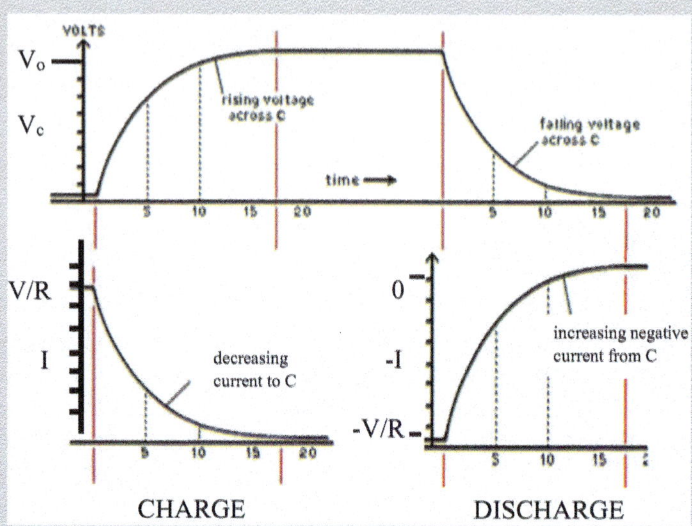

Fig. 8.10 Charging and discharging of an RC circuit

When the capacitor is fully charged and the switch is closed in the middle arm position, the battery is out of the circuit and the capacitor discharges through the resistor. Starting at $t = 0$ s, the current from the capacitor decreases exponentially in the negative direction (opposite from the charging direction) until it goes to zero when the capacitor is fully discharged. The voltage across the capacitor also decreases exponentially to zero. This is shown in the right side of Fig. 8.10.

Again the time constant for the exponential changes is given by the product of the resistance and capacitance in the circuit.

Circuits like this are useful when there is a temporary need for a current. Examples are pacemakers and camera flash bulbs. They can also act as filters as we will discuss in the next section.

8.5 AC Current

In the introduction to this chapter, we mentioned that there are two types of currents. The situations we have discussed in the previous sections utilize direct current (DC) which always flows in the same direction. The other type of current is alternating current (AC) which periodically reverses direction. AC electricity is better than DC electricity for transferring power over long distances with low power loss. To distribute enough electricity to power a large region requires a power plant to generate the electricity and a grid of power lines to transmit it to the end user. Since power is the product of voltage and current the transmission over long distances of high power with low losses is best done with high voltages and low currents. This is because the resistance with high currents causes significant loss through heating. Also, AC voltage can be stepped up or stepped down easily with transformers (as discussed in Chap. 10) while DC voltage cannot. This voltage change is critical in going from transmission voltage to usage voltage. Thus it is common for major power generation stations to produce AC electricity, step up the voltage, and transmit this on high-power lines to residential areas where step-down transformers bring it to wall outlets at a lower voltage. The wall electrical outlets in the United States provide electricity alternating direction at 60 cycles per second. The frequency of AC electricity in other countries is generally either 50 or 60 cycles per second. The source of AC electricity is a device such as a generator which can be wired to produce either AC or DC electricity instead of a battery that provides DC electricity only.

Both the voltage and the current obtained from an AC electrical outlet vary with the pattern of a sinewave. The expressions for instantaneous voltage and current are given mathematically as

$$v = V_{\text{max}} \sin(2\pi f t) \tag{8.13}$$

$$i = I_{\text{max}} \sin(2\pi f t) \tag{8.14}$$

where f is the frequency of the alternating electricity and t is the time that the instantaneous values are being measured. Note that both v and i vary with time as a sine function as shown in Fig. 8.11. This shows one cycle of current and voltage traveling in phase with each other. For a typical outlet in the U.S., the voltage varies between + 170 V and – 170 V. The voltage for the outlet is given as root mean square (rms) average value. To calculate the rated voltage of the outlet, the instantaneous voltage is squared, the average of many cycles is found, and the square root taken. This is equivalent to

$$V_{\text{rms}} = V_{\text{max}}/\sqrt{2} \tag{8.15}$$

$$I_{\text{rms}} = I_{\text{max}}/\sqrt{2}. \tag{8.16}$$

This turns out to be 120 V. Most devices in this country are rated for 120 V sources although some require 240 V.

It is important that the voltage and current sinewaves remain in-phase with each other. We are interested in delivering electrical power which we learned is defined as the product of current and voltage, $P = IV$. This product is at its highest when AC current and voltage are in phase with each other and it decreases as they get more and more out of phase.

A typical AC circuit has as source of power along with elements such as resistors, capacitors and inductors as shown in Fig. 8.12. AC power sources and inductors are discussed in more detail in Chap. 10 because they involve the interaction between electricity and magnetism. The resistor acts to impede current flow

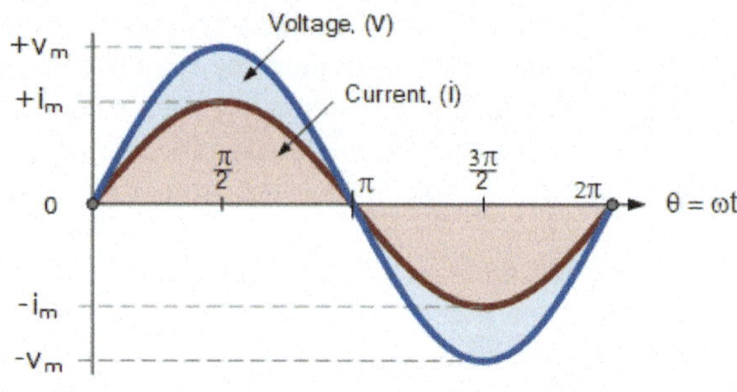

Fig. 8.11 AC current and voltage sine waves

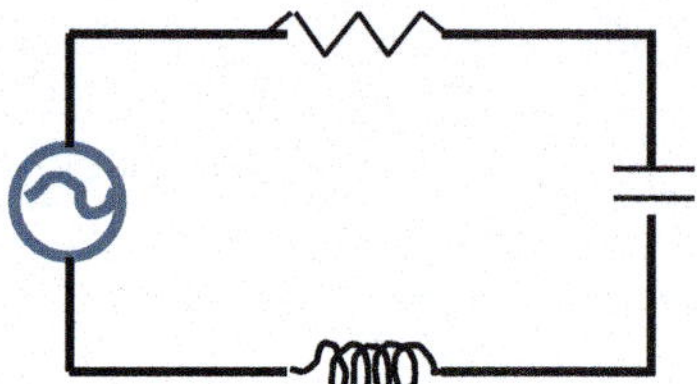

Fig. 8.12 AC circuits

just like it does in a DC circuit. Its effect on the current flow is the same no matter which direction the AC current is flowing.

As the AC current changes direction, a capacitor changes from charging to discharging. Because of its time dependence in charging and discharging as discussed in the previous section, a capacitor in an AC circuit acts like a resistor and impedes the flow of current. This effect is described by a property called reactance which is given by

$$X_{\mathrm{C}} = (2\pi f C)^{-1} \tag{8.17}$$

where C is the capacitance of the capacitor and f is the frequency of the AC electricity. The units of reactance are ohms.

An important difference between AC and DC circuits are the frequency dependences of the AC circuit components. The resistance of a resistor has no frequency dependence. However, at low frequencies, a capacitor has a high reactance, so AC current is blocked from flowing. At high frequencies the reactance of a capacitor is low so AC current can flow. The reactance's of more than one capacitor connected in series add together to give an effective reactance,

$$X_{\mathrm{C}} = \Sigma_i X_{\mathrm{C}_i}. \tag{8.18}$$

If capacitors are connected in parallel, their total effective reactance is given by

$$1/X_{\mathrm{C}} = \sum_i (1/X_{\mathrm{Ci}}). \tag{8.19}$$

Another component used is AC circuits is called an inductor. In a circuit it is shown as a coiled wire. Because its operation involves the interaction of electricity and magnetism, it is described in more detail in Chap. 10. In an AC circuit an inductor has a reactance given by

$$X_L = 2\pi fL \tag{8.20}$$

in units of Ohms. L is the inductance of the component defined in Chap. 10. The unit of inductance is the henry, designated by H. $1H$ is 1 $\text{kgm}^2/(\text{s}^2\text{A}^2)$. In this case the reactance is directly proportional to frequency. Therefore, an inductor blocks the flow of high frequency AC current and does not block the flow of low frequency current. The reactance's of more than one inductor connected in series add together for an effective reactance of

$$X_L = \Sigma_i X_{Li}. \tag{8.21}$$

If they are connected in parallel, the total effective reactance is given by

$$1/X_L = \sum_i (1/X_{Li}). \tag{8.22}$$

The total reactance of a circuit is a combination of the capacitive reactance and inductive reactance is given by

$$X = |X_C - X_L|. \tag{8.23}$$

where the vertical lines indicate that this expression gives the magnitude of the reactance without a plus or minus sign.

The reason for the negative sign is that the capacitive reactance and the inductive reactance have opposite effects on the phase relationship between the sinewave describing the current and the sinewave describing the voltage. When a switch closes on an AC circuit containing only a capacitor, the current immediately starts to flow at its maximum value to charge the capacitor but the voltage starts from zero and builds up to its maximum value occurring a quarter cycle later. Thus the sinewave describing the voltage lags the current sinewave by a phase factor of 90° or $\pi/4$.

We will learn in Chap. 10 that an inductor creates a magnetic field when current flows through it and this produces an emf that opposes any change in current. When a switch is closed on an AC circuit containing only an inductor, the voltage starts at its maximum value and decreases while the current starts at zero and increases. Thus the voltage leads the current by a phase factor of 90° or $\pi/4$.

The fact that capacitive and inductive reactance have exactly the opposite effect on the phase shift between the current and voltage sinewaves is the reason for the negative sign in Eq. 8.23. For capacitors in an AC circuit, the peak of the current curve leads the peak of the voltage curve by 90°, while for inductors current lags voltage by 90°. For a resistor, the current and voltage sinewaves remain in phase.

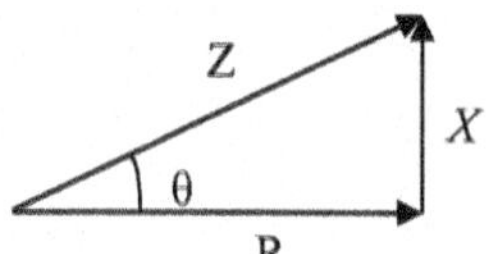

Fig. 8.13 Impedance

If X_C and X_L balance each other out, the dependencies on frequency of the total reactance of the circuit is zero, and the current and voltage curves stay in phase with each other.

The sum of the resistance and reactance in an AC circuit at a specific frequency is called impedance. It is designated as Z and has units of ohms. Because of the differences in their phases, the R and X components cannot be simply added together. The magnitude of the impedance is expressed as Ohm's law for AC circuits,

$$Z = V_{\text{max}}/I_{\text{max}} = \sqrt{X^2 + R^2}. \tag{8.24}$$

The phase difference between current and voltage with an impedance Z can be represented vectorially as shown in Fig. 8.13. A horizontal line represents the I and V sinewaves being in phase with each other, and the angle θ for the impedance shows how much they are out of phase.

One interesting aspect of a circuit with an AC source that has a resistor, capacitor, and inductor, is its resonant frequency. In Chap. 6, Sect. 6.4, we learned about the importance of a resonance frequency in an oscillating system. For this case Eq. 8.24 becomes

$$Z = \sqrt{X^2 + R^2} = \sqrt{\left(2\pi fL - \frac{1}{2\pi fc}\right)^2 + R^2}. \tag{8.25}$$

This frequency dependent impedance will be minimum when the reactance term is zero. This occurs at a frequency of

$$\mathrm{f_r} = \frac{1}{2\pi\sqrt{LC}}. \tag{8.26}$$

This is the resonant frequency for the circuit. It provides the largest current per voltage. This is important when many frequencies are present and you want to select just one, such as tuning to a specific radio station. The resonant frequency of this circuit helps you select just that frequency.

These concepts are summarized in Fig. 8.14.

Now try the following problems.

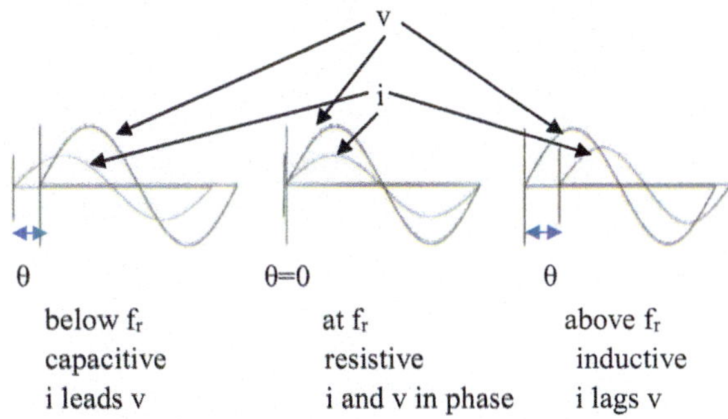

Fig. 8.14 Phase versus impedance

Student

8.11 A capacitor is measured to have a capacitive reactance of 25 Ω when connected to an AC circuit operating at 400 cycles per second. What is its capacitance? If the resistance in the circuit is 30 Ω, what is the phase angle between the current and voltage?

8.12 The voltage in an AC circuit with a resistance of 30 Ω varies sinusoidally with time with a maximum of 170 V. What is the instantaneous voltage when it has reached 45° in its cycle. What is the current at that point?

Summing Up

You now know how to analyze electric circuits using Ohms' Law and Kirchhoff's laws. This is useful in understanding how many electrical appliances work. It involves knowing about resistors, capacitors, and inductors. In a later chapter we will add transistors to this list of components. For A.C. circuits you learned about controlling the impedance in order to keep the current and voltage in phase with each other. In Chap. 10 we will apply some of this knowledge to devices like motors and generators.

Answers to the Problems

8.1 If you want the maximum current, you want the least resistance. Thus, you want to use the material with the lowest resistivity. Of the materials listed in the table, silver has the lowest resistivity. Of course, this is not practical in most applications, because silver is expensive. Thus, copper is more often used, because it has a very low resistivity but is not as expensive as silver.

8.2

Knowns: $l = 1.0\text{m}$; $\rho = 1.72 \times 10^{-8}\Omega\text{m}$; $R = 8.6 \times 10^{-5}\Omega$; $A = d^2$

Unknowns: d

We know that resistance depends on resistivity (which we have in the table), length (given), and cross-sectional area. We can calculate that:

$$R = \rho L/A$$

$$A = \rho L/R = \left(1.72 \times 10^{-8}\ \Omega\text{m}\right)(1.0\ \text{m})/\left(8.6 \times 10^{-5}\ \Omega\right) = 0.00020\,\text{m}^2$$

Assuming the width and height are equal, the length of each is simply the square root of the area,

$$d = \sqrt{A} = \sqrt{0.00020\ \text{m}^2} = 0.014\ \text{m}.$$

8.3

Knowns: $P = 14.5$ W; $R = 25.0\ \Omega$

Unknowns: V

To answer this question, we first need to know what the current is. We can get that from Eq. 8.4:

$$P = I^2R$$

$$I = \sqrt{P/R} = \sqrt{\frac{14.5\ \text{J/s}}{25.0\ \text{V/A}}} = 0.762\ \text{A}$$

Notice how the units work out. Since a Volt is a J/C, Joules cancel, and we are left with A^2 under the square root. Now that we have current, we can use Eq. 8.5 to get voltage:

$$P = IV$$

$$V = P/I = \frac{14.5\ \text{J/s}}{0.762\ \text{C/s}} = 19.0\,\text{V}$$

8.4

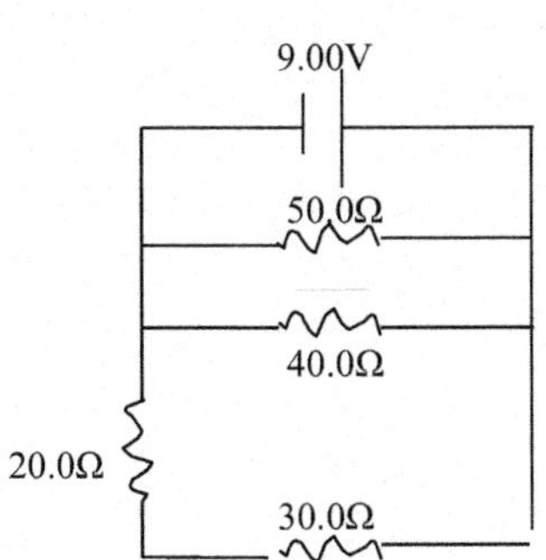

Knowns: shown in figure.

Unknowns: P

The first thing we need to do is to replace all of the resistors with an equivalent resistor. To do that, we need to determine which resistors are in parallel and which are in series. The 30.0 Ω and 20.0 Ω resistors are in series, because any current which goes through one must go through the other. Thus, we can add their resistances and come up with an equivalent resistor for these two of

$$R'_{\text{eff}} = 20.0\ \Omega + 30.0\ \Omega = 50.0\ \Omega$$

This is in parallel with the other two resistors. Since all resistors are in parallel, we can find an equivalent resistor for the total circuit as

$$\begin{aligned} 1/R_{\text{eff}} &= 1/R_1 + 1/R_2 + 1/R'_{\text{eff}} = 1/50.0\ \Omega + 1/40.0\ \Omega + 1/50.0\ \Omega = 0.0650\ \Omega^{-1} \\ R_{\text{eff}} &= 15.4\ \Omega \end{aligned}$$

This takes us down to the battery and one 15.4 Ω resistor. To determine the power that this circuit draws, we must first determine the current:

$$\begin{aligned} V &= I \cdot R \\ I &= V/R = 9.00\ \text{V}/15.4\ \Omega = 0.584\ \text{A} \end{aligned}$$

Now we can get power from either Eq. 8.4 or 8.5:

$$P = IV = (0.584\ \text{A})(9.0\ \text{V}) = 5.26\ \text{W}$$

8.5

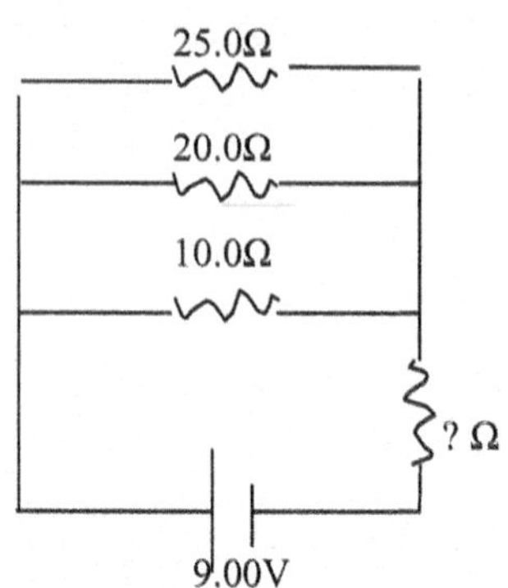

Knowns: $I = 0.500$ A

Unknowns: ? Ω

Note that the resistor with the question mark is not a part of the parallel circuit. The other resistors are in parallel, but *all* of the current must travel through the unknown resistor. We can reduce the three parallel resistors to one effective resistor:

$$1/R_{\text{eff}} = 1/R_1 + 1/R_2 + 1/R_3 = 1/25.0\ \Omega + 1/20.0\ \Omega + 1/10.0\ \Omega = 0.190\ \Omega^{-1}$$
$$R_{\text{eff}} = 5.26\ \Omega$$

Since the resistors that are left in series, they add. Thus, if we could find some way to get the total resistance, getting the unknown resistance is easy. We know the voltage and the current, so Ohm's Law will tell us the resistance:

$$V = IR$$
$$R = V/I = 9.00\ \text{V}/0.500\ \text{A} = 18.0\ \Omega$$

That's the total effective resistance. Thus, we can now calculate the unknown resistance:

$$R_{\text{eff}} = R_1 + ?$$
$$? = R_{\text{eff}} - R_1 = 18.0\ \Omega - 5.26\ \Omega = 12.7\ \Omega$$

8.6

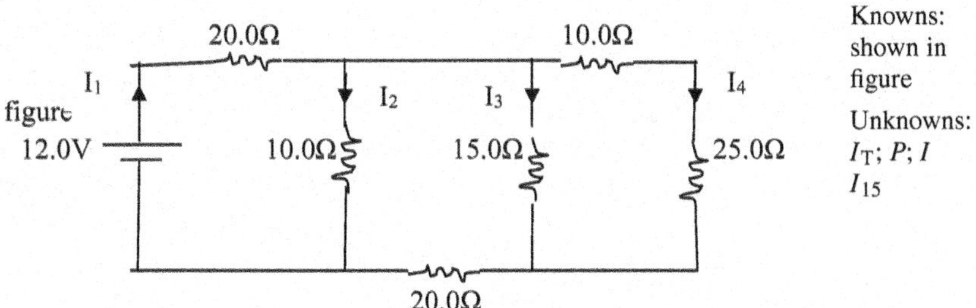

To solve a problem like this, we use Kirchhoff's Laws. There is another way to solve the problem, but it is no easier, so we might as well stick with the process we know. Let's start with the first loop to the left. In this loop, we have I_1 going from the plus side of the battery through the 20.0 Ω resistor, but then the current has a choice. Thus, only part of the current, I_2, will go through the 10.0 Ω resistor. Then, we encounter the battery, negative side first. Thus the voltages in the loop are:

$$12.0\ \text{V} - I_1(20.0\ \Omega) - I_2(10.0\ \Omega) = 0$$

That gives us our first equation, but we have two unknowns. Thus, we have to continue.

The next loop to the right has I_1 going through the 20.0 Ω resistor, but then the current has a choice. Thus, only part of the current, I_3, will go through the 15.0 Ω resistor. Look at the other 20.0 Ω resistor on this loop, however. We cannot say that I_3 runs through it, because there is a junction behind it. Whatever is running through the 10.0 Ω and 25.0 Ω resistors on the far right of the circuit will also run through this 20.0 Ω resistor. Let's call that I_4. This means that *both* I_3 and I_4 run through the 20.0 Ω resistor. Then, we encounter the battery, negative side first. Thus:

$$12.0\ \text{V} - I_1(20.0\ \Omega) - I_3(15.0\ \Omega) - (I_3 + I_4)(20.0\ \Omega) = 0$$

Now we have two equations, but four unknowns. Thus, we have go on to the next loop.

In this loop, we have I_1 going through the 20.0 Ω resistor, but then the current has a choice. We already called the current that runs through the far right of the circuit I_4, and it will go through the 10.0 Ω and 25.0 Ω resistors. Then, we encounter the 20.0 Ω resistor. We already determined that both I_3 and I_4 will run through that one. Finally, we encounter the battery, negative side first. Thus:

$$12.0\ \text{V} - I_1 \cdot (20.0\ \Omega) - I_4 \cdot (10.0\ \Omega) - I_4 \cdot (25.0\ \Omega) - (I_3 + I_4) \cdot (20.0\ \Omega) = 0$$

Finally, we know that all of the current must recombine once it comes back from the last junction. Thus:

$$I_1 = I_2 + I_3 + I_4$$

We now have a system of 4 equations and 4 unknowns.

Let's get I_2 in terms of I_1 using the first equation:

$$I_2 = \frac{12.0\ \text{V} - I_1(20.0\ \Omega)}{10.0\ \Omega}$$

In the second loop equation we can use the current equation to replace $I_3 + I_4$ with $I_1 - I_2$ and solve it to get an expression for I_3.

$$I_3 = \frac{12.0\ \text{V} - I_1(40.0\ \Omega) + I_2(20.0\ \Omega)}{15.0\ \Omega}$$

Since we have an expression for I_2 in terms of I_1, we can substitute it into this equation to get an expression for I_3 in terms of I_1.

$$I_3 = \frac{36.0\ \text{V} - I_1(80.0\ \Omega)}{15.0\ \Omega}$$

Now using the third loop equation and the expressions for I_2 and I_3 in terms of I_1, we can get an expression for I_4 in terms of I_1.

$$I_4 = \frac{36.0\ \text{V} - I_1(80.0\ \Omega)}{35.0\ \Omega}$$

Now we can put all of these expressions for I_2, I_3, and I_4 into the total current equation:

$$\begin{aligned} I_1 &= I_2 + I_3 + I_4 \\ &= \frac{12.0\ \text{V} - I_1(20.0\ \Omega)}{10.0\ \Omega} + \frac{36.0\ \text{V} - I_1(80.0\ \Omega)}{15.0\ \Omega} + \frac{36.0\ \text{V} - I_1(80.0\ \Omega)}{35.0\ \Omega} \\ &= \frac{42.0\ \text{V} + 84.0\ \text{V} + 36.0\ \text{V}}{372\ \Omega} = 0.435\ \text{A} \end{aligned}$$

Since that's the sum of all the individual currents, it is the total current. Thus, the total current is 0.435 A.

The power drawn is easy:

$$P = IV = (0.435\ \text{C/s})(12.0\ \text{J/C}) = 5.22\ \text{W}$$

Finally, the problem asks for the current in the 15.0 Ω resistor. According to our definitions, that was I_3.

$$I_3 = \frac{36.0\ \text{V} - (0.435\ \text{A})(80.0\ \Omega)}{15.0\ \Omega} = 0.080\ \text{A}$$

Therefore, 0.080 A flow through the 15.0 Ω resistor.

8.7

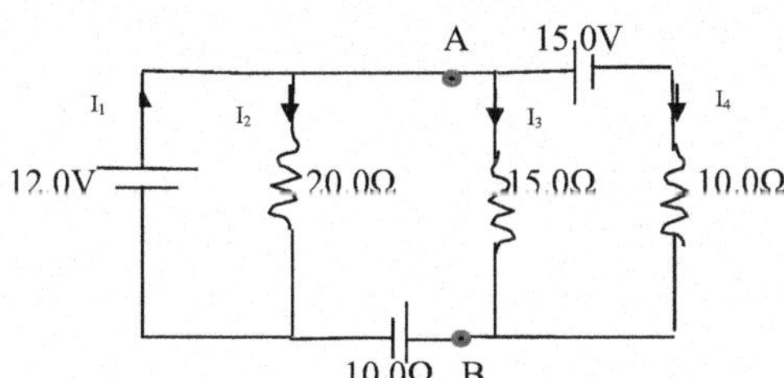

Knowns: listed in figure

Unknowns: I_R; V_{AB}

Once again, we have to use Kirchhoff's Laws. The first loop is on the left of the circuit through the 20.0 Ω resistor. In this loop, I_1 leaves the battery, but it soon has to make a choice. Only part of the current (I_2) will go through the 20.0 Ω resistor. Then, we encounter the negative side of the 12.0 V battery. Thus, this loop tells us:

$$12.0\ \text{V} - I_2(20.0\ \Omega) = 0$$

We can solve this equation for I_2.

$$I_2 = 12.0\ \text{V}/20.0\ \Omega = 0.600\ \text{A}$$

We can now go to the next loop to the right. In this loop, I_1 leaves the battery, but it soon has to make a choice. Only part of the current (I_3) will go through the 15.0 Ω resistor. Then, we encounter the positive side of the 10.0 V battery. Once we pass through that battery, we encounter the negative side of the 12.0 V battery. Thus, this loop tells us:

$$12.0\ \text{V} - 10.0\ \text{V} - I_3(15.0\ \Omega) = 0$$

$$I_3 = 2.0\ \text{V}/15.0\ \Omega = 0.13\ \text{A}.$$

We can now go to the next loop to the right. In this loop, I_1 leaves the battery, but it soon has to make a choice. Only part of the current (I_4) will go through the 10.0 Ω resistor. Before reaching the resistor, however, it encounters the positive side of the 15.0 V battery. Then, it encounters the resistor, then it encounters the positive side of the 10.0 V battery. Then it encounters the negative side of the 12.0 V battery. This loop, then, tells us:

$$-15.0\ \text{V} - 10.0\ \text{V} + 12.0\ \text{V} - I_4(10.0\ \Omega) = 0$$

$$I_4 = (13.0\ \text{V})/(10.0\ \Omega) = -1.30\ \text{A}$$

The negative sign tells us that the direction of the current is opposite of what we drew. This tells us, then, that 0.600 A run through the 20.0 Ω resistor; 0.13 A run through the 15.0 Ω resistor; and 1.30 A run through the 10.0 Ω resistor.

We are not quite done, however. We also need to know the potential difference between points A and B. Let's assume we are taking the second loop. At point A, I_3 and I_4 are running through the wire. If we follow the second loop, I_3 runs down the 15.0 Ω resistor. Then, it encounters point B. The difference in potential, then, is just the drop in voltage across that resistor.

$$V = I_3R = (0.13\ \text{A}) \cdot (15.0\ \Omega) = 2.0\ \text{V}$$

Thus, point B is 2.0 V lower in potential than point A. Please note that had we chosen the third loop instead, we would have gotten the same answer. In the third loop, the current passes through the 15.0 V battery encountering the positive side first. That drops the potential by 15.0 V. Then, I_4 passes through the 10.0 Ω resistor. However, we have the direction backwards, so we are traveling opposite the current, which raises the potential by IR, or 13.0 V. In the end, then, that loop also indicates a 2.0 V drop from point A to point B, as it should.

8.8

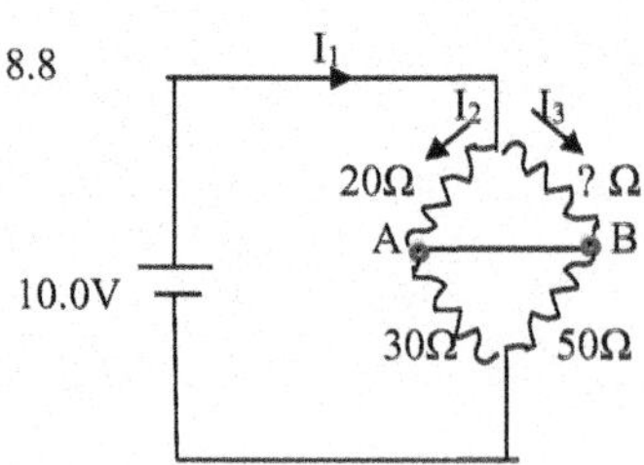

Knowns: shown in figure; $V_{AB} = 0\text{V}$

Unknowns: ?Ω

Since there is no current flowing from point A to point B, there are only two possible loops in which to travel. The first is the left branch. I_1 leaves the battery but reaches a choice. Part of the current (I_2) travels through the 20.0 Ω and 30.0 Ω resistors, and returns to the battery, encountering the 10.0 V battery negative side first. Thus:

$$10.0\ \text{V} - I_2(20.0\ \Omega) - I_2(30.0\ \Omega) = 0$$

This can be solved to show that $I_2 = 0.200$ A.

The other loop is the right branch. A part of I_1, called I_3, will flow through the unknown resistor and the 50.0 Ω resistor, and then it will encounter the negative side of the battery. Thus:

$$10.0\ \text{V} - I_3 \cdot (?\Omega) - I_3 \cdot (50.0\ \Omega) = 0$$

This doesn't help, however, because there are two unknowns. We could say that $I_1 = I_2 + I_3$, but that doesn't help, either, since it just adds one more unknown. Remember, there is no potential difference between point A and point B. That means whatever potential drop I_2 experiences as it travels through the 20.0 Ω resistor, I_3 must experience that exact same drop as it passes through the unknown resistor. After all, both currents are at the same potential before they encounter the resistors. To get to point A, I_2 must experience a drop as it passes through the 20.0 Ω resistor.

To get to point B, I_3 must experience a drop as it passes through the unknown resistor. If those drops are not identical, there will be a potential difference between A and B. However, we know that there is not, so the drops must be the same. Thus:

$$I_2 \cdot (20.0\ \Omega) = I_3 \cdot (?\ \Omega)$$

We already know I_2, so we can put it in:

$$(0.200\ \text{A})(20.0\ \Omega) = I_3 \cdot (?\ \Omega)$$
$$I_3 \cdot (?\ \Omega) = 4.00\ \text{V}$$

Well, we can take the equation for loop 2 and just stick 4.00 V in for $I_3 \cdot (?\ \Omega)$.

$$10.0\ \text{V} - 4.00\ \text{V} - I_3(50.0\ \Omega) = 0$$

This tells us that $I_3 = 0.12$ A. With that, we can now get the resistance:

$$I_3 \cdot (?\ \Omega) = 4.00\ \text{V}$$
$$(0.12\ \text{A}) \cdot (?\ \Omega) = 4.00\ \text{V}$$

The unknown resistance, then, is 33 Ω.

8.9

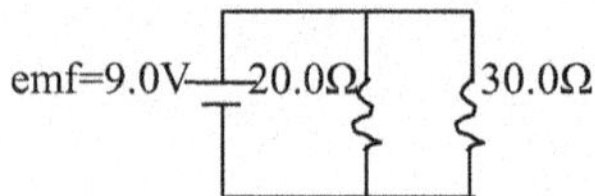

Knowns: emf = 9.0 V; $V = 8.5$ V

Unknowns: r

First get the effective resistance of the external resistors,

$$1/R_{\text{eff}} = 1/20.0\ \Omega + 1/30.0\ \Omega = 0.0833\ \Omega^{-1}$$
$$R_{\text{eff}} = 12\ \Omega$$

Now the total resistance must be *more* than 12.0 Ω, because of the internal resistance of the battery.

We know that after the internal resistor of the battery, the potential drops 0.41 V, because the battery has an emf of 9.00 V but delivers only 8.59 V. That means the internal resistor causes it to lose 0.41 V. Thus, we know that

$$0.41\ \text{V} = I \cdot r$$

where "r" is the internal resistor. The problem is, this has two unknowns. However, we do know something else. We know that the total current is simply the emf divided by the total resistance:

$$I = 9.00\ \text{V}/(12.0\ \Omega + r)$$

Using this for the current in the previous expression gives

$$0.41\ \text{V} = \frac{9.00\ \text{V}}{12.0\ \Omega + r} r$$

$$r = 4.9\ \text{V}\Omega/8.59\ \Omega = 0.57\ \Omega$$

8.10

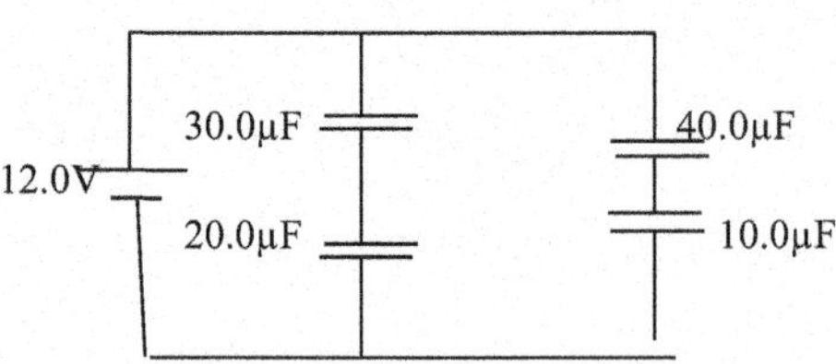

Knowns: given in figure

Unknowns: Q_i

We can first calculate an effective capacitance for the capacitors in series, using Eq. 8.7

$$1/C_{\text{eff1}} = 1/C_1 + 1/C_2 = 1/30.0\ \mu\text{F} + 1/20.0\ \mu\text{F}$$
$$C_{\text{eff1}} = 12.0\ \mu\text{F}$$

$$1/C_{\text{eff2}} = 1/C_1 + 1/C_2 = 1/40.0\ \mu\text{F} + 1/10.0\ \mu\text{F}$$
$$C_{\text{eff2}} = 8.00\ \mu\text{F}$$

This reduces the circuit to these two effective capacitors in parallel.

Capacitors in parallel are subject to the same potential, so each effective capacitor has 12.0 V across its plates. That means the first branch stores:

$$\Delta V = Q_1/C_{\text{eff1}}$$
$$Q_1 = \Delta V_{\text{eff1}} = (12.0\ \text{V})\left(1.2 \times 10^{-5}\ \text{C/V}\right) = 1.44 \times 10^{-4}\ \text{C}$$

The second branch stores:

$$\Delta V = Q_1/C_{\text{eff2}}$$
$$Q_2 = \Delta V C_{\text{eff2}} = (12.0\ \text{V})\left(8.00 \times 10^{-6}\ \text{C/V}\right) = 9.60 \times 10^{-5}\ C$$

Since capacitors in series store the same amount of charge, the 20.0 μF and 30.0 μF capacitors store 1.44×10^{-4} C, and the other two capacitors store 9.60×10^{-5} C.

8.11

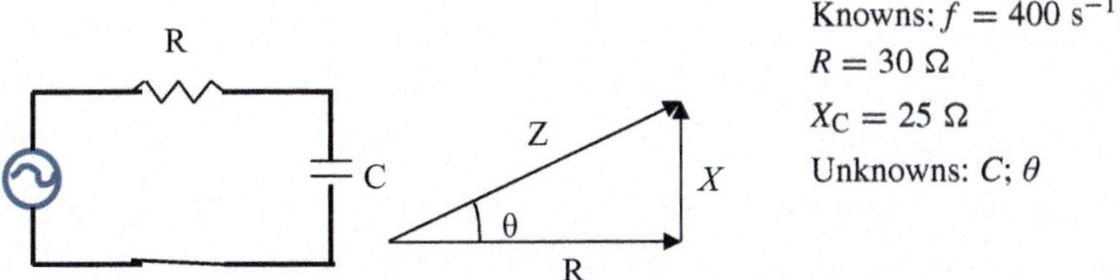

From Eq. 8.16, the capacitance is

$$X_C = (2\pi f C)^{-1}$$
$$C = \frac{1}{2\pi f X_C} = \frac{1}{2\pi\left(400\ \text{s}^{-1}\right)(25\ \Omega)} = 1.6 \times 10^{-5}\ \text{f} = 16\ \mu\text{f}$$

Note that $1\ f = 1$ coulomb/volt = 1 coulomb/(amp ohm) = 1 sec/ohm .

From the impedance diagram, the phase angle is

$$\tan^{-1}\theta = X/R = 25/30 = 39.8°$$

Since the reactance is capacitive, i leads v.

8.12

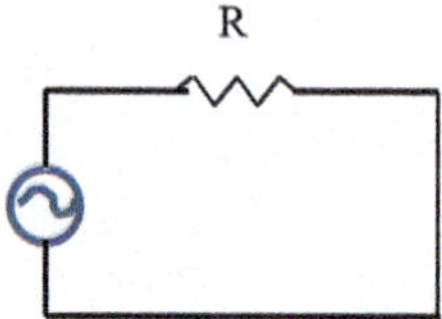

Knowns: $R = 30\ \Omega$; $v_m = 170$ V; $\theta = 45°$

Unknowns: v; i

$$v = v_m \sin\theta = 170\ \text{V}\ \sin 45° = 120\ \text{V}$$

Since the impedance is only from the resistor, the current is in phase with the voltage so we can use Ohm's law to get the current

$$i = v/R = 120\ \text{V}/30\ \Omega = 4.0\ \text{A}$$

Study Guide for This Chapter

1. Two conductors identical in length and shape are connected to both sides of a battery. The first is made of lead, and the second is made of iron. What is the ratio of the current in the first conductor to the current in the second conductor? (See Table 8.1)
2. When a light bulb burns out, it fails to conduct electricity. In the following circuit, all light bulbs start out lighting up. If light bulb #2 suddenly burns out, what other light bulb(s) will go out?

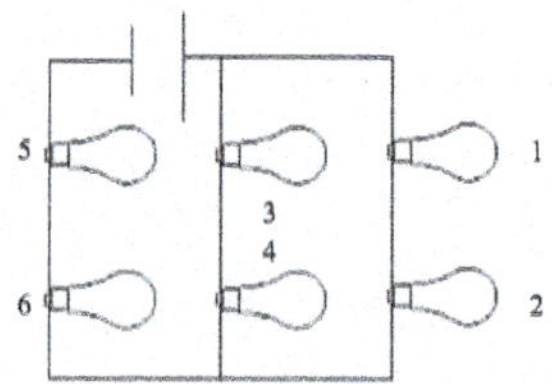

3. A "critical" bulb is a bulb that will cause the entire circuit to go dark if it burns out. List any critical bulbs in the circuit diagram above.
4. A circuit consists of a battery and a 15 Ω resistor. The power output of that circuit is measured and recorded as "P." The 15 Ω resistor is then replaced with a 45 Ω resistor. What is the new power output, in terms of "P?"

5. A battery is rated to produce 9.0 V of potential difference. When put in a circuit that has a total resistance of 10 Ω, however, it delivers only 8.5 V. If the same battery is put in a circuit whose total resistance is 100 Ω, will the potential it delivers increase, decrease, or stay the same?

6. In the circuit diagrammed to the right, switch S_1 and switch S_2 are initially open. S_1 is then closed while S_2 remains open. After a long time, S_1 is opened and then S_2 is closed. After a long time, what will be the potential across both capacitors?

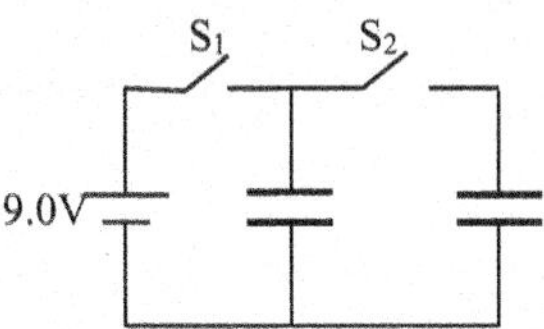

7. You have the simple circuit that is drawn to the right. You also have another 15 Ω resistor which you can place anywhere you would like in this circuit. Where would you place it to decrease the current flowing from the battery?

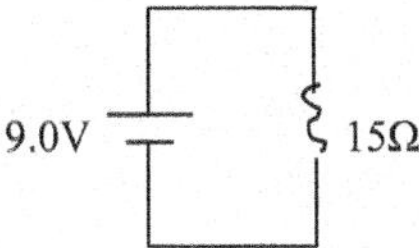

8. What three things make up the impedance of an AC circuit?
9. What determines the resonant frequency and an AC circuit?
10. A 5.00 W electric motor runs on a 9.00 V battery. What is the resistance of the motor?
11. An Ohm-meter is a device that measures the resistance between two points. If an Ohm-meter was hooked up between points A and B in the diagram below, what resistance would it read?

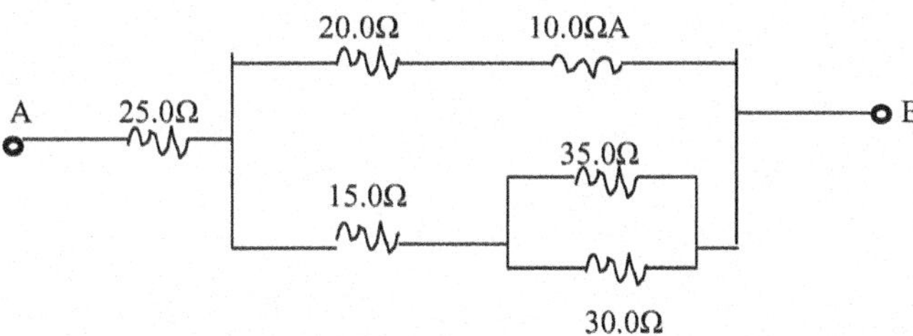

12. Suppose 5.00 A of current entered the circuit drawn above at point A. How much current would flow through the 35.0 Ω resistor? (HINT: You calculated the resistance from point A to point B. Think of the voltage drop that occurs between these two points as a result of the 5.00 A current running through that resistance.)

Questions 13–15 refer to the following circuit diagram:

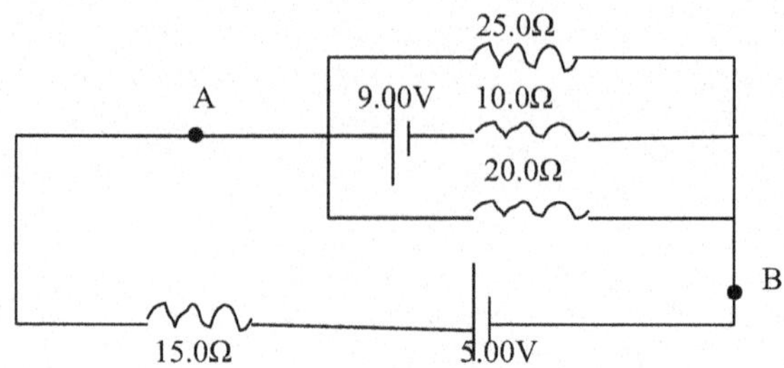

13. Draw the current as it flows in this circuit, labeling the direction as well as the amount of current in each part of the circuit.
14. Calculate the potential difference between points A and B.
15. Calculate the amount of energy dissipated by the 20.0 Ω resistor in 1.00 minute.
16. In the circuit diagram below, what is the potential difference between points A and B?

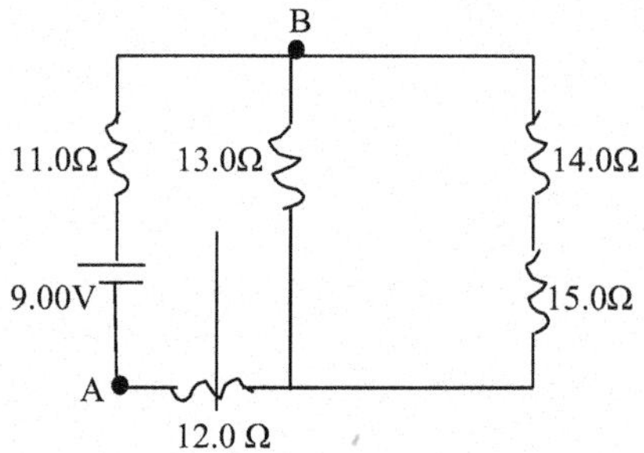

17. Work problem #16 again, assuming that the battery has an internal resistance of 3.0 Ω.
18. In the circuit below, how much charge is stored on each capacitor?

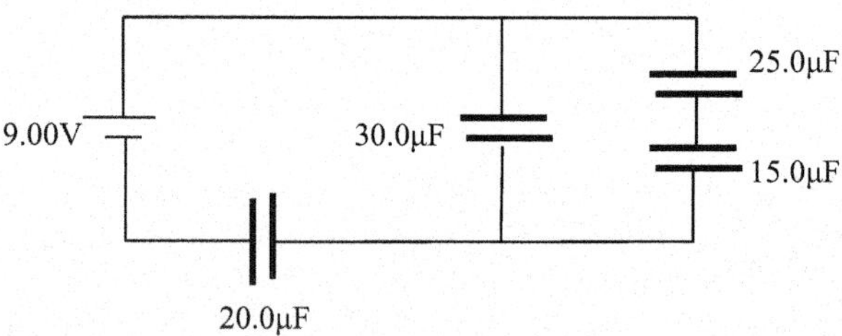

19. A radio station is broadcasting at 98.1 kHz. You have a tuning circuit with an inductor of 2 μH. What is the capacitance of the capacitor you need to tune your circuit to this radio station?
20. An AC circuit operating at a frequency of 60 s^{-1} and $V_{max} = 220$ V, has a resistor, capacitor, and inductor in series with $R = 10.0\ \Omega$, $C = 4.00\ \mu F$, and

$L = 2.00\ \mu H$. What is the impedance of the circuit, the maximum current in the circuit, and the resonant frequency of the circuit?

Next Level

21. Consider the RC circuit shown in Fig. 8.9 with $V_s = 10\,V$, $R = 2\Omega$, and $C = 5\ \mu f$. What is the time constant for the circuit? What is the initial current that flows when the charging switch is closed? What is the charge on the capacitor when it is fully charged.
22. You have an application that requires a burst of voltage that is equal to or greater than 2 V for a time of 0.1 s. You are given a 5 V battery and a 10 μf capacitor. What must the time constant of your circuit be and what other element do you need to get to complete the circuit?

Magnetism 9

Chapter Summary

Magnetism is not a new phenomenon to men. Chinese writings mentioning magnetism date to 645 B.C. and magnetic artifacts have been found in Egyptian pyramids and ancient Greek archeological sites. However, it wasn't until the late 1500s that the English scientist William Gilbert investigated magnetism using scientific methods. He discovered that the Earth itself is a magnet. Later, in the early part of the 1800s, Danish scientist Hans Christian Oersted first suggested the relationship between electricity and magnetism. In this module we will review the basic physical properties of magnetism and in the following module we will discuss how we use electro-magnetism. Some of the themes of the chapter are represented in Figs. 9.1, 9.2 and 9.3. Figure 9.1 shows a 4-pole magnet. Figure 9.2 shows birds using the earth's magnet field to migrate. Figure 9.3 shows a magnet used to accelerate electrons.

Main Concepts in This Chapter

- Permanent Magnets
- Charged Particle in a Magnetic Field
- Motional EMF
- Force on a Current Carrying Wire

R. C. Powell, *Physics*, Undergraduate Lecture Notes in Physics,
https://doi.org/10.1007/978-3-032-09361-5_9

Fig. 9.1 Magnetic quadrupole. *Credit* K. Aainaqatai, public domain

Fig. 9.2 Migrating birds. *Credit* Handigumus, Creative Commons 0

Fig. 9.3 Cyclotron. *Credit* NASA Glenn Research Center, public domain

9.1 Introduction

Magnetism is closely related to electricity which is why they are grouped together in our study of physics. Magnetism is the force that one magnetic object exerts on another one through its magnetic field. We will review the basic physics of magnetism in this chapter and discuss the relationship between electricity and magnetism in Chap. 10.

Like gravity and electrostatic fields, magnetic fields cause a force to act at a distance without two objects needing to touch each other. Thus the first thing we need to understand is the source of a magnetic field.

All materials are made up of atoms which have electrons orbiting around their nuclei (see Chap. 16). The moving charge results in a local magnetic field surrounding each atom. However, remember that a field is a vector quantity. In most materials these microscopic magnetic fields point in all directions and their superposition cancels each other out so the material has no net magnetic field. In some materials, there are regions called domains where all the atoms are aligned so their magnetic fields reinforce each other instead of canceling each other. This results in a measurable local magnetic field in that region of the material. Each domain can have dimensions of tens of micrometers and contain many billions of atoms. However, a sample of material has many different domains with their magnetic field vectors pointing in random directions so their magnetic fields still cancel each other. In some cases, the magnetic fields of many domains become aligned in the same direction so the material can have a net magnetic field surrounding it. If this magnetic field remains permanently, these types of materials are called ferromagnetic. One common example is iron.

There is another type of material called paramagnetic. In these materials, the microscopic magnetic fields generally cancel out. However, if a paramagnetic material is placed in an external magnetic field its microscopic magnetic fields will become aligned to produce a net magnetic field in the direction of the external field. When the external field is removed, the net field of the paramagnetic material disappears due to the thermal motion of the atoms. Aluminum is a paramagnetic material.

A third type of material is called diamagnetic. Like paramagnetic materials, diamagnetic materials are non-magnetic until they are exposed to an external magnetic field. Then they become magnetic with their field in the opposite direction to the external field. Their magnetic field disappears when the external field is taken away due to the thermal motion of the atoms. Copper is a diamagnetic material.

The parameter describing the magnetic properties of a type of material is its magnetic permeability. This is represented by the Greek letter μ. Its units are henries per meter which are equivalent to newtons per ampere squared (N/A^2). It is a measure of the amount of magnetization a material acquires in response to an external applied magnetic field. The permeability of free space is an important scientific constant given by $\mu_o = 4\pi \times 10^{-7}\ N/A^2 = 12.57 \times 10^{-7}\ N/A^2$. Ferromagnetic and paramagnetic materials have values of permeability greater than μ_o while diamagnetic materials have values of permeability less than μ_o.

We will discuss permeability more in the next chapter. For now, perform the following experiment to demonstrate the importance of permeability.

Experiment 9.1

Magnetic Permeability

(The idea for this experiment came from the Exploratorium Teachers Institute.)

Supplies

- Magnet (the stronger the better; using a stack of three disk magnetics 2 cm in diameter works well)
- Two pieces of thin (2 mm) cardboard of equal size, about 13×4 cm
- Two pieces of construction paper, each 5×10 cm
- Large, 5 cm metal paper clip
- Glue
- Stack of books
- Five long, thin knives made out of different materials (steel, silver, copper, plastic, wood, etc.)

Introduction—This experiment will test different types of materials to determine whether they're permeable or nonpermeable.

Procedure:

1. Build a hollow sandwich by placing the two pieces of cardboard near each edge and between the construction paper and glue in place.

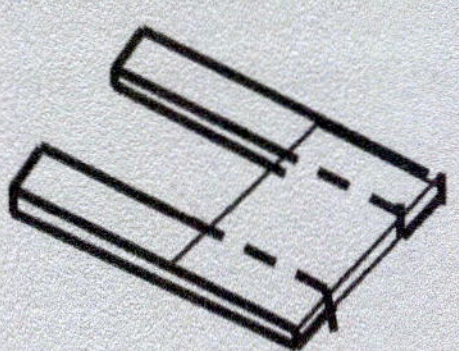

2. Place the extended ends of the cardboard between the books in the stack to suspend the sandwich in the air.
3. Place the magnet on top of the top piece of the sandwich.
4. Suspend the paperclip by the magnetic attraction on the bottom of the sandwich.
5. One at a time, insert the thin objects into the opening in the sandwich and see what happens to the paperclip.

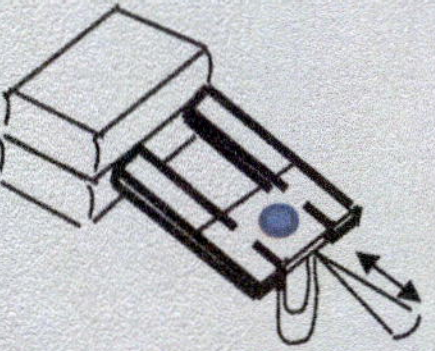

6. In your lab notebook, record an entry for each object you used stating what it was made of and what happened to the paperclip.
7. Put everything away.

Conclusions:

Explain what happened to the paperclip when each object was inserted in the sandwich opening. For your five objects, which were made of magnetic permeable material and which were not. If you wanted to shield a sensitive instrument from an external magnetic field, which of your materials would you use for the shield?

So what happened in the experiment? The magnet field lines from the magnet pass through the paper and air to the paperclip. Since the paperclip is made of a paramagnetic material, the magnetic force holds it in place. Putting nonpermeable materials between the magnet and the paperclip does nothing to change this because magnetic fields do not form inside them. However, putting permeable materials in the sandwich opening blocks the magnetic field lines from getting to the paperclip so the magnetic force holding the paperclip up goes to zero. Instead, the magnetic field lines are deflected inside the permeable material and travel back to the other pole of the magnet

9.2 Permanent Magnets

Permanent magnets are ferromagnetic materials with their domains aligned in the same direction. This creates a dipole as shown in Fig. 9.4 with the two poles designated as north and south. In some ways this is analogous to an object with positive electrical charge concentrated at one end and negative charge at the other end.

However, there is an important difference between these two situations. Positive and negative charged particles can each exist as individual entities. North and south magnetic poles only exist in pairs.

Special Topic
In today's society, the word *magnetism* is synonymous with a positive attraction. Saying that someone has a magnetic personality means that others are attracted to them. In physics, we know that magnetism has both properties of attraction and of repulsion and that they both exist together in one entity called a dipole. Maybe if we look closely enough at a person or organization we consider to be magnetically attractive we will find that they also have opposite characteristics. Society does recognize a condition called bipolar disorder that some people suffer from that changes their personality from positive to negative. When we use scientific terms like magnetism it is important that we understand the complete meaning.

The lines of the magnetic field of a magnetic dipole are shown in Fig. 9.4. This field is designated as **B** and has units of Tesla designated as T. For unit analysis, a Tesla is a N/(A m). The direction of the field is out of the north pole of the magnet and into the south pole. In regions parallel to the dipole, the direction of the field is from the north pole to the south pole. The field strength weakens as the distance from the magnet increases. Any magnetic material placed in the magnetic field of this magnetic will have a magnetic force exerted on it. The direction of the force

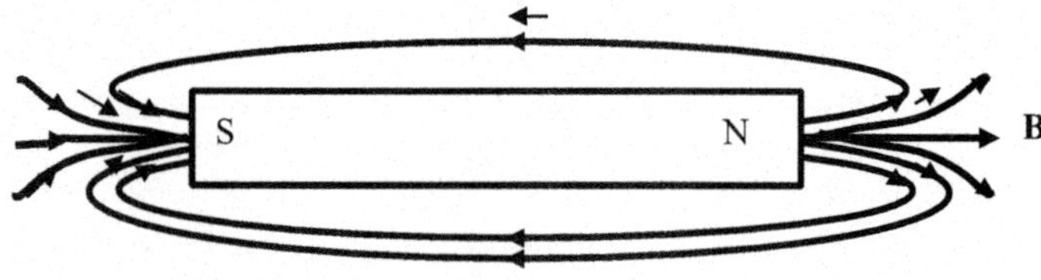

Fig. 9.4 Magnetic field around a permanent magnet

depends on the alignment of the two magnetic dipoles. Two similar poles (both north poles or both south poles) will repel each other. Two opposite poles (one north and one south) will attract each other. Note this is similar to the electrostatic charges we studied in Chap. 7 where like charges repelled each other and unlike charges attracted each other.

The most prominent permanent magnetic is the earth itself. Geologists believe that the central core of the earth is molten metal that is rotating in a way that it creates a magnetic dipole. The magnet pole in the northern hemisphere is located 11.5° away from the north pole of the axis of rotation. Because it is in the northern hemisphere, we call it the magnetic north pole even though it is actually the south pole of the earth's magnetic field. That is, the earth's magnetic field lines are directed inward at this pole. The magnetic field lines point outward from the earth's magnetic pole in the southern hemisphere making it the north pole of the magnet.

A compass has a permanent magnetic needle on an axis that is free to rotate. It aligns itself with the field lines of the earth's magnetic field and points to the earth's magnetic north. Some birds and fish have built-in ways to detect the earth's magnetic field and this helps them with their direction when they migrate. The fact that the earth is surrounded by magnetic field lines is important in protecting life on earth. The sun emits charged particles like electrons and protons which enter the earth's atmosphere along with sunlight. These charged particles could cause radiation damage to living creatures as well as disrupting communication signals. In the next section, we will learn about the motion of a charged particle in a magnetic field. In this case, the earth's magnetic field causes the charged particles to spiral toward the north and south magnetic poles where they do little damage. When they collide with particles in the earth's atmosphere, they create colorful lights called auroras.

For the remainder of this chapter we will be dealing with uniform magnetic fields. Just as we can produce a uniform electric field between two oppositely charged parallel plates, we can produce a uniform magnetic field between two parallel, opposite magnetic poles. As long as you are not near the edges of the poles, the magnetic field between the two poles will be uniform as shown if Fig. 9.5. As stated previously, the field strength is represented by the vector **B** whose magnitude is given in Teslas (N/Am) and direction from N to S.

As a further review, work the problem below.

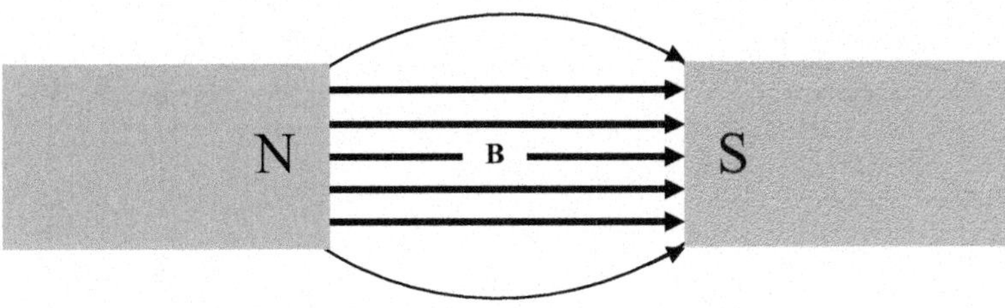

Fig. 9.5 Uniform magnetic field

Student

9.1 Suppose a substance was placed directly in between the north and south poles pictured in Fig. 9.5. If the substance was diamagnetic, which way would the magnetic field inside the substance point? What if the substance was paramagnetic? What if the substance was ferromagnetic?

9.3 Charged Particle in a Magnetic Field

Since the origin of a magnetic field is associated with the motion of electrically charged particles, it is not surprising that when a charged particle moves in a magnetic field it experiences a force. This situation is shown in Fig. 9.6 where q is the charge on the particle, **v** is its velocity vector, and **B** is a uniform magnetic field in the direction of the arrows.

Experiments show that this force is proportional to the component of the strength of the magnetic field, B, perpendicular to the particle's velocity, as well as the particle's charge and its velocity. Mathematically this is expressed as

$$\boldsymbol{F} = q(\mathbf{v} \times \mathbf{B}) \tag{9.1}$$

This is a vector with magnitude

$$F = qvB\sin\theta. \tag{9.2}$$

and direction given by the right-hand rule. In your study of vectors in mathematics, you learned the right-hand rule for vector cross products. When you point the fingers of your right hand in the direction of the first vector and curl them toward the direction of the second vector, your thumb points toward the direction of the resultant vector. In this case, point the fingers of your right hand in the direction of the velocity and curl them toward the direction of the magnetic field. Then your thumb points in the direction of the force. For the situation shown in Fig. 9.6 the direction of the force is out of the page. The angle θ is the angle between the direction of the velocity and the direction of the magnetic field. If the particle is traveling in the direction of the magnetic field, it feels no force from the field.

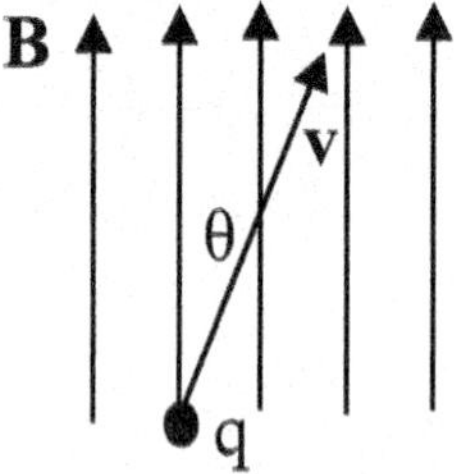

Fig. 9.6 Charged particle moving in a magnetic field

It feels a maximum force when it travels perpendicular to the field and thus cuts across the magnetic field lines.

Example 9.1
A charged particle ($q = -0.561$ mC) is moving with a velocity of 3.40×10^4 m/s at an angle of 45.0° in a uniform magnetic field of 0.350 Teslas at 180.0°. (Assume the normal directions of an x–y coordinate system.) What is the force experienced by the particle?

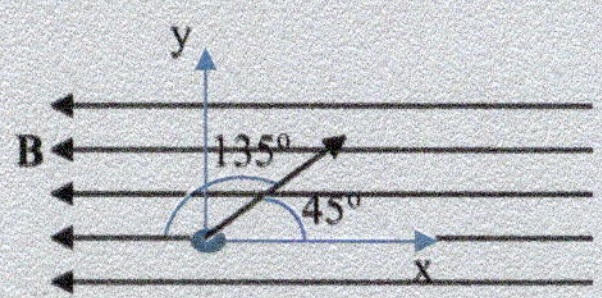

Knowns: $q = -0.561$ mC; $v = 3.40 \times 10^4$ m/s
$B = 0.350$ T; $\theta = 135°$
Unknowns: **F**

According to Eq. 9.2, the magnitude of the force is

$$F = qvB\sin\theta = (-0.000561\ \text{C})\left(3.40 \times 10^4\ \text{m/s}\right)(0.350\ \text{N/Am})\sin 135°$$
$$= -4.72\ \text{N}$$

Notice that the units do work out to give us Newtons in the end.

The force vector has a direction which is given by the right-hand rule. Since the magnitude came out negative, it means that the direction of the force vector is *opposite* that which is given by the right-hand rule. In the right-hand rule, we take the fingers of our right hand and point them in the direction of the first vector (**v**). Then, we curl our fingers to the second vector (**B**), along the arc of the angle between the two vectors. When we are finished, our thumb points in the direction of the cross product. When we do that here, our thumb points in the direction out of the plane of the paper. Thus, the cross product is directed perpendicular to the paper, pointing out of it. Remember, however, that the magnitude came out negative. Thus, the direction of the force is opposite that of the cross product. In other words, it is pointing into the plane of the paper. The force, then, is 4.72 N directed perpendicular to and pointing into the plane of the paper.

When you use the right-hand rule, you will always come up with a direction that is perpendicular to both vectors. Thus, since the magnetic force depends on the cross product between the velocity and magnetic field vectors, the magnetic force experienced by a moving charged particle in a magnetic field will always be perpendicular to the field and the particle's velocity vector. What happens when an object experiences a force which is perpendicular to its velocity? It moves in a circle, because a force perpendicular to the velocity acts as a centripetal force as discussed in Chap. 5.

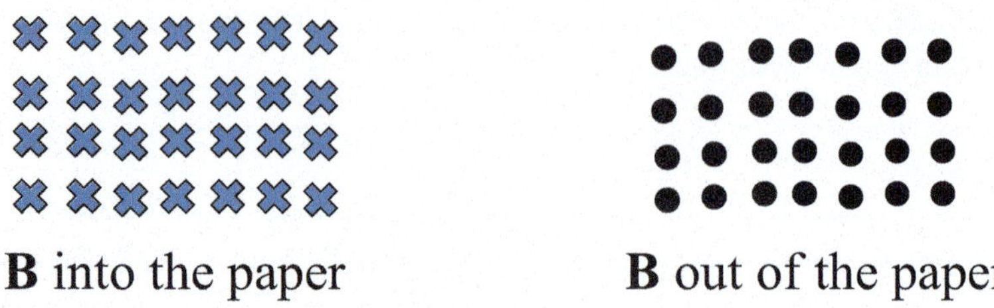

Fig. 9.7 Magnetic fields perpendicular to the paper

Before we look at this in detail, we need to think about how we can illustrate uniform magnetic fields perpendicular to the plane of the paper. This is shown in Fig. 9.7.

When you see a series of X's, then, you will know that the field lines are pointing into the plane of the paper. If you were to point your fingers in the direction of the magnetic field, you would hold your hand above the paper and point directly down at the paper. On the other hand, a series of dots means that the magnetic field is perpendicular to and pointing out of the plane of the paper. If you were to point in the direction of the magnetic field, then, you would hold your hand above the paper and then point straight up.

One important type of problem occurs when a charged particle is injected into a magnetic field traveling in a direction perpendicular to the field as shown in Fig. 9.8. In this case, the uniform magnetic field **B** is pointed into the page and a particle of mass m and charge q is accelerated to a velocity $\mathbf{v}$ in the plus x direction. When it enters the magnetic field, the right-hand rule tells us that it experiences an initial force in the plus y direction. As the particle changes to go in the plus y direction the force changes to the negative x direction. As the particle changes to the negative x direction the force changes to the negative y direction. As the particle changes to the negative y direction, the force changes to the plus x direction. As shown in Fig. 9.8 the particle goes around in a circle with the magnetic force always acting radially toward the center of the circle.

In Chaps. 4 and 5 we learned that the force needed to keep an object moving in uniform circular motion is called centripetal force. According to Eq. 4.7, the magnitude of the centripetal force is directly proportional to the mass of the object and the square of its velocity and is inversely proportional to it radius. In this case Eq. 9.2 tells us the strength of the magnetic force. Equating these two expressions gives

$$qvB = mv^2/r$$

Then the radius of the circular orbit is given by

$$r = (mv)/(qB). \tag{9.3}$$

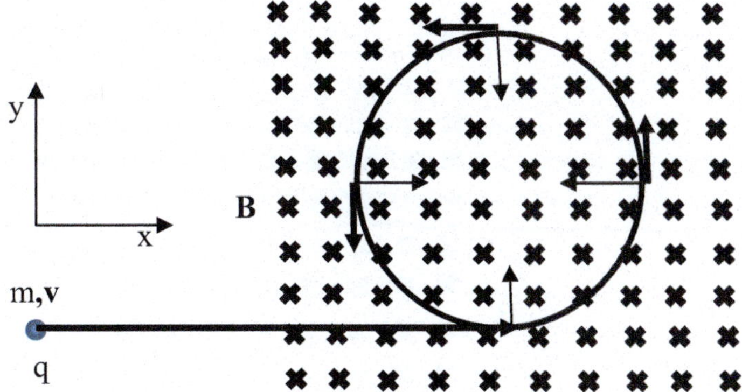

Fig. 9.8 Centripetal force on charged particle in a magnetic field

Note that the greater the particle's momentum, the greater the radius of the circle. The greater the strength of the magnetic field and the charge on the particle, the smaller the radius will be.

The motion of a charged particle in a magnetic field has been the basis for some very important applications. For example, this configuration is important in the construction of particle accelerators called cyclotrons that are critical to nuclear physics as discussed in Chap. 16.

One important application is the operation of mass spectrometers that are used to identify atomic size particles. This is illustrated in Fig. 9.9. Particles with the same charge and velocity but different masses will have a different radius of curvature in a magnetic field. The number of particles with a specific mass will be counted by a particle detector at located at that radius. Using this information the mass of the particle can be determined. This is demonstrated in the following example.

Example 9.2

In a mass spectrometer, a particle of unknown mass is given an extra electron, so that its electric charge is -1.6×10^{-19} C. The particle is then accelerated to 1.3×10^{4} m/s and sent into a perpendicular magnetic field of strength 0.050 T. The particle's final positionindicates that while it was in the magnetic field, it moved in a circle with a radius of 1.3 cm.

What is the mass of the particle?

This is a simple application of Eq. 9.3. Remember, we do not use the sign of the charge in the equation.

$$r = (mv)/(qB)$$

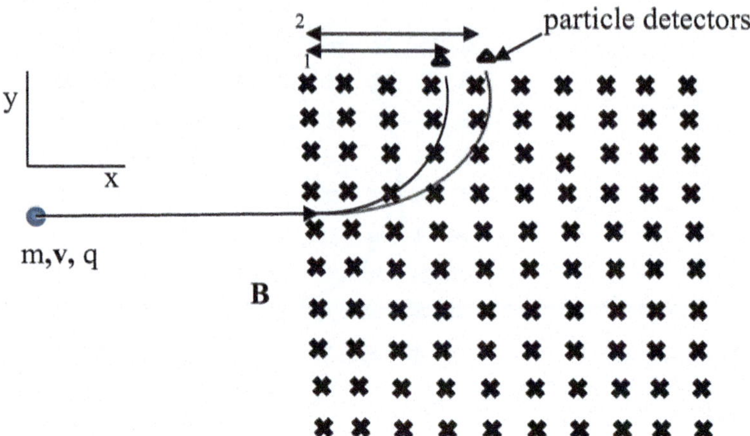

Fig. 9.9 Mass spectrometer

So,

$$m = qBr/v = \left(1.6 \times 10^{-19}\ \text{C}\right)(0.050\ \text{N/Am})(0.013\ \text{m})/\left(1.3 \times 10^{4}\ \text{m/s}\right)$$
$$= 8.0 \times 10^{-27}\ \text{kg}$$

Notice how the units work out. An Ampere is a Coulomb per second. Thus, the Coulombscancel, leaving s^2 N/m, which is a kg. Notice also how small this mass is. Clearly, you could never measure such a tiny mass on a balance. Thus, a mass spectrometer can be used to measure the mass of very tiny things.

Now work the following problems.

Student

9.2 A positive particle is shot between the plates of a parallel- plate capacitor, which is also in a magnetic field, as shown to the right. The electric field between the plates is 3014 N/C, and the speed of the particle is 3210 m/s. What must the strength of the magnetic field be in order for the particle to travel straight through both fields with no deflection?

9.3 If the electric field in the previous problem was shut off, but the magnetic field that you calculated in your answer was used, what would be the radius of the circle in which the particle would travel while in the magnetic field? The mass of the particle is 15.0 g, and the charge is 0.340 C.

9.4 A mass spectrometer shows that a sample contains two types of particles. Type 1 particles travel through a radius that is half as large as the radius of travel for Type 2 particles. What is the charge-to-mass ratio of Type 1 particles compared to that of Type 2 Particles?

Next Level

Cyclotrons

Another instrument based on these physics principles is a cyclotron. This is a device that accelerates charged particles to high speeds for collisions that produce elementary particles discussed in Chap. 16. A cyclotron is shown schematically in the diagram in Fig. 9.10.

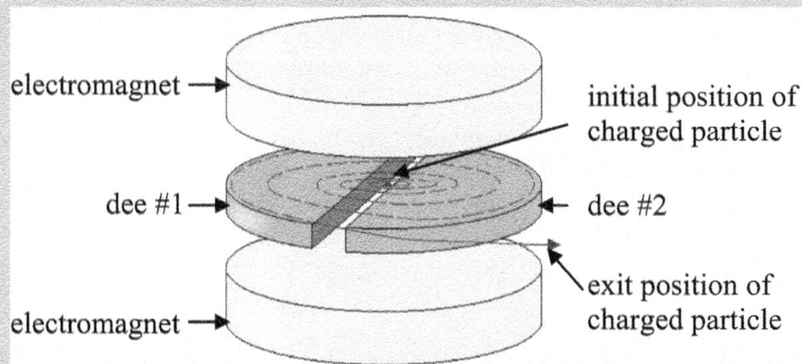

Fig. 9.10 Schematic of a cyclotron

A cyclotron is composed of two electromagnets that create a uniform magnetic field perpendicular to the plane in which the charged particle moves. In between these electromagnets are two hollow, D-shaped conductors called **dee's**. The dee's are connected to an alternating voltage supply. A charged particle starts at the center of the cyclotron, and one dee is given the same charge as the particle while the other dee is given the opposite charge. In the Fig. 9.10, for example, assume that the particle is negative. Dee #2 would be charged positive and dee #1 would be charged negative. As a result the particle will start accelerating away from dee #1 and towards dee #2. Once it reaches dee #2, it will enter the hollow conductor. Since you learned in Chap. 7 that inside the conductor it *will not* feel the effects of the electric field from the dee's at all, because electric fields cannot exist inside a conductor. Thus the particle does not increase its speed. However, it is already traveling at a given speed, which means it will continue to travel at that speed. Since it still feels the effect of the magnetic field, it will travel in a circle.

Eventually, its motion will result in the charged particle leaving dee #2. Before that happens, however, the voltage source will switch potential. That way, dee #1 will be positive and dee #2 will be negative. As a result when the charged particle leaves dee #2, it will suddenly feel the effects of the electric field again, and it will accelerate towards dee #1. Once it enters dee #1, it no longer feels the effects of the electric field and begins to travel in a circle at constant speed again but larger radius than before because of the larger velocity. Before it leaves dee #1, the voltage source switches potential again, so that when the particle leaves dee #1, it sees that dee #2 is positively charged and begins accelerating towards that dee.

Each time it leaves a dee, the voltage has been switched so that it is always nearest to the negatively charged dee and farthest from the positively charged dee. Thus, every time it leaves a dee, it is accelerated. Once in the dee, it travels in a circle. As Eq. 9.3 tells you, however, the faster it is traveling, the larger the radius of the circle. Thus, each time the particle accelerates, it will travel in a larger circle. Eventually, the circle becomes as large as the cyclotron, and the particle is traveling at a high rate of speed.

An interesting aspect of a cyclotron is that until the particle reaches speeds near the speed of light, the frequency at which the voltage must switch is constant. Each time the particle leaves a dee, it is accelerated. Thus, once it enters the next dee, it travels in a larger circle. You might think, therefore, that it takes more time for the particle to leave the dee, since it must travel in a larger circle. However, it is also traveling faster. These two effects cancel one another out. The distance the charged particle travels increases every time it enters a dee, but its speed also increases, which allows it to travel that longer distance in the same amount of time. Thus, the time that the charged particle spends in a dee is constant. As a result, the frequency with which the voltage must change is constant as well.

Notice that the frequency at which the voltage changes is constant, as long as the particle has not reached a speed near the speed of light. Once the particle reaches a speed near that of light, special relativity starts to play a role, and that throws off the timing. As a result, a cyclotron is a useful tool for accelerating particles, but it cannot accelerate particles to speeds near that of light, because of the effects of special relativity discussed in Chap. 15. This fact is actually another of the many experimental observations that prove the theory of special relativity.

9.4 Motional EMF

So far we have been discussing free charged particles in a magnetic field. Next we want to consider charges that are free to move inside a conducting wire traveling in a magnetic field as shown in Fig. 9.11. In this figure there is a closed electrical circuit of conductors. The left branch of the circuit is moving to the left at a velocity v. The magnetic field creates a force on the mobile electrons in the branch of the circuit that is moving. Since they are moving at the same speed in the same magnetic field, all the electrons experience the same force. According to the right-hand rule, Eq. 9.1 tells us the magnetic force on the moving charges pushes the negatively charged electrons up. This creates a current flow in the opposite direction of the electron motion because of the negative electrical charge on electron. The arrows in the figure show the direction of current flow around the circuit.

The force on the electrons causing them to move as a current is a similar result to adding a battery to the circuit. Using Eqs. 7.5 along with Eq. 9.1, the magnetic force on the electrons is equivalent to the force of an electric field given by

$$F = qE = qvB \tag{9.4}$$

where the angle between **v** and **B** is 90°. Then from Eq. 7.6 the potential difference in the circuit is

$$\Delta V = LvB. \tag{9.5}$$

where L is the length of the moving wire in the magnetic field. Thus, the moving wire acts like a battery that drives a current in the circuit. The induced voltage is called motional emf.

Motional emf—A potential difference established by the motion of a conductor in a magnetic field

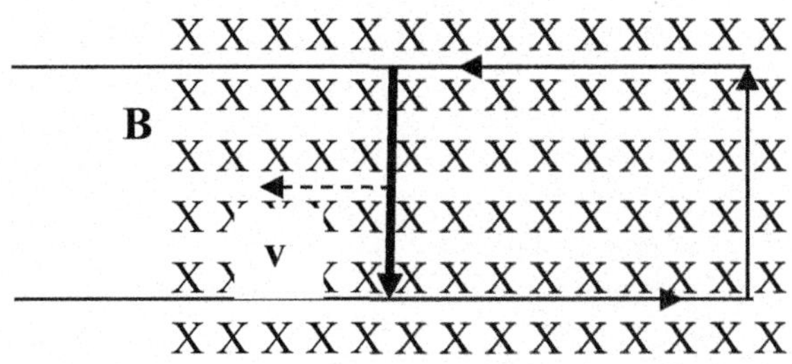

Fig. 9.11 Moving conductor in a magnetic field

Example 9.3

A 50.0 cm conductor is placed on stationary, conductive bars which have a total resistance of 2.0 Ω. If the conductor moves with a speed of 2.0 m/s in a 0.067 T magnetic field as illustrated in Fig. 9.11, what is the current that moves in the conductive bars? What is the direction of the current?

The figure is given above. Knowns: $L = 50.0$ cm; $R = 2.0\ \Omega$; $v = 2.0$ m/s; $B = 0.067$ T.
Unknowns: I

When the conductor moves, it will generate a potential difference. Thus, it will act like a battery. Since it is in contact with conductive bars, the voltage will cause current to flow. If we figure out the voltage, we can use Ohm's Law to figure out the current.

$$\Delta V = vBL = (2.0\ \text{m/s})(0.067\ \text{N/Am})(0.500\ \text{m}) = 0.067\ \text{V}$$

Notice how the units work out. Since an Ampere is a C/s, a Joule is a N m, and a Volt is a J/C, the potential difference comes out with the unit of Volt, which it should.

Now that we have the potential difference, Ohm's Law can give us the current.

$$V = IR$$
$$I = V/R = 0.067\ \text{V}/2.0\ \Omega = 0.034\ \text{A}$$

The problem also asked for the direction of the current. To determine that, we just need to determine which side of the "battery" (the moving conductor) is positive. If we apply the right hand rule to determine $\mathbf{v} \times \mathbf{B}$, we get that the cross product is pointed down the conductor. However, electrons experience a force (**Fe**) opposite the cross product, because of their negative charge. Thus, the electrons move up the conductor. That means the top of the conductor is negative, and the bottom is positive. Thus, conventional current flows counterclockwise around the bars.

It may appear that in a situation such as the one discussed above, electricity is being produced with no energy input. However, that is obviously not true. It takes energy to move the conductor. That mechanical energy is being converted (via the magnetic field) into electrical energy. Ideally, the energy used to move the conductor would be equal to the electrical energy produced. You will actually demonstrate that in the problem that follows. However, because of both friction and the law of conservation of energy, the amount of electrical energy produced will be less than the energy it takes to move the conductor. In the next Chapter we will learn how useful motional emf can be. For now, try the following problems.

Student

9.5 For the situation given in the example above, demonstrate that the power required to move the conductor is equal to the power dissipated in the resistance of the bars. (HINT: The moving conductor acts like a current-carrying wire. Think about the force exerted by the magnetic field on the current-carrying wire.)

9.6 A 75.0 cm conductor moves on stationary, conductive bars in a 0.0557 T magnetic field as shown to the right. The bars have essentially zero resistance, but there is a 15.0 Ω resistor connecting them. If 0.0433 A of current run through the resistor, what is the speed at which the bar is moving? What is the direction of the current?

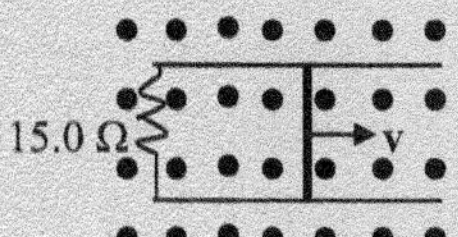

9.5 Force on a Current Carrying Wire

Next we can extend our discussion to current carrying wires in a magnetic field as shown in Fig. 9.12. Assuming that the current is made up of flowing electrons in the wire, the magnetic field exerts a force on each electron given by Eq. 9.2. The current is the charge per unit time flowing past a point in the wire so the charge in Eq. 9.2 can be replaced by $q = It$. The time it takes a charge to travel the length L in the magnetic field is $t = L/v$. Thus the product qv in Eq. 9.2 can be replaced by IL. Then the magnitude of the force on a current carrying wire in a magnetic field is

$$F = ILB \sin \theta \tag{9.6}$$

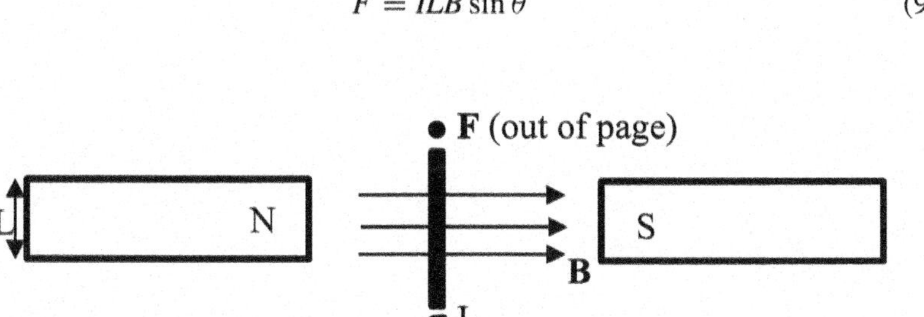

Fig. 9.12 Force on a current carrying wire in a magnetic field

θ is the angle between current direction and the magnetic field direction. The direction of the force is again found from the right-hand rule. Pointing the fingers of your right hand in the direction of the current and rotating them toward the direction of the magnetic field, your thumb points in the direction of the force on the wire. In Fig. 9.12 this is out of the page. Remember that an electron's velocity is opposite to the direction of current flow but the negative charge on an electron reverses the direction of the force given by the right-hand rule. Therefore the force of the magnetic field on each of the electrons in the wire is also out of the page. Note that L is a vector whose magnitude is the length of the wire in the magnetic field and whose direction is the direction of the current. In vector terms, the above equation can be rewritten as

$$\mathbf{F} = \mathrm{I}(\mathbf{L} \times \mathbf{B}) \tag{9.7}$$

where again the direction of the vector **L** is in the direction of current flow.

Consider the following two examples.

Example 9.4

A wire carrying 0.0450 A of current is placed in a 0.0512 Tesla magnetic field as illustrated to the right. The length of the wire in the magnetic field is 35.0 cm. What are the magnitude and direction of the magnetic force on the wire?

Figure is given. Knowns: $I = 0.0450$ A; $B = 0.0512$ T; $L = 35.0$ cm
Unknowns: F

This is a simple application of Eq. 9.6. First, we can get the magnitude using the equation that gives us the magnitude of the cross product:

$$F = ILB \sin \theta$$

Since the external magnetic field is coming out of the plane of the paper, and since the current is in the plane of the paper, the angle between them is 90.0°. So,

$$F = IBL = (0.0450 \text{ A})(0.350 \text{ m})(0.0512 \text{ N/Am}) = 8.06 \times 10^{-4} \text{ N}$$

To determine the direction of the force, we take our right hand and point our fingers in the direction of the current. Then we curl them out of the plane of the paper (in the direction of the magnetic field). That means the force is directed straight down. Thus, the force is 8.06×10^{-4} N directed straight down.

Example 9.5

A "U"-shaped wire is suspended in a uniform 0.456 T magnetic field as shown to the right. What are the magnitude and direction of the net force exerted on the "U?" The current is 0.0654 A, and the length of each side of the "U" is 10.0 cm.

Figure given. Knowns: $B = 0.456$ T; $I = 0.0654$ A; $L_1 = L_2 = L_3 = 10.0$ cm.
Unknowns: F

We can treat this situation as if there were three individual wires carrying current. The "wire" on the left carries current down. The "wire" on the right carries current up, and the wire running across the bottom carries current to the right. If we think about the cross product that determines the force, we see something that simplifies the situation enormously. The "wire" on the left carries current down. If we take our right hand and point our fingers down, and then curl them into the page (the direction of the magnetic field), we see that the force on that "wire" is pointed to the right. The "wire" on the right carries current up. If we take our right hand, point our fingers up, and then curl them into the page, we see that the force is pointed to the left. Since both "wires" carry the same current and are exposed to the same magnetic field, the magnitude of the forces on each "wire" are the same, but the directions are opposite. Thus, *these forces cancel out, and the only force remaining is the force on the bottom "wire."*

The bottom wire is perpendicular to the magnetic field, so the angle between **L** and **B** is 90.0°. Thus,

$$F = ILB = (0.0654\ \text{A})(0.100\ \text{m})(0.456\ \text{N/Am}) = 2.98 \times 10^{-3}\ \text{N}$$

The direction is found from the right-hand rule. The current is carried to the right, and the magnetic field is pointing into the paper. If we take our right hand, point the fingers to the right and curl them into the paper, our thumb points up. Thus, the force is 2.98×10^{-3} N pointing straight up.

Now try the following problems.

Student

9.7 A square of wire that is free to rotate is placed in a magnetic field as shown to the right. The current in the wire is 0.500 A, and the magnetic field strength is 0.126 T. If each side of the wire is 15.0 cm long, what is the total torque experienced by the square? Ignore the tiny gap on the left. Treat this as a complete square.

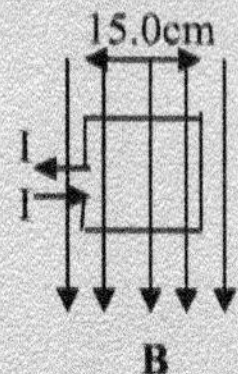

9.8 A knight wants to send a message to a fair maiden at the window to her castle 50 ft above the ground. He puts two parallel metal poles 10 cm apart at an angle against the window. On these poles he mounts a current carrying wire with a permanent magnetic below it that provides a magnetic field of 0.049 T downwards perpendicular to the plane of the rods. The moving assembly has a mass of 0.25 kg. He passes a current of 10 A up one rod, through the wire, and down the other rod. Neglecting friction, the wire assembly moves up the rods at a constant velocity and delivers the note to the maiden. What must the lengths of the rods be?

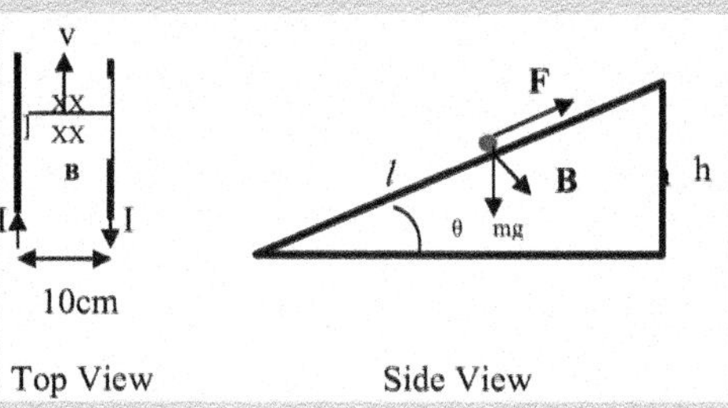

Summing Up

You know now about permanent magnets and the source of magnetic field. You learned about the magnetic force on an electrically charged particle in a magnetic field and how this is expanded to mobile electrons in a conductor. In the next chapter we will learn about the direct connection between electricity and magnetism and how it is useful for making devices such as motors and generators.

Answers to the Problems

9.1

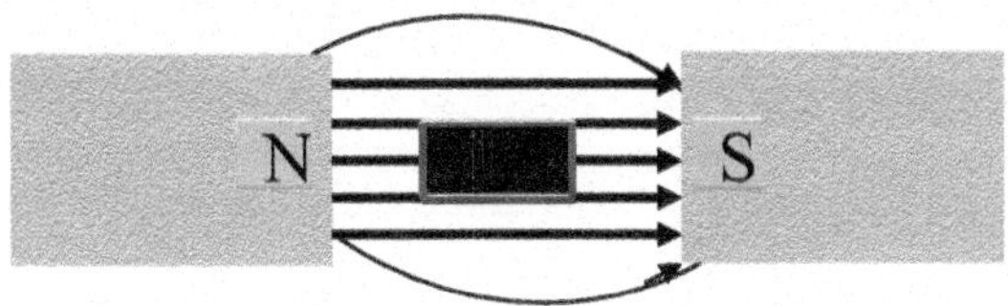

In diamagnetic substances, the induced magnetic field is weak and opposes the applied magnetic field. Thus, in the diamagnetic substance, the field would point to the left in the picture. In paramagnetic substances and ferromagnetic substances, the magnetic atoms align with the field. Thus, in the paramagnetic and ferromagnetic substances, it would point to the right.

9.2

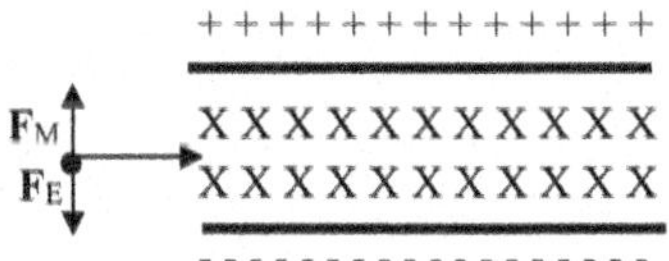

Knowns: $E = 3014$ N/C; $v = 3210$ m/s

Unknowns: B

When the positive particle enters the plates, it will be pulled down by the electric field.

Equation 7.4 gives us the force in terms of E and q:

$$\mathbf{F} = \mathrm{q} \cdot \mathbf{E}$$

Based on the right hand rule, the magnetic field will pull up on the particle (point your right hand's fingers in the direction of the velocity and curl them into the paper (the direction of the magnetic field), and your thumb points up. We also know that force, from Eq. 9.1

$$\mathbf{F} = \mathrm{q}(\mathbf{v} \times \mathbf{B})$$

For no deflection, the forces must be equal in magnitude and in the opposite direction. If we therefore set the magnitudes of the forces equal to one another, we get:

$$\mathrm{qE} = \mathrm{qvB}$$

$$\mathrm{B} = \mathrm{E/v} = (3,014\mathrm{N/C})/(3,210\mathrm{m/s}) = 0.939\mathrm{T}$$

Notice that since an Ampere is a C/sec, the units work out to give us Teslas.

9.3 Without the electric field, the particle will travel in a circle, with a radius given by Eq. 9.3:

$$r = (mv)/(qB) = (0.0150\ \text{kg})(3210\ \text{m/s})/(0.340\ \text{C})(0.939\text{N/Am}) = 151\text{m}$$

Notice that because an Ampere is a C/s and a Newton is a kg m/s^2, the units work out to give us meters.

9.4

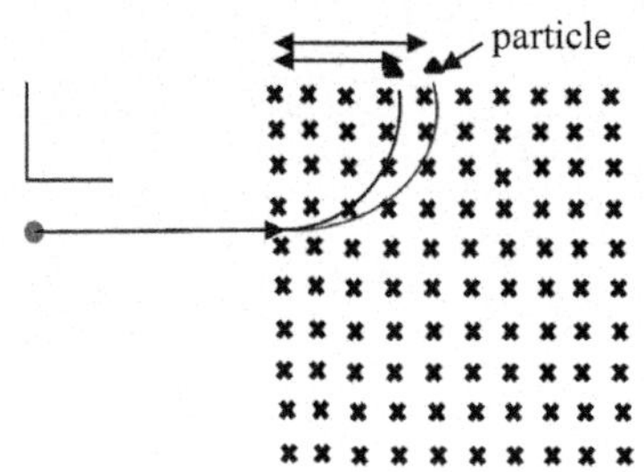

Knowns: $r_1 = 1/2r_2$

Unknowns: $(q/m)_2$

$r = (m/q)(v/B)$

$(m/q)_1(v/B) = (1/2)\ (m/q)_2(v/B)$

$(q/m)_1 = 2(q/m)_2$

9.5

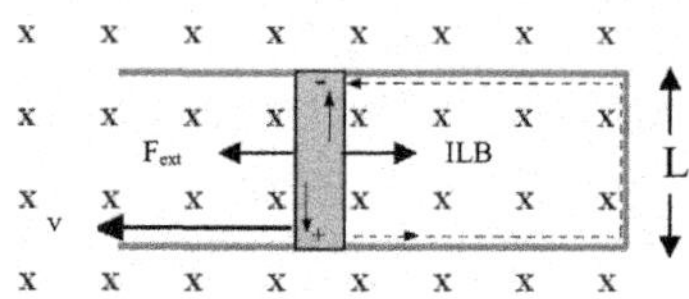

Knowns: $L = 50$ cm; $R = 2.0\ \Omega$; $v = 2.0$ m/s; $B = 0.067$ T

Unknowns: I

Since the moving conductor is like a current-carrying wire, we know that it experiences a force from the magnetic field, given by Eq. 9.7,

$$\mathbf{F} = I(L \times \mathbf{B})$$

The angle between the magnetic field and the current is 90°, so the magnitude of the force is:

$$F = ILB \sin(90) = ILB$$

The direction is given by the right-hand rule. Point your fingers in the direction of the conventional current in the conductor (down), and curl them into the page (the direction of the magnetic field), and you find that the force is *opposite* the motion.

Thus, there must be an external force ($\mathbf{F}_{ext}$) being exerted on the conductor in order to keep it moving. Since the conductor moves at constant velocity, the sum of the forces must be zero. Thus:

$$F_{ext} = ILB$$

According to Eq. 3.9, $P = \mathbf{F} \cdot \mathbf{v}$ when F and v are constant. Thus:

$$P = Fv = ILBv$$

That's the power required to move the conductor. Well, according to Eq. 8.4:

$$P = IV$$

Since, according to Eq. 9.5, motional emf is $B \cdot L \cdot v$, that means the power produced by the current is:

$$P = IBLv$$

The power required to pull the conductor, then, is the same as the power produced by the electrical current, in compliance with the conservation of energy.

9.6

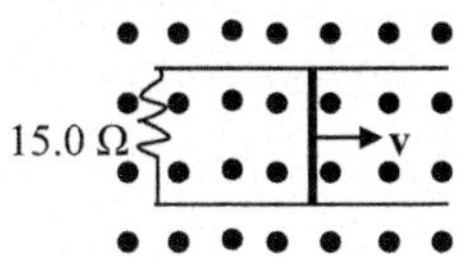

Knowns: $L = 75$ cm; $B = 0.0557$ T; $R = 15.0\ \Omega$; $I = 0.0433$ A

Unknowns: v; I direction

We know the resistance and the current, so if we go to Ohm's Law, we can get the voltage:

$$V = IR = (0.0433\text{ A})(15.0\ \Omega) = 0.650\text{ V}$$

That's the potential difference supplied by the emf, which is equal to:

$$\Delta V = vBL$$

$$v = \Delta V / BL = (0.650\text{ V})/(0.0557\text{ N/Am})(0.750\text{ m}) = 15.6\text{ m/s}$$

Notice that since a Volt is a Joule per Coulomb, an Amp is a Coulomb per second, and a Joule is a Newton·meter, the units work out to give us m/s.

The problem also asks for the direction of the current. The electrons are moving to the right, along with the conductor. When we take $\mathbf{v} \times \mathbf{B}$, we point our fingers to the right (direction of the velocity) and curl them out of the page (the direction of the magnetic field). As a result, our thumb points down. Since electrons are negative, however, that means they move opposite, or up. Thus, the top of the conductor is negative, and the bottom is positive. Conventional current runs from the positive end of the "battery," around the circuit, and to the negative end.

Thus, the current runs clockwise.

9.7

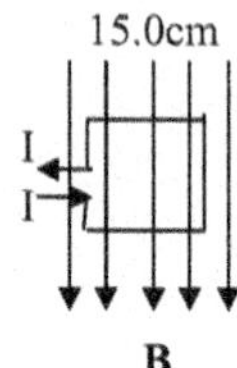

Knowns: $I = 0.500$ A; $B = 0.126$ T; $L = 15$ cm

Unknowns: τ

The force that a current-carrying wire experiences in a magnetic field is given by Eq. 9.7. We can treat the square of wire as four individual straight wires, so we really just apply Eq. 9.7 four times. Two of those applications are easy, however. As you can see in the figure, the current (shown by the black arrows) is parallel to the magnetic field on the left side of the square and directly opposite the magnetic field on the right. Since the force exerted by the magnetic field depends on the cross product between the direction of the current and the magnetic field, the force on both sides is zero. Thus, the only sides that experience a force are the top and bottom of the square.

To determine the direction of the force experienced by the top and bottom of the square, we can use the right-hand rule. The current in the top of the square is moving to the left.

Pointing the fingers of your right hand to the left and curling them down (in the direction of the magnetic field), you find that the force is directed out of the plane of the paper. The current on the bottom is traveling to the right. Doing the same thing there, you find that the force is pointing into the plane of the paper. Thus, we have a force on top of the square pushing out of the plane of the paper and a force on the bottom pushing into the plane of the paper. What will this do? It will cause the square to *rotate* around the axis shown by the dashed line in the drawing.

The loop, then, experiences a torque, which is equal to $\mathbf{r} \times \mathbf{F}$, or $rF\sin\theta$. The "r" is the lever arm, which is the distance between the axis of rotation and the force. In this case, since each side of the square is 15.0 cm long, the distance from the force to the axis is 7.50 cm. The "θ" is the angle between the lever arm and the force, which is 90.0°:

$$\tau = rF\sin\theta = rF$$

Since Eq. 9.7 allows us to calculate the magnitude of the force:

$$\tau = rILB\ \sin\theta$$

This "θ" is the angle between the current and the magnetic field, which is also 90.0°:

$$\tau = rILB = (0.0750\ \text{m})(0.500\ \text{A})(0.150\ \text{m})(0.126\ \text{N/Am}) = 0.000709\ \text{Nm}$$

That is the torque experienced by one of the two wires. Since both wires will have the same r, I, L, and B, their torques will each have the same magnitude. Also, since they both promote rotation in the same direction, they are both pointing in the same direction. Thus, we can just take the number above and multiply by 2. The total torque, then, is 0.00142 N m.

9.8

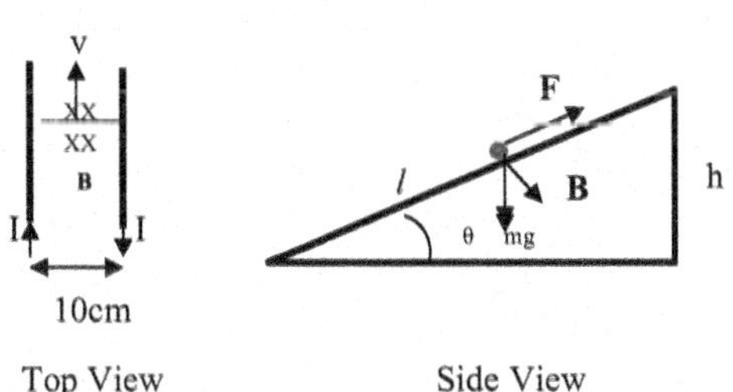

Knowns: $h = 50$ ft; L = 10 cm; $B =$ 0.049 T m = 0.25 kg; $I = 10$ A; a = 0 m/s^2

Unknowns: l

The force on the wire with the message is given by Eq. 9.7 to be

$$F = ILB\sin 90^\circ = (10\ \text{A})(0.10\ \text{m})(0.49\ \text{N/Am}) = 0.49\ \text{N}$$

The component of the weight of the wire acting down the incline is

$$F = mg\sin\theta$$

Since there is now acceleration, these two forces must be equal in magnitude and opposite in direction. So,

$$\sin\theta = F/mg = (0.49\ \text{N})/(0.25\ \text{kg})\left(9.81\ \text{m/s}^2\right) = 0.02$$

But from the figure,

$$\sin\theta = h/l$$

so $l = h/\sin\theta = 50\ \text{ft}/0.02 = 2500\ \text{ft}$

Study Guide for This Chapter

$$(\mu_o = 4\pi \times 10^{-7}\text{Tm/A})$$

1. Is it possible for a charged particle to move through a magnetic field without experiencing a force? If so, how?
2. Is it possible for the kinetic energy of a charged particle to change as it moves through a magnetic field? If so, how?
3. A negatively charged particle moves in a magnetic field. As shown to the right, the magnetic field lines point to the left and the charge experiences a magnetic force directed downwards. What is the direction of the particle's velocity?

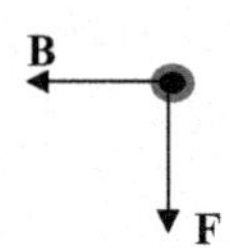

4. A conductor is moved through a magnetic field so that it develops a potential difference between the ends. However, it is not attached to any circuit, so current does not flow. When the conductor is stopped (assume the magnetic field is still present), what will happen to the potential difference?
5. Imagine two charged particles moving directly up this page. Suddenly, a magnetic field is turned on. One particle is deflected out of the page, while the other is deflected into the page. What can you say about these two charged particles?
6. What is the source of the magnetic properties of a sample of material?
7. Which particles have a greater orbital radius in a mass spectrometer, those with larger mass or these with smaller mass? Which particles have a greater orbital radius in a mass spectrometer, those with greater charge or these with smaller charge?

Problems 8–10 refer to the following situation: A charged particle is traveling at a speed of 3014 m/s and enters a perpendicular magnetic field of 0.0456 T, as shown to the right. Its mass is 4.60 mg, and its charge is – 2.50 C.

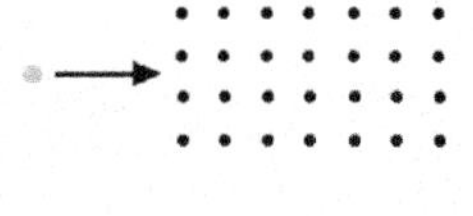

8. What are the magnitude and direction of the force that the particle will experience the instant it enters the magnetic field?
9. The charged particle exits the magnetic field on the same side which it entered, with a velocity pointed in exactly the opposite direction. How far is its point of exit from its point of entry? Is its point of exit above or below its point of entry?
10. How much time elapses from the moment it enters the field until the moment it leaves the field?

11. A portion of a current-carrying circuit is suspended (without support) in a magnetic field, as shown to the right. The circuit has a mass of 50.0 g, and it carries a current of 3.40 A. What must be the strength and direction of the magnetic field if the loop simply "floats" in that position?

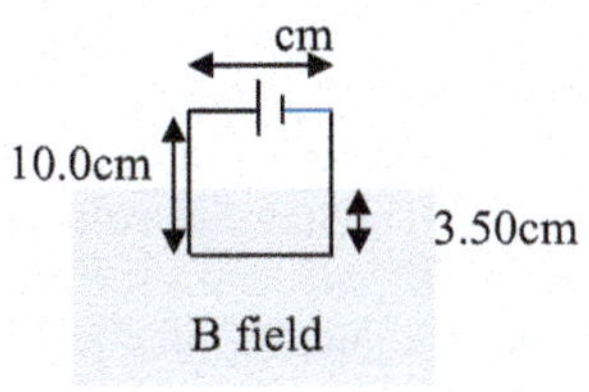

12. Suppose, in the situation described above, you suspended the circuit with a string so that the circuit did not have to be held up by the magnetic field. Then, instead of pointing the field as you determined in problem #4, you pointed it directly to the right. What would happen to the circuit? What is the net torque that the circuit would experience?

13. A 25.0 cm conductor is moved at a constant velocity in a 0.861 T magnetic field as illustrated in the drawing to the right. The potential difference between the ends of the conductor is measured to be 1.86 V. What is the speed at which the conductor is moved, and which end (top or bottom) of the conductor is at the higher potential?

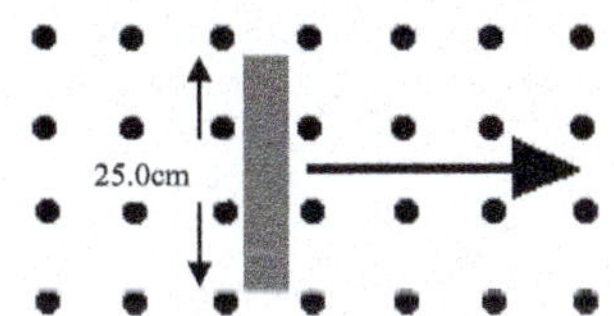

14. Find the force on an oil drop with 2×10^{-17} C of electrical charge moving at a speed of 1×10^{2} m/s across a perpendicular magnetic field of 2 T

15. As shown, a 30 cm long wire carrying a current of 25 A at an angle of 60° to a magnetic field of 8.0×10^{-4} T. What is the force on the wire?

16. Two wires are placed parallel and 15cm apart from one another. They each carry 15 A of current in the same direction. What is the magnetic field at a point directly in between the two wires?

Next Level

17. If a cyclotron works to accelerate electrons, what A.C. frequency should be used to switch the voltage if the magnetic field used is 0.09 T? ($q_e = 1.602 \times 10^{-19}$ C; $m_e = 9.11 \times 10^{-31}$ kg)
18. For the cyclotron in problem 17, if you want the electrons to emerge with a kinetic energy of 3.0×10^{-12} J, what is the maximum radius of the dees?

Electromagnetism

10

Chapter Summary

The development of electromagnetic theory was one of the great accomplishments of physics in the 1800s. Learning how to use electromagnetism has given us the ability to make hundreds of different types of devices as well as provide electrical power to every home and business. In the history of civilization, we are currently in the age of electricity. Some of the themes of the chapter are represented in Figs. 10.1, 10.2, 10.3 and 10.4. Figure 10.1 shows an electric power plant. Figure 10.2 shows a meter for measuring current and voltage in an electrical circuit. Figure 10.3 shows part of an electrical power delivery system. Figure 10.4 shows a bandpass filter.

Main Concepts in This Chapter

- Magnetic Flux
- Faraday's Law and Lenz's Law
- Meters, Motors, and Generators
- Inductors, Filters, and Transformers

10.1 Introduction

In the introduction to the last chapter we mentioned that electricity and magnetism are closely related. Although all areas of physics are related to each other to some degree in different ways, the relationship between electricity and magnetism is special. This will be demonstrated explicitly in the current chapter. The fact that electricity and magnetism are related was first observed by Danish physicist Hans

R. C. Powell, *Physics*, Undergraduate Lecture Notes in Physics,
https://doi.org/10.1007/978-3-032-09361-5_10

Fig. 10.1 Electric power plant. *Credit* CEphoto, Uwe Aranas. Creative Commons Attribution Share Alike 3.0

Fig. 10.2 Amp and volt meter. *Credit* Islander6, Creative Commons Attribution-Share Alike 4.0

Christian Ørsted in 1820. However it was 53 years later that James Clerk Maxwell published *A Treatise on Electricity and Magnetism* that provided a mathematical description of electromagnetic theory. Maxwell's mathematical model is described by four partial differential equations. The calculus required to use these equations is beyond the mathematics in this text, but the physics they describe can still be understood.

Fig. 10.3 Inductor in electrical transmission line. *Credit* Mario Sediak, Creative Commons Attribution-Share Alike 3.0

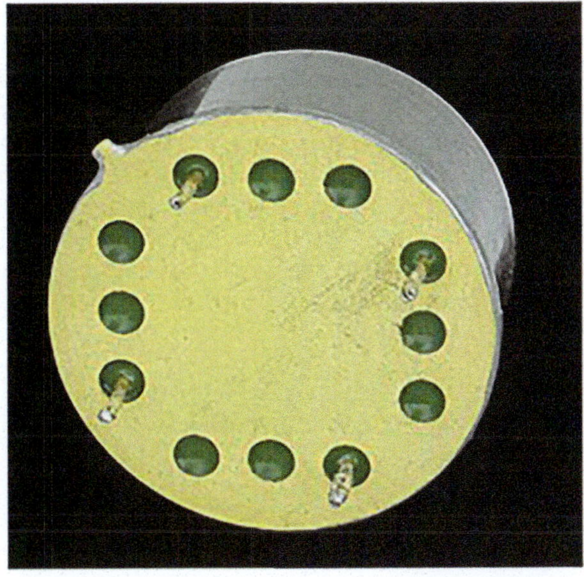

Fig. 10.4 Bandpass filter combining inductors and capacitors. *Credit* Mister rf, Creative Commons Attribution-Share Alike 4.0

The connections between electricity and magnetism are important because they form the basis for constructing electromagnetic devices. The basic electromagnetic principles and devices are discussed in this chapter. Later in Chap. 14 we will extend this discussion to the basis of microelectronic devices.

10.2 Magnetic Flux

We begin our discussion of electromagnetism by considering the magnetic field surrounding a current-carrying wire. In discussing permanent magnets in Sect. 9.1, we said that the origin of magnetism is the motion of electrons around the nuclei of atoms. This is like a microscopic current. Therefore, it should not be surprising that when electrons move in a macroscopic current, they also create a magnetic field.

Remember that the direction of the magnetic field surrounding an electrical current is always perpendicular to the direction of current flow. Thus, for a straight wire conductor, the lines of the magnetic field are circles around the wire. The direction of the field can be determined using the "right-hand rule". This is shown in Fig. 10.5. For a straight length of wire, when the thumb of your right hand points in the direction of the current in the wire, your fingers curl around the wire in the direction of the magnetic field as shown in (a). If the wire is shaped into a closed loop like the circuit shown in (b), applying the right-hand rule to each of the four sides of the loop shows that the magnetic field always into the page on the inside of the loop and out of the page on the outside of the loop. Figure 10.5c shows what happens when the wire has multiple loops in the circuit. The multiple loops create an enhanced magnetic field with a distinct north pole and south pole as shown. This creates a device called a solenoid. Sometimes a permeant magnet is inserted in the core of the solenoid to enhance the magnetic field further. Solenoids are important elements in many electromagnetic circuits and are discussed further below.

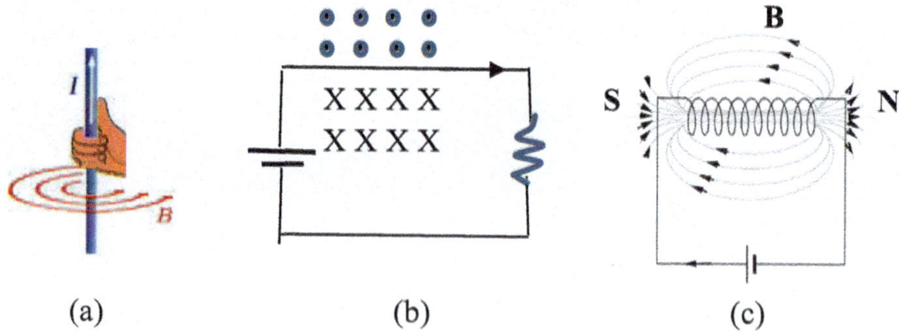

Fig. 10.5 Right hand rule for magnetic field around a current in a wire

Now that we know the direction of the magnetic field vector **B** around a current-carrying wire, we can determine the magnitude of B. Through making experimental measurements, the magnitude of the magnetic field surrounding a current-carrying wire has been found to be given by the equation

$$B = \left(\frac{\mu_o}{2\pi}\right)\left(\frac{I}{r}\right). \tag{10.1}$$

In this expression, I is the current in the wire, r is the radial distance out from the wire, and μ_o is the permeability of free space which we defined in the last chapter to be a constant $\mu_o = 4\pi \times 10^{-7}$ Tm/A. Note that the strength of the magnetic field increases as the current in the wire increases and it decreases as the distance from the wire increases.

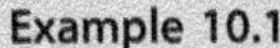

Example 10.1

What is the strength of the magnetic field a distance of $r = 0.1$ m away from a wire carrying a current of $I = 2$ A.

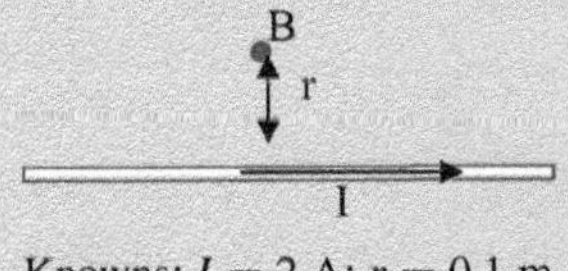

Knowns: $I = 2$ A; $r = 0.1$ m

Unknowns: **B**

Using Eq. 10.1 gives

$$B = \left(\frac{4\pi \times 10^{-7}\ \text{Tm/A}}{2\pi}\right)(2\ \text{A}/0.1\ \text{m}) = 4 \times 10^{-6}\ \text{T}.$$

Try the following problem.

Student

10.1 A physicist wants to use a current-carrying wire to generate a magnetic field. If she wants to produce a magnetic field of strength 4.32×10^{-5} T at a distance of 10.0 cm from the wire, what current will she need to pass through the wire? Based on the illustration to the right, will the magnetic field be pointed into or out of the plane of the paper at the point which she is studying?

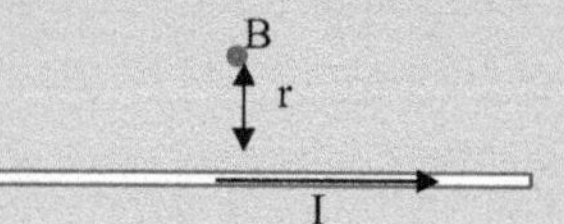

Perform the following experiment to demonstrate the magnetic fields around a current-carrying wire.

Experiment 10.1

Magnetic Field Around a Current-Carrying Wire

Supplies

- a 1.5 V battery
- a length of insulated wire with ends stripped for contacts
- a compass
- an iron nails
- a few metal paper clips

Introduction—A magnetic field is created when an electric current flows through a wire. A single wire does not produce a strong magnetic field, but a coiled wire around an iron core does. We investigate this behavior.

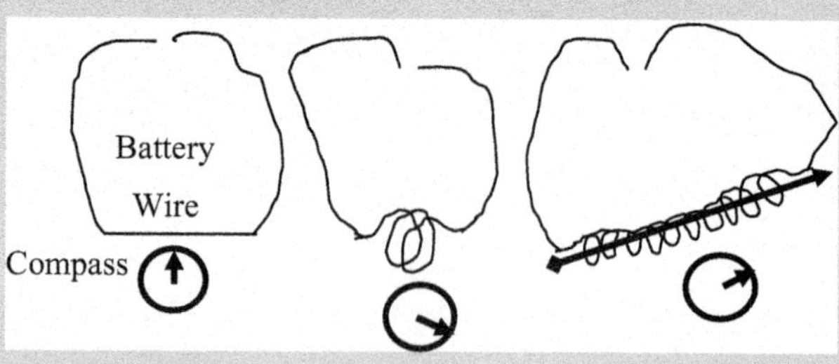

Procedure:

1. Lay the wire on the table and move the compass around it to make sure there are no magnetic fields present.
2. Hook the ends of the wire up to the terminals of the battery and repeat step 1 to detect the direction of any magnet field. Record the results in your lab notebook. Warning: Touching a bare conductor in a live circuit can be lethal if you do not know what you are doing! You should never do that unless someone who knows exactly what he or she is doing tells you!
3. Unhook the wire from the battery and bend it into two loops.
4. Re-hook the wire to the battery and repeat step 2.
5. Insert the nail into the loops of the wire and repeat step 2.
6. See how many paperclips the magnetic field of the nail can hold and record this in your lab notebook.
7. Unhook the wire from the battery and remove the nail.
8. Bend the wire into 10 loops and re-hook the wire to the battery.
9. Repeat step 2.
10. Insert the nail into the loops of wire and repeat steps 5 and 6.
11. Clean up and return everything to its proper place.

This experiment may be used to demonstrate the difference in magnetism of different types of materials. The experiment shows the effect on the magnetic field of making a solenoid out of the wire. Think about how the number of coils in the solenoid and the type of core in the solenoid change the magnetic field.

Since a current-carrying wire can generate its own magnetic field, it ought to interact with other magnetic fields. Consider, for example, a current-carrying wire that is placed in an external magnetic field. The magnetic field that is generated by the current in the wire will interact with the external magnetic field. This interaction is the basis of the relatively simple equation that we developed in the last chapter that allows us to calculate the force caused by that interaction:

$$\boldsymbol{F} = I(\boldsymbol{L} \times \boldsymbol{B}) \tag{9.7}$$

In this equation, I is the current, **B** is the external magnetic field vector. The **L** in the equation is a vector whose magnitude is equal to the length of the wire that is in the magnetic field, and the direction is equal to the direction of the conventional current going through the wire. In Chap. 9 we worked several problems using this equation.

In Chap. 9 we also discussed the motional emf associated with a current-carrying wire moving in a magnetic field. Now we want to extend this discussion to see what happens to a closed circuit loop in a changing magnetic field. If a permanent magnetic is moved in or out of a closed circuit loop of conducting wire, a current will be induced to flow in the wire even though it is not connected to

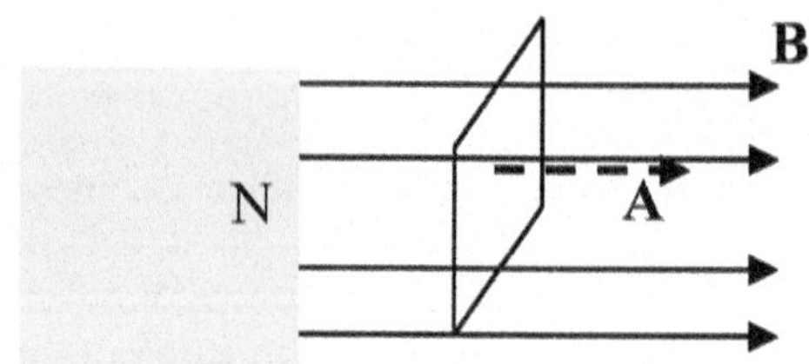

Fig. 10.6 Conducting loop in a magnetic field

a battery or other source of potential difference. This is shown in Fig. 10.6. The amount of current that flows depends on the strength of the magnetic field and the geometry of the magnet compared to the loop of wire. How effective the magnetic field is in inducing a current to flow is dependent on its magnetic flux, ϕ, defined as

$$\phi = \mathrm{BA}\cos(\theta) \tag{10.2}$$

or in vector form,

$$\phi = \boldsymbol{B} \cdot \boldsymbol{A} \tag{10.3}$$

where the magnitude of **A** is the area enclosed by the loop of the conductor, its direction is perpendicular to the plane of the loop, and θ is the angle between the direction of the magnetic field and a vector normal to the plane of the loop. The units of magnetic flux are T m^2 and are called Webers (Wb) in honor of Wilhelm Eduard Weber, a German physicist whose work in electrostatics and electrodynamics was very important in the development of Maxwell's equations. The magnetic flux describes the amount of magnetic field lines in a specific area. ϕ is maximum when **B** and **A** are parallel so $\theta = 0°$ and $\cos\theta = 1$ and can be thought of as the perpendicular component of a magnetic field through a given area.

10.3 Faraday's Law and Lenz' Law

Faraday's Law of Electromagnetic Induction tells us that the electromotive force that induces the current to flow in the conductor loop is equal to the time rate of change of the magnetic flux through the loop,

$$\mathrm{emf} = \Delta V = \Delta\phi/\Delta t \tag{10.4}$$

This causes a current to flow just like attaching a battery to the circuit. The faster the change in the magnetic flux, the larger the potential difference it produces in the loop. If there are multiple loops involved such as in the solenoid shown in Fig. 10.5c, a factor of N is included in Eq. 10.4 for the number of loops involved in creating the emf.

Note that according to Eq. 10.2 the magnetic flux can change if the magnetic field strength changes or the area of the conducting loop changes or the angle between the field and the perpendicular to the area of the loop changes. All three of these can be important in using Faraday's law of induction.

As an example of Faraday's law, consider an example of the changing area of the loop.

Example 10.2

Consider the situation in which a moving conductor slides along stationary bars in a magnetic field as shown.
In Chap. 9 we treated this as motional emf. Now let's see how we can use Faraday's law to find the induced emf.

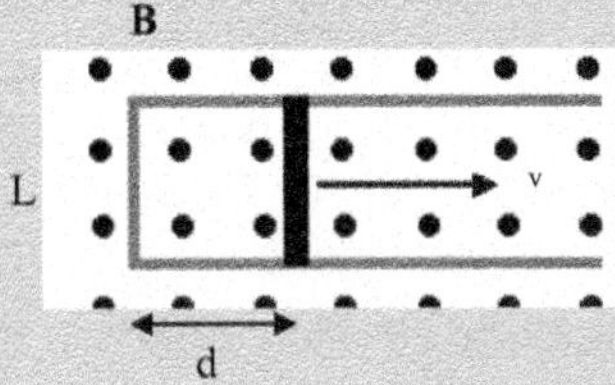

Notice that the moving conductor and the stationary bars form a rectangle. Since there is a magnetic field here, that means there is a magnetic flux through the rectangle. We can even calculate that flux. The height of the rectangle is L, the length of the conductor, and the width is d. That means the magnitude of the vector **A** that we must define is Ld. Since the vector's direction is perpendicular to the face of the rectangle, it is pointing out of the paper.

That is the same as the magnetic field. Thus, **B** and **A** are parallel, so the angle between them is zero. That means:

$$\phi = \boldsymbol{B} \cdot \boldsymbol{A} = BLd \cdot \cos(0) = BLd$$

Think for a moment about what happens to the area of the rectangle as the conductor moves. L doesn't change while the conductor moves, but d does. As d changes, the flux changes. Thus, we could say:

$$\Delta\phi = BL\Delta d$$

That's how the flux changes, but how does it change over time? If we divide both sides of the equation by Δt, we would get the average rate at which the flux changes:

$$\Delta\phi/\Delta t = BL\Delta d/\Delta t = BLv$$

where we have set $v = \Delta d/\Delta t$, the speed at which the wire is moving. The right-hand side of this equation is the motional emf according to Eq. 9.5.

Thus we have the same emf

$$\Delta V = \Delta\phi / \Delta t$$

from changing flux, which is Faraday's law.

This is an incredibly important equation. It tells us that if magnetic flux changes in a conductor, an emf is produced. That, of course, will lead to a current. It is important to note that since Δt is a finite amount of time, the emf calculated in this way is simply the *average* emf produced over the time interval. Just as the speed you calculate from $v = \Delta x/\Delta t$ is the *average* speed over the time interval Δt, the emf you calculate from this equation is the average emf over the time period of Δt. According to Eq. 10.2, similar types of emfs can be introduced in the circuit by changing the magnitude of B with time ($\Delta B/\Delta t$) or by changing the angle θ with time. Applying Faraday's law to these cases will be discussed below.

Note that Faraday's law gives us a way to calculate the magnitude of the induced emf which will result in a current flow according to Ohm's law. It does not give us the direction of the current. However, the current is created in the conductor by changing the magnetic flux surrounding the conductor. But we know a current flowing in a conductor creates a magnetic field around it.

Lenz's Law states that the induced current will create a magnetic field that opposes the change in the magnetic flux that caused it.

This gives us a way to determine the direction of the current.

Consider Example 10.2 above. The magnetic flux is coming out of the page and is increasing as the area of the rectangular loop increases. This time dependent increase in flux induces a current which has a magnetic field surrounding it. If the current flows in a clockwise direction, the induced magnetic field lines around each branch of the circuit point inward within the loop. This opposes the increasing lines of flux out of the loop and thus satisfies Lenz law.

Example 10.3

A rectangular loop of wire (20.0 cm x 50.0 cm) is placed in a perpendicular 0.176 T magnetic field as illustrated to the right. The strength of the magnetic field is then increased to 0.500 T over a period of 20.0 s. What is the average emf produced in this way? If the resistance of the loop is 1.50 Ω, what is the average current produced? What is the direction of the current in the loop?

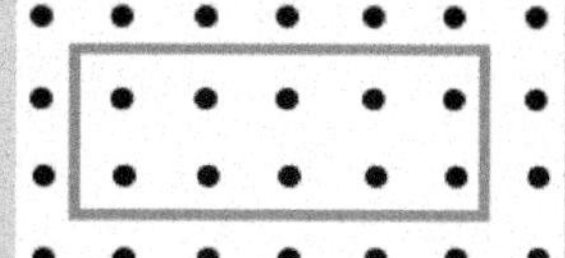

Figure is given.	Knowns: $A = 20.0 \times 50.0$ cm; $B_i = 0.176$ T; $B_f = 0.500$ T; $t =$ 20.0s; $R = 1.50\ \Omega$ Unknowns: emf; I

Let's start with the emf. The area of the rectangle is (0.200 m)·(0.500 m), or 0.100 m^2. If we define the vector **A** as coming straight out of the paper towards us, then **A** and **B** are parallel. Thus, the angle between them is zero. The initial magnetic flux, then, is:

$$\phi_i = \boldsymbol{B} \cdot \boldsymbol{A} = BA\cos(0) = (0.176\ \text{T})\left(0.100\ \text{m}^2\right) = 0.0176\ \text{T m}^2 = 0.0176\ \text{Wb}$$

Once the magnetic field is increased to 0.500 T, the magnetic flux is:

$$\phi_f = \boldsymbol{B} \cdot \boldsymbol{A} = BA\ \cos(0) = (0.500\ \text{T})\left(0.100\ \text{m}^2\right) = 0.0500\ \text{T m}^2 = 0.0500\ \text{Wb}$$

This change in flux happened in 20.0 s, so the average flux rate of change is equal to:

$$\text{emf} = \Delta\phi/\Delta t = \left(0.0500\ \text{Tm}^2 - 0.0176\ \text{Tm}^2\right)/20.0\ \text{s} = 0.00162\ \text{V}$$

Notice that when we plug in the definitions of T and A, the units cancel to leave Volts. That's the answer to the first question. The second question just requires Ohm's Law:

$$V = IR$$
$$I = V/R = 0.00162\ \text{V}/1.50\ \Omega = 0.00108\ \text{A}$$

The third question requires the use of Lens's Law. The current will flow so that its magnetic field opposes the change in flux. The magnetic field inside the rectangle is pointing out of the page, and it is increasing. The current, then, will produce a magnetic field inside the rectangle that is increasing *into* the page. If we look at the bottom of the rectangle and point my your right hand's thumb so that the current travels *clockwise*, the resulting magnetic field goes *into* the paper inside the rectangle, which is what we need to oppose the change in the flux. As a result, the current runs clockwise so that the resulting magnetic field opposes the change in flux.

We need to do one more example, because there is another way that flux can change. If the magnetic field stays constant, the flux can vary if the loop of conductor moves so that the angle between the surface of the loop and the magnetic field changes. That will produce an emfas well.

Example 10.4

A square is made out of 200 individual wire loops which are each 15.0 cm x 15.0 cm. This square is placed between two poles of a permanent magnet so that it experiences a uniform magnetic field of 0.0145 T. The loop is initially placed so that its surface is perpendicular to the magnetic field. In 0.500 s, the loop is twisted so that its surface is parallel to the magnetic field lines. When this rotation occurs, which point (A or B) has the higher potential, and how much potential difference exists between them?

Figure is given.

Knowns: $N = 200$; $A = 15.0 \times 15.0\ \text{cm}^2$; $B = 0.0145$ T; $t = 0.500$ s; $\theta_i = 0°$; $\theta_f = 90°$

Unknowns: ΔV_{AB}.

The dashed arrow is **A**, the vector that represents the surface area. At the beginning, **A** is parallel to **B**, so the angle between them is zero. Remember, this square actually contains 200 individual squares. They *each* have their own flux. As a result, the *total* flux will be the sum of all of their individual fluxes. According to Eq. 10.2, the initial flux is:

$$\phi_i = \text{NB} \cdot \boldsymbol{A} = 200(0.0145\ \text{T})(0.150\ \text{m})(0.150\ \text{m})\cos(0°) = 0.0653\ \text{Wb}$$

At the end, the vector **A** is perpendicular to the magnetic field. Thus, the angle between **A** and **B** is 90°:

$$\phi_f = NB \cdot \mathbf{A} = 200BA\ \cos(90°) = 0$$

The change in flux over time, then is given by Eq. 10.4 to be

$$\Delta\phi/\Delta t = \Delta V = \frac{0\ \mathrm{Tm}^2 - 0.0653\ \mathrm{Tm}^2}{0.500\ \mathrm{s}} = -0.131\ \mathrm{V}$$

This means that the flux *decreased* over time which is important because we need to know that to determine the direction of the current that would be produced.

To determine the direction of the current (which will also tell us which point has the higher potential), let's look at the first drawing. The flux is to the right, because the magnetic field lines pass through the rectangle, pointed to the right. However, the flux to the right *decreases* over time. The current would produce a magnetic field that is opposed to that change.

Thus, the current will produce a magnetic field that *increases* to the right as it passes through the loop. Using the right-hand rule, then, the current travels from point B to point A, because that current will produce a magnetic field that points to the right inside the loop. Since current flows from point B to point A, that means B has the highest potential. Thus, point B is at a potential that is 0.131 V higher than point A.

This is how you analyze situations that produce electricity through changes in magnetic flux. The change in flux can be due to one of three things, a change in the magnitude of the magnetic field, a change in the area of the flux, or a change in the angle between the field and the area. You can calculate the emf by using Eq. 10.4, and you can determine the direction of the current by using Lens's Law. If the flux *increases* to the right, for example, the current will be produced so that the current's magnetic field will point to the left at all points inside the loop. If the flux *decreases* to the right, however, the current will be produced so that the current's magnetic field will point to the right at all points inside the loop. See if you can do this problem.

Student

10.2 A circle of radius $R = 1.00$ cm is made up of 500 individual circles of wire. The total resistance of this circle is 0.0251 Ω. It is placed $r = 15.0$ cm from a long, straight wire, as shown to the right.
The current in the wire is uniformly increased from 0 A to 10.0 A in 30.0 s. What is the average current in the circle? In which direction does it travel?

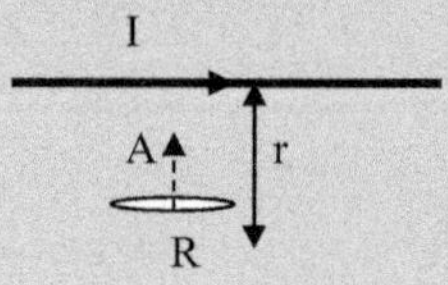

10.3 A bar magnet is shoved into a circle of wire as shown to the right. What is the direction of the current produced? If the magnet were pulled away from the wire rather than shoved into it, what would be the direction of the current produced?

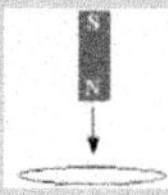

10.4 Meters, Motors, and Generators

Faraday's Law and Lenz's Law form the basis of important electromagnetic devices such as the meters, motors, and generators, discussed in this section, as well as the devices discussed in the next section. Since the devices discussed in this section are all based on a conducting loop in a magnetic field, we consider the situation shown in Fig. 10.7. With the current flowing in the direction shown in the figure, according to Eq. 9.7 the left side of the loop will experience a force in the downward direction and the right side of the loop will experience a force in the upward direction. If the loop is mounted on an axis that is free to rotate, the torque provided by these forces will rotate it in a counterclockwise direction. If the magnitudes of B and L (the size of the loop) are constant, the strength of the torque is proportional to the magnitude of the current and the orientation of the loop.

The situation shown in Fig. 10.7 can be made into a meter by attaching the loop to a rotating axis with a needle on the end and calibrating the rotation of the needle with the magnitude of a known current. A spring is used to provide reverse torque so the needle doesn't spin too far. This is called a galvanometer. If it is hooked into one of the circuits shown in Chap. 8 in series, it will measure the amount of current flowing in that branch of the circuit.

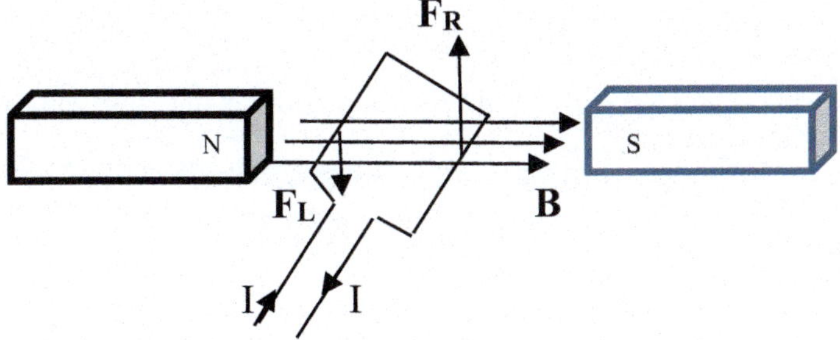

Fig. 10.7 Forces on a conducting loop in a magnetic field

Example 10.5

A galvanometer with a torsional restoring spring is shown in the figure. What is the deflection of the pointer θ if there are 75 windings around an armature of with a loop area of 2 cm^2 rotating in a radial magneticfield of 0.1 T. Assume the current is 10 μA and the restoring force on the spring is 10^{-6} Nm/rad. (For a torsional spring, $\tau = -k\theta$.)

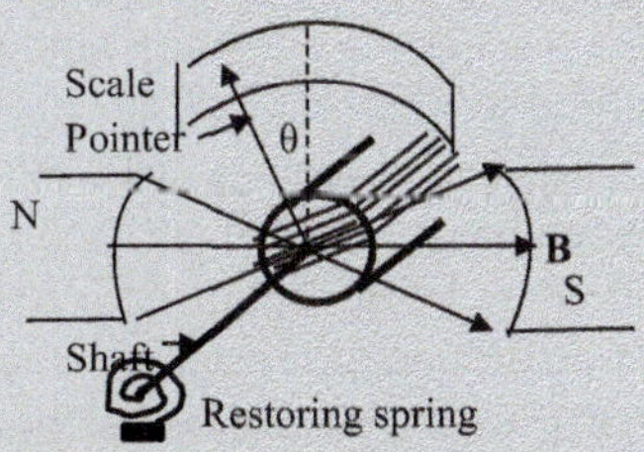

Figure is given.

Knowns: $N = 75$; $A = 2 \times 10^{-4}$ m^2; $B = 0.1$T; $\mathrm{I} = 10^{-6}$ A; $k = 0.510^{-6}$ Nm/rad

Unknowns: θ

At maximum deflection the shaft stops rotating because the torque from the current-carrying wire in the magnetic field balances the restoring spring torque in the opposite direction.

$$\tau_{\mathrm{B}} + \tau_{\mathrm{s}} = 0$$

$$k\theta = NAIB$$

So

$$\theta = NAIB/k = (75)\left(2 \times 10^{-4}\ \mathrm{m}^2\right)\left(10^{-5}\ \mathrm{A}\right)(0.1\ \mathrm{T})/\left(0.510^{-6}\ \mathrm{Nm/rad}\right)$$
$$= 0.03\ \mathrm{rad}.$$

Note we have used the fact that the angle between the magnetic force and the moment arm is 90° and the angle between the current and magnetic field is 90°. Also, using the units for a tesla as N/Am lets the units work out for the angle in radians.

Galvanometers are usually very sensitive and can measure microamps of current. To measure larger amounts of current a shunt resistor is wired in parallel to the resistance of the loop. The total current is then the sum of the current through the shunt resistor plus the current through the loop. Since the voltage across the two parallel branches of the circuit is the same,

$$I_T = V/R_L + V/R_s = (V/R_L)(1 + R_L/R_s) = I_L(1 + R_L/R_s)$$

So if the ratio of the loop resistance to the shunt resistance is equal to 9, the total current is 10 times the current recorded by the loop. An instrument like this with shunt resistors is called an ammeter. Usually an ammeter has a switch that can be turned to different values of shunt resistors so the calibrated dial can read different magnitudes such as 0–10 milliamps or 0–100 milliamps.

A galvanometer can also be used to measure the potential difference between two points in a circuit. In this case a multiplier resistor is wired in series with the resistance of the loop. The galvanometer measures the current flowing through it as $I = V/R_{eff}$ where the voltage across the meter is what we want to measure and the effective resistance is the sum of the multiplier resistor and the resistance of the loop. By changing the magnitude of the multiplier resistance the voltage reading for a specific amount of current is changed. This instrument is called a voltmeter. Again the scale can be calibrated so that by changing the multiplier resistor it can read 0–1 V or 0–10 V, etc.

Try the following problem.

Student

10.4 A galvanometer has an internal resistance of 10Ω and a full scale deflection when it has a current of 2 mA. How can you change it to give a full-scale deflection for a current of 0.25A?

Electric motors work on the same principle of torque on a current carrying loop of wire like the meters discussed above. However, in this case there is no spring to limit the amount of rotation of the loop so it can continue for 360°. This allows the electrical energy of the current to be converted into mechanical rotational kinetic energy. A schematic picture of an electric motor is shown in Fig. 10.8.

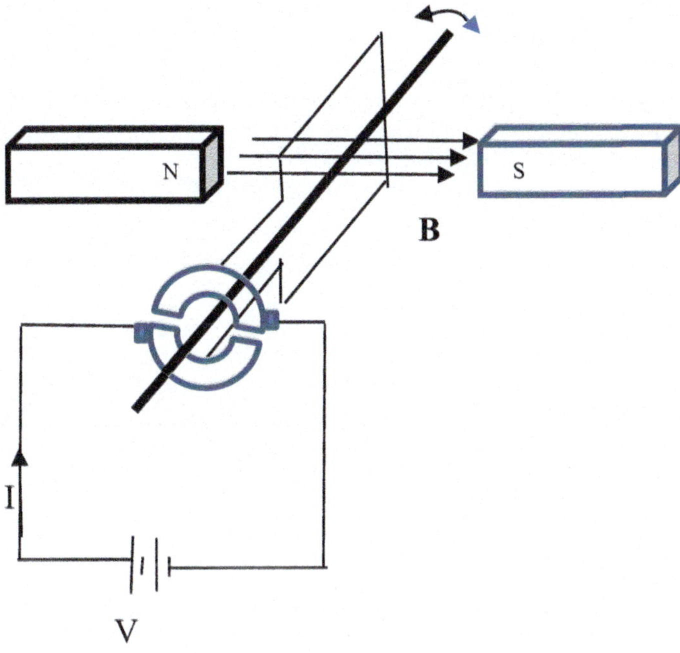

Fig. 10.8 Electric motor

The problem of maintaining a loop rotation of 360° can be seen by looking at Fig. 10.7. Once the loop reaches a vertical position where its area is perpendicular to the magnetic field, the forces on the two sides of the wire are parallel to their moment arms so they exert zero torque. If the loop coasts through the vertical position, the torque created will oppose the direction of motion and stop it from rotating. In order to get around this problem, the current in the loop must reverse direction every time the loop reaches a vertical position. Then when the loop coasts past the vertical position the torque will switch directions and work to increase the rotation in the same direction.

One way to cause this to happen with DC electricity is shown in Fig. 10.8. This is based on a split-ring commutator that is a circular conductor that rotates with the shaft the loop is on. It is split into two semicircular parts as shown. The wire from one side of the conducting loop is attached to one part of the split-ring commutator and the wire from the other side is attached to the other part of the commutator. The current is provided by a battery making contact with the two parts of the commutator through brushes. The brushes are not fastened to the commutator but instead make a pressure contact with whichever part of the split-ring that is rotating closest to it at a specific point in time. As the loop rotates the current reverses direction every time the brushes change the part of the ring with which they are in contact. In Fig. 10.8 the shaft supporting the wire loop and commutator is rotating in the counterclockwise direction and has just reached a vertical position. The current flowing from the battery is about to shift from making contact through the lower ring of the commutator to the upper ring. As the loop coasts through the vertical position the current starts to flow into the upper wire on the loop providing a torque in the counterclockwise direction. Meanwhile, the current flows out of the bottom of the wire loop into the lower semicircle of the commutator. This also provides a torque to the shaft in the counterclockwise direction. The speed of the rotation of the shaft is controlled by changing the strength of the current flowing through the loop. Most electric motors use many coils of wire to enhance the strength of the motor. These groups of coils are called armatures.

It is also possible to produce an electric motor using AC current where the current automatically reverses direction in every half cycle. In this case the split ring commutator shown in Fig. 10.8 is replaced by two solid circular slip rings, one attached to the upper wire of the loop and one attached to the lower wire. They both rotate with the shaft holding the loop and are connected to the AC electricity source through brush contacts.

A generator can be thought of as an electric motor acting in reverse. In other words, mechanical rotational kinetic energy is converted into electrical energy. For example, in Fig. 10.8, instead of a battery providing current to the armature, a steam turbine can provide the torque required to rotate the armature shaft. As the conducting loops rotate in the magnetic field, the amount of flux through the loop varies with time and Faraday's Law tells us that an emf is produced in the circuit causing a current to flow. If the output is connected directly to an electrical circuit with two solid slip rings, AC current is produced. This is called an alternator. By using a split-ring commutator the current produced is always in the same direction (DC) but varies in amplitude like the positive part of a sinewave being repeated over and over.

It is also possible to construct an alternator by keeping the armature fixed and rotating the magnet around it. This has the same effect of producing a changing magnetic flux in the armature coils and thus producing AC current output. The advantage of this design is having the output wires of the armature fixed in space so there is no need for a rotating ring and brush connector that can cause problems such a sparking.

Example 10.6

In a generator, the magnetic field is provided by wire coils called "field coils" and the electricity is generated by wire coils called "armature coils". If they are wired in series, and a load wired in parallel with the field coils, the circuit looks like the figure to the right. If this generator has a terminal voltage of $V_L = 120$ V and delivers 1.50 kW of power to the line, what is the emf of the generator and its efficiency, if $R_F = 200\ \Omega$ and $R_A = 0.500\ \Omega$?

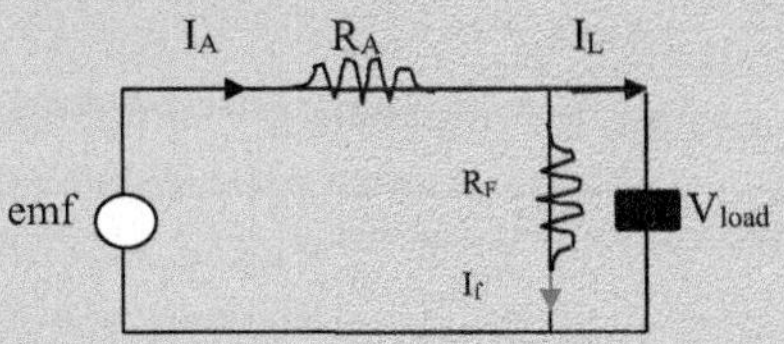

Knowns: $V_L = 120$ V; $P_L = 1.50$ kW; $R_F = 200\ \Omega$; $R_A = 0.500\ \Omega$.
Unknowns: emf; efficiency.
The power delivered to the load is

$$P_L = I_L V_L$$

So,

$$I_L = P_L/V_L = 1500\ \text{W}/120\ \text{V} = 12.5\ \text{A}$$

The field coils have the same voltage as the load. Using Ohm's law

$$V_L = I_F R_F$$

So,

$$I_F = V_L/R_F = 120\ \text{V}/200\ \Omega = 0.600\ \text{A}$$

Now using Kirchhoff's current law,

$$I_A = I_L + I_F = 12.5\ \text{A} + 0.600\ \text{A} = 13.100\ \text{A}$$

Then using Kirchhoff's voltage law,

$$\text{emf} = I_A R_A + I_F R_F = (13.1\ \text{A})(0.5\ \Omega) + 120 = 126.55\ \text{V}$$

The answer to the first part of the question is that the generator produces an emf of 126.55 V. Next we need to find how efficient the generator is. To do this we need to compare the power it produces to the power it delivers to the load.

The generator produces a power of

$$P_{\mathrm{i}} = (I_{\mathrm{A}})\mathrm{emf} = (13.1\ \mathrm{A})(126.55\ \mathrm{V}) = 1657.8\ \mathrm{W}$$

Then the efficiency is

$$\eta = P_{\mathrm{o}}/P_{\mathrm{i}} = 1500\ \mathrm{W}/1657.8\ \mathrm{W} = 0.905 = 90.5\%$$

So the generator is over ninety percent efficient.

Next Level

Back EMF

When a motor is first turned on, there is a spike in the current flow which decreases when the motor is operating at a constant speed. This is due to back emf. When the current starts to flow in the armature wire of a motor, the effective emf driving the current is the external emf, V. However, once the current is flowing in a moving wire in a magnetic field, Lenz's law tells us that an emf is created to oppose the change in current. Thus the effective emf is the external emf minus the back emf, $(V - E_{\mathrm{b}})$. This reduced potential difference results in a smaller current flow. If the motor on an appliance wired in series with a light source, the lights will have a high initial brightness when the circuit is turned on but quickly return to their lower brightness when the motor is running at a constant speed. The strength of E_{b} depends on the change in the number of magnetic lines of flux cut by the armature coils per unit time as the motor spins. This depends on the size of the armature coils, the number of coils, and the rotation speed of the motor. The faster the motor turns, the greater the back emf and thus the lower the current in the circuit.

Before going on to the next section, try the following problem.

Student

10.5 A two horsepower electric motor is connected to a 120 V line. If it operates for a half an hour, how much current does it draw and how much electrical energy does it use? (Assume no friction losses.)

10.5 Inductors, Filters, and Transformers

The back emf discussed above is also responsible for the self-inductance L present in an inductor coil in an AC circuit as discussed in Sect. 8.5. An inductor is a passive electrical device used in AC electrical circuits. It is made in the form of a solenoid which has a coil of conducting wire wrapped around a core that can be either air or a ferromagnetic material. Electrical current through the conductor creates a magnetic flux proportional to the current. This is shown in Fig. 10.5c. The lines of flux go through the center of the coil of wire, out one end, around the side, and in the other end. A change in the current creates a change in magnetic flux. According to Faraday's law, this generates an emf and according to Lenz's law this emf acts to oppose the change in current that caused it.

The important property of an inductor is its inductance, designated L, whose units are Henries. (Remember a Henry is a $kgm^2/(s^2A^2)$.) The magnitude of the inductance for an inductor is determined by its shape and the material of the core. For a cylindrical solenoid

$$L = \mu A N^2 / l \tag{10.5}$$

where μ is the permeability of the core material, A is the cross-sectional area of the solenoid of length l, and N is the number of turns in the coil. Remember that permeability describes the magnetization developed in the material produced by an external magnetic field. If the change in magnetic flux is produced by an external magnetic field, the response is called inductance. If the change in magnetic flux is produced by the current flowing through the device itself, it is called self-inductance.

We learned in Sect. 8.5 that an inductor in the circuit inhibits the flow of AC electricity by providing a reactance proportional to its inductance (Eq. 8.20)

$$X_{\mathrm{L}} = 2\pi f L$$

where f is the frequency of the AC electricity. When the AC voltage changes direction, the back emf delays the time of the current buildup so the phase of the current sinewave lags the voltage sinewave.

Remember from Sect. 8.5 that capacitors also provide a reactance in an AC circuit. The capacitive reactance is given by (Eq. 8.17)

$$X_{\mathrm{C}} = 1/(2\pi f C)$$

where C is the capacitance of the capacitor. The effect of capacitive reactance is to cause the phase of the current sinewave to lead the voltage sinewave. This is because when the AC current changes direction the current starts to flow immediately while the voltage across the capacitor is dependent of the buildup of charge on the plates.

Consider the following example and problem.

Example 10.7

Consider the following circuit. When it is powered by a DC source of 120 V it draws 0.80 A of current. When it is powered by a 60 cycle AC source it draws 0.60 A of current. What are the values of the resistance, inductance, reactance, impedance, and phase angle?

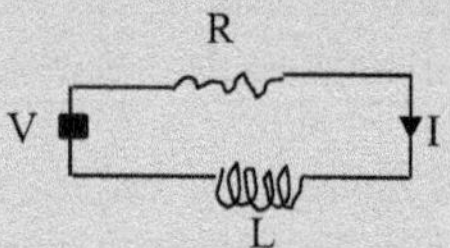

For the DC voltage source, $f = 0$ so $X_L = 2\pi fL = 0$.

That means that the impedance is just equal to the resistance, $Z = R$, and the resistance can be found from Ohm's law,

$$R = V/I = 120\ \text{V}/0.80\ \text{A} = 150\ \Omega = Z$$

For the AC voltage, this will also be the resistance. In this case,

$$Z = V/I = 120\ \text{V}/0.60\ \text{A} = 200\ \Omega$$

The phase diagram looks like:

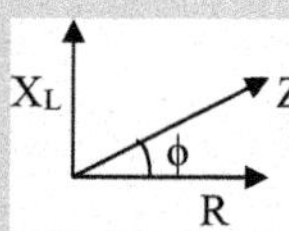

So,

$$Z^2 = R^2 + XL^2$$

or

$$X_L{}^2 = 40,000\ \Omega^2 - 22,500\ \Omega^2 = 17,500\ \Omega^2$$
$$X_L = 132.29\ \Omega$$

Then the inductance is found from,

$$X_L = 2\pi fL$$

$$L = X_L/(2\pi f) = 132.29\ \Omega/\left(2\pi 60\ \text{s}^{-1}\right) = 0.35\ \text{H}$$

The phase angle is given by.

$\phi = \sin^{-1}(X_L/Z) = \sin^{-1}(132.29\ \Omega/200\ \Omega) = 41.38^\circ$ voltage leads current.

Student

10.6 Consider the following AC circuit. If $R = 500\ \Omega$, $L = 0.2$ H, and $C = 3.0\ \mu$f, calculate the impedance and draw the phase diagram for a frequency of $500\ \text{s}^{-1}$.

Because inductive reactance is directly proportional to frequency while capacitive reactance is inversely proportional to frequency, circuits using inductors and capacitors can be made into frequency filters. Inductors will block AC signals with high frequencies because they have high reactance and pass low frequencies because they have low reactance. Instead, capacitors will block AC signals with low frequencies because of their high reactance and pass high frequencies because of their low reactance.

A typical high pass filter is shown in Fig. 10.9. A circuit like this might be used in a music player. The music source provides an AC signal with many different frequencies. The filter controls the amount of high frequency signals reaching the speaker. This controls the sound quality making it more of a treble sound. Figure 10.10 shows the circuit for a low pass filter. It allows low frequencies from the source to reach the load. This gives it a stronger bass sound.

Remember from Sect. 8.5 that if we choose L and C for a circuit to give a zero reactance, we can use the circuit to select a resonant frequency signal from a multi-frequency source. These types of frequency filter devices have many applications when we are dealing with multi-frequency electromagnetic signals.

Try your hand with the following problem.

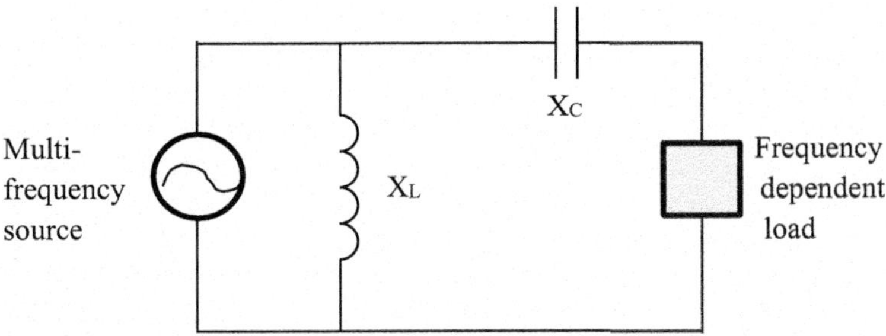

Fig. 10.9 High pass filter

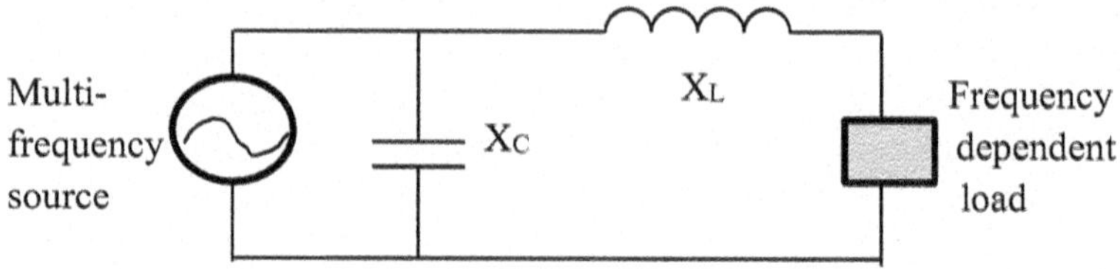

Fig. 10.10 Low pass filter

Student

10.7 If a circuit with and impedance of 140 Ω will block a signal. Does a 10 μf capacitor or an inductor with 0.2 H block a signal with a frequency of 120 Hz?

Another important device that depends on magnetic flux is a transformer. This is used to increase or decrease AC voltages. For example, most electrical power plants generate AC electricity and send it through a step-up transformer to increase its voltage before being transmitted to customers. When we use electrical power in our home, most devices operate on lower voltages, so they need a step-down transformer.

Figure 10.11 shows a typical transformer. It is made up of an iron core that efficiently transmits magnetic flux. The primary side has the input voltage V_p and number of windings N_p. The alternating current in the primary circuit creates a magnetic flux ϕ_p in the transformer. Because of the permeability of the core material, this fluctuating flux is transmitted through the iron core to the secondary arm of the magnet as ϕ_s. The secondary side has N_s windings that experience the changing magnetic flux and this creates an emf V_s in the secondary circuit. This is called mutual induction. The voltages in the primary and secondary circuits are given by Faraday's Law as

$$V_p = N_p \Delta\phi_p / \Delta t$$

Fig. 10.11 Transformer

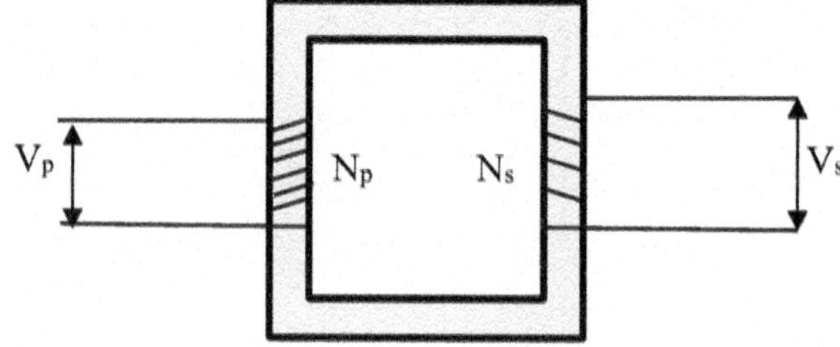

and

$$V_s = N_s \Delta\phi_s / \Delta t$$

Assuming that the transformer has perfect coupling between the primary and secondary arms, the magnetic flux in the two arms is the same. Thus the ratio of secondary to primary voltage is equal to the ratio of the secondary to primary number of coils,

$$V_s / V_p = N_s / N_p. \tag{10.6}$$

Example 10.8

If you plug a device into a wall socket with 120 V and it has a transformer with 100 coils on the primary arm and 10 coils on the secondary arm, what is the voltage used by the device?

Using 10.11 for the figure, the knowns and unknowns are
Knowns: $V_p = 120$ V; $N_p = 100$; $N_s = 10$.
Unknowns: V_s.
Then using Eq. (10.6), the secondary voltage is given by

$$V_s = V_p (N_s / N_p) = (120 \text{ V})(10/100) = 12 \text{ V}.$$

This is a step-down transformer that reduces the input voltage to one-tenth its original amount.

In other cases, we might need a step-up transformer that has more coils on the secondary than on the primary, so the voltage is increased. Note that transformers are useful for AC circuits but not for DC circuits since they require a changing current to work.

To finish our discussion on electromagnetism, work the following problem.

Student

10.8 A power transmission line operating at 240 kV delivers 2400 V through a step-down transformer. If it is 100% efficient and the primary of the transformer has 10,000 loops of wire, how many loops are on the secondary arm of the transformer? If the current in the secondary is 500 A, what is the current in the primary?

Special Topic
The power distribution grid in the United States has been referred to as one of the Wonders of the Modern World. It consists of many power generation sources and power delivery lines spread through the entire country. There are a variety of sources for powering the electrical generators including burning fossil fuel (coal, oil, gas), hydroelectric, nuclear, geothermal, solar and wind to name a few. The generators produce AC electricity at a low voltage. To transmit the high power that is needed to serve many customers, it is better to use high voltage with low current. This is because transmitting the same power with high current and low voltage creates heating loss due to resistance in the lines. Thus a step-up transformer is used to put the power on the transmission lines with high voltage. The local power companies have sophisticated control rooms where their operators determine the load requirements of their customers and take power from the grid sources to meet this load. Balancing the source with the load minimizes the transmission losses due to impedance in the circuit. Variable inductors and capacitors are also wired into the circuit to keep the current and voltage in-phase during transmission. The end user requires low voltage, so a step-down transfer is required at the customer's site. Note that all of the devices discussed in this chapter are relevant to the power grid. The reliable delivery of electricity is challenged by anything that changes in this complicated distribution system. Society is currently moving toward the greater use of renewable energy sources such as solar and wind. These have an attractive benefit of being clean sources of energy production, but they have an issue of being intermittent which provides a challenge to the grid operators.

Summing Up

Now that you know about magnetic flux along with Faraday's Law and Lenz's Law, you know how electrical motors and generators work. In addition, you know how A.C. circuit elements like inductors and transformers work and how useful they are in power transmission and tuning filters. In the following chapter we will discuss the propagation of electromagnetic waves and in Chap. 15 we will talk about the quantum mechanical aspects of electromagnetic propagation. In addition, we will learn more about the electromagnetic properties of solids in Chap. 14 and electromagnetic properties of atoms in Chap. 16. In other words, electromagnetism is an important part of many different areas of physics.

Solutions to the Problems

10.1 Figure is given. Knowns: $B = 4.32 \times 10^{-5}$ T; $r = 10.0$ cm

Unknowns: I, direction of B.

Equation 10.1 will give us the strength:

$$B = \mu_0 I/(2\pi r)$$

$$I = B2\pi r/\mu_o = \left(4.32 \times 10^{-5}\ \mathrm{T}\right)(2\pi)(0.100\ \mathrm{m})/\left(4\pi \times 10^{-7}\ \mathrm{Tm/A}\right) = 21.6\ \mathrm{A}$$

That's a *lot* of current. Most household circuits run less than 20 A, so even a large amount of current produces a small magnetic field. Also, notice that since the value for μ_o is *exact* (as stated in the text), it does not enter into consideration for significant figures.

The problem also asks for the direction of the magnetic field. Pointing your right thumb in the direction of the conventional current, your fingers curl out of the page above the wire and into the page below the wire. Thus, the magnetic field points out of the page at the point the physicist is studying.

10.2 Knowns: $R = 1.00$ cm; $N = 500$ individual; $\Delta I_w =$ 10.0A $\Delta t = 30$ s; Resistance $= 0.0251\Omega$; $r =$ 15.0 cm

Unknowns: I_c

The current-carrying wire will produce a magnetic field, which will produce a flux through the circles of wire. Initially, there is no flux, because there is no magnetic field. At the end, the magnetic field in the center of the circle is:

$$B = \mu_0 I/(2\pi r) = \left(4\pi x 10^{-7}\mathrm{Tm/A}\right)(10.0\mathrm{A})/(2\pi 0.150\mathrm{m}) = 1.33 \times 10^{-5}\mathrm{T}$$

Note that the r in this equation is *not* the radius of the circle, but the distance from the wire.

There are 500 individual circles, so they all experience a change in magnetic field from 0 to 1.33×10^{-5} T. They experience the following rate of change in flux:

$$\Delta\phi/\Delta t = N\Delta BA/t = (500)\left(1.33 \times 10^{-5}\ T\right)(\pi)(0.0100\ \mathrm{m})^2/(30.0\ \mathrm{s})$$
$$= 6.96 \times 10^{-8}\ \mathrm{V}$$

That's the emf produced by the changing magnetic flux. Using Ohm's Law, we can calculate the resulting current:

$$V = I\mathrm{Resistance}$$
$$I = V/\mathrm{Resistance} = 6.96 \times 10^{-8}\ \mathrm{V}/0.0215\ \Omega = 2.77 \times 10^{-6}\ \mathrm{A}$$

The current must be produced so that its magnetic field opposes the change in flux that caused it. According to the right-hand rule (point your thumb in the direction of the conventional current and the fingers curl in the direction of the field), the magnetic field goes into the paper at the point where the circles of wire are. The flux increases over time, so the current in the circles must produce a magnetic field which comes up out of the paper. Using your right hand, then, your fingers must curl out of the paper inside the circle. Your thumb indicates that the current must therefore travel counterclockwise.

10.3

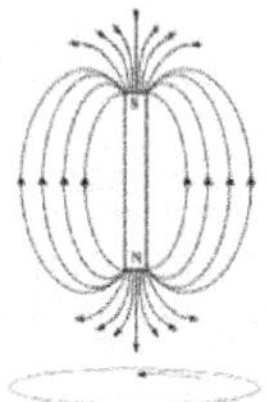

Knowns: magnet shoved in; magnet pulled out

Unknowns: current directions in loop

To solve this correctly, you first must think of the direction of the magnetic field lines. They come out of the magnet's north pole and end up in its south pole. Thus, as the magnet is pushed into the loop of wire, the magnetic field points *down* through the circle, and the flux is increasing. The current must produce a magnetic field which opposes this, so the current's magnetic field would have to point *up* out of the circle. Curling your right-hand fingers up out of the circle makes your thumb point so that the current flows counterclockwise around the circle.

If the magnet were pulled away from the circle, the magnetic field lines would still be pointed down, but the flux would *decrease* in that direction. If the flux decreased in that direction, the current's magnetic field would oppose it by pointing in the same direction (so as to "fight" the decrease). The right-hand rule tells us that to produce a magnetic field which travels down through the circle, the current would have to flow clockwise. Thus, if the magnet were pulled away, the current would flow clockwise.

10.4

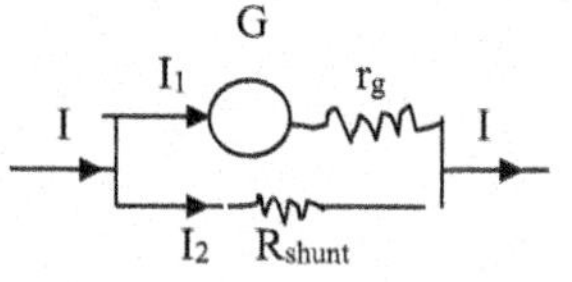

Knowns: $r_g = 10\ \Omega$; $I_1 = 2$ mA; $I = 0.25$ A;

UUnknowns: R_{shunt}

For I_1 to be 2×10^{-3} A when $I = 0.25$ A means

$$I_2 = I - I_1 = 250\ \text{mA} - 2\ \text{mA} = 248\ \text{mA}$$

The voltage across the galvanometer and the shunt resister is the same so,

$$I_{11}r_g = I_2R_{\text{shunt}}$$

$$\text{So } R_{\text{shunt}} = rgI_1/I_2 = (10\ \Omega)(2\ \text{mA})/(248\ \text{mA}) = 0.08\ \Omega.$$

Therefore adding a resistor of 0.08 Ω makes the maximum deflection of this galvanometer $I = 0.25$ A.

10.5

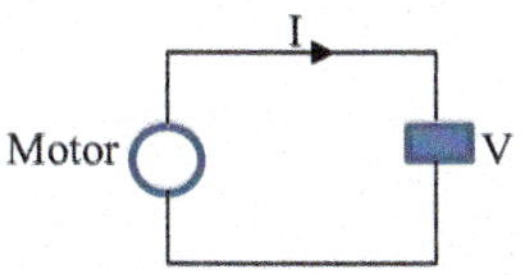

Knowns: $P = 2$ hp; $V = 120$ V; $t = 0.5$ h
Unknowns: I; E_{elec}

Power is voltage times current, $P = VI = 2$ hp $= 1492$ W

$$I = P/V = 1492\ \text{W}/120\ \text{V} = 12.43\ \text{A}$$

The electrical energy is power times time,

$$E_{\text{elec}} = Pt = 1492\ \text{W}(0.5\ \text{h}) = 0.746\ \text{kW-h}$$

We have expressed the electrical energy in the units of kW-h because that is the common unit use by electrical companies.

10.6

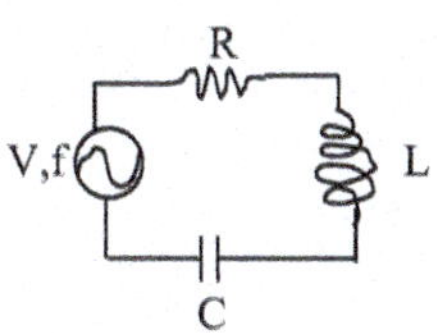

Knowns: $R = 500\ \Omega$; $L = 0.2$ H; $C = 0.3$ μf; $f = 500\ \text{s}^{-1}$
Unknowns: Z; vector phase diagrams

Equation 8.25 tells us that the impedance is related to the capacitive and inductive reactances,

$$Z = \sqrt{X^2 + R^2} = \sqrt{\left(2\pi fL - \frac{1}{2\pi fC}\right)^2 + R^2}.$$

At a frequency of 500 Hz,

$$X_{\mathrm{L}} = 2\pi fL = 2\pi\left(500\ \mathrm{s}^{-1}\right)(0.2\ \mathrm{H}) = 628\ \Omega$$

where we have used the fact that one henry is one kgm^2/sec^{-2}A^{-2} so the units of inductive reactive are Ohms.

$$X_{\mathrm{C}} = (2\pi fC)^{-1} = \left[2\pi\left(500\ \mathrm{s}^{-1}\right)\left(0.3\times10^{-6}\ \mathrm{f}\right)\right]^{-1} = 1060\ \Omega$$

According to Eq. 8.24, the total reactance is the difference of inductive and capacitive reactance,

$$X = X_{\mathrm{L}} - X_{\mathrm{C}} = 628 - 1060 = -432\ \Omega$$

The impedance is then,

$$Z = \sqrt{X^2 + R^2} = \left[(-432\ \Omega)^2 + (500\ \Omega)^2\right]^{1/2} = 661\ \Omega$$

The phase diagram for $f = 500$ Hz is with the values of R, X, and Z given above ant the phase angle.

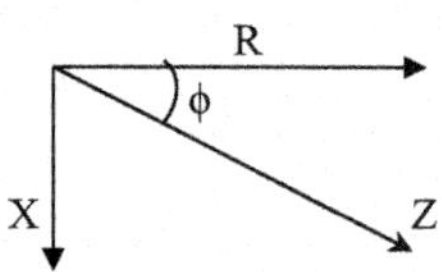

$$\phi = \sin^{-1} X/Z = -432/661 = -40.8^\circ \quad \text{Current lags voltage.}$$

10.7

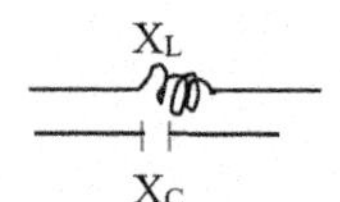

Knowns: $C = 10\ \mu$f; $L = 0.2$ H; $f = 120$ Hz; $Z = 140\ \Omega$

Unknowns: X_{L}; X_{C}

$$Z^2 = \left(R^2 + X^2\right)^{1/2}$$

If there is no resistor present, the impedance equals the reactance.

$$X_{\mathrm{L}} = 2\pi fL = 2\pi\left(120\ \mathrm{s}^{-1}\right)(0.2\ \mathrm{H}) = 150.8\ \Omega > 140\ \Omega$$

$$X_{\mathrm{C}} = \frac{1}{2\pi fC} = \left[2\pi\left(120\ \mathrm{s}^{-1}\right)\left(10^{-5}\ \mathrm{f}\right)\right]^{-1} = 132.6\ \Omega < 140\ \Omega$$

This shows that a 0.2 h inductor has enough impedance to block the signal at 120 Hz while the impedance of a 10 μf capacitor will not block this signal.

10.8

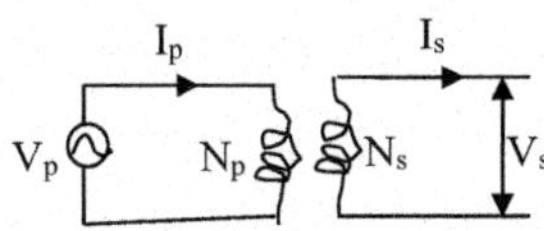

Knowns: $V_P = 20{,}000$ V; $V_S = 2000$ V; $N_P = 12{,}000$; $I_S = 400$ A

Unknowns: N_S; I_P

According to Eq. 10.6, $V_P/V_S = N_P/N_S$ so

$$N_S = N_P V_S / V_P = (1.2 \times 10^4)(2.0 \times 10^3 \text{ V})/\left(2.0 \times 10^5 \text{ V}\right) = 120$$

For 100% efficiency, conservation of energy requires power out equals power in and $P = VI$. So,

$$V_P I_P = V_S I_S$$

$$I_P = I_S V_S / V_P = (400 \text{ A})(2000 \text{ V})/\left(2.0 \times 10^5 \text{ V}\right) = 4.0 \text{ A}$$

Study Guide for This Chapter

$$\left(\mu_o = 4\pi \times 10^{-7} \text{Tm/A}\right)$$

1. A permanent magnet is pulled through a loop of wire as shown to the right. The current in the loop is measured during the entire process. Does the direction of the current change? If so, why?

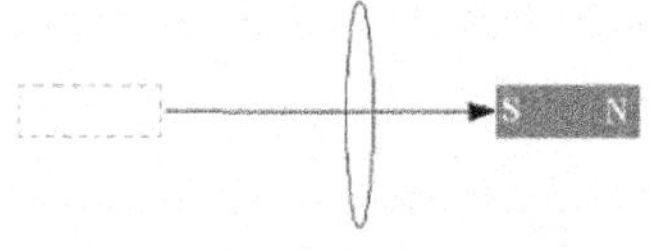

2. In the situation above, what would happen to the current if, once the magnet was inside the loop, it was held stationary?
3. A permanent magnet is repeatedly pushed in and out of a loop of wire.
   a. Will the current produced be alternating current or direct current?
   b. As the frequency with which the magnet is pushed in and out increases, what happens to the magnitude of the current produced?
4. State Faraday's law and Lenz's law in words and discuss their relationship.
5. What are three ways to change the amount of magnetic flux in a loop of wire?
6. How do you change the sensitivity scale on a voltmeter or ammeter?
7. For AC circuits, when does voltage lead current and when does voltage lag current?

8. Why aren't transformers found in DC circuits?

9. A loop of wire ($R = 0.0340\ \Omega$) is placed in between two current carrying wires which each carry 1.00 A to the right, as illustrated in the drawing to the right. The radius of the loop of wire is 0.500 cm. Over a period of 0.500 s, the current in the bottom wire is reduced to zero. What are the magnitude and direction of the current that will develop?

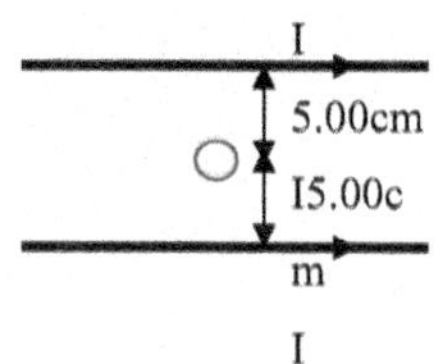

10. In the situation described above, what will be the magnitude and direction of the current in the loop once the current in the bottom wire is reduced and held at zero?
11. A square made of 500 individual squares of wire is turned in a magnetic field as illustrated below. The squares are all 50.0 cm × 50.0 cm, and the strength of the magnetic field is 0.987 T.

If the square is turned as illustrated over a period of 0.300 s, what is the average total current?

In what direction does it flow? The total resistance of all of the loops combined is 1.20 Ω.

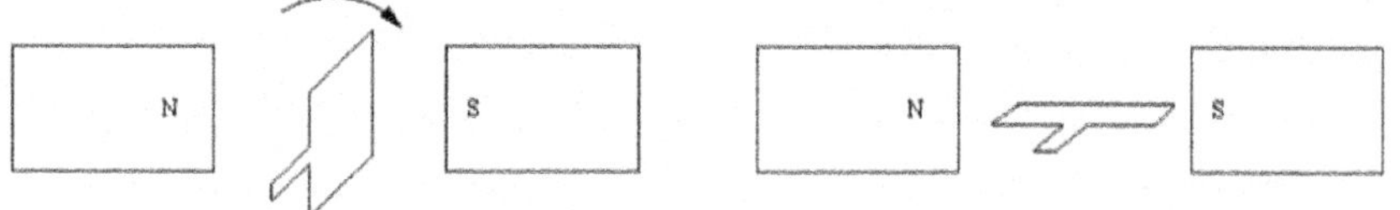

12. For the galvanometer in the problem 10.4, what modification must be made so it gives a full scale deflection for a potential difference of 5.0V?
13. What is the impedance of a 5H inductor at frequencies of 120 Hz and 220 Hz?
14. A low-pass filter has a 1 μf capacitor and an inductor with a variable inductance. To effectively work at a maximum frequency of 1000 Hz, the inductive reactance must be less than or equal to one-tenth of the capacitive reactance. What is the maximum amount the inductance can be?
15. A transmission line uses a transformer to step down the voltage from 2000 V to 100 V. If it is 90% efficient with an output power of 10.0 kW, and a primary winding with 5000 turns, what are the currents in the primary and secondary arms of the transformer, the power on the primary side of the transformer, and the number of turns in the secondary?

Next Level

16. A DC motor with an internal resistance of 0.200 Ω is wired in series with a 120 V source. At its rated speed, it produces a current of 10.0 A. What is the back emf of the armature?
17. A light source is on the same electrical circuit as an appliance with an electric motor. The light's brightness is directly proportional to the current flowing through it. The circuit is driven by a 10 V battery. If the light dims to half its initial brightness when the motor is turned on, what is the back emf of the motor?

Waves I: Optics

11

Chapter Summary

How do signals such as electromagnetic fields, music, or water disturbances propagate from one place to another? The answer is by waves and this makes wave motion an important area to understand in physics. All waves have certain characteristics in common. However, other characteristics can be very different from one type of wave to another. In this chapter we use light as an example of one type of wave. Its physical properties are significantly different from those of sound waves discussed in the following chapter. Some of the themes of the chapter are represented in Figs. 11.1, 11.2, 11.3, 11.4, 11.5 and 11.6. Figure 11.1 shows circular waves in a liquid. Figure 11.2 shows a set of mirrors. Figure 11.3 shows a child looking at her reflection in a mirror. Figure 11.4 shows dispersion and reflection of white light by a prism. Figure 11.5 shows an interference pattern of light going through a single slit. Figure 11.6 shows an interference pattern of light by a thin film of oil.

Main Concepts in This Chapter

- Waves and Pulses
- Boundary Conditions and Interference
- Geometric Optics—Mirrors and Lenses
- Physical OpticsDispersion, Interference and Polarization

R. C. Powell, *Physics*, Undergraduate Lecture Notes in Physics,
https://doi.org/10.1007/978-3-032-09361-5_11

Fig. 11.1 Surface waves. *Credit* Roger McLassus, Creative Commons Attribution-Share Alike 3.0

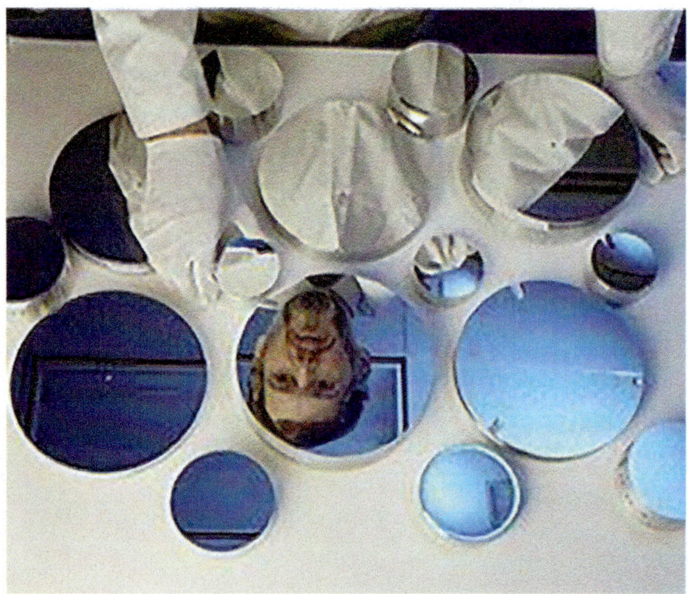

Fig. 11.2 Optics—mirrors and lenses. *Credit* doe oakridge, public domain

11.1 Introduction

The next area of physics we will study deals with wave motion. There are two important applications of waves in physics: the fields of optics and acoustics. The general concepts of wave motion are important in both of these fields, but we learned that optic waves and acoustic waves are really quite different. Of course, there are other situations where waves play an important role in the world around us (such as ocean waves), and what we learn in the following chapters can be applied to these other situations. We will begin by reviewing the general physical properties of waves and then switch to their specific applications in optics. The next chapter will cover acoustics.

Fig. 11.3 Mirror reflection. *Credit* Psychopoesie, Creative Commons Attribution-Share Alike 3.0

Fig. 11.4 Dispersion from prism. *Credit* Spigget, Creative Commons Attribution-Share Alike 3.0

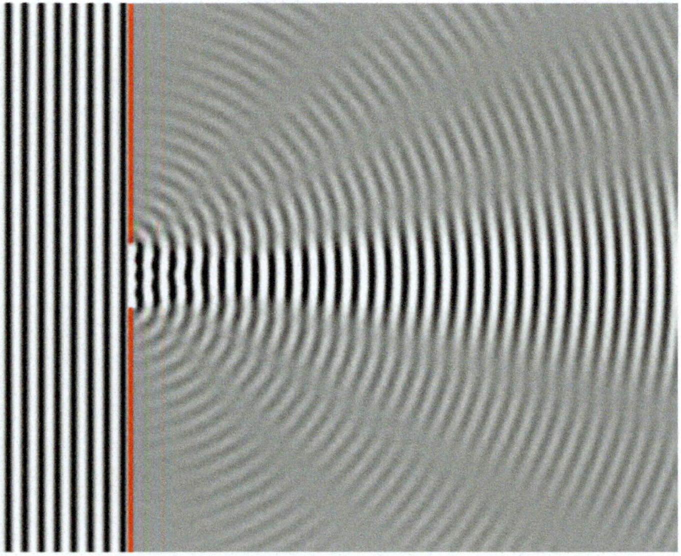

Fig. 11.5 Single slit diffraction pattern. *Credit* Dicklyon, public domain

Fig. 11.6 Oil spill on the road. *Credit* John, Creative Commons Attribution-Share Alike 2.5

11.2 Waves and Pulses

A wave is a periodic, continuous disturbance that transmits energy through matter or space. A pulse is a single disturbance that moves in time and space. A wave can be thought of as a train of pulses. Pulses and waves have similar physical properties. We will discuss these properties in terms of waves because of their important applications but you should remember that the same properties apply to pulses.

In many ways the properties of waves are similar to the periodic motion of mechanical systems that we discussed in Chap. 6. The basic concepts of mechanics such as conservation of energy apply to pulses and waves. However, in the case of classical mechanics, mass is transmitted along with energy whereas in wave or pulse motion only energy is transmitted. Some pulses and waves require the presence of matter such as air or water to transmit energy. These are called mechanical waves and sound waves are one example of a mechanical wave. Other waves such as optical waves can transmit energy in empty space. One example is sunlight traveling through outer space from the sun to the earth.

Pulses and waves are classified into two types depending on how the disturbance causing it is created. The first type is a transverse pulse or wave where the disturbance is perpendicular to the direction of its propagation as shown in Fig. 11.7a. One example is plucking a string on a musical instrument. The second type is a longitudinal pulse or wave where the disturbance is parallel to the direction of propagation as shown in Fig. 11.7b. One example is squeezing a coiled wire in the form of a slinky. In both cases the disturbance that is created propagates in both directions along the medium supporting it with a velocity $v = \Delta x/\Delta t$. Note that the strength of the disturbance that is created determines the amplitude of the pulse or wave. If more work is done in creating the disturbance, the amplitude is greater and the pulse or wave transmits more energy. The disturbance creating the wave also is responsible for its frequency. The velocity of the propagation depends on the medium the wave is in.

The general properties of transverse and longitudinal pulses or waves are the same. Since it is easier to visualize transverse waves, we will use this type of wave as an example of these properties. Figure 11.8 shows a diagram of a transverse wave with its various properties. Part (a) shows the spatial propagation of the wave while part (b) shows its evolution in time. In both cases, the y-axis shows the amplitude of the wave at different points in space and time. Remember from your introductory course that the peak amplitudes for transverse waves are called crests and the minimum amplitudes are called troughs. For longitudinal waves, the terms compressions and rarefactions are used.

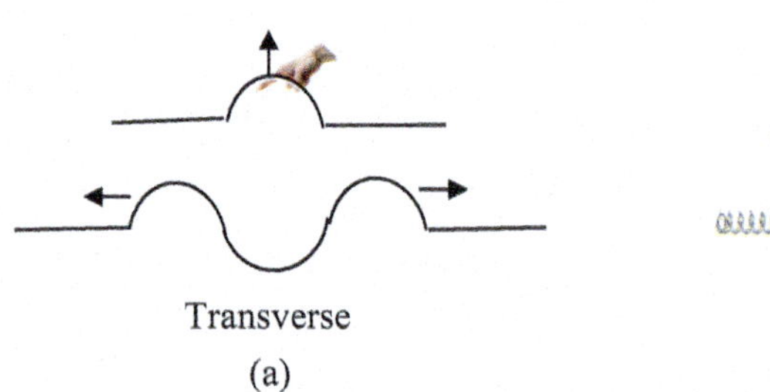

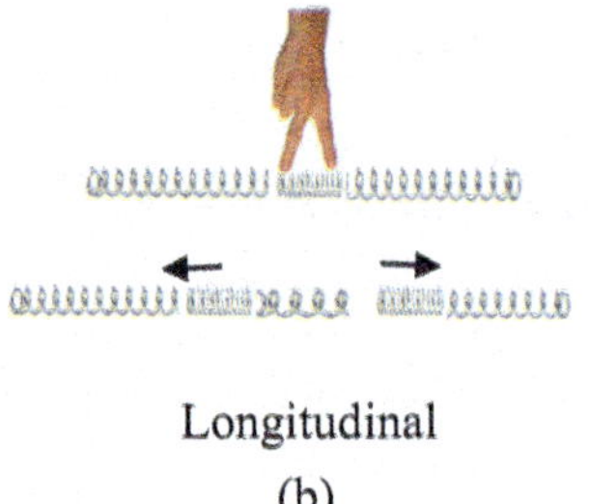

Fig. 11.7 Types of waves

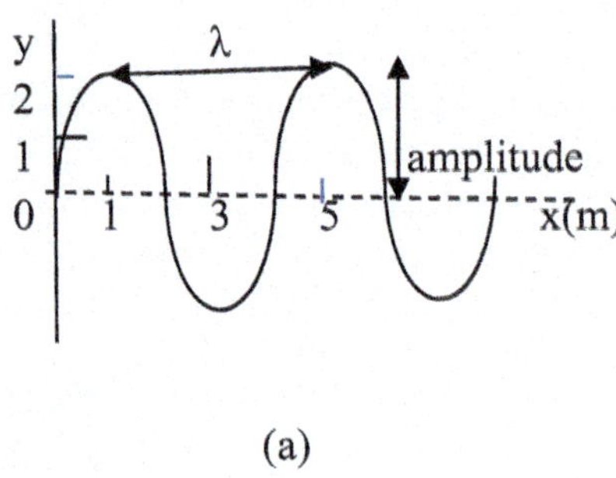

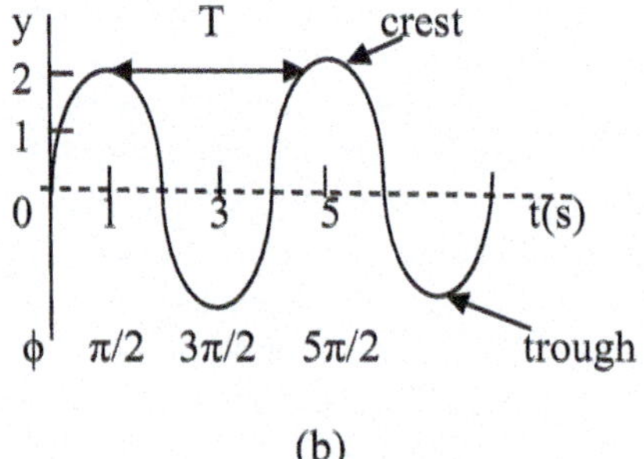

Fig. 11.8 Wave properties

In Fig. 11.8a, the shortest distance between two similar points on the wave (i.e., two crests, two troughs, etc.) is called the wavelength designated by λ. In Fig. 11.8b, the shortest time between two similar points on the wave (i.e., two crests, two troughs, etc.) is called the period designated by T. The period is the time for the wave to travel a distance of one wavelength which is called one cycle of the wave. The frequency of the wave, f, is the number of cycles per unit time passing a specific point. Its units are Hertz, Hz. Frequency depends on the source that created the wave and not on the medium it is in. These properties are related by the following equations:

$$f = 1/T; \tag{11.1}$$

$$\lambda = vT = v/f. \tag{11.2}$$

The second of these two equations is sometimes referred to as the wave equation.

For the wave shown in Fig. 11.8, the graphs show that the amplitude is $y_{max} =$ 2 m, the wavelength is $\lambda = 4$ m, and the period is $T = 4$ s. Then using Eqs. 11.1 and 11.2, the velocity is $v = 1$ m/s and the frequency is $f = 0.25\ s^{-1} = 0.25$ Hz.

Another way of describing points on a wave is by their phase, ϕ. This is an angular measurement that repeats itself every 2π or 360°. The importance of phase is describing the difference in the positions between two points on the wave pattern so the origin of $\phi = 0°$ is arbitrary. However, it is common to choose $\phi = 0°$ to be at the origin of the coordinate systems shown in Fig. 11.8. Then the first crest occurs at $\phi = 90°$ ($\pi/2$) and at $\phi = 180°$ (π) the wave is crossing the equilibrium line. At $\phi = 360°$ (2π) the wave has completed one complete cycle. Points separated by an amount $n\lambda$ are said to be "in phase" with each other. A crest and trough are 180° (π) out of phase with each other.

Example 11.1

As an example, consider a wave at a frequency of 192 Hz that travels 20 m in 0.06 s. What are the velocity, wavelength, and period of the wave? If the frequency is changed to 440 Hz, how do the wavelength and period of the wave change?

Use Fig. 11.8 as the picture of the problem.

Knowns: $f_1 = 192$ Hz; $f_2 = 440$ Hz; $\Delta x = 20$ m, $t = 0.06$s.

Unknowns: v; λ; T

This calls for a simple application of Eqs. 11.1 and 11.2. For the first part of the problem,

$$v = \Delta x/t = 20\ \text{m}/0.06\ \text{s} = 333\ \text{m/s}$$

$$\lambda = v/f = (333\ \text{m/s})/192\ \text{s}^{-1} = 1.73\ \text{m}$$

and

$$T = 1/f = 1/192\ \text{s}^{-1} = 5.2 \times 10^{-3}\ \text{s}.$$

When the frequency is changed to 440 Hz, the wavelength and the period are

$$\lambda = v/f = (333\ \text{m/s})/440\ \text{s}^{-1} = 0.76\ \text{m}$$

and

$$T = 1/f = 1/440\ \text{s}^{-1} = 2.3 \times 10^{-3}\ \text{s}$$

Another important characteristic of a wave is its speed of propagation. For a wave that is not a mechanical wave, this is a constant that we will discuss later in the chapter. For a mechanical wave its speed is determined by the properties of the medium supporting it. The best way to understand this is to consider a wave traveling along a string. Look at the forceswhich are at play.

Take a small section of the wave at the top of the crest and look at the forces acting on it. Ignoring gravity, the only force is the tension in the string (T). It is trying to straighten the string out, so it is pulling down on the crest of the wave from both sides. Now, if we split these tensions into horizontal ($T \cdot \cos\,\theta$) and vertical ($T \cdot \sin\,\theta$) components, the horizontal components are equal but opposite and cancel. However, the vertical components point in the same direction. Thus, they add to give the thick arrow, which is $2 \cdot T \cdot \sin\,\theta$. Since θ is small, however, $\sin\,\theta \approx \theta$, so the total force acting on the crest of the wave is approximately $2 \cdot T \cdot \theta$.

The force is pointed directly to the center of a circle whose arc length is "s." That's a centripetal force, which you learned about in Chap. 5. Thus, the small portion of the wave (the arc noted as "s") is in circular motion with the centripetal force equal to $2 \cdot T \cdot \theta$. Equation 5.8 gave the centripetal force as $F_c = mv^2/r$ where m is the mass of the object in circular motion, v is its speed, and r is the radius of the circle. Setting this expression for centripetal force equal to $2 \cdot T \cdot \theta$ gives

$$2 \cdot T \cdot \theta = mv^2/r$$

As you should remember from geometry, the length of an arc (s) is equal to the radius of the circle (r) times the angle subtended by the arc (θ). Since the arc in Fig. 11.9 subtends an angle of $2 \cdot \theta$

$$r = s/2 \cdot \theta$$

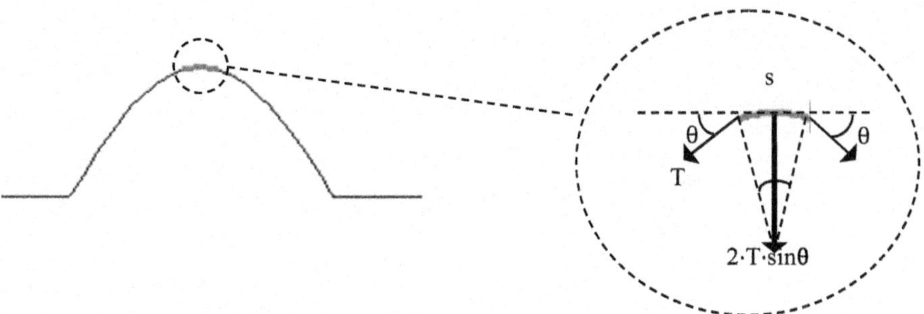

Fig. 11.9 Wave moving on a string

Substituting this expression for r in the centripetal force equation and solving for velocity gives

$$v^2 = Ts/m.$$

The mass of the arc length (m) divided by the length of the arc (s) is defined as the linear mass density (μ) of the string. Mathematically:

$$\mu = m/s \tag{11.3}$$

This has units of kilograms per meter. Using this in the equation for velocity,

$$v = \sqrt{\frac{T}{\mu}} \tag{11.4}$$

This tells us that the speed of a wave on a stretched string depends only on the tension in the string and the linear mass density (mass divided by length) of the string. The equation for the velocity of waves traveling in other types of media will be different. However in each case the velocity depends only on the properties of the media.

Example 11.2

A string is stretched with a tension of 25.0 N. What is the speed of a wave on the string if it is 45.0 cm long and has a mass of 17.5 g?

Knowns: $T = 25.0$ N; $L = 45.0$ cm; $m = 17.5$ gm

Unknowns: v

This is a straightforward application of Eq. 11.4. However, we need to keep our units consistent. Since the tension is given in Newtons, the length must be in meters and the mass in kilograms.

$$v = \sqrt{\frac{T}{\mu}} = v = \sqrt{\frac{25.0\ \mathrm{N}}{0.0175\ \mathrm{kg}/0.45\ \mathrm{m}}} = 25.4\ \mathrm{m/s}$$

Notice how the units work out. When you substitute the definition of the Newton into the equation, kilograms cancel and you are left with $\mathrm{m}^2/\mathrm{sec}^2$ under the square root sign. When you take the square root, then, you get the unit for speed, m/s.

Now try the following problem.

Student

11.1 A string ($m = 50.0$ g, length $= 1.0$ m) is stretched over a pulley and suspends a mass of 2.5 kg. What is the speed of a wave on the string?

Waves are many times three-dimensional instead of just the two-dimensional pictures shown in Figs. 11.7 and 11.8. Like a pebble thrown into water in a pond, they travel out in all directions from the source. This is shown in Fig. 11.10. A line drawn through all the crests of the wave at a given distance from the source is called a wavefront. As shown in the figure, the wavefront is spherical near to the source. However, at long distances from the source and with a detector that sees only a small part of the entire wave, the wavefront looks like a plane. Many of the concepts we will discuss later in this chapter deal with plane waves. A ray is an arrow perpendicular to the wavefront pointed in the direction the wave is traveling. A ray will continue going in a straight line until it hits some type of boundary discussed below. Using rays is critical for the study of geometric optics treated in a later section of this chapter.

Waves with the sine wave shape shown in Fig. 11.7 are called harmonic waves. We discussed harmonic motion in Chap. 6. Harmonic waves can be described by the equation

$$y = y_{\max} \sin\left[\frac{2\pi}{\lambda}(vt \pm x)\right] \tag{11.5}$$

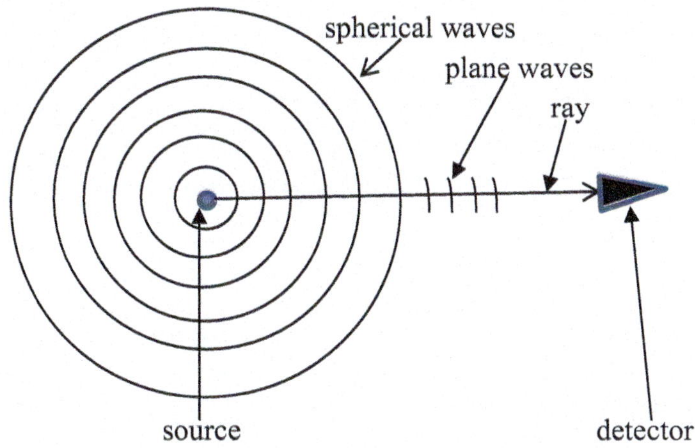

Fig. 11.10 Wavefronts

where the plus sign represents a wave moving to the left and the minus sign rep resents a wave traveling to the right. y_{max} is the amplitude of the wave. Again using Fig. 11.8 as an example, at point x = 1.5 m and t = 1 s, y = 2sin[(2π/ 4 m)(1.0 m–1.5 m)] = – 0.027 m. This equation is important if we need to know the size of a wave at a specific point in time and space.

Example 11.3

A harmonic wave travels to the right along a rope that has a linear mass density of 0.310 kg/m and is under 25.0 N of tension. The wave has an amplitude of 10.0 cm and a wavelength of 50.0 cm. What is the vertical displacement of the rope at x = 1.10 m and t = 1.00 s?

Knowns: $\mu = 0.310$ kg/m; $T = 25.0$ N; $y_m = 10.0$ cm; $\lambda =$ 50.0 cm; $x = 1.10$ m;

$t = 1.00$ s

Unknowns: y

To answer the question, we have to determine the equation for the harmonic wave.

We were given y_m and λ and we can calculate v, because we know the tension and the linear mass density of the rope. From Eq. 11.4

$$v = \sqrt{\frac{T}{\mu}} = \sqrt{\frac{25.0\ \text{N}}{0.310\ \text{kg/m}}} = 8.98\ \text{m/s}$$

The harmonic equation for a wave traveling to the right is,

$$\begin{aligned} y &= y_m \sin\left[\frac{2\pi}{\lambda}(vt - x)\right] \\ &= y_m \sin\left[\frac{2\pi}{0.500\text{m}}\{(8.89\ \text{m/s})(1.0\ \text{s}) - 1.10\ \text{m}\}\right] \\ &= -0.0990\ \text{m} \end{aligned}$$

If you were to look at the point 1.10 m to the right of the origin at a time of 1.00 s after the wave began, you would see that the rope would be displaced 9.98 cm *downwards.* Please note that to get the proper answer, your calculator *must* be in radians.

Now try the following problem.

Student

11.2 A harmonic wave moving to the left is generated on a rope. The speed of the wave is 2.00 m/s, and the wavelength is 32.1 cm. At $t = 2.00$ s and 1.25 m to the right of the origin, the wave displaces the rope 3.44 cm downwards. What is the amplitude of the wave?

11.3 Boundary Conditions and Interference

When a medium has a place where its properties that support wave transmission change, this acts like a boundary for the wave propagation. When a wave hits a boundary, it is called an incident wave. Part of it may continue across the boundary with different wave properties and is called a transmitted wave. Part of the incident wave may bounce off the boundary and create a reflected wave.

As examples, consider the boundaries for a wave on a string shown in Fig. 11.11. If the string is attached to a fixed point on a solid wall, a wave on the string incident on the boundary is completely reflected. As a crest in the wave approaches the fixed point, the force from the wall pulls it downward. This drives the reflected wave into a trough so it is 180° out of phase with the incident wave as shown in Fig. 11.11a. However, if the boundary is movable, this phase shift does not occur. For example, if the string is attached to a ring on a pole that can slide up and down, when the crest of a wave on the string reaches the pole the ring moves up the pole. As it slides back down the pole it sends a reflected wave back down the string going in the opposite direction but in phase with the incident wave as shown in Fig. 11.11b. In addition, if there is another string attached to the ring on the other side of the boundary, the oscillation of the ring caused by the incident wave generates a transmitted wave in this region. The transmitted wave has a velocity and wavelength dependent on material on the right side of the boundary but its frequency is the same as that of the incident wave. In summary:

> Waves reflected off a stationary boundary are inverted. Waves reflected off a moving boundary are not inverted.

Sometimes a boundary is simply going from one transmission medium to another. For example, going from air to water or going from a thin string to a thick rope. When the wave crosses the boundary into the second media, its velocity will change since it depends on the medium of propagation. The velocity of the transmitted wave can be either greater or less than the velocity of the incident wave depending on how the microscopic media of the two regions move to support the waves. However, the frequency of the wave will not change since it depends only on the source and not the medium of propagation. If the speed of the transmitted wave is less than the speed of the incident wave, the reflected wave will be inverted. If the speed of the transmitted wave is greater than the speed of the incident wave, the reflected wave will not be inverted.

The wave equation, Eq. 11.2, relates the velocity to the frequency and wavelength of the wave through $\lambda = v/f$. Thus, if the velocity changes and the frequency is constant, the wavelength must change,

$$\lambda_2/\lambda_1 = v_2/v_1 \tag{11.6}$$

where the 1 and 2 subscripts refer to the two regions across the boundary.

If the incident wave hits the wall at an angle θ_i to the direction perpendicular to the boundary, the reflected wave leaves the wall at an angle θ_r to the same direction as shown in Fig. 11.11c. The arrows in this figure represent rays perpendicular to the wavefront in the direction of wave velocity. Remember, the wavefront is a

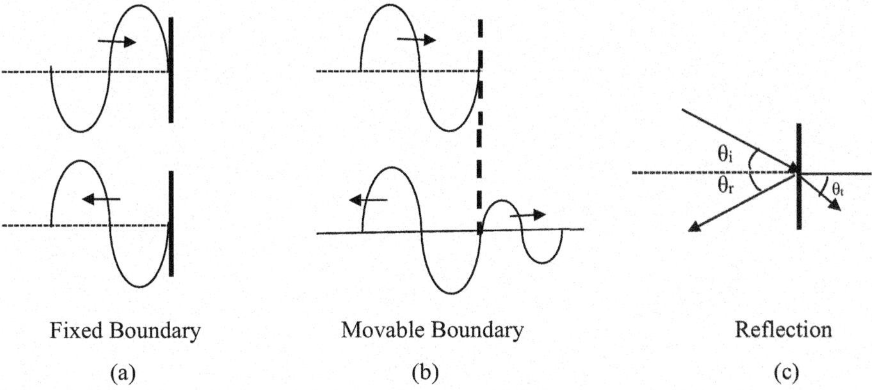

Fig. 11.11 Waves at a boundary

line representing the crest of the wave. The Law of Reflection states that angle of reflection equals the angle of incidence,

$$\theta_r = \theta_i. \tag{11.7}$$

This will be discussed more in a later section. When the incident wave hits the boundary at an angle, the transmitted wave moves in a different direction. This effect is called refraction and is discussed later in this chapter.

The discussion above indicates that it is possible for more than one wave to be present in the same media at the same time. When this occurs, the Principle of Superposition states that the displacement of a media caused by two or more waves is the algebraic sum of the displacements of the individual waves. Because one of the waves in a superposition can have a different phase with respect to the other wave, the result is called interference between the two waves. If the two waves are in phase with each other, constructive interference occurs. The peaks of the two waves are added together to get the peak of the superposition wave and the two troughs are added together to get the trough of the superposition wave. If the two waves are 180° out of phase with each other, destructive interference occurs. In this case the peak of one wave lines up with the trough of the other wave and they tend to cancel each other out. This is shown in Fig. 11.12.

Note that if waves A and B have a phase difference somewhere between 0° and 180° the maximum of their interference wave is less than it is for complete constructive interference. This loss in maximum is what occurs for reactive losses in the transmission of AC electricity discussed in Chap. 10.

Try these following problems.

Student

11.3 A wave travels along a string whose linear mass density is 0.050 kg/m at a speed of 3.00 m/s. Another string whose linear mass density is 0.100 kg/m is connected to the first string.

Both strings have the same tension. When the wave encounters the second string, what will the speed of the transmitted wave be? What

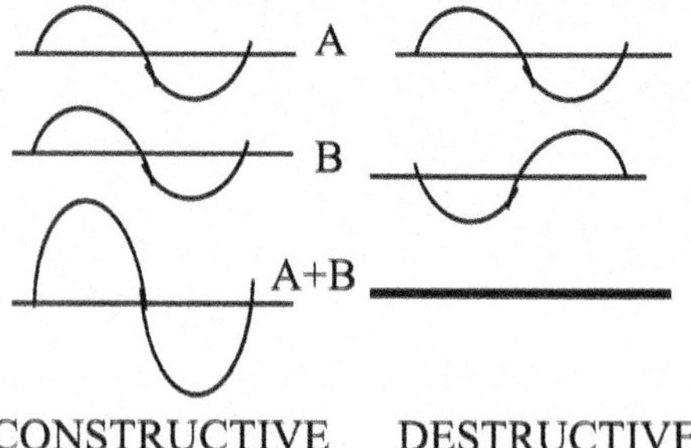

Fig. 11.12 Interference between waves A and B

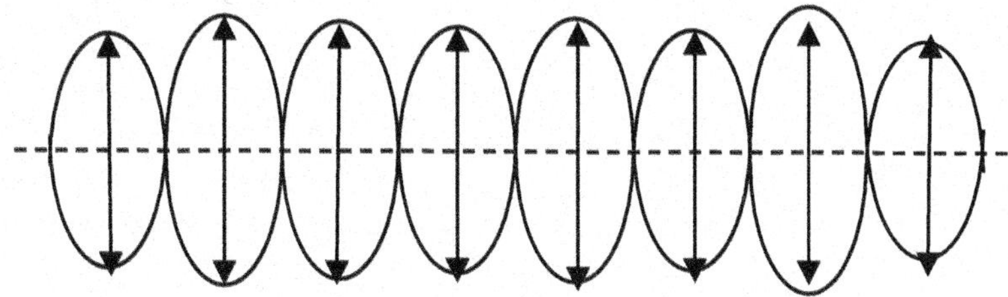

Fig. 11.13 Standing wave

will the speed of the reflected wave be? Will the reflected wave be inverted?

11.4 Two waves approach one another. The first has an amplitude of 15.0 cm and the second has an amplitude of 5.0 cm. What is the minimum and maximum displacement experienced by the medium once the superposition begins?

One very interesting case of interference is when there are two waves with the same frequency and wavelength traveling in opposite directions. Their interference pattern sets up a standing wave as shown in Fig. 11.13. In this case the wave is stationary in position, x, but oscillates in time, t. The nodes are fixed points in time and space where the wave disturbance is zero because of destructive interference. The antinodes are points that oscillate between being crests and troughs through constructive interference.

A standing wave can be described using Eq. 11.5. The waves going to the left and to the right are described as

$$y_{\mathrm{r}}(x,t) = y_{\mathrm{m}} \sin(2\pi x/\lambda - 2\pi ft)$$

$$y_{\mathrm{l}}(x,t) = y_{\mathrm{m}} \sin(2\pi x/\lambda + 2\pi ft)$$

so the standing wave is

$$y_{\mathrm{s}}(x,t) = y_{\mathrm{r}}(x,t) + y_{\mathrm{l}}(x,t) = y_{\mathrm{m}} \sin(2\pi x/\lambda - 2\pi ft) + y_{\mathrm{m}} \sin(2\pi x/\lambda + 2\pi ft)$$

Using the trig identity, $\sin a + \sin b = 2\sin\left(\frac{a+b}{2}\right)\cos\left(\frac{a-b}{2}\right)$, the standing wave equation is

$$y_{\mathrm{s}}(x,t) = 2y_{\mathrm{m}} \sin(2\pi x/\lambda)\cos(2\pi ft)$$

The strength of the wave at any point x is given by

$$y(x) = 2y_{\mathrm{m}} \sin(2\pi x/\lambda).$$

Nodes where the strength of the wave is zero for all times occur when x is an even multiple of $\lambda/4$,

$$x_{\text{node}} = 0, \lambda/2, \lambda, 3\lambda/2, \ldots, n\lambda/4 \; (n \text{ even number})$$

since that is where the sine is zero. The maxima in the wave strength occur when x is an odd multiple of $\lambda/4$

$$x_{\max} = \lambda/4, 3\lambda/4, 5\lambda/4, \ldots, n\lambda/4 (n \text{ odd number})$$

There will be n antinodes in a standing wave and $n + 1$ nodes.

The positions of the nodes in the equation for a standing wave give us a relationship between the length of the medium supporting the wave, L, and the wavelength of the standing wave,

$$\lambda_n = 2L/n \quad (n = 1, 2, 3, \ldots) \tag{11.8}$$

The wavelength for $n = 1$ is called the fundamental wave while the wavelengths for higher n's are called overtones or harmonics.

According to Eq. 11.2, if we know the wavelength and velocity of a wave in a medium, we can determine its frequency. The frequencies of standing waves supported by a medium are sometimes referred to as the resonant frequencies of the medium.

Standing waves are especially relevant in acoustics. Their application in music will be discussed in the next chapter. For now, try the following problems.

Student

11.5 A string of length 15.0 cm is stretched between two supports. A standing wave is developed. The standing wave has a total of 4 antinodes. What is the wavelength of this wave?

11.6 A string of length 25.0 cm is stretched between two supports with a tension of 50.0 N. The mass of the mass of the string is 7.50 g. What is the resonant frequency of the fundamental vibration of this string?

11.4 Geometric Optics–Mirrors and Lenses

Optics is the study of the physical properties of light and other types of radiation. Because of the invention of the laser with all its many applications, optics has become one of the most important areas of physics. Optics can be approached in two ways. One is geometric optics which is based on the fact that light tends to travel in a straight line until it hits something. Geometric optics uses the concept of rays to describe how light behaves while ignoring its fundamental physical nature. If light hits an opaque object it is reflected. If it hits a transparent object it is transmitted. If it hits a translucent object it transmitted but not in a clear fashion. You can't "see through" a translucent object like you can a transparent object.

Special Topic
Optics is an enabling science. New developments in optics provide important support for many other disciplines. For example, telescopes for astronomers, microscopes for biologists, optical fibers for telecommunications, imaging techniques for medicine, and lasers for surgery. Light emitting diodes are revolutionizing the lighting industry. Thus learning about the physics of optics is not only important for itself, it is critical for understanding every area of science and technology.

The other approach to optics is called physical optics. It accounts for the basic physical properties of light. It is important to put light in the context of the rest of nature. We will learn in our study of modern physics that everything has a dual nature. It sometimes acts as a wave and sometimes as a beam of particles. This is especially important in our study of light. Light exhibits wave properties such as interference but delivers energy in small packets called photons. To fully understand light, we must understand both its wave and particle properties. In this chapter we will focus on the wave nature of light. The particle nature of light is discussed further in Chaps. 14 and 15 of this book.

The geometrical and physical approaches to optics are reviewed in the following sections. Before discussing optical properties, it is important to understand the fundamental nature of light. Light is not a mechanical wave. Since it isn't based on a disturbance in a media, it can travel in free space. Instead, light is the oscillation of coupled electric and magnetic fields. This is shown in Fig. 11.14. A source emits **E** and **B** fields that oscillate with the same wavelength and phase in directions perpendicular to each other and to the direction of propagation.

Formally, the properties of these coupled electric and magnetic fields can be described by a set of equations called Maxwell's Equations plus the wave equation. Among other things, these equations show that when a charged particle is accelerated, it emits an electromagnetic wave. For example, a radio antenna is a metal rod and when the electrons in the rod are accelerated back and forth at a specific frequency they emit electromagnetic waves in the radio frequency range.

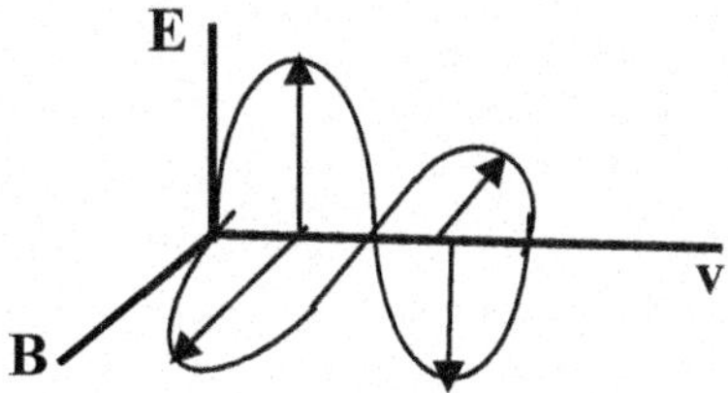

Fig. 11.14 Electromagnetic waves

However, the math required to use Maxwell's Equations involves vector calculus which is beyond the scope of this book. The important thing to remember is that the source of electromagnetic radiation is an electron undergoing acceleration. We can still learn much about the properties of light without using such sophisticated mathematics. For example, it has been shown experimentally that transverse electromagnetic waves can travel in a vacuum (so no medium is required to support this type of wave). The speed of an electromagnetic wave in a vacuum is a constant, $c = 2.998 \times 10^8$ m/s. This is one of the important physical constants.

It is important to remember that light is one small part of a very broad spectrum of electromagnetic waves. This is shown in Fig. 11.15. Radiation in the electromagnetic spectrum varies from very high energy, short wavelength gamma rays to very low energy, long wavelength radio waves. What we call visible light is in the middle of this range. It is defined by what our eyes are able to see. Human vision can respond to waves ranging from blue light at about $\lambda = 400$ nm to red light at about $\lambda = 700$ nm. The rest of the radiation in the electromagnetic spectrum is invisible to human vision. Of course, other spectral regions can still be very useful such as low energy waves for radio transmission and high energy waves for medical imaging. Our study of optics will focus on the properties of visible light, but the results can be applied to other types of electromagnetic waves.

Let's begin our study of geometric optics by reviewing the concepts of ray tracing when light hits mirrors and lenses. A ray representing an electromagnetic wave will travel in a straight line until it hits something. When it hits an object that does not permit it to travel across the boundary, it is reflected. There are two types of reflection. If the object is smooth, the wave undergoes specular reflection and the ray is changed to one specific new direction. The angle of reflection equals the angle of incidence. The plane of incidence is the plane containing the incident ray and the normal to the surface of the reflecting object. The ray of the reflected wave is in this plane. If the reflecting object has a rough surface, the wave undergoes diffuse reflection. In this case, the reflected wave goes in many different directions depending on the parts of the surface that each part of the wavefront hits.

A mirror is an object that produces specular reflection of light. There are three types of mirrors shown by their side views in Fig. 11.16. The horizontal line through their centers is called the principal axis. In each case, the incident light rays are traveling from left to right and the reflected rays from right to left. The

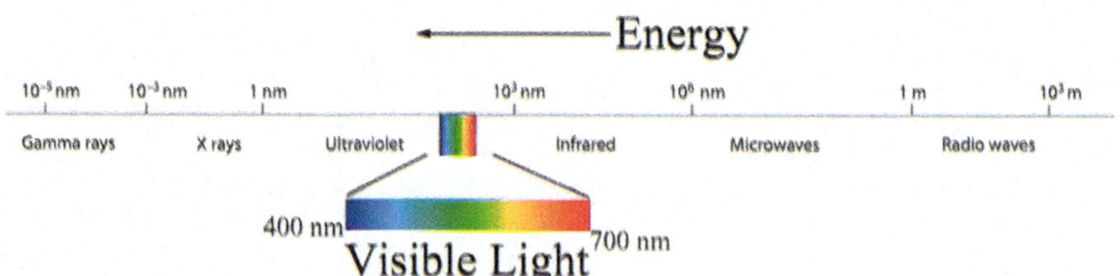

Fig. 11.15 Electromagnetic spectrum

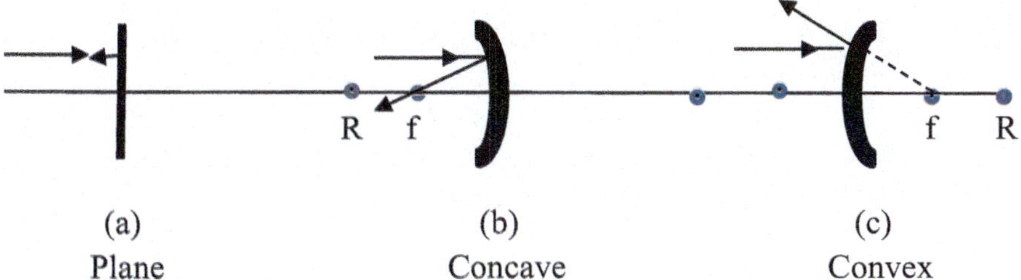

Fig. 11.16 Types of mirrors

first type is a plane mirror where the reflecting surface is flat. The next two types have spherically curved reflecting surfaces with a radius of curvature a distance R along the central axis and a focal point f at half that distance. In Fig. 11.16b the reflecting surface is curved inward at the center and this is called a concave mirror. In Fig. 11.16c the reflecting surface is curved outward at the center and this is called a convex mirror. Each of these types of mirrors reflect light rays very differently as discussed below.

Ray tracing is a technique for finding the image made by a mirror or lens. In ray tracing with mirrors, we generally start with an object that is emitting light and determine the properties of the image that is formed. An image is the representation of the object formed by the mirror (or lens as we will discuss in the next section). In general, we are interested in the position and size of the image and whether it is upright or inverted compared to the object. In addition, we need to know if the image is real or virtual. For a real image, the light is actually at that point so it can be projected on a screen. A virtual image appears to be at a point where there really is no light so it cannot be projected. The magnification of a mirror is the image height divided by the object height.

To begin with, consider an object in the form of an arrow emitting light that hits a plane mirror as shown in Fig. 11.17. To determine properties of the image we can trace rays coming from the point of the arrow. The first ray travels parallel to the primary axis and hits the mirror perpendicular to its plane. This will reflect directly back on itself. To make it useful in finding the image, we must project its travel as if it kept going through the mirror in a straight line. The second ray we can use comes from the same point at the tip of the object and travels to the center of the mirror on the main axis. It is reflected with $\theta_r = \theta_i$ to obey the law of reflection. Again to make this useful in defining the image we must project rays backwards on the right side of the mirror. Now where these two projected rays cross is the position of the image of the arrow point. An observer looking into the mirror will see the image at this point. It will look like it is behind the mirror. Because the triangles on each side of the mirror are equivalent, the image distance d_i is equal to the object distance, d_o. Also, the image is upright and the same size as the object. Obviously it is a virtual image because light does not really travel through the mirror.

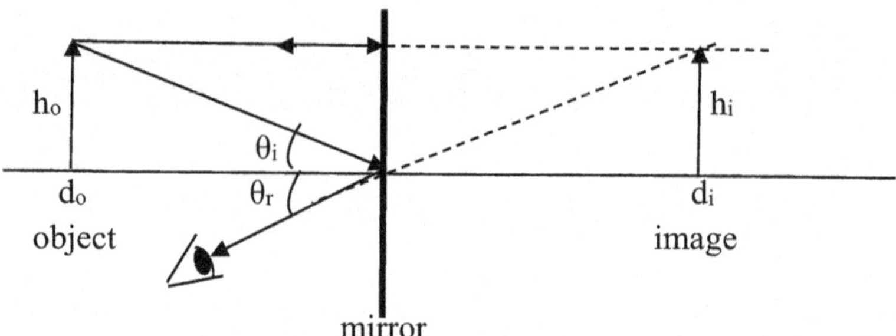

Fig. 11.17 Reflection from a plane mirror

Next consider an example of a concave mirror, sometimes called a converging mirror. For this case any ray parallel to the principal axis will reflect off the mirror and converge to a focal point, f, on the principal axis located at half the distance to the radius of curvature of a spherical mirror. Figure 11.18 shows the ray tracing needed to find the image formed by a concave mirror. Three rays coming from the top of the object can be used although only any two of these are needed. The first ray is incident on the mirror parallel to the primary axis so its reflected ray comes back through the focal point f. The second ray goes from the top of the object through the focal point so its reflected ray is parallel to the primary axis. The point where these two rays cross is the top of the image. The third ray shown in the figure goes through the radius of curvature. Since this always hits perpendicular to the surface of the mirror, it is reflected directly back on itself. As seen from the figure the image is located at position d_i along the primary axis, is h_i high, and is inverted. It is real and in front of the mirror. Its magnification is $m = h_i/h_o$.

Finally, consider the reflection from a convex mirror as shown in Fig. 11.19. This is sometimes called a diverging mirror. In this case the focal point and radius of curvature are behind the reflecting surface of the mirror so their positions are negative. Again, two out of three rays coming from the top of the object are used

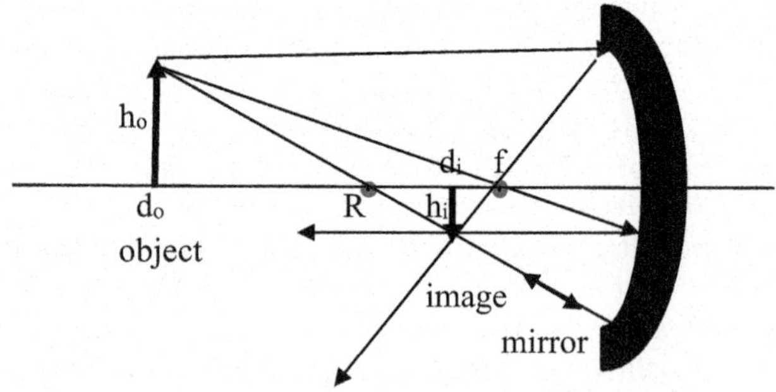

Fig. 11.18 Reflection from a concave mirror

to define the image. The first of these hits the mirror surface traveling parallel to the principal axis. This means its reflected ray can be extended to go through the focal point on the right side of the mirror and projected back from there. The second ray is going toward the focal point and when it hits the mirror it is reflected in a direction that is extended as if it is coming from the focal point. The third ray is headed towered its extension going through the radius of curvature. When it hits the mirror it is reflected back on itself. The figure shows that the image is virtual, upright, and behind the mirror by distance d_i along the principal axis. Its height is h_i so the magnification is $m = h_i/h_o$. The rules for forming images with concave and convex mirrors are summarized in Table 11.1. Note how the image properties change for a concave mirror when the object is between the focal distance and the mirror.

There is an important equation that can be used along with ray tracing to characterize images formed by mirrors. This is called the mirror equation,

$$\frac{1}{d_o} + \frac{1}{d_i} = \frac{1}{f} \tag{11.9}$$

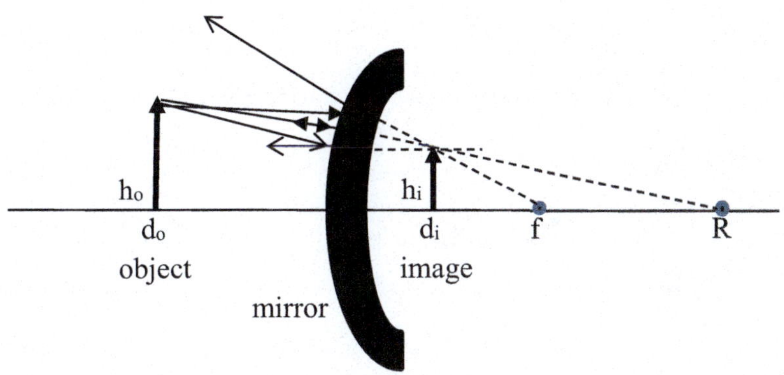

Fig. 11.19 Reflection from a convex mirror

Table 11.1 Image properties of mirrors

Mirror	d_o	d_i	m	Type
Plane (no f)	> 0	$-d_o$	1	Virtual upright
Concave (f is +)	> 0			
	$> f$	$d_i > 0$	$-\frac{d_i}{d_o}$	Real inverted
	$< f$	$d_i < 0$	$-\frac{d_i}{d_o}$	Virtual upright
Convex (f is −)	> 0	$d_i < 0$	$-\frac{d_i}{d_o}$	Virtual upright

The size of the image is given by

$$m = -d_i/d_o \tag{11.10}$$

This quantifies the distances along the primary axis of the object and image with respect to the focal length. Table 11.1 summarized the sign conventions to use in this equation. The mirror is at position 0. The focal length is either plus or minus depending on whether the mirror is concave or convex. The object distance is always positive but the image is positive for a concave mirror and negative for a convex mirror. The magnification factor m is the ratio of the image height to the object height. As shown in the table, this is related to the positions of the image and object. The minus sign is present so that m is positive when the image is upright and m is negative when the image is inverted.

Example 11.4

An object is placed 10.0 cm from a concave mirror whose radius of curvature is 25.0 cm.

On which side of the mirror is the image formed? Is it real or virtual? Is it upright or inverted? What is the magnification?

Use the figure given below. Knowns: $d_o = 10.0$ cm; $R = 25.0$ cm.
Unknowns: Image characteristics.

Let's first solve this with the mirror equation and then show that ray tracing gives the same result. Since this is a concave mirror, "f" is positive. Since the focal length is half the radius of curvature, $f = 12.5$ cm. The mirror equation, then, gives us:

$$\frac{1}{d_o} + \frac{1}{d_i} = \frac{1}{f}$$

$$\frac{1}{10.0\ \text{cm}} + \frac{1}{d_i} = \frac{1}{12.5\ \text{cm}}$$

$$d_i = -50.0\ \text{cm}$$

Since the image distance is negative, the image is on the side of the mirror opposite of the object and is virtual. To determine the magnification and whether the image is upright or inverted, we must use the magnification equation.

$$m = -\frac{d_i}{d_o} = -\frac{-50.0\ \text{cm}}{10.0\ \text{cm}} = 5.00$$

Thus, the image is upright and five times as large as the object.

Ray tracing will give us all of that information except the numbers. If we follow the ray tracing rules, we get

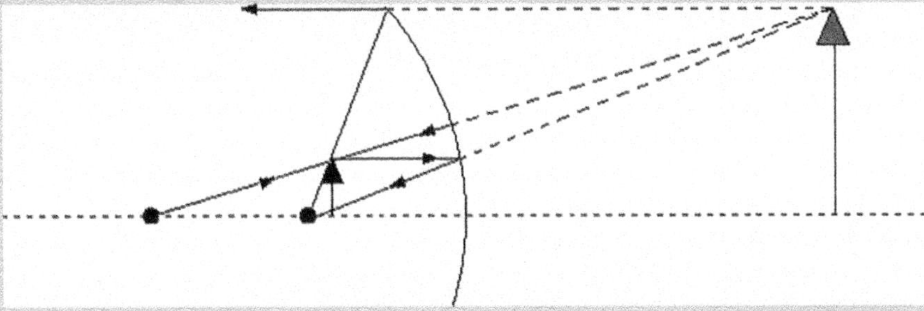

Notice that in this picture, the image is on the opposite side of the mirror relative to the object; it is virtual; it is magnified; and it is upright.

Now answer all of the same questions if the situation above used a *convex* mirror instead.

The only mathematical difference between this problem and the one before is the fact that since the mirror is convex, the focal length is negative. The mirror equation becomes

$$\frac{1}{10.0\ \text{cm}} + \frac{1}{d_\text{i}} = \frac{1}{-12.5\ \text{cm}}$$
$$d_\text{i} = -5.56\text{cm}$$

This tells us that the image is on the opposite side of the mirror relative to the object and is virtual. The magnification equation gives us:

$$m = -\frac{d_\text{i}}{d_\text{o}} = -\frac{-5.56\ \text{cm}}{10.0\ \text{cm}} = 0.556$$

This tells us that the image is upright and only 0.556 times as tall as the object. If you went through the ray tracing steps for this situation, you would find that it gives consistent results.

Try the following problem.

Student

11.7 An object is placed 15.0 cm from a convex mirror whose radius of curvature is 20.0 cm.

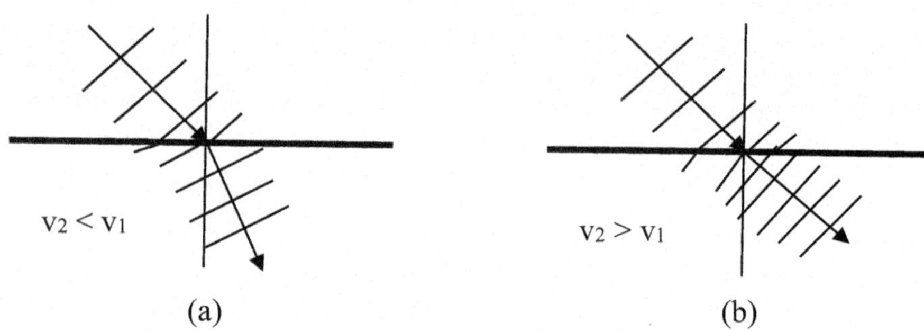

Fig. 11.20 Refraction of light at a boundary

What is the magnification of the image relative to the object? Is the image real or virtual? Is it upright or inverted? Repeat the problem for a concave mirror.

Now let's turn to the interaction of light with lenses. In the previous section we mentioned refraction, or the bending of light when it crosses a boundary between two media in which light travels at different speeds. Figure 11.20 shows the refraction of a plane wave front as it travels through a boundary. If the speed of the wave in the second region is faster than it is in the first region, the ray representing the transmitted wave will be bent away from the perpendicular to the boundary of the surface. If the wave moves slower in the second region, the transmitted ray will be bent toward the perpendicular line.

The speed of an optical wave in a medium is determined by its index of refraction n defined as the ratio of the speed of light in a vacuum to the speed of light in the medium,

$$n = c/v. \tag{11.11}$$

Remember that the speed of light in a vacuum is one of the fundamental constants in physics, $c = 2.998 \times 10^8$ m/s. Nothing can be accelerated to a speed faster than c. Thus $n \geq 1$.

In Fig. 11.20a, media 2 has a refractive index n_2 that is greater than n_1 so the speed of the wave is less. Also, because of the relationships in Eqs. 11.2 and 11.5, its wavelength is smaller in the second media. Its frequency does not change. The angles of incidence θ_i and refraction θ_r are measured between the rays representing the incident and refracted waves and the perpendicular to the surface at the point of contact. Snell's law of refraction states that

$$n_1 \sin\theta_i = n_2 \sin\theta_r. \tag{11.12}$$

In Fig. 11.20a $\theta_i > \theta_r$. In Fig. 11.20b, media 2 has a refractive index n_2 that is less than n_1 so the speed of the wave is greater and its wavelength is greater in this media. In this case $\theta_i < \theta_r$. As before, the frequency of the wave is the same in media 1 and 2.

The bending of light rays traveling across a boundary between two different media has many important consequences. Do the following problem.

Student

11.8 Suppose a light ray is traveling in air ($n = 1.00$) and hits a layer of glass, as shown in the figure. The incident angle is 60.0° relative to the perpendicular.

a. Draw five light rays that will result, filling in their angles.
b. If the reflected light is allowed to hit a screen above the glass layers, an interference pattern will be observed. Why?
c. Consider the two light rays that are interfering with one another as they travel to the screen. They each travel a different distance to the screen. If the difference between the distances they travel is Δl, what is the relationship between Δl and λ that will result in destructive interference?

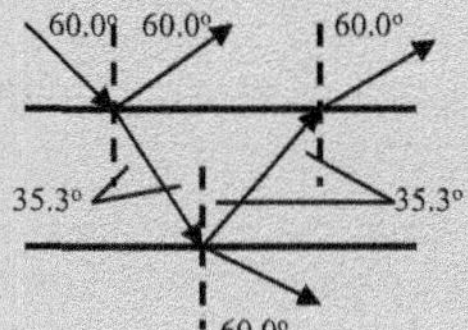

One of the great inventions of the twentieth century was fiber optics. Scientists learned how to draw glass into fibers that were the diameter of a human hair and many miles in length. Single frequencies of laser light can travel the length of the fiber with very small loss of intensity. This ability to accurately deliver light to precise remote locations is the basic technology behind optical telecommunications, medical imaging, laser surgery and machine vision. The reason fiber optics works so well is the concept of total internal reflection. This is shown schematically in Fig. 11.21 where light is going from a substance with lower refractive index to one with a higher refractive index. As the angle of incidence of the light is increased, it reaches an angle where the angle of refraction is equal to 90°. At that angle and at any greater angle of incidence, there is no light transmitted into the region with n_2. All of the light is reflected back into the region with n_1. This

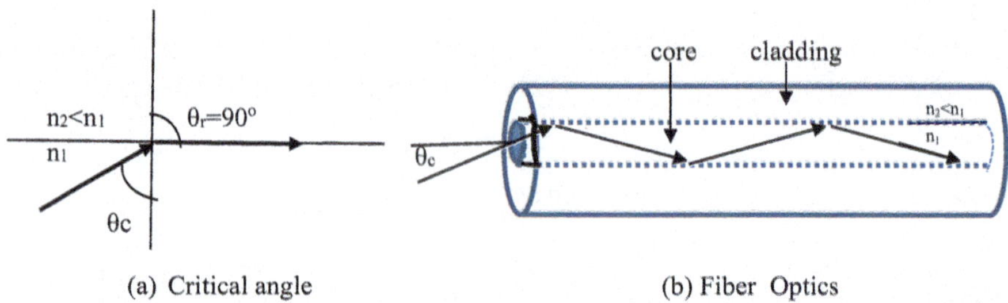

Fig. 11.21 Total internal reflection

situation is shown in Fig. 11.21a and this special angle of incidence is called the critical angle.

A mathematical expression for the critical angle can be calculated from Snell's law in Eq. 11.12 as

$$n_1 \sin \theta_c = n_2 \sin 90^\circ .$$

Thus,

$$\theta_c = \sin^{-1}(n_2/n_1). \tag{11.13}$$

The way the concept of total internal reflection is used in fiber optics is shown in Fig. 11.21b. An optical fiber has an inner core of a material with a refractive index n_1 and an outer cladding of material with a refractive index n_2 where $n_1 > n_2$. When light is incident on the inner core at an angle greater than the critical angle, the light wave propagates down the fiber by undergoing total internal reflection inside the core every time it hits the cladding. For high quality glass, there is very little loss of light over the entire length of the fiber.

For telecommunications, light pulses travel at the speed of light in the glass fibers which is much faster than the speed of electrical signals on metal wires. Using slightly different angles of incidence of the light going into the fiber allows simultaneous transmission of different signals. A single optical fiber is capable of handling on the order of 3000 transmission signals at once which is much greater than an electrical cable. Repeater stations are used when necessary to maintain the strength of the optical signals.

Fiber optics has also become important in medical applications. Fiber optic endoscopes can be inserted into the body in a very non-invasive way. Sending light through the endoscope allows a doctor to look at the inside of a patient and, using lasers, to perform some surgical procedures with less trauma to the body than normal surgery.

Example 11.5

A light beam traveling in a pool of water ($n_w = 1.33$) hits an air interface at an angle of incidence of $\theta_i = 20°$. Can someone standing at the side of the pool see it?

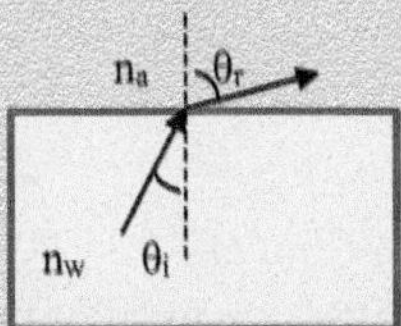

Knowns: $\theta_i = 20°$; $n_w = 1.33$; $n_a = 1$

Unknowns: θ_c

Another way of asking this question is whether the angle of incidence is equal to or greater than the critical angle. We can use Eq. 11.13 to determine θ_c

$$\theta_c = \sin^{-1}(n_2/n_1) = \sin^{-1}(1/1.33) = 48.8°$$

Since the angle if incidence of $\theta_i = 20°$ is less than the critical angle of $\theta_c = 48.8°$ the refracted light wave will be visible in the air.

Now try your hand at working the following problem.

Student

11.9 A diamond is acting as a point source of light at the bottom of a pool of water 2 m deep. In order to stop anyone from seeing it, you want to cut a circular piece of wood to float on the surface of the water to block the light. What is the smallest radius you need for the wood? ($n_w = 1.33$)

The most useful application of refraction is the focusing of images by lenses. The technique of ray tracing applied to mirrors in the previous section can also be used to analyze the effect of a lens on a light wave. Mirrors manipulate light through reflection while lenses manipulate light through refraction. A lens is a piece of transparent material shaped in a way that refraction causes parallel light waves that go through it to converge to a focal point or to diverge away from a focal point. Figure 11.22 shows two common types of lenses. The double convex lens is thicker in the middle than at the edges and it causes light waves to converge to the focal point on the opposite side of the lens. The double concave lens is thicker at the edges than in the middle and it causes light waves to diverge away from the focal point in the incident side of the lens. Ray tracing can be used to

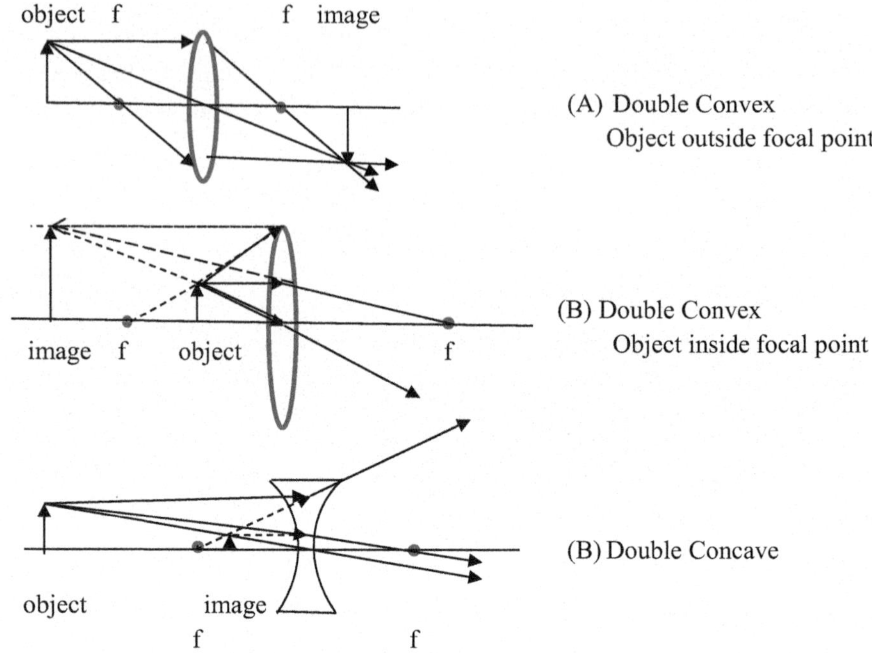

Fig. 11.22 Common lenses

determine the properties of the image created by each type of lens. The thickness of the lens is considered to be so small it can be ignored.

The following summary should remind you of the ray tracing rules for lenses that you learned in your introductory course:

Ray Tracing Rules for a Converging Lens

1. **A light ray that travels horizontally will be refracted so that it goes through the focal point.**
2. **A light ray that passes through the center of the lens experiences no net deflection.**
3. **A light ray that travels towards the focal point will be refracted so that it travels horizontally.**

Ray Tracing Rules for a Diverging Lens

1. **A light ray that travels horizontally will be refracted so that it appears to have originated at the focal point.**
2. **A light ray that passes through the center of the lens experiences no net deflection.**

3. **A light ray that travels towards the focal point on the other side of the lens will be refracted so that it travels horizontally.**

Note that the image properties of a converging lens depend on the position of the object with respect to the focal point. If $d_o > 2f$, then $d_i < 2f$ and the image is smaller than the object. If $d_o = 2f$, then $d_i = 2f$ and the image is the same size as the object. If $f < d_o < 2f$, then $d_i > 2f$ and the image is bigger than the object. If $d_o = f$ no image is formed. The final case is $d_o < f$ In this case the image is virtual, upright, larger than the object, on the same side of the lens as the object, and further from the lens than the object.

In the previous section we referred to Eq. 11.9 as the mirror equation. The same equation can be applied to lenses if they are thin enough that their thickness is not a factor. It is sometimes called the thin lens equation. It can be used to determine the image position

$$1/f = 1/d_o + 1/d_i. \quad (11.9)$$

The size of the image is again given by the magnification

$$m = -d_i/d_o \quad (11.10)$$

In these expressions, d_i is positive for a real image and negative for a virtual image. The magnification factor is negative for an inverted image and positive for an upright image. The lens is located at position zero on the primary axis. The object distance is always positive. The focal point for a converging lens is positive and for a diverging lens is negative. If the image distance is positive, the image is real and is on the opposite side of the lens from the object. If the image distance is negative the image is virtual and is on the same side as the object. Table 11.2 summarizes these results and the following example shows the use of the thin lens equation.

Example 11.6
Consider the converging lens in Fig. 11.22a. The drawing is consistent with an object being placed 14.0 cm from a lens whose focal length is 10.0 cm.

Table 11.2 Image properties of a lens

Lens	d_o	d_i	m	Type
Convex (f +)	$> 2f$	$2f > d_i > f$	< 1	Real inverted
	$2f > d_o > f$	$> 2f$	> 1	Real inverted
	$f > d_o > 0$	$\|d_i\| > d_o$ Neg.	> 1	Virtual upright
Concave (f–)	> 0	$\|f\| > d_i > 0$ Neg	< 1	Virtual upright

Use Eqs. 11.9 and 11.10 to determine whether the image is real or virtual, upright or inverted, and what the magnification of the image is relative to the size of the object.

From the ray tracing diagram, we know the answers to all of these questions except the last one. However, it is instructive to see that Eqs. 11.9 and 11.10 give us the same answers. Since this is a converging lens, "f" is positive. Also, the object's position is

$$1/f = 1/d_o + 1/d_i.$$
$$1/10.0 \text{ cm} = 1/14.0 \text{ cm} + 1/d_i.$$
$$d_i = 35.0 \text{ cm}$$

Thus, the image is real, because the image distance is positive. Also, it is on the opposite side of the lens as compared to the object, but the question did not ask about that. To determine the answer to the other questions, we must use the magnification equation:

$$m = -d_i/d_o = -35.0 \text{ cm}/14.0 \text{ cm} = -2.50$$

As shown in the ray tracing diagram, then, the image is inverted. The equation tells us that the image is 2.5 times larger than the object. So our results of using the thin lens and magnification equations are consistent with the results of ray tracing.

Now its time for you to try the following problem.

Student

11.10 An object is placed 10.0 cm from a diverging lens that has a focal length of 15.0 cm. On what side of the lens is the image produced? Is it real or virtual? Is it upright or inverted? What is the size of the image relative to the object?

Ray tracing techniques are extremely important for designing optical instruments like microscopes and telescopes that involve multiple mirrors and lenses. The individual elements can be more complex than the simple examples given here. For example, a lens can be concave on one side and convex on the other. Sophisticated software programs have been developed that use ray tracing for designing complicated optical systems that minimize image aberrations and unwanted scattered light.

Next Level

Ray Tracing for Compound Lenses

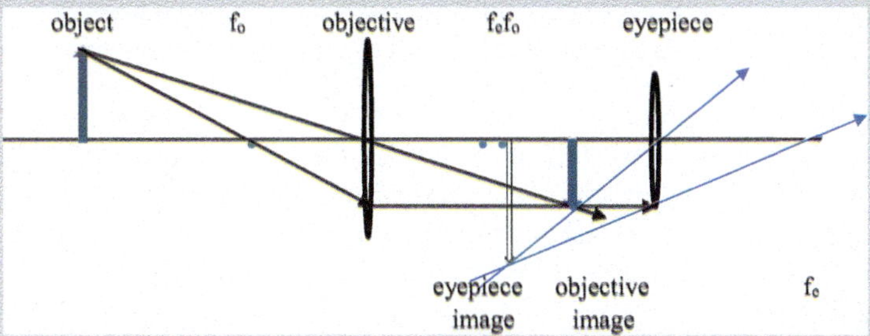

The figure shows a simple refractor telescope with a double convex objective lens and a double convex eyepiece. For a two-lens system, the image of the first lens becomes the object for the second lens. As shown in the figure, the objective lens has a real image which the eyepiece sees as a magnified virtual image. The thin lens equation can be applied sequentially to the objective lens and the eyepiece to find image properties. The magnification of the pair can be found as the product of the magnifications of each lens from Eq. 11.10.

11.5 Physical Optics–Dispersion, Interference and Polarization

Some properties of light cannot be explained by the use of rays and simple geometric techniques. Instead, we have to consider the fact that light is made up of transverse electromagnetic waves. These phenomena like interference and polarization fall in the category of physical optics and are considered in this section.

Before we talk about specific phenomena of light, we need to we need to think about what we mean by light. Light is made up of electromagnetic waves in the visible region of the electromagnetic spectrum. It is emitted by a source such as the sun or a light bulb and reaches a detector like our eyes either directly or by reflecting off some object. A common light source emits "white light" which is made up of many different individual light waves covering all the different wavelengths of visible light and going out in all directions. An important property of each of the waves is the direction of vibration of their electric field. This is called the polarization direction for the wave. For a typical white light source all directions of wave polarizations are present so the light is called unpolarized. In addition, the phases of the individual waves that are emitted by the source are uncorrelated. This type of light is called incoherent light. Some of the phenomena we will discuss require special types of light. This includes monochromatic light in which all the waves have the same wavelength, polarized light in which all of the waves have the same direction of the vibration of their **E** fields, or coherent light in which all the light waves go in the same direction in phase with each other. There are various ways to obtain these special types of light which we will discuss. In Chap. 16 we will learn about lasers that are good sources for coherent, monochromatic, polarized light.

One important property of a light source is the amount of light it emits. For a white light source, this is called its luminous flux, P, measured in lumens (lm). The number of lumens emitted by a source is the total amount of visible light it emits. For example, a 60 W incandescent light bulb emits about 800 lumens. The illuminance of the light we measure from the source (also called intensity) is the luminous flux per unit area,

$$I = P/A. \tag{11.14}$$

Its units are lux (equivalent to lumens per meter squared). For a point source emitting in all directions of a sphere,

$$I = P/\left(4\pi r^2\right) \tag{11.15}$$

where r is the radius of a sphere from the source to the point of measurement. This is an important relationship referred to as the inverse squared law for light sources. It is important in applications such as astrophysics where the luminous flux of certain objects such as specific types of supernova are known. Using this known value of P and the measured value of I in Eq. 11.15 gives the distance r from the earth to the object. If a light source emits light in a direction with a limited solid angle, the denominator is Eq. 11.15 is modified to account for this geometry. The luminous intensity of the source is then given in units of candelas (cd) which are equal to lumen per solid angle.

First let's talk about how to get monochromatic light. Normal white light from the sun is a mixture of waves with all visible frequencies. Since the angle of refraction depends on the wavelength of the light, refraction can be used to separate white light into its different color wavelengths. This is called dispersion. One way of doing this is by using a prism as shown in Fig. 11.23. When the white light enters a glass prism the frequencies of each of its waves stays the same but the wavelengths of the different waves change since their speed in the glass depends on wavelength. According to Eq. 11.2, $v = \lambda f$. Substituting the definition of the refractive index in terms of wave velocity into Snell's law from Eq. 11.12, the angle of refraction for a specific wave is given by

$$n_1 \sin\theta_\mathrm{i} = n_2 \sin\theta_\mathrm{r}, \quad n = c/v$$

$$\theta_\mathrm{r} = \sin^{-1}\left[\frac{v}{c}\sin\theta_\mathrm{i}\right]. \tag{11.16}$$

where we have assumed the incident light is traveling in air with a speed c and incident angle θ_i for all wavelengths. However, since the velocity of waves with different wavelengths in the glass prism is different, their angle of refraction will be different. Red light has the longest wavelength and thus the greatest velocity. Violet light has the shortest wavelength and thus the smallest velocity. Because of this, the angle of refraction for red light is greater than for violet light. Coming out the other side of the prism this wavelength selective refraction has separated the white light into its color components. This is one way to get a monochromatic beam of light.

Another property of light that is important is polarization. The direction in which the electric field oscillates in an electromagnetic wave is called the polarization direction for the wave. The plane in which **E** is polarized is always perpendicular to the direction of the ray representing the direction of travel of the wave front. Typical light sources emit unpolarized light. That is, all directions of polarization can be found in the waves they are emitting. It is common to characterize the polarization of a wave by the combination of the horizontal and vertical components of the **E** vector. This is shown in Fig. 11.24.

There are several ways to obtain a beam of light with one direction of polarization. One way is by transmission through a polarization filter. When light hits an object of matter, its oscillating electric field vector causes the electrons of the matter to oscillate. This can result in absorption or transmission of the light. A polarizer is a piece of matter made up of long chain molecules aligned in one

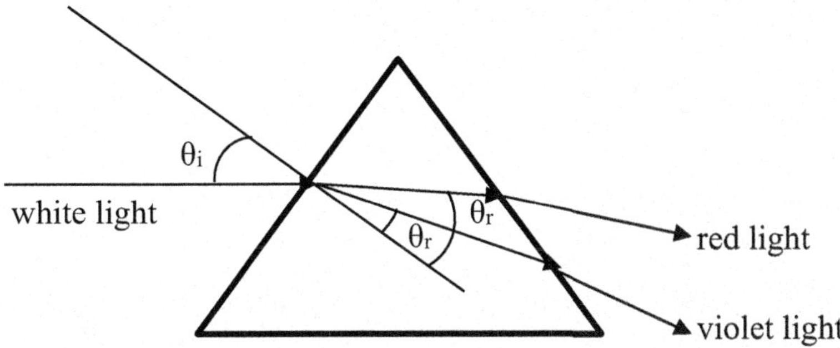

Fig. 11.23 Dispersion of light with a prism

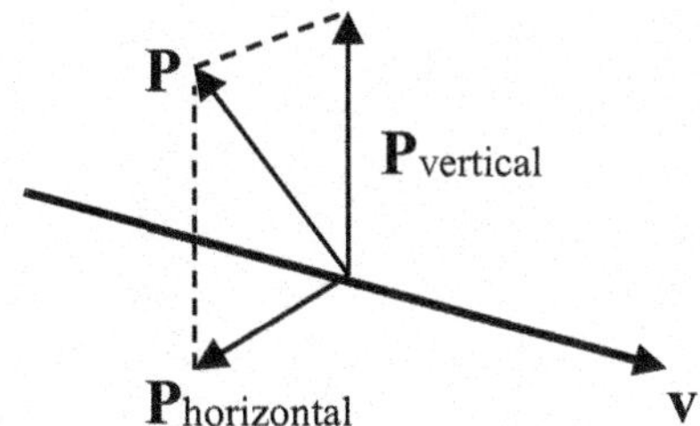

Fig. 11.24 Polarization components

direction. A light wave with its **E** field polarized parallel to the direction of these molecules is absorbed while a light wave with its **E** field polarized perpendicular to the direction of these molecules is transmitted. Using a polarizing filter such as this can produce a beam of light polarized in one direction. If a second polarizer if placed in the path of the polarized beam of light, the amount of light that it transmits is given by Malus's equation

$$I_2 = I_1 \cos^2 \theta \tag{11.17}$$

where θ is the angle between the polarization directions of the first and second polarizing filters and I_1 is the intensity of the incident light. This is useful in limiting the amount of light hitting a detector.

Another way of creating polarized light is by reflection. When unpolarized light is reflected from the surface of an object to our eyes or some other optical instrument, it tends to be polarized in the direction parallel to the plane of the surface it hit. This is because the waves from the source that are polarized parallel to the surface cause the electrons in the material to vibrate in that plane and the reflected wave that is radiated by these electrons has to have its polarization (**E** vector) in this direction and perpendicular to the direction of the reflected ray. The incident waves with other directions of polarization cause the electrons of the material to oscillate in directions that radiate a refracted wave that is unpolarized. At a specific angle of incidence, the angle between the reflected and refracted waves is 90° and the reflected wave is completely polarized. This is called the Brewster angle, θ_B, and is shown in Fig. 11.25. Using Snell's law and the law of reflection, an expression for the Brewster angle can be derived. From the figure, $\theta_1 = \theta_3$ and $\theta_1 + \theta_2 = 90°$. Then Eq. 11.12 can be written as

$$n_1 \sin \theta_B = n_2 \sin\left(90^\circ - \theta_B\right)$$

so

$$\tan \theta_B = (n_2/n_1) \tag{11.18}$$

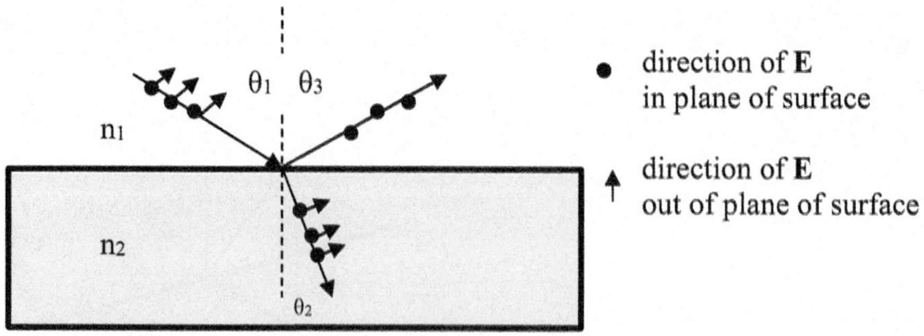

θ1 θ3
n1
n2
θ2
direction of **E** in plane of surface
direction of **E** out of plane of surface

Fig. 11.25 Brewster's angle

where we have used the trig identities $\sin(90° - \theta) = \cos\theta$ and $\tan\theta = \sin\theta/\cos\theta$. This is called Brewster's law. At angles other than θ_B or 0° the reflected wave is only partially polarized but has more light with its polarization parallel to the interface than any other direction. Since most sunlight is reflected from the earth's surface in this manner, glasses made with polaroid filters oriented to block light that is polarized in the horizontal direction are effective in decreasing the amount of scattered light hitting our eyes.

It is also possible to create polarized light through refraction. This is done in certain types of crystals where the index of refraction and thus the speed of the light wave depends on the direction of polarization. The horizontal and vertical components of an unpolarized light beam are split into two beams traveling in different directions because of refraction.

This is shown in Fig. 11.26.

Perhaps the most important phenomenon associated with the wave nature of light is the interference patterns associated with diffraction of light through small openings. This was discussed in your introductory course but to remind you of the properties of diffraction, perform the following experiment.

Experiment 11.1

Interference of Light Waves

Supplies:

- A comb

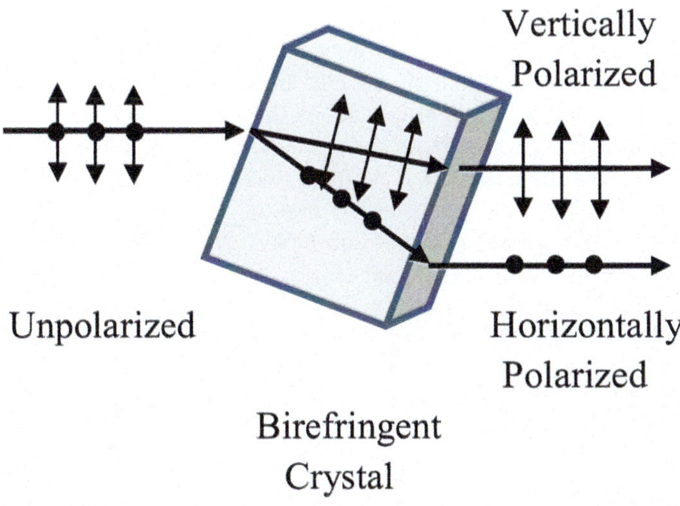

Fig. 11.26 Birefringent polarization

- A cardboard tube (A tube from a roll of toilet paper works best. If you have a longer tube, cut it down to the length of a roll of toilet paper.)
- Black construction paper
- Tape
- A flashlight (It needs to be a small flashlight that can fit inside the tube.)
- A dark room

Introduction—As you already know, waves can constructively and destructively interfere with one another. In this experiment, you will see how light can constructively and destructively interfere with itself, indicating that light is a wave.

Procedure:

1. Most combs have teeth that are spaced farther from one another at one end and closer to one another at the other. You want to use the construction paper and tape to cover the entire comb *except for one tooth and the spaces on each side of the tooth.* Use a tooth on the end where the teeth are close together. Your comb should look like the following drawing:

Comb

Comb that is mostly covered with black paper—one tooth and the space on each side is exposed.

2. Put the comb covered in paper across the end of the tube. The exposed tooth on the comb should be near the center of the tube opening. Use tape to attach the comb to the tube.
3. If there are any open spaces between the comb and the tube, cover them with black paper and tape. In the end, the only way light should be able to pass through that end of the tube is through the two spaces on either side of the exposed tooth.
4. Take the flashlight and the comb/tube assembly into a dark room. Turn on the flashlight and stick it in the tube.
5. Point the tube to a blank wall and look at the light that comes out. It will probably look a little strange.
6. With the tube still pointing to the wall, tilt the flashlight so that it shines mostly at the side of the tube.
7. Looking at the light cast on the wall, move the flashlight back and forth (and perhaps play with the tilt of the flashlight) until you see several bars of light on the wall.
8. Now remember, there are only two spaces through which the light can travel (the spaces on either side of the exposed tooth). However, you

should be able to position the flashlight so that you see *several* bars of light on the wall. How does that happen? You will learn the answer to that question in a moment.

9. Before you finish, play around with the tilt and position of the flashlight to see what kinds of patterns you can make.
10. Clean everything up

You saw the wave nature of light in this experiment. Imagine that the flashlight was putting out waves. The waves hit the comb, which has just two slits through which the waves could travel. As the waves began traveling through the slits, eventually, the waves from one slit began to overlap with the waves from the other slit. The result was a pattern of constructive and destructive interference that produced a pattern of light and dark spots on the wall. This situation is depicted in the figure below.

This experiment, (shown in Fig. 11.27), was originally performed by Thomas Young in 1801. He considered it solid evidence that light is, indeed, a wave. That view of light remained unchanged until quantum mechanics came on the scene in the early 1900s. You will learn about what happened then in Chap. 15.

Let's look at the details of what happens in a diffraction experiment. First, notice that when the light passes through the slits, the direction of its travel changes. Prior to hitting the slits, the waves were traveling straight from the left side of the figure to the right side of the figure. Once the waves hit the slits, however, they "bent" around the slits and started radiating outwards from the slits. This is consistent with Huygen's principle that you learned about in your first physics course. This phenomenon is called diffraction, and all waves behave this way when they travel through small openings or encounter obstacles. In the experiment, diffraction makes the two slits act like individual light bulbs whose light radiates in all directions on the other side of the slits. Unlike two individual light sources, however, the light from the two slits is coherent, which means that the waves all have the same phase angle.

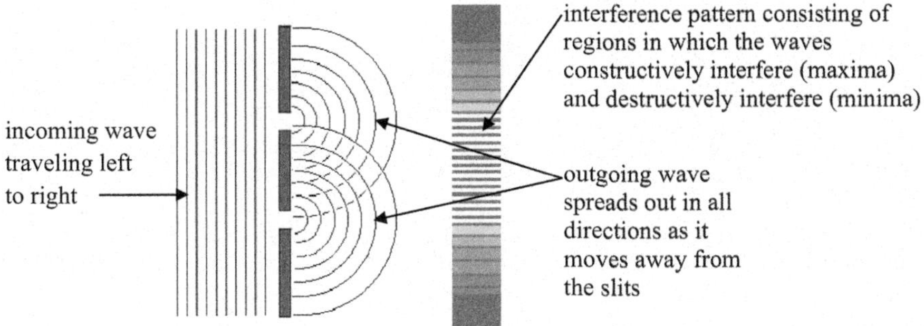

Fig. 11.27 Interference of waves passing through two slits

If two waves that have exactly the same equation (the wavelength, speed, amplitude, and phase angle are all the same), it is easy to determine where constructive and destructive interference will occur. Constructive interference will occur when the light from one slit travels a distance that is λ (or any multiple of λ) farther than the light from the other slit travels. When that happens, the wave crests overlap and the wave troughs overlap, making constructive interference. That's where you will see bright spots. When the light from one slit travels ½·λ (or any odd multiple of ½·λ) more than light from the other slit, the crests of one wave will line up with the troughs of the other, making destructive interference. The interference pattern seen on the screen, then, depends on the wavelength of light involved. This geometry is shown in Fig. 11.28.

Young's double slit experiment is a demonstration of diffraction. Young used coherent, monochromatic light of wavelength λ. He sent it through two small slits separated by a distance d as shown in Fig. 11.28. The interference pattern created by the waves emanating from the two slits is project on a screen a distance L away. The interference pattern has a bright central region and with alternating light and dark regions on either side. These show up as light and dark fringes on the screen. The first bright fringe on the side of the central maximum can be calculated as shown. Since $d << L$ we have very small angles which gives us two similar triangles,

$$\tan\theta = y/L \text{ and } \sin\theta = \lambda/d.$$

Since $\lambda << d$, we are dealing with very small angles where $\tan\,\theta = \sin\,\theta$. Therefore the first bright fringe occurs at $y = \lambda L/d$ on either side of the central maximum. In general, the light fringes occur at

$$y_{\mathrm{m}} = m\lambda L/d \tag{11.19}$$

where $m = 0, \pm 1, \pm 2, \pm 3, \ldots\ldots$

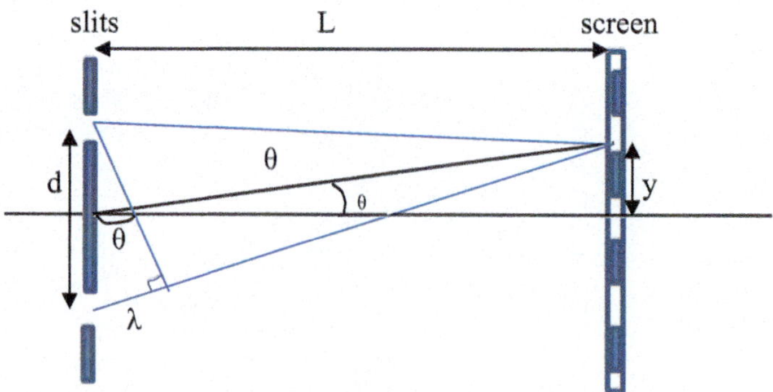

Fig. 11.28 Double slit interference pattern

Consider the following example which shows how a two-slit diffraction experiment can be used to measure the wavelength of light.

Example 11.7

In a two-slit diffraction experiment the first maximum fringe above the central maximum appears at $y_{+1} = 2 \times 10^{-2}$m. If the screen is 0.8 m away from two slits separated by $d = 2 \times 10^{-5}$ m, what is the wavelength of the light?

Figure 11.27 can be used for the figure.

Knowns: $y_{+1} = 2 \times 10^{-2}$ m; $L = 0.8$ m; $d = 2 \times 10^{-5}$ m.
Unknowns: λ

This is just a simple application of Eq. 11.19

$$\lambda = y + 1d/L = \left(2 \times 10^{-2}\ \text{m}\right)\left(2 \times 10^{-5}\ \text{m}\right)/0.8\ \text{m} = 5 \times 10^{-7}\ \text{m} = 500\ \text{nm}$$

A small circular opening can diffract light the same way a narrow slit does. The diffraction pattern of a circle appears as light and dark interference rings. The central bright spot is called Airy's disk and the bright rings are Airy's rings. The exact mathematical expression to describe this circular diffraction pattern requires the use of Bessel functions which is beyond the math used in this book. However, the reasoning for getting bright and dark fringe rings is exactly the same as it was for light going through a small slit. That is, the waves from one part of the opening interfere constructively and destructively with the waves from another part of the opening. Using the small angle approximation, the equation for the width of the central bright spot in a single slit pattern Eq. 11.16 is modified for circular geometry to be the radius of the central bright circle r_1,

$$r_1 = 1.22\lambda L/d \tag{11.20}$$

where d is the diameter of the aperture and L is the distance to the screen.

Many important optical instruments like telescopes have circular apertures that will produce diffraction patterns. If all the other optical aberrations present in producing an image through a telescope are very small, the resolution of two closely spaced objects is limited by the size of the diffraction patterns produced by light from the objects. For a diffraction limited instrument, the ability to resolve an image of two closely spaced objected is called the Rayleigh resolution criterion. The limit of resolution occurs when the bright spot of one image falls on the first dark ring of the second image. Objects closer than this have a blurred image that is not resolved. According to Eq. 11.20 the minimum resolvable separation between two objects is

$$x = 1.22\lambda L_{\text{obj}}/d \tag{11.21}$$

where L_{obj} is the distance to the two objects. It is usual to convert this to the angular separation between the two objects by dividing both sides by L_{obj},

$$\Delta\theta = 1.22\lambda/d \tag{11.22}$$

This Rayleigh resolution criterion is important to know for astronomical telescopes and other optical instruments that form images.

There are a couple of more types of interference phenomena that are important to review. The first is associated with the light hitting a diffraction grating. A diffraction grating is made of many single slits that diffract light and form the superposition of many individual diffraction patterns. The slit spacings can be as small as 1 μm. A precision ruling of slits on the grating can either go clear through the material for a transmission grating or only on the surface for a reflection grating. Both types of gratings disperse incident white light into wavelengths of different colors. The wavelength of light equals the distance between the slits, d, times the sine of the angle to the first order bright line in the diffraction pattern,

$$\lambda = d \sin\theta. \tag{11.23}$$

The diffraction angle is different for each wavelength of light which is why a diffraction grating disperses the colors of white light like a prism. This is useful in instruments like spectrometers that analyze the wavelengths of the light emitted or absorbed by a substance.

The final interference phenomenon to consider is the reflection of light from a thin film. You may be familiar with multiple colors coming from a thin film of oil on a garage floor. Figure 11.29 shows how light reflected by a thin film can be color selective. When light in air is incident on a thin film of material with $n_{film} > n_{air}$, part of the incident light is reflected without change the phase of the light wave. The other part of the wave is transmitted through the film and reflected off the back surface. Since $n_{film} > n_{air}$, this reflection inverts wave so the phase changes by $\lambda_f/2$ where λ_f is the wavelength of the light in the material of the film. If the thickness of the film d is equal to $\lambda_f/4$, the wave travels a distance of $\lambda_f/2$ in the film and then exits the front surface of the film. Adding this phase change to the phase shift at the reflection in the film makes the transmitted wave exit the film in phase with the initial reflected wave. Constructive interference between these two waves gives a strong reflected light intensity. If the incident beam is white light, the reflected beam has an enhanced color at this specific wavelength. This can also occur for film thicknesses of $d = 3\lambda_f/4$, $5\lambda_f/4$, etc. If the film varies in thickness, different wavelengths of light will show constructive interference in their reflections from different parts of the film. This makes the reflection from the film looks like it is multi-colored.

Summing Up

In this Chapter you learned about the physical properties of light as an electromagnetic wave from both a geometrical and physical optics perspective. This

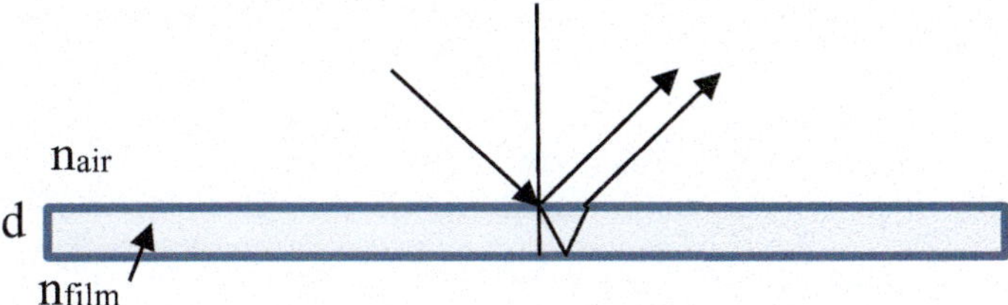

Fig. 11.29 Thin film interference

demonstrated how important optics is to many different scientific applications through optical instruments such as microscopes, telescopes, and other imaging devices. You should be able to identify some of the properties of optical waves in the world around you, such as polarization, dispersion, and thin film interference effects. We will return to the subject of the optical properties of solids in Chap. 14, the particle nature of light in Chap. 15 and light absorption and emission in Chap. 16.

Solutions to the Problems

11.1

Use the figure given in the problem. Knowns: $m = 50.0$ g; $l = 1.0$ m; $M = 2.5$ kg
Unknowns: v

Using what you have learned in the past about Newton's Laws, you can quickly determine that the tension in the string is given by:

$$T = mg = (2.5\ \text{kg})\left(9.81\ \text{ms}^{-1}\right) = 25\ \text{N}$$

Now that you know the tension, the speed is a simple application of Eq. 11.4.

$$v = \sqrt{\frac{T}{\mu}} = \sqrt{\frac{25\ \text{N}}{0.0500\ \text{kg}/1.0\ \text{m}}} = 22\ \text{m/s}$$

11.2

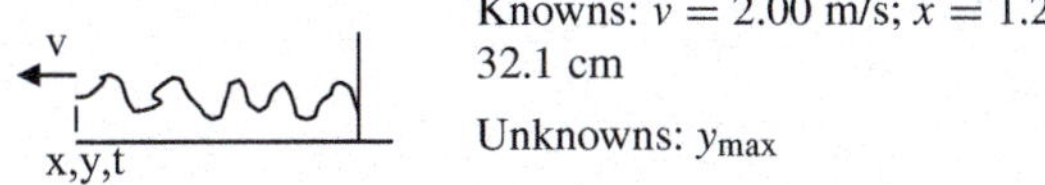

Knowns: $v = 2.00$ m/s; $x = 1.25$ m; $y = 3.44$ cm; $t = 2.00$ s; $\lambda =$ 32.1 cm

Unknowns: y_{max}

Since we know the speed of the wave and the wavelength, we can get most of Eq. 11.3 describing the wave

$$\begin{aligned} y &= y_{\max} \sin\left[\frac{2\pi}{\lambda}(vt \pm x)\right] \\ &= y_{\max} \sin\left[\frac{2\pi}{0.321\ \text{m}}\{(2.00\ \text{m/s})(2.00\ \text{s}) + 1.25\ \text{m}\}\right] \\ &= -0.0344\ \text{m} \end{aligned}$$

We use the plus sign because the wave is moving to the left (the negative x direction). Since this equation tells us the vertical displacement of the medium for any time and position, we can put that information in and solve for $y_{\max}$:

$$y_{\max} = \frac{-0.0344\ \text{m}}{\sin\left[\frac{2\pi}{0.321\ \text{m}}\{(2.00\ \text{m/s})(2.00\ \text{s}) + 1.25\ \text{m}\}\right]} = -0.0436\ \text{m}$$

The negative just means that the medium is displaced downwards. Thus, the amplitude is 0.0436 m.

11.3

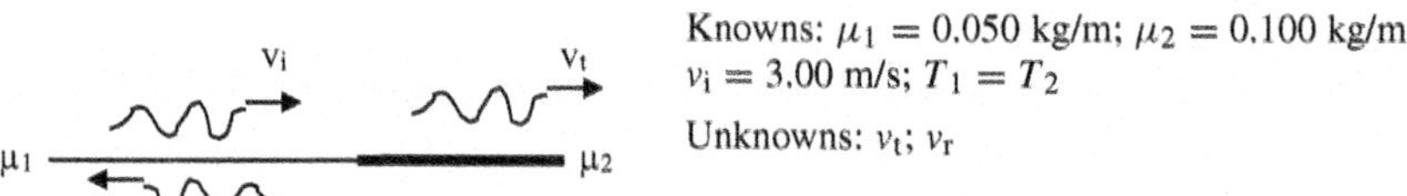

Knowns: $\mu_1 = 0.050$ kg/m; $\mu_2 = 0.100$ kg/m; $v_i = 3.00$ m/s; $T_1 = T_2$

Unknowns: v_t; v_r

To figure out the speed of the wave in the second string, we need to figure out the tension. We can do that from the speed of the wave in the first string:

$$v_i = \sqrt{\frac{T}{\mu_1}} = \sqrt{\frac{T}{0.050\ \text{kg/m}}} = 3.00\ \text{m/s}$$

$$\text{T} = (3.00\ \text{m/s})^2(0.050\ \text{kg/m}) = 0.45\ \text{N}$$

where we have used the equivalent unit 1 N = 1 kg m/s^2.

Now we can figure out the speed in the second string:

$$v_t = \sqrt{\frac{T}{\mu_2}} = \sqrt{\frac{0.45\ \text{N}}{0.100\ \text{kg/m}}} = 2.1\ \text{m/s}$$

The transmitted wave will have a speed of 2.1 m/s. The reflected wave will have the same speed as the original wave, 3.00 m/s. Remember, the speed of a wave on a string depends only on the tension and the linear mass density of the string. The reflected wave will be inverted, because it encountered a medium in which it moved more slowly than the medium it was originally in.

11.4 The waves will both destructively and constructively interfere with one another, depending on the overlap. As they approach each other and continue moving, all possible overlaps will eventually happen. When the waves constructively interfere, the medium will experience the greatest oscillation, which will be the sum of the two amplitudes. The least displacement of a medium in *any* wave is zero. Thus, the greatest displacement will be 20.0 cm and the least displacement will be 0 cm.

11.5

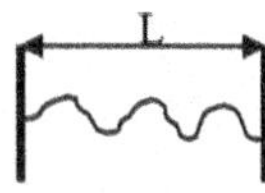

Knowns: $L = 15.0$ cm; 4 antinodes

Unknowns: λ

When there are 4 antinodes, $n = 4$. The equation for the wavelength of a standing wave, Eq. 11.8, is

$$\lambda_n = 2L/n \quad (n = 1, 2, 3, \ldots)$$

Thus

$$\lambda_4 = 2(15.0\text{cm})/4 = 7.50\text{cm}$$

11.6

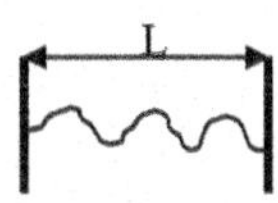

Knowns: $L = 25.0$ cm; $T = 50$ N; m = 7.50 gm; $n = 1$

Unknowns: f_{fund}

The wavelength is easy to calculate:

$$\lambda_1 = 2\text{L}/1 = 2(25.0\text{cm})/1 = 50.0\text{cm}$$

To get the speed of the wave we can use Eq. 11.4,

$$v = \sqrt{\frac{T}{\mu}} = \sqrt{\frac{50.0\ \text{N}}{0.00750\ \text{kg}/0.250\ \text{m}}} = 40.8\ \text{m/s}$$

Now that we know the velocity and wavelength, we can find the frequency from Eq. 11.2,

$$f_{\text{fund}} = v/\lambda = \frac{40.8\ \text{m/s}}{0.500\ \text{m}} = 81.6\ \text{Hz}$$

11.7

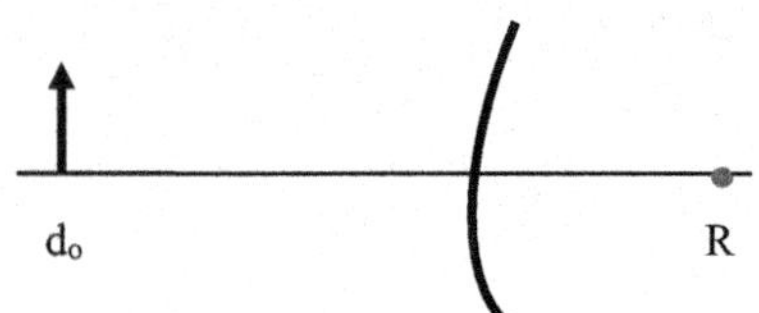

Knowns: $d_o = 15.0$ cm; $R = 20.0$ cm

Unknowns: image

The focal length (half the radius of curvature) of a convex mirror is negative. Thus the mirror equation (Eq. 11.9) is

$$\frac{1}{d_o} + \frac{1}{d_i} = \frac{1}{f}$$
$$\frac{1}{15.0\text{cm}} + \frac{1}{d_i} = \frac{1}{-10.0\text{cm}}$$
$$\text{d}_\text{i} = -6.00\text{cm}$$

This tells us that the image is virtual. To get the magnification we can use Eq. 11.10,

$$m = -d_\text{i}/d_\text{o} = -(-6.00\text{cm})/(15.0\text{cm}) = 0.400$$

The image is 40.0% of the object's height and is upright.

The only difference for a concave mirror is that the focal length is positive.

$$\frac{1}{d_o} + \frac{1}{d_i} = \frac{1}{f}$$
$$\frac{1}{15.0\text{cm}} + \frac{1}{d_i} = \frac{1}{10.0\text{cm}}$$
$$\text{d}_\text{i} = 30.0\text{cm}$$

This tells us that the image is real. To get the magnification we can use Eq. 11.10,

$$m = -d_{\mathrm{i}}/d_{\mathrm{o}} = -(30.0\ \mathrm{cm})/(15.0\ \mathrm{cm}) = -200$$

The image is 2.00 times the object's height and is inverted.

11.8

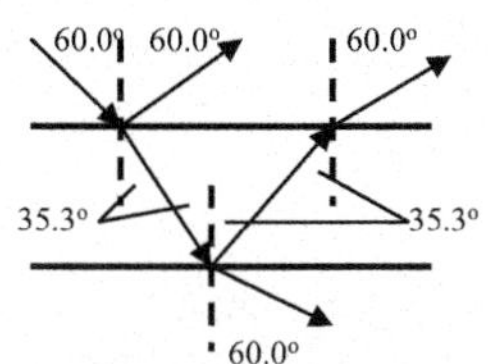

Knowns: $n_{\mathrm{a}} = 1.00$; $n_{\mathrm{g}} = 1.50$; $\theta_{\mathrm{i}} = 60°$

Unknowns: destructive interference

(a) The reflected rays are shown in the figure. The light will be reflected at each boundary, and it will also be refracted. When the light moves into a medium of higher index of refraction, the light will refract towards the perpendicular. When it moves into a medium of lower index of refraction, it will bend away from the perpendicular. In the reflected rays, the angle of incidence equals the angle of reflection. In the refracted rays, Snell's Law can be used to determine the angle of the first refracted ray:

$$n_1 \sin\theta_1 = n_2 \sin\theta_2$$

$$\sin\theta_2 = (n_1/n_2)\sin\theta_1 = (1.00/1.50)\sin 60.0°$$

$$\theta_2 = 35.3°$$

To determine the angle of the other refracted rays, we can also use Snell's Law, or the trigonometry of the triangles in the figure.

(b) An interference pattern will be observed because there are two light sources: the reflection off of the top layer of glass and the reflection off of the bottom layer. These light sources are coherent, so they will produce an interference pattern. Note that the interference pattern will not be very strong, since the ray reflected off of the bottom of the glass and then refracted through the top of the glass is less intense than the ray that reflects off of the top of the glass.

(c) Remember that a wave will invert when reflected off of a boundary when the wave travels slower in the new medium than in the old medium. Thus, the wave reflected from the top of the glass will be inverted. However, the wave reflecting from the bottom of the glass will *not* be inverted. Thus, everywhere the wave reflected from the top of the glass has a crest, the other wave will have a trough. Thus, to get destructive interference, these waves just have to "line up." As a result, when Δl is any integral multiple of the wavelength, destructive interference will result.

11.9

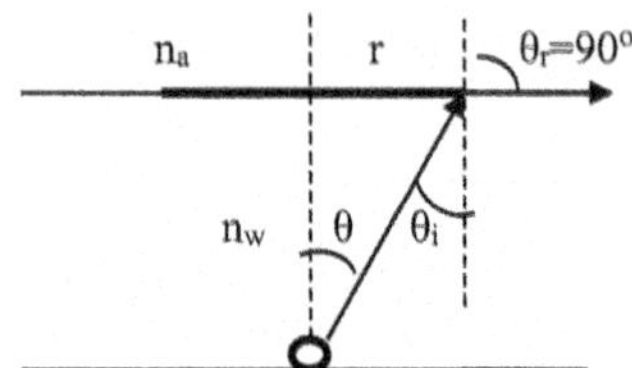

Knowns: $n_a = 1$; $n_w = 1.33$; $\theta = \theta_i$; $\theta_r = 90°$

Unknowns: r

$\theta = \theta_i \geq \theta_c$ for no light rays to be refracted into the air. For the smallest value of θ we use the equal sign. Using Snell's law and the definition of critical angle,

$$n_w \sin\theta_i = n_a \sin\theta_r = 1 \sin 90° = 1$$

Using the triangle with the radius of the wood circle and the incident angle to define the sine,

$$n_w \frac{r}{\sqrt{d^2 + r^2}} = 1$$

$$r = \sqrt{\frac{d^2}{n_w^2 - 1}} = \sqrt{\frac{(2m)^2}{(1.33)^2 - 1}} = 2.28\text{m}$$

11.10

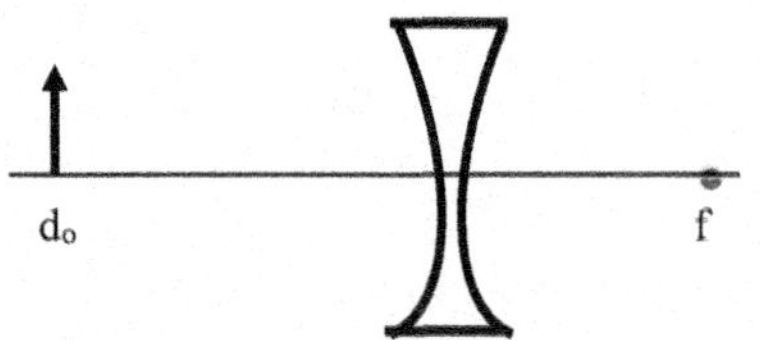

Knowns: $d_o = 10$ cm; $f = -15$ cm

Unknowns: Image

In a diverging lens, the focal length is negative. Equation 11.9 gives

$$\frac{1}{d_o} + \frac{1}{d_i} = \frac{1}{f}$$

$$\frac{1}{10.0\text{ cm}} + \frac{1}{d_i} = \frac{1}{-15.0\text{c}}$$

$$d_i = -6.00\text{ cm}$$

Based on the sign conventions for lenses, the image is virtual and is on the same side of the lens as is the object. The magnification equation (Eq. 11.10) tells us the rest of what we need to know:

$$m = -d_\mathrm{i}/d_\mathrm{o} = -(-6.00\ \mathrm{cm})/(10.0\ \mathrm{cm}) = 0.600$$

The image is 60.0% the size of the object and is upright.

Study Guide for This Chapter

1. The speed of waves moving along a stretched string is measured. If the researcher wanted to double the speed of the waves in the same string, by what factor would the tension of the string need to be changed?
2. The end of a stretched rope is oscillated at a certain frequency in order to create waves. If the frequency is suddenly doubled, what happens to the wavelength of the waves? What happens to the speed of the waves?
3. A wave on a string encounters a new string. The reflected portion of the wave is inverted compared to the original wave. Compare the linear mass densities of the strings.
4. Two transparent pieces of plastic are stacked on top of one another, and a beam of light is directed through one layer. The top piece of plastic has an index of refraction equal to 2.5, and the bottom piece has an index of refraction equal to 1.4. If the experimenter wants to observe total internal reflection, should the beam of light originate in the top piece of plastic or the bottom piece of plastic?
5. When you look at something that is underwater, the depth at which you observe it is not the proper depth. Why? Is the observed depth more or less than the proper depth?
6. Suppose you set up a two-slit diffraction experiment. If you looked at the interference pattern produced and then increased the distance between the screen and the slits, what would happen to the distance between the bright lines?
7. The diagram to the right shows where the image appears for a distant in an eye that is nearsighted. In order for the image to be seen clearly, it should be focused right on the retina. If you were to use glasses to correct this problem, would the glasses be made of converging or diverging lenses?

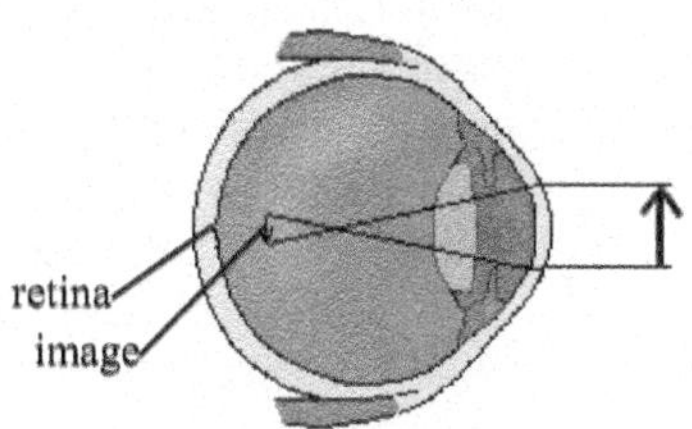

8. What is the significance of Brewster's angle?
9. A rope (length = 4.00 m, mass = 0.750 kg) is stretched with a tension of 85.0 N. A wave ($\lambda = 45.0$ cm) travels down it. What is the frequency of the wave?

10. A wave travels along a string according to the following equation:

$$y = (0.111\ \text{m}) \sin\left[\frac{2\pi}{0.350\ \text{m}}(\{35.0\ \text{m/s}\}t + x)\right]$$

   a. If motion to the right is positive, which way is the wave propagating?
   b. What is the amplitude of the wave?
   c. If the tension in the string is 341 N, what is the linear mass density of the string?
11. The speed of a wave traveling down a stretched rope (length = 5.50 m) is measured to be 45.0 m/s. What is the frequency of a standing wave on that rope if the wave has 4 nodes?
12. A rope ($\mu = 0.350$ kg/m) is stretched between two fixed points with a tension of 45.0 N. An experimenter notices that the rope has a fundamental resonant frequency of 1.13 Hz. How long is the rope?
13. In Young's double-slit experiment, the interference pattern produced had a bright spot on the screen right in the center between the two slits. This is called the "central maximum." How much farther (in terms of λ) did the light from one slit travel as compared to light from the other slit for the bright spot that is right above the central maximum?
14. A fisherman drops his flashlight in a lake. He doesn'trealize it for a moment, however, so he keeps traveling in his boat. Eventually he notices that he doesn't have his flashlight anymore. However, he is not worried. He knows that the light was on, and since it is night, he decides that he can see his flashlight shining underwater. However, when he turns around, he does not see the flashlight. He starts heading back the way he came and eventually sees the light shining underneath the water. Assuming that the flashlight is shining at a slight angle relative to the perpendicular, at what depth in the water ($n = 1.4$) is the flashlight given the fact that the fisherman must be within 5.0 m of the flashlight's horizontal position before he sees it?

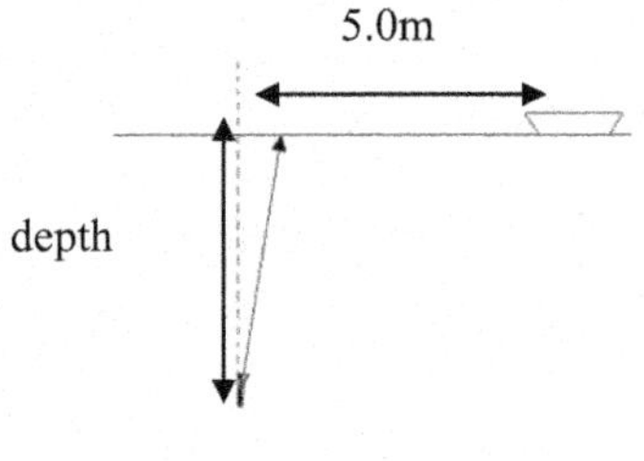

15. When white light passes through a prism, it is often separated into its colors. In the situation depicted below, which color is on top? Which color is on the bottom?

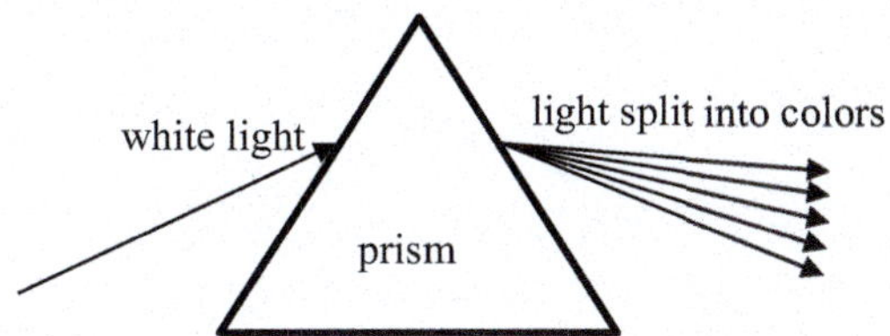

16. An object of height 15 cm is placed 7.5 cm in front of a convex mirror. An image forms 5.0 cm on the opposite side of the mirror. What is the radius of curvature of the mirror? Is the image real or virtual? Is it upright or inverted? What is the size of the image?
17. You want to magnify an object 15.0 cm away using a converging lens. If you want to magnify the object so that it is upright and twice as large, what focal length will you need for the lens?
18. A beam of sunlight is hitting a sheet of bright metal ($n_m = 1.5$). How do you orient the sheet so the reflected light is completely polarized in the horizontal direction?

Next Level

19. A simple microscope is shown in the figure. It is made of two double convex lenses of focal lengths $f_e = 26.0$ mm and $f_o = 11.0$ mm of the eyepiece and objective, respectfully. For an object at 12.0 mm from the objective lens, a distance image is formed 252 mm from the eyepiece. Use the thin lens equation to find the distance between the lenses and the magnification of the microscope? Is the image virtual or real?

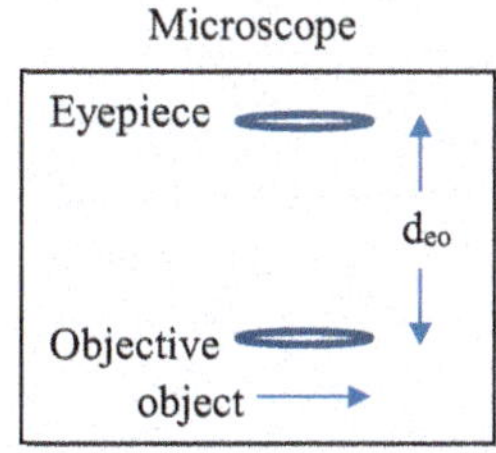

20. Use ray tracing to show the position, size and nature of the image in problem 19.

Waves II: Acoustics

12

Chapter Summary

Our lives are full of sounds ranging from annoying noise to beautiful music. The sound of speech is an important way we have to communicate. Because of the critical nature of sound in our lives it is important that we understand what sound is and learn about its physical characteristics. We already know a lot about waves but sound waves are very different from light waves. We focus on these differences in this chapter. Some of the themes of the chapter are represented in Fig. 12.1 which shows an airplane creating a sonic boom.

Main Concepts in This Chapter

- Sound Waves
- Effect of Moving Sources
- Music
- Sonar, Ultrasound, and Infrasound

12.1 Introduction

Now we want to discuss sound waves. This branch of physics is called acoustics. Sound waves exhibit some of the same wave-like properties of the light waves we have been discussing but they are basically quite different. For example, a sound wave exhibits the same periodic properties as a light wave such as frequency,

R. C. Powell, *Physics*, Undergraduate Lecture Notes in Physics,
https://doi.org/10.1007/978-3-032-09361-5_12

wavelength, and period. Its speed is still given by Eq. 11.2, $v = f\lambda$. The amplitude of the wave is its maximum displacement from an equilibrium position. Multiple sound waves can exhibit constructive and destructive interference like light waves. All of these general wave characteristics were discussed in the previous chapter. However, sound waves are longitudinal waves whereas light waves are transverse waves. Sound waves are mechanical waves that require a media of some type for propagation whereas light waves are not mechanical waves and therefore can travel in a vacuum. These characteristics lead to some important differences in physical properties.

Sound waves are pressure waves. A source drives the atoms or molecules of the surrounding media to move so that there are some regions where they are much closer together than they are in equilibrium and there are some regions where they are much further apart. The former areas are called areas of compression and the latter areas of rarefaction. The particles in the areas of compression collide with neighboring particles and cause the compression area to move and an area of rarefaction to form behind it. The wavelength is the distance between two areas of compression.

Remember that light is the part of the electromagnetic spectrum that we can see. Similarly, sound is the part of the spectrum of acoustic waves that we can hear. These are mechanical waves in the approximate frequency range from 20 to 20,000 Hz. Waves with higher or lower frequencies are called ultrasound and infrasound, respectively. These types of waves have important applications that will be discussed later.

Detectors convert the kinetic energy of vibrating particles in sound waves to other forms of energy. We refer to hearing sound waves when they are detected by our ears and the signal transmitted to the brain. The brain interprets the sound in terms of its pitch, loudness, tone and quality. These are terms used in describing the sound we hear. In terms of wave terminology, pitch is the frequency of the sound wave, loudness is related to the amplitude of the sound wave, and tone refers to the number of frequencies that are present in the sound you hear. The quality or timbre of the sound is determined by the shape of the wave and how close it is to a harmonic wave form. These concepts are discussed further in a later section of this chapter.

Let's begin our discussion of acoustics by performing the following experiment that demonstrates the properties of mechanical waves.

Experiment 12.1

Wave Motion and Standing Waves

Supplies:

1. Slinky (A metal one works best, but you can use a plastic one.)
2. A person to help you
3. Table with a smooth surface that will not get scratched up.

Fig. 12.1 Breaking the sound barrier. *Credit* Ensign John Gay, U.S. Navy, public domain

Introduction: This experiment will give you some experience with the propagation of transverse and longitudinal waves. It will also give you some experience with wave interference and standing waves, which will be discussed in upcoming sections of this chapter.

Procedure:

1. Hand one end of the slinky to your helper and take the other end yourself.
2. Stand so that you and your helper are 3.5–4.0 m apart from one another and the Slinky is stretched out in between both of you.
3. Have your helper hold his or her end still.
4. Generate a wave on the slinky by raising your end up and then bringing it down quickly. Notice how the wave travels down the Slinky, reflects off of the end that is being held still, and travels back towards you. This is a transverse wave because the wave moves from you to your helper and back (horizontally), but the Slinky oscillates up and down (vertically). Notice that the Slinky does not move horizontally. Only the wave does. Thus, the Slinky is the medium for the wave. The wave you created is called a **mechanical wave**, because it was the result of the mechanical motion of your arm.
5. Allow the Slinky to become reasonably still again. Once again, generate a wave on the Slinky by raising your end up and then bringing it down quickly. Make sure your helper continues to hold his end still. This time, notice *how* the wave bounces off of your helper. When the crest of the

wave hits your helper, it bounces back as a trough. That means the wave is **inverted** when it is reflected.

6. Allow the Slinky to become reasonably still again. Now, start creating many waves by moving your end up and down at a constant rate. Make sure that your helper holds his or her end still.
7. Notice that the waves you are creating begin overlapping with each other and creating different patterns. This is called **wave interference**. Since waves move the Slinky up and down, if the waves overlap, they will affect one another. If a trough of one wave, for example, overlaps with the crest of another, the trough and crest will cancel each other out.
8. Let the Slinky settle again. Now, have your helper move his or her end up and down while you do the same. You are now both generating waves which are interfering with each other.

If you both move your ends up and down at the right frequency, you can eventually make the Slinky look like this:

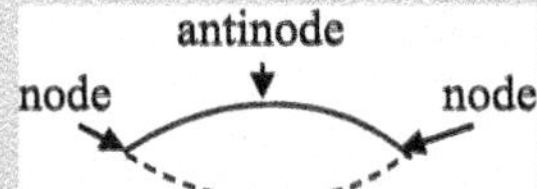

When this happens, the waves are interfering with each other in just the right way so that the entire Slinky is either moving up or moving down. This is called a **standing wave**. This particular standing wave is called the **fundamental frequency** (also called the **first harmonic**) of the Slinky. The point on the Slinky that experiences the greatest oscillation amplitude is called the **antinode**, while the points which experience no oscillation at all are called **nodes**. In this situation, then, there are two nodes and one antinode.

9. Now speed up the frequency at which you both move the slinky up and down. Once again, if you both start moving your ends up and down with exactly the right frequency, you will make the Slinky look like this with the nodes and antinodes as designated above:

This is another standing wave, and it is typically called the **second harmonic**. Notice that in this standing wave, there are three nodes and two antinodes. Compare the frequency with which you needed to shake the Slinky in this step to the frequency in the previous step.

10. If you can do it, try and form a standing wave with three antinodes and 4 nodes. This is the **third harmonic**, and it would look like this:

Compare the frequency with which you needed to shake the Slinky in this step to the frequencies in the previous two steps.

11. Thank your helper. He or she can leave now. For the next part of the experiment, find a clear tabletop that you will not scratch with the Slinky.
12. Place the Slinky on the tabletop.
13. Stretch it out so that it is about three-quarters of a meter long.
14. Hold one end still and push the other end in and out once. Notice what happens. A portion of the spring compresses (forming the compression) while the portion behind stretches out (forming the rarefaction). The compressed area moves to the end that is being held still. It then bounces off and comes back. This is a longitudinal wave traveling down the Slinky. It will not travel nearly as well as the transverse waves did, since there is a lot of friction between the Slinky and the table.
15. Continue to hold one end still and begin moving the other end back and forth rapidly. See if you can set up a standing wave so that the compression areas do not move. This will be much harder because friction reduces the waves so much. Nevertheless, you should at least be able to see something that is close to a standing wave.
16. Put the Slinky away.

12.2 Sound Waves

As you know, the frequency of a wave is related to the speed of the wave divided by its wavelength (Eq. 11.2). Thus, if we know the pitch of a sound wave as well as its speed, we can determine the wavelength. Not surprisingly, the speed depends on the medium that is transmitting the sound. In general, the speed of sound is given by:

$$v = \sqrt{\frac{\kappa}{\rho}} \tag{12.1}$$

where v is the speed, and ρ is the density of the medium. The parameter κ (the Greek letter kappa) is the bulk modulus of the medium. The bulk modulus is a measure of how easy it is to compress the medium. If the medium is easily compressed, its bulk modulus is low; if the medium is hard to compress, its bulk modulus is high.

Equation 12.1 is not all that useful because the bulk modulus of a medium depends on lots of things. Thus, you have to have many tables in order to use this equation properly. However, there is one thing that you can learn from this equation. In general, the compressibility of a gas is lower than that of a liquid, which is, in turn, lower than that of a solid.

Thus, the speed of sound in a gas is lower than the speed of sound in a liquid, which is lower than the speed of sound in a solid. This is an important point:

Sound waves travel slower in gas than they do in liquid. They also travel slower in liquid than they do in solid.

Since the speed of a sound wave changes when it changes the medium through which it travels, it is important to note that the frequency does not change. If a sound wave is traveling in air and then suddenly hits water, the wave will travel more quickly in the water, but with a longer wavelength. It will have the same frequency in the water as it had traveling in the air.

Since we mostly deal with sound waves traveling through air, it would be nice to have Eq. 12.1 evaluated for the specific medium of air. Both the bulk modulus as well as the density of air depend on the temperature. When these effects are taken into account, the speed of sound in air is given by the equation:

$$v = (331.5 + 0.606T)\,\mathrm{m/s} \tag{12.2}$$

where v is the speed of sound in air, and T is the temperature expressed in degrees Celsius.

When using this equation, use the temperature in degrees Celsius but just put it into the equation as a number with no unit. If you do that, then the equation is defined so that the speed comes out in m/s.

Use this equation in the following problem.

Student

12.1 If the temperature drops from 30 to 10 °C, how much does the speed of sound in air change?

12.2 What is the speed of sound in glass if it has a bulk modulus of 1000 N/mm^2 and a density of 2500 kg/m^3.

The loudness of sound is determined by the amplitude of the wave. The tighter the compressions in the sound wave, the louder the sound. It turns out to be a bit easier to measure the intensity of a sound wave than it is to measure the amplitude, so we will discuss intensity rather than amplitude.

Intensity of a sound wave—The rate at which sound energy flows through a given area.

The intensity of a wave is proportional to the square of its amplitude. This is determined by the source of the vibration causing the wave. Intensity is the energy in the wave per unit time per unit area,

$$I = P/A \tag{12.3}$$

where we have used the fact that energy per unit time is power. For sound waves the intensity is measured in watts per meter squared. Human ears have a hearing threshold of about 10^{-12} W/m^2 so sounds with lower intensities cannot be heard. Sounds with intensities above 1 W/m^2 are so loud they cause discomfort and can damage the ear. This level of intensity is called the threshold of pain. Because of the wide range of intensities that can be heard, the intensity of sound can be expressed in decibels (dB) defined as

$$\beta = 10\log(I/I_o) \tag{12.4}$$

where I_o is the threshold for hearing ($I_o = 1.0 \times 10^{-12}$ W/m^2). Thus in decibels the range of hearing runs from intensities of $\beta = 0$ dB to 120 dB. (Remember that log A/B = log A – log B.) Normal conversation, for example, typically takes place at a loudness of 40 dB. From Eq. 12.4, when the intensity of the sound waves is equal to Io, the loudness of the sound is 0 dB. Now, since the equation involves taking a log of a ratio, we say that the loudness scale is a logarithmic scale. That means an increase in 1 dB is an increase of a factor of 10 in intensity. Thus, a 1 dB sound wave is ten times as intense as a 0 dB sound wave, but a 2 dB sound wave is 100 times as intense as a sound wave at the threshold of hearing. Then, 40 dB for normal conversation involves sound waves which are 10,000 times more intense than the sound waves that are at the threshold of human hearing.

Of course, these sound levels depend on where you are relative to the sound. As you travel away from the source of a sound, the loudness of the sound decreases. As the sound wave travels, it spreads out in a spherical wave front whose surface area is proportional to the square of the distance traveled. This is the same as Eq. 11.15 for light waves. It means that:

The intensity of a sound wave is inversely proportional to the square of the distance that the sound wave travels.

Table 12.1 lists the loudness of common sounds in decibels.

Table 12.1 Loudness of some common sounds

Sound	Decibels	Sound	Decibels
Soft whisper	20	Powered mower	95
Normal conversation	40	Typical rock concert	115
Busy traffic	70	Physical pain to ears	120
Pneumatic drill	80	Physical damage to ears	130

Suppose you are standing next to a gasoline-powered mower and do not like the loud noise. If you increase the distance between you and the mower by a factor of 10, the intensity of the wave will decrease by a factor of 10^2, or 100. That corresponds to a drop of 20 dB. Thus, the noise level from the mower will drop from 95 to 75 dB.

To give you a little experience with sound waves, perform the following experiment.

Experiment 12.2
Sound Waves in a Bottle and the "Bottle Paradox"

Supplies:

- Two identical empty glass bottles
- Water
- Vinegar (Any kind)
- Baking soda
- A butter knife
- A funnel
- Paper towels
- A countertop or desktop

Introduction: This experiment will help you learn about the relationship between wavelength and pitch of a sound wave, how the bulk modulus of the medium affects the speed of sound, and why the pitch of a sound depends on *what* is actually vibrating.

Procedure:

1. Take a glass bottle and place your lower lip on the edge of the bottle's opening. Then, blow into the bottle. With a bit of practice, you should eventually hear a nice sound coming from the bottle. That sound is the result of a longitudinal standing wave that you are producing in the air inside the bottle.
2. Next, put the bottle on a countertop or desktop and strike it gently with the knife. You should hear a sound that is markedly different in pitch than the sound you heard in Step #1. Why is it different? In this step, the *glass bottle* is the thing that is vibrating. Those vibrations then push air back and forth, making a sound wave. Thus, you are not making the air vibrate initially. Instead, you are making the glass bottle vibrate, which then makes the air vibrate. This is the reason that the sounds are different.
3. Fill the other bottle one-third of the way full with water. Now, blow into this bottle the same way you did in Step #1.

4. Blow into the bottle that is still empty. Notice the difference in the pitch. The pitch of the bottle one-third full of water should be *higher* than the pitch of the empty bottle. Why? The bottle with water has less air in it. The distance from your mouth to the end of the air in the bottle is lower than it is in the empty bottle. As a result, the standing wave you can produce in the bottle with water has a *smaller wavelength* than does the wave in the empty bottle. As a result, the wave in the bottle with water has a *higher frequency*. Thus, it makes a higher pitched sound.
5. Put both bottles on the countertop or desktop.
6. Hit each bottle with the knife again. Now which one produces the higher-pitched sound? The *empty bottle* produces the higher-pitched sound here. Why? In this case, the bottle with the water can support the longer wavelength, because of the water you added. Thus, it produces a *lower* frequency. This is often called the "Bottle Paradox," because the bottle with water produces a higher pitch when blown and a lower one when struck.
7. Leaving the bottle with water in it alone, fill the other bottle two-thirds with water and repeat Steps 3–6. Once again, you should hear that the bottle with the higher level of water produces a higher pitch when blown and a lower pitch when struck.
8. Empty both bottles.
9. Fill each bottle one-third full of vinegar.
10. Put about a teaspoon of baking soda into one of the two bottles. You will see a reaction occurring, because the acetic acid in the vinegar is reacting with the sodium bicarbonate in the baking soda to make carbon dioxide gas. Allow the solution in the bottle to settle. This will take a while.
11. Once the solution in the bottle has settled, wipe the tops of both bottles with paper towels to get rid of any vinegar that may be on the lips of the bottles.
12. Blow into the bottle that contains just vinegar in order to remind yourself of the pitch it produces. Try not to inhale while your mouth is over the bottle, as vinegar fumes are nasty.
13. Blow into the bottle to which you added baking soda. Note the difference in pitch.
14. Blow into that bottle several times. Note that the pitch changes until eventually, it sounds like the other bottle. Why did this happen? The bottle to which you added baking soda was no longer full of air. It was full of the carbon dioxide produced in the reaction. The bulk modulus of carbon dioxide is lower than that of air. Thus, the speed of sound in the carbon dioxide is lower than the speed of sound in air. For a given wavelength (determined by the water level), then, the bottle with carbon dioxide will produce a lower pitch. Why did the pitch change as you

kept blowing? Well, as you blew, the carbon dioxide in the bottle was replaced by the mixture of gases you exhaled, which is close to that of air. Thus, the pitch increased until it sounded essentially like the pitch from the other bottle.

Now try the following problems.

Student

12.3 An amplifier increases the loudness of a sound by increasing the amplitude and thus the intensity of the sound wave. A certain amplifier increases the loudness of a sound from 30 to 100 dB. By what factor did the amplifier increase the intensity of the sound wave?

12.4 Suppose you were to heat up an empty bottle to about 90 °C and cool down an identical bottle to about 5 °C. Then, suppose you blew on each bottle like you did in Experiment 12.2. Would the pitch of the sounds produced be the same? If not, which bottle would produce the lower pitch?

12.5 You move from being close to a pneumatic drill where the noise level is 80 dB to twice the distance away. What is the noise level you then hear?

12.3 Effect of Moving Sources

An interesting property of waves occurs when there is relative motion between the source and the observer. This applies to all waves including both light waves and sound waves. It is described in this chapter because some of the most common examples of this property occur with sound waves.

The Doppler effect is the apparent change in frequency of a wave traveling from a source to an observer when the two of them are moving either toward or away from each other. This effect is named after Austrian physicist Christian Doppler. The Doppler effect is best explained with the figure below. Consider a car that is traveling down the street. The driver sees you standing on the sidewalk. He honks his horn in one steady blast as he drives by you. He hears the horn with a pitch that does not change. What do you hear?

The horn from a car produces sound waves with a constant frequency. When the car moves, however, the horn emits those waves as the car travels. Thus, after it has emitted one wave, it moves forwards to emit the next. This causes the waves to be bunched up in front of the car and stretched out behind the car. Since the sound waves get bunched up in front of the car, the wavelength seems shorter, so the frequency is higher. If the car is heading towards you, then, you will hear a pitch that is higher than the true pitch of the horn. When the car passes you,

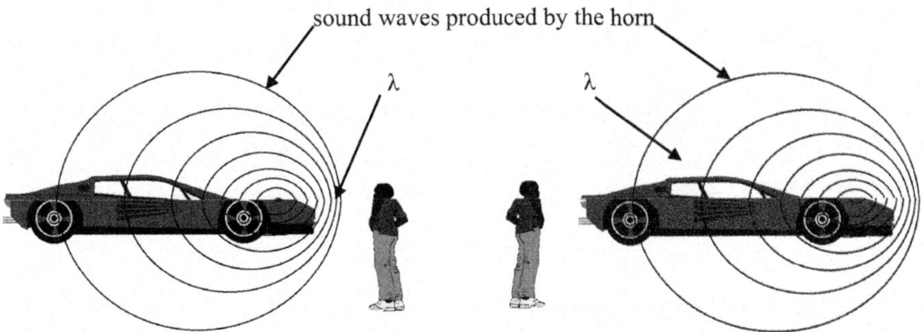

Fig. 12.2 The Doppler effect

the sound waves that reach your ears are the ones traveling behind the car. These sound waves are stretched out, which gives them a longer wavelength and thus a lower frequency. This results in a pitch that is suddenly lower than that of the true pitch. Obviously, the Doppler effect is a result of the wave nature of sound (Fig. 12.2).

This effect depends on the relative motion between the source and the observer. You can produce the same effect by moving towards or away from a fixed object which is emitting sound waves. If the object is stationary, its sound waves do not get "bunched up" or "stretched out" like those of a moving object. However, if you move towards the object, you will start encountering its sound waves at a high frequency, because the relative velocity between you and the sound waves has decreased. Thus, moving towards an object generates the same Doppler effect that you experience when a moving object moves towards you. In the same way, if you run away from a stationary object that is emitting a sound, the sound waves will hit you with a lower frequency, because the relative velocity between you and the sound waves has decreased. Thus, moving away from a fixed object results in the same Doppler effect that you experience when a moving object is traveling away from you.

The Doppler effect is not difficult to calculate. It depends only on the true frequency of the sound wave (in other words, the pitch you would hear if neither you nor the source of the sound were moving), the speed of the observer, and the speed of the sound'ssource:

$$f_{\text{observed}} = \left[\frac{v_{\text{sound}} \pm v_{\text{observer}}}{v_{\text{sound}} \mp v_{\text{source}}}\right] f_{\text{true}} \tag{12.5}$$

The upper signs are used when the motion between the source and observer is toward each other and the lower signs when the motion is away from each other. For example, suppose the source is moving. If it is moving *away from* the observer, you must use a *plus* in the denominator of the equation. After all, when the source moves away from the observer, the frequency observed will be *lower* than the true frequency. The way you will get an observed frequency lower than the true frequency is if the number in the denominator is increased. Thus, you use the plus sign so that the denominator gets bigger. In the same way, suppose the observer is moving away from the source. If that's the case, then once again you expect the observed frequency to be lower than the true frequency, so you would use the *minus* sign in the numerator of the equation so that the numerator (and thus the observed frequency) gets smaller.

Example 12.1

A horn emits a sound with a frequency of 355 Hz when the car is at rest. If a car is traveling towards a person on a bicycle and beeps the horn, what frequency will the bicyclist hear? The car is traveling at a speed of 21.0 m/s, and the bicyclist is traveling towards the car with a speed of 4.2 m/s. Assume that the speed of sound is 343.0 m/s.

Knowns: $f_{true} = 355$ Hz; $v_{source} = 21.0$ m/s
$v_{observer} = 4.2$ m/s; $v_{sound} = 343.0$ m/s
Unknowns: $f_{observed}$

This is a direct application of Eq. 12.5. Since the car is traveling towards the observer, that will increase the horn's frequency. Thus, we must use a minus in the denominator of the equation to make the denominator smaller. Since the observer is moving towards the source, that will also increase the frequency. Thus, we must use a plus sign in the numerator of the equation to make the numerator bigger. The equation, then, is:

$$f_{observed} = \left[\frac{v_{sound} \pm v_{observer}}{v_{sound} \mp v_{source}}\right] f_{true}$$
$$= \left[\frac{343.0\,\text{m/s} + 4.2\,\text{m/s}}{343.0\,\text{m/s} - 21.0\,\text{m/s}}\right] 355\,\text{Hz} = 383\,\text{Hz}$$

Student

12.6 A train is traveling north at 45.0 m/s. A man is behind the train in a car. He is also traveling north at 25.0 m/s. If the train blows its horn (frequency = 411 Hz when the train and observer are both stationary), what is the frequency that the man will hear? ($T = 25.0$ °C).

Monitoring the change in frequency or wavelength of waves due to the relative motion between the source and detector can be a very sensitive way to measure motion. This is true for light waves as well as sound waves. One example is in astronomy where the fluorescent light emitted by a specific chemical element is known to be at a specific wavelength on earth. When the source of light comes from the same type of gas in a star in a distant galaxy, the wavelength is measured to be smaller than it is on earth. The Doppler effect has caused a "red shift" in the spectral line meaning that the source is moving away from the observer. This is evidence that the outer galaxies are moving away from us and therefore the university is expanding.

An interesting effect happens when the source of the wave is faster than the propagation velocity of the wave it emits. Of course, this cannot happen with light waves since no source can travel faster than the speed of light. However, this can happen for sound waves if the source can travel at extremely high speeds like a jet airplane. A supersonic jet can travel at a velocity greater than the speed of sound. Speeds of this magnitude can be designated by their Mach number,

$$\text{Mach number} = v/v_{\text{sound}}. \tag{12.6}$$

Thus a jet traveling at Mach 1 is going the speed of sound. Mach numbers less than one indicate speeds slower than the speed of sound while Mach numbers greater than one indicate speeds greater than the speed of sound.

When the source moves faster than the sound waves it is creating, the waves it creates interfere with each other and the points of maximum interference form a cone as shown in Fig. 12.3a. Since these are pressure waves, the surface of this constructive interference cone is a region of very high pressure. When this cone reaches an observer on the ground, the High pressure acts like a shock wave which sounds like a thunderclap and is called a sonic boom. The angle of the cone is shown in Fig. 12.3b to be given by

$$\sin\theta = v_{\text{sound}}/v_{\text{source}} \tag{12.7}$$

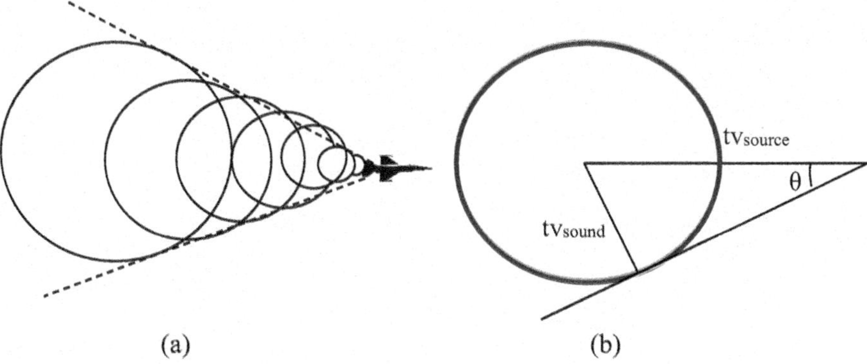

Fig. 12.3 Sonic boom

So the faster the source the smaller the cone angle and the further away the source is by the time the observer hears the sonic boom.

Now work the following problem.

Student

12.7 A jet plane flying at Mach 1.5 = 514.5 m/s passes over you and 3 s later you hear a sonic boom. How far has the jet traveled between the time it emitted the sound wave and the time you hear the sonic boom?

12.4 Music

One of the most pleasant aspects of acoustics is music. Music is basically a series of pressure waves with different frequencies that are organized together or in unison to create a specific pattern of sound. Each individual musical note comes from a standing wave resonance of either a vibrating surface or column of air. We learned about the wave characteristics of both of these in the last chapter and we can use them to model some musical instruments. The vibrating surface can be activated either by striking a drumhead or by plucking, bowing, or strumming a taut string. The air wave can be created either by directly blowing into a tube of a specific length or by blowing through a vibrating reed into a tube. The acoustic properties of these two types of resonances are discussed below.

In discussing standing waves in Chap. 11 we learned that a resonance is a frequency at which a standing wave pattern is formed. A standing wave is formed by two waves of the same frequency traveling at the same speed in opposite directions in the same medium. The waves with resonant frequencies supported by a string or pipe are amplified while waves with other frequencies die out. An example of a standing wave pattern is shown in Fig. 11.7. This is for a transverse wave on a string. A tube with a longitudinal pressure wave looks more like Fig. 12.4. Since it is easier to visualize transverse waves, we will use their shapes to show standing wave patterns even when dealing with longitudinal sound waves.

Think for a moment about the bottles you used in Experiment 12.2. They were open on one end and closed on the other. This tells you something about the standing waves you were able to set up in the bottle. There is no way that the air could vibrate past the closed end of the bottle. Thus, the end of the bottle (or

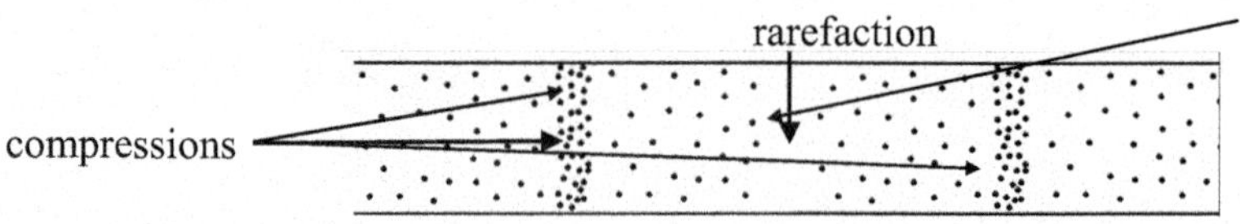

Fig. 12.4 Sound wave in a tube

the level of the water, which was the end of the air available to the sound wave) must be a *node* of the standing sound wave. However, the open end of the bottle was where the vibrations originated. The vibrations, then, were strongest at that point. Thus, the open end of the bottle was an *antinode* in the experiment. In other words, there was always either a compression or rarefaction at the open end of the bottle.

As was discussed in Chap. 11, the type of standing wave pattern that is established depends on the properties of the medium and the boundary conditions at the ends of the medium.

This is true for both waves on a string and waves in a tube. We showed that the strength of the standing wave at any point x is given by

$$y(x) = 2y_{\mathrm{m}} \sin(2\pi x/\lambda).$$

Nodes where the strength of the wave is zero for all times occur when x is an even multiple of $\lambda/4$,

$$x_{\mathrm{node}} = 0, \lambda/2, \lambda, 3\lambda/2, \ldots N\lambda/4 \ (N \text{ even number})$$

since that is where the sine is zero. The maxima in the wave strength occur when x is an odd multiple of $\lambda/4$

$$x_{\mathrm{max}} = \lambda/4, 3\lambda/4, 5\lambda/4 \ldots N\lambda/4 \ (N \text{ odd number})$$

Thus, there will always be N antinodes in a standing wave and $N + 1$ nodes.

The distance between the first and third nodes in a standing wave equals the wavelength of the wave. If a medium of length L supports a wave with $N + 1 = 3$ nodes its length is equal to the wavelength of the standing wave. Then in general the length of the medium supporting the wave, L, and the wavelengths of the resonant waves the medium supports is given by

$$\lambda_N = 2L/N \ (N = 1, 2, 3, \ldots) \tag{11.8}$$

The wavelength for $N = 1$ is called the fundamental wave. The frequency of the wave can be found from knowing that $f_N = v/\lambda_N$ where v is the velocity of sound. Using $N = 1$ gives the fundamental frequency. Frequencies with higher values of N are called overtones. $N = 2$ is the second overtone, etc. If an overtone is an integral multiple of the fundamental frequency, it is called a harmonic. Sometimes the fundamental is referred to as the first harmonic. Now let's see how this applies to string and wind musical instruments.

First consider a stringed instrument such as guitar or violin. The source of the sound originates from a string that has been made to vibrate by bowing or plucking. The string is fixed at both ends creating closed boundary conditions for particle displacement waves at these points. The vibrations travel in the string with a speed that depends on its tension, its diameter, and its type of material

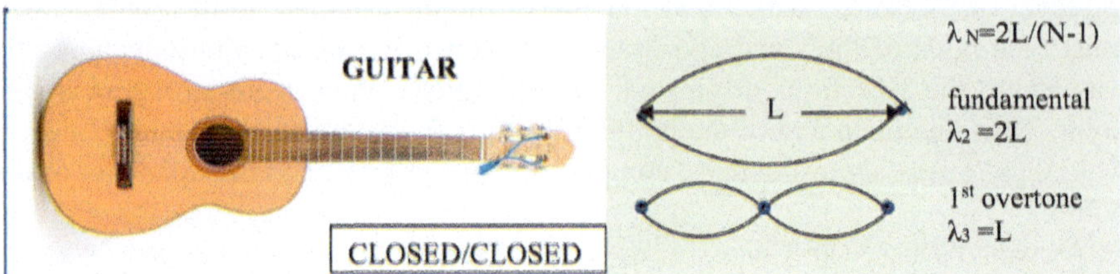

Fig. 12.5 Standing waves for a string instrument

as given in Eq. 12.1. Typical materials for musical strings are nylon, steel, and gut. The displacement in the string travels to the end points where it is reflected, resulting in two waves traveling in opposite directions with the same frequency. This establishes a standing wave at one of the resonant frequencies of the string. The vibrations of the string create pressure waves in the surrounding air that are transmitted to the resonating cavity of the instrument which amplifies sound at all frequencies.

The wavelengths of the standing waves that are supported by one of the strings are determined by the length of the string as shown in Fig. 12.5. Note that these are waves of displacement and therefore have nodes for the closed end boundary conditions. The equation relating the resonant wave wavelength to the length of the string has been modified so in this case n is the number of nodes.

The longest resonant wavelength is called the fundamental tone. As shown in the figure, it has only the two end nodes and its wavelength is $\lambda = 2L$. The next highest resonance has one additional node as shown giving a wavelength of $\lambda = L$. This is known as an overtone. Standing waves with additional nodes produce higher order overtones whose wavelengths are given by the equation $\lambda_n = 2L/(n - 1)$ where n is the number of nodes.

Since $f = v/\lambda$, strings with the same wavelengths can produce different frequency tones if the velocities of the waves in the two strings are different. The velocity of a wave in a string was given by Eq. 11.4 to be

$$v = \sqrt{\frac{T}{\mu}} \tag{11.4}$$

where T is the tension of the string and μ is the mass per unit length of the string. Thus the velocity can be adjusted by changing the tension in the string or by using different types of strings with different diameters or different materials. This creates resonant waves with lower or higher frequencies for strings of the same length.

Example 12.2
Consider a guitar string in which the wave velocity is $v = 143$ m/s and its length is $L = 0.65$ m. Calculate its fundamental and first three overtone frequencies.

Using the expression for the wavelength in terms of the number of nodes gives

$$\lambda_n = 2L/(N-1) = 2(0.65\,\text{m})/(N-1)$$

where $N = 2, 3, 4, 5$. Thus, $\lambda = 1.30$ m, 0.65 m, 0.433 m; and 0.325 m for the wavelengths of the fundamental and first three overtones. Then,

$$f = v/\lambda = (143\,\text{m/s})/\lambda = 110\,\text{Hz},\ 220\,\text{Hz},\ 330\,\text{Hz},\ \text{and}\ 440\,\text{Hz}$$

for the frequencies of the fundamental and first three overtones of this string. Note that the three overtone frequencies are integral multiples of the fundamental frequency so they are called harmonics. Each of these frequencies are assigned to a name of a musical note using the letters A through G. For example, the third harmonic at $f_3 = 4f_0 = 440$ Hz is the A above middle C.

A specific type of string of a given length on a stringed instrument has a fundamental frequency of vibration that is changed by adjusting its tension with the tuning peg. This changes the velocity of sound in the string and allows the various strings on the instrument to be tuned to frequencies with respect to each other and to other instruments. Once a string has been tuned to a specific fundamental frequency, its resonant frequencies can be changed by changing the length of the string. This is done by pressing the string down against the neck of the instrument thus shortening the fundamental wavelength of the string's resonance and creating a higher frequency note. The quality of the tone that is produced by plucking or bowing the string is determined by the mixture of harmonics that are present. This is generally associated with the amplitude of the distortion produced in the string when it is plucked or bowed.

The vibrating string is not the sound that you hear. Its vibrations create pressure waves surrounding it that come to your ears. In general, the coupling between the string and the air would not be strong enough to create a very loud sound. Because of this, the strings on stringed instruments are attached to the body of the instrument which is a resonant cavity. The body of an acoustic (non-electric) instrument is a hollow cavity made of thin wood. The strings are directly attached to it through a device called a bridge. The bridge supports the strings and transmits the vibration to the top surface of the instrument which serves as a sound board. The hollow cavity amplifies the sound as it vibrates and specially designed holes let the sound out at a much higher amplitude than the sound of the original vibrating string. Other stringed instruments like the violin, viola, and cello operate in the same way as described here for a guitar. The piano is a type of stringed instrument where the strings each have a fixed pitch.

Wind instruments like a clarinet and a flute are very close to being cylindrical in shape. Thus, they can be modeled as standing waves in cylindrical tubes. The player pushes air into the tube, and standing wave patterns are formed, depending on the tube length and boundary conditions. First consider a clarinet which has one closed end and one open end. The closed end is the mouthpiece which holds a reed. When the player blows into the mouthpiece it causes the reed to vibrate and sends the vibrating pressure wave down the instrument. At the open end, the wave hits a discontinuity in its transmission (a boundary) which results in a reflected wave and establishes the standing wave pattern.

As shown in Fig. 12.6, the fundamental resonant frequency for the clarinet has a particle displacement node at the mouthpiece and an antinode at the open end. This results in a wavelength of $\lambda = 4L$ where L is the length of the instrument with all the tone holes closed. The first overtone has an additional node in the center as shown resulting of a standing wave with $\lambda = 4L/3$. The equation for the wavelengths of different harmonic modes is $\lambda_n = 4L/(2N - 1)$ where N is the number of nodes. The wavelength and frequency for the fundamental and first three overtones of a clarinet that is 60 cm long when the velocity of sound in air is 345 m/s are

$$\lambda_n = 4L/(2N - 1) = (2.4\,\mathrm{m})/(2N - 1)$$

where $N = 1, 2, 3, 4$. Thus, $\lambda = 2.40$ m, 0.80 m, 0.48 m; and 0.34 m for the wavelengths of the fundamental and first three overtones. Then,

$$f = v/\lambda = (345\,\mathrm{m/s})/\lambda = 144\,\mathrm{Hz},\ 431\,\mathrm{Hz},\ 719\,\mathrm{Hz},\ \text{and}\ 1015\,\mathrm{Hz}$$

are the frequencies for the fundamental and first three overtones of this clarinet with all the tone holes closed. Note that in this case the first overtone is three times the fundamental frequency, the second overtone is five times the fundamental, and the third overtone is seven times the fundamental. Thus, in this case only the odd harmonics are supported as resonant sound waves.

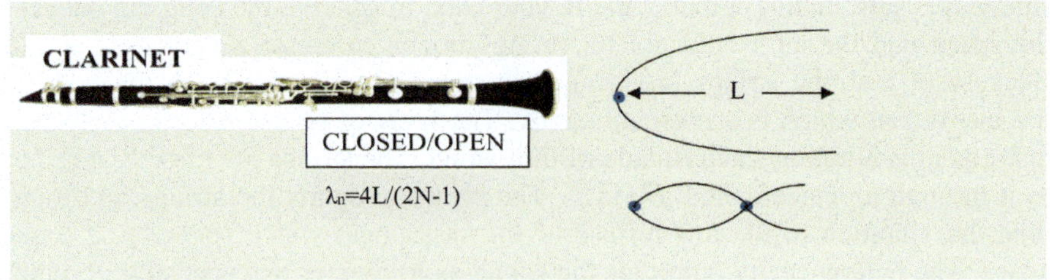

Fig. 12.6 Standing waves for a clarinet

The flute is another wind instrument that can be analyzed as a cylindrical tube. In this case both ends are open and therefore it produces standing waves of particle displacement with antinodes at both ends. The initial pressure wave is created by blowing air across a hole in the head joint called the embouchure hole. Since there are antinodes at both ends of the wave, the fundamental resonance has one node in the center as shown in Fig. 12.7. That means the wavelength of the fundamental is twice the length of the flute. Adding an additional node produces the first overtone at a wavelength equal to the length of the flute. The equation for the wavelengths of the overtone wave with n nodes is $\lambda_n = 2L/N$. The wavelength for the fundamental and first three overtones of a flute that is 66 cm long when the velocity of sound in air is 345 m/s are

$$\lambda_n = 2L/N = 1.32\,\text{m}/N$$

where $N = 1, 2, 3, 4$. Thus, $\lambda = 1.32$ m, 0.66 m, 0.44 m; and 0.33 m for the wavelengths of the fundamental and first three overtones. The frequencies associated with these wavelengths are

$$f = v/\lambda = (345\,\text{m/s})/\lambda = 261\,\text{Hz},\ 522\,\text{Hz},\ 784\,\text{Hz},\ \text{and}\ 1045\,\text{Hz}$$

for the fundamental and first three overtones of this flute with all the tone holes closed. In this case, the overtone frequencies are integral multiples of the fundamental frequency so both odd and even harmonics are supported.

The clarinet and flute are constructed with several sections. The upper joint can be moved in or out to produce a fine adjustment to the length of the instrument, which changes the resonant frequencies. This allows these instruments to be tuned to the same frequency as other instruments for playing together in an ensemble. Opening different tone holes changes the effective length of either instrument resulting in a different resonant frequency and therefore a different musical note. As mentioned previously, different frequencies are assigned different musical note names. For example, a middle C is 261.6 Hz and the A above middle C is 440 Hz. For a group of instruments playing together in an ensemble, it is important that they are all in tune. The 440 Hz A is generally the frequency they all tune to.

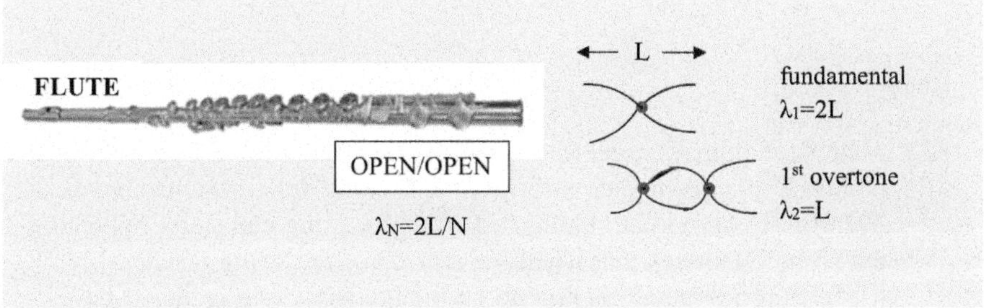

Fig. 12.7 Standing waves for a flute

Other woodwind instruments such as oboe, bassoon and saxophone have the same boundary conditions as the clarinet. However, their operation cannot be modeled as a cylindrical tube. Their shape is conical, so the analysis of their resonances is more complicated. The sound wave traveling down a conical tube loses intensity as the tube diameter increases and this changes the shape of the resonances. The result is that these instruments can support both even and odd harmonics. Organ pipes have the same boundary conditions as the clarinet and thus support only odd harmonics.

Example 12.3
If an organ pipe is closed on one end and is 33.0 cm long, what is its fundamental frequency at 24.0 °C?

Since the pipe is closed on one end and open on the other, it can be modeled like a clarinet:

$$\lambda_n = 4L/(2N - 1)$$

For the fundamental there is one node so $N = 1$

$$\lambda_1 = 4L = 4(0.330\,\text{m}) = 1.32\,\text{m}$$

We can determine the frequency if we know the speed. For that, we must use Eq. 12.2:

$$v = (331.5 + 0.606T)\text{m/s} = [3315 + 0.606(24.0)]\text{m/s} = 346.0\,\text{m/s}$$

Now we can determine the frequency:

$$f = v/\lambda = (346.0\,\text{m/s})/(1.32\,\text{m}) = 262\,\text{Hz}$$

That's the frequency for middle "C."

Now try an problem.

Student

12.8 A physics student stretches a string ($\mu = 0.0150\,\text{kg/m}$) between two points. The student then strikes a "G" tuning fork, which vibrates at 392 Hz. He brings the tuning fork near the string and starts tightening the string. He notes that when the tension of the string reaches 144 N, it begins to vibrate at its second harmonic. How long is the string?

12.9 A physics student makes two tubes: one that is closed on one end and one that is open. What is the ratio of the lengths of the tubes if they produce the same pitch when air oscillates inside?

Why do harmonics matter? When a person blows into a wind instrument, several of the harmonics are set up as standing waves. Since each harmonic has its own distinct wavelength, each harmonic has its own frequency (pitch) as well. Thus, when a person blows into a flute or clarinet, several pitches are produced. However, the harmonic with the largest amplitude (and thus the loudest volume) is the fundamental (first) harmonic. The other harmonics are there, each producing its own pitch, but because they mix so well with the sound of the first harmonic, and because they are softer than the first harmonic, they cannot be distinguished as separate notes. Thus, the harmonics "add" to the fundamental to make what is generally called the **timbre** (tam'ber) of the instrument. This is what typically separates the sound of one instrument from the sound of another. If each instrument always played only the fundamental harmonic of a given note, they would all sound the same. This is true of all instruments, including your voice. When a guitar string is plucked, for example, several harmonics are set up on the guitar string. Thus, several different sound wave frequencies are formed. Once again, however, the fundamental harmonic is the loudest, so that's the pitch you hear.

The timbre of a note depends on the number and amplitude of the harmonics that are mixed with the fundamental frequency in the sound that you hear. This is generally controlled by the way the player creates the initial vibration through blowing into the instrument or plucking the string. However, the same note played on the flute and the clarinet will sound quite different since the harmonics supported by the two instruments are different. The more harmonics that are present the more the simple sinewave form of the sound wave is distorted thus producing a more complex sound. This is shown in Fig. 12.8. The composite waveform is a mixture of the fundamental and several of the lower order harmonics mixed together with different amplitudes. Note that the envelope of the composite waveform is essentially the same as the fundamental waveform but with a significant amount of structure. The structure depends on the number and amplitude of the harmonics that are present. The multiple frequencies in the structured form of the soundwave allows us to hear a much richer sound than the tone of the single fundamental frequency.

Musical notes are often played together simultaneously to produce a chord or tone cluster. The frequencies of notes heard at the same time produce a wide variety of timbres, or sound colors, ranging from harmonious to discordant. These can be combined to evoke a wide range of different emotions, thoughts and ideas.

An interesting phenomenon occurs when two sound waves that have frequencies very close to each other are superimposed. Their combined wave pattern alternates in time between constructive and destructive interference as shown in Fig. 12.9. This produces a periodic change in the amplitude of the sound they produce and thus the volume you hear. This effect is called a beat and the beat frequency is

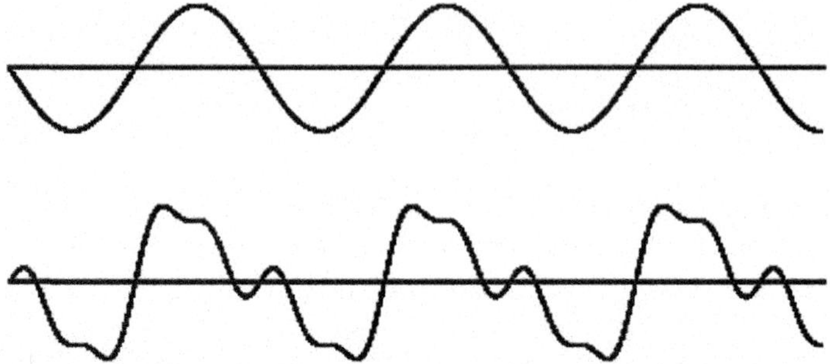

Fig. 12.8 Waveforms of a fundamental note and a composite with harmonics

given by the difference in the two interfering frequencies,

$$f_{\text{beat}} = |f_a - f_b|. \tag{12.8}$$

Eliminating beat frequencies is another reason it is important for all instruments in an ensemble to be tuned to the exact same frequency.

Here is one final problem to try.

Student

12.10 Your flute is tuned to 440A but when you play with the flutist next to you there is a beat frequency with a period of about a third of a second. What frequency is the other flute tuned to?

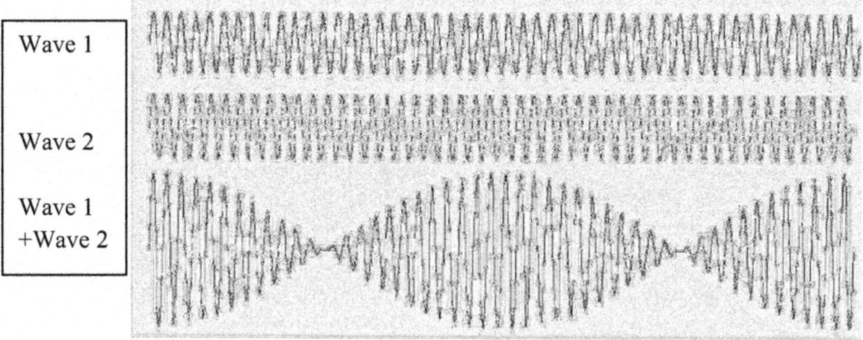

Fig. 12.9 Beat frequency

Next Level

Reverberation

Music is often heard in a performance venue and the quality of the sound you hear can be significantly affected by the architectural acoustics of the venue itself. The intensity of the sound emitted on stage decreases with the $1/r^2$ law as it travels throughout the venue. At the same time, it is reflected, diffused, and absorbed by the material making up the venue. The sound that reaches you in the audience is a combination of direct and reflected sound. This affects the quality of the sound you hear. An example of an important architectural acoustic characteristics is reverberation time which is essentially the amount of time a sound can be heard in a venue after it is created and before it dies out to an amplitude less than our threshold for hearing. This depends on the size, shape and type of material of the venue. It is defined as the amount of time it takes sounds at a specific frequency to decay in intensity by 60 dB. This is a somewhat arbitrary amount but was chosen because a typical orchestra crescendo is around 100 dB and a typical background noise in an auditorium is 40 dB. If sound decays from 100 to 40 dB in 2.0 s, the reverberation time is $T_{60} = 2.0$ s.

The equation to calculate reverberation time is called Sabine's equation,

$$T_{60} = 0.16\,\mathrm{V}/(aA)$$

V is the total volume of the venue, A is the surface area of the room, and a is the sound absorption coefficient expressed as the ratio of the percentage of sound absorbed to percentage of sound reflected when a sound wave hits the material. The denominator is actually the sum of the products of the areas and absorption coefficients of each different component of the room (i.e. floor, ceiling, walls, chairs, etc.). The numerical factor was derived using the speed of sound at 20 °C and has units of s/m. The absorption coefficient is given is units of Sabine's defined as the percent of sound absorbed per percent reflected when a sound wave hits the material. One Sabine is 100% absorption of 1 m^2 of material.

If the reverberation time is too short, the sound of music is dry but the sound of speech is very clear. If the reverberation is too long, the sound of music is rich but speech is difficult to understand. General use venues usually have reverberation times between 1.5 and 2.5 s. A reverberation of 1.0 s is better for lectures and a reverberation time of 2.0 is better for orchestras.

Many venues have developed ways to alter their reverberation times so both lectures and concerts can be supported. One important variable is the presence of an audience because bodies absorb sound. This differs for each frequency and requires testing of each type of material used in the construction at each location in the venue. Architectural acoustics remains a very complicated field.

We have been treating music as a science. However, bringing creative structure and design to combinations of acoustical sounds is an art form. Understanding the scientific basis of acoustics is helpful to fully appreciate this art. The discussion above is meant to provide a few examples of the sound waves that we hear.

Special Topic

Music is an integral part of our culture that can evoke a broad spectrum of different emotions. But music is more important than just a form of enhancement to our quality of life. Music is now being used for therapeutic treatment in many areas of medicine backed by clinical, evidence-based research. Here are some examples where music therapy has been found to be effective.

- It has been used to lower blood pressure, reduce heart rate and relax muscle tension. Listening to music has been found to reduce the anxiety associated with chemotherapy and radiotherapy cancer treatments. It can also decrease nausea and vomiting for patients receiving these treatments.
- Music therapy decreases pain perception which reduces the amount of pain medication needed and helps relieve depression in pain patients.
- For mental health, this form of therapy reduces the negative effects of stress, including emotional and behavioral problems. It can also improve the quality of life for people with dementia by helping to evoke memories, reduce agitation, assist communication, and improve physical coordination.
- Music therapy can help people who are recovering from a stroke or traumatic brain injury that has damaged their left-brain region responsible for speech. Because singing ability originates in the right side of the brain, people can work around the injury to the left side of their brain by first singing their thoughts and then gradually dropping the melody. A great example of this is former U.S. Representative Gabby Giffords who used this technique to enable her to testify before a Congressional committee two years after a gunshot wound to her brain destroyed her ability to speak.

The physics of acoustics is the foundation for all of these important applications.

12.5 Sonar, Ultrasound, and Infrasound

The technique of sound navigation and ranging is called sonar. This represents an important application of acoustic waves for both human beings and different types of animals. Many times ultrasound or infrasound waves are used instead of sound waves within the human hearing range for sonar applications.

One important use of sonar is "seeing" things underwater. This is shown schematically in Fig. 12.10. Ultrasound waves are sent out from a source in a ship and when they reflect off of an object their acoustic echo comes back to the ship and is detected. Knowing the speed of sound in seawater, the distance of the object can be determined from the time it takes the echo to return. For example, the speed of sound waves in sea water is 1450 m/s. Thus if it takes 2 s for a sound wave to leave a source, bounce off an object, and return to a detector, the distance of the object from the source can be calculated to be $d = vt = (1450\ \text{m/s}) \times (1.0\ \text{s}) = 1450\ \text{m}$. Ultrasound frequencies are generally used for underwater sonar applications because their high frequencies and small wavelengths generally allow them to travel farther underwater than lower frequency sound waves. However, the speed of sound in sea water depends on the temperature, pressure, and salinity of the water. Sonar is a standard tool for the navy in detecting submarines, mines, and other underwater obstructions. It is also used by scientists to map the ocean floor, by sailors to detect reefs, and by fishermen to detect schools of fish.

Another application of ultrasound is in medicine where it is used for both imaging and treatment. Ultrasound imaging can accurately detect shapes inside the human body without the use of dangerous radiation. One important example

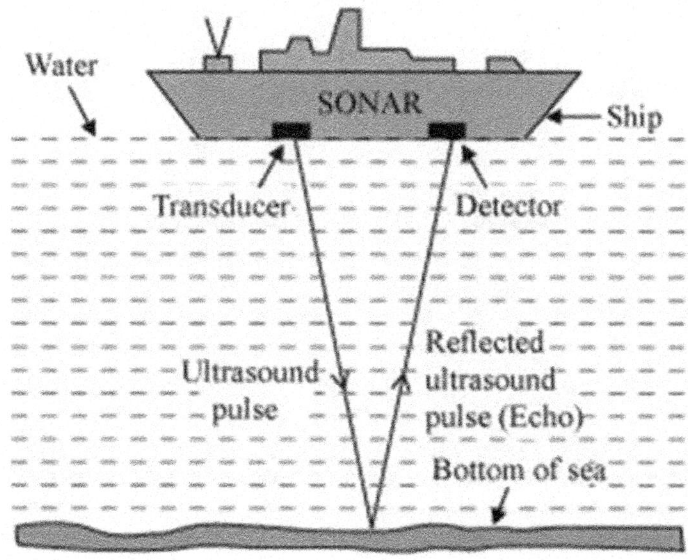

Fig. 12.10 Sonar

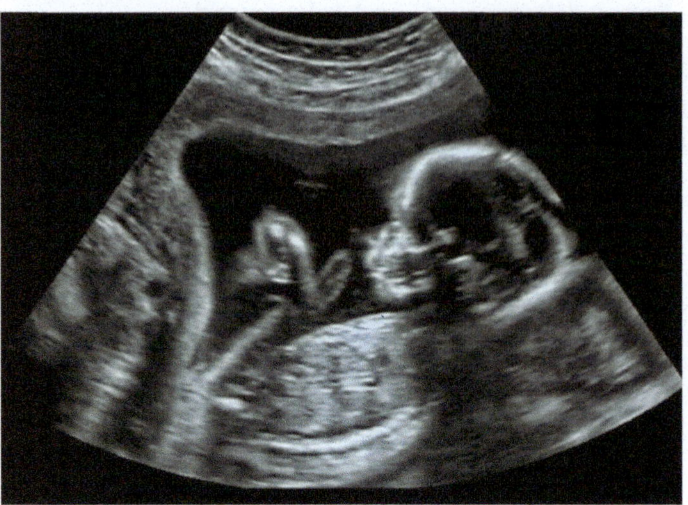

Fig. 12.11 Ultrasound image

of this is in obstetrics where sonograms can show the details of the fetus in the womb as shown in Fig. 12.11. This type of imaging is also useful for looking for tumors, blood clots and other obstructions. Doppler ultrasound is useful for detecting moving targets such as blood flow and heart movement. An echocardiogram creates an image of blood flow and movement as the heart beats and detects any abnormalities and leaking.

As well as providing images, acoustics is also useful in treating some medical problems. An example of this is lithotripsy to break up kidney stones that are too large to pass normally. Acoustic shock waves at about 25 kHz and 98 dB are directed onto a large stone and the vibration that this produces causes the stone to break up into small fragments that can pass.

Many animals can hear sound waves outside of the human frequency range. For example, the upper end of a dogs hearing range is somewhere between 47 and 65 kHz, much higher than the 20 kHz cutoff for human hearing. Some animals' also use acoustic sonar for navigating and communicating. Bats use this echolocation technique to "see" objects as they are flying at night. To do this, bats produce sounds between 9 and 200 kHz. This is mostly in the ultrasound region above the limit for human hearing. Bats' acoustic system is so sensitive they use the Doppler frequency shift to detect motion of objects. This is useful in detecting flying insects for them to eat.

Another example of an animal use of acoustics is the voice of the whale. Whales use sound for echolocation and for communicating with other whales. The sound range they produce is from 30 Hz to 8 kHz. Most of this is in the infrasound region below the range of human hearing. The interesting thing about whale acoustics is

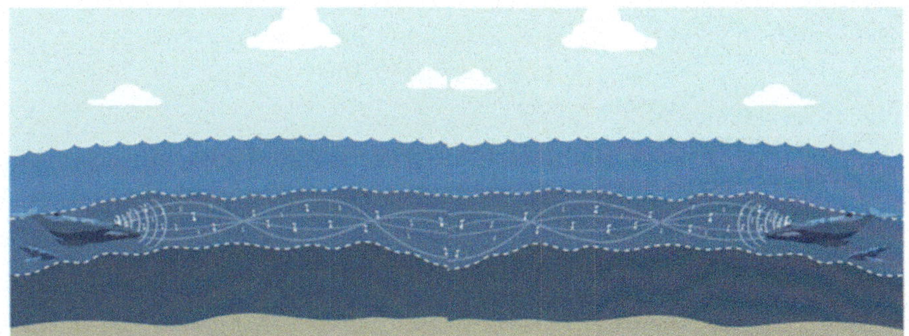

Fig. 12.12 Voice of a whale

that they can communicate with other whales that are thousands of miles away. They do this with a technique called SOFAR (sound fixing and ranging) which is illustrated in Fig. 12.12. The ocean has layers where the temperature, pressure, and salinity are different from other layers. As the depth of the ocean increases the temperature of the water decreases causing the velocity of sound to decrease. However, at the same time the pressure of the water increases causing the speed of sound to increase. At a specific depth in the ocean there is a SOFAR channel layer between two other layers. In the SOFAR layer the speed of sound is a minimum and the channel acts like a waveguide reflecting low frequency sound waves back into the channel. This allows them to travel long distances with little amplitude attenuation. Whales use the SOFAR channel for their long distance communication.

Summing Up

With your knowledge of acoustics, you can now explain the Doppler effect, sonic booms, and musical notes. You also know how bats and whales communicate. Hopefully this discussion of acoustics has stimulated your interest to learn more about this very complex and important area of physics.

Solutions to the Problems

12.1

$v_{sound}(T_1)$

$v_{sound}(T_2)$

Knowns: $T_1 = 30\ ^\circ\text{C}$; $T_2 = 10\ ^\circ\text{C}$

Unknowns: Δv_{sound}

This is a simple application of Eq. 12.2;

$$v = (331.5 + 0.606T)\,\mathrm{m/s}$$

$$v_1 = (331.5 + (0.606)(30))\mathrm{m/s} = 349.68\,\mathrm{m/s}$$
$$v_2 = (331.5 + (0.606)(10))\mathrm{m/s} = 337.56\,\mathrm{m/s}$$
$$\Delta v_{\mathrm{sound}} = 12.12\,\mathrm{m/s}$$

12.2

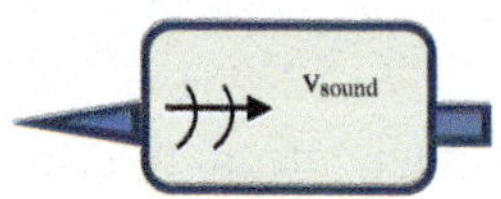

Knowns: $\kappa = 1000\ \mathrm{N/mm^2}$; $\rho = 2500\ \mathrm{kg/m^3}$
Unknowns: v_{sound}

This is a straight forward application of Eq. 12.2:

$$v = \sqrt{\frac{\kappa}{\rho}}$$

$$v_{\mathrm{sound}} = \sqrt{\frac{1000\,\mathrm{N/mm^2}\ \left(10^6\,\mathrm{mm^2/m^2}\right)}{2500\,\mathrm{kg/m^3}}} = 40\,\mathrm{m/s}$$

To make the units come out right, we must change millimeters to meters and use the equivalent units for Newtons of kg m/s^2.

12.3

Knowns: $\beta_1 = 30$ dB; $\beta_2 = 100$ dB
Unknowns: ΔI

Equation 12.4 relates the loudness in decibels to the intensity.

$$\beta = 10\log(I/I_o)\beta = 10\log(I/I_o)$$

Whenever the intensity is changed by a power of 10, $\beta/10$ changes by one unit. Going from 30 to 100 dB is a change of 70 dB. Thus $\beta/10$ changes by 7 units which means the amplifier increased the intensity of the sound wave by a factor of 10^7.

12.4

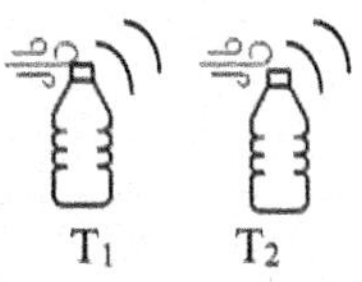

Knowns: $T_1 = 90\ °\mathrm{C}$; $T_2 = 5\ °\mathrm{C}$

Unknowns: $f_1; f_2$

The pitches would not be the same. Since the bottles are identical, the *wavelength* is the same in each sound wave. However, the *speed* of a sound wave depends on temperature. The sound will travel more slowly in the cold air. The wavelength and speed of a wave are related to its frequency by Eq. 11.2,

$$f = v/\lambda$$

you see that a lower speed will mean a lower frequency and thus a lower pitch. As a result, the cold bottle will produce a lower pitch than will the warm bottle.

12.5

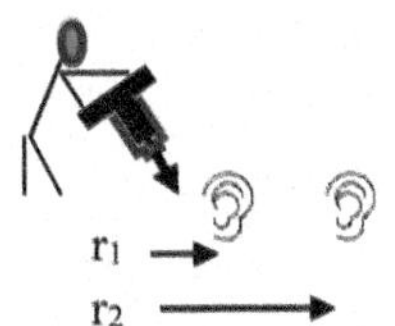

Knowns: $\beta_1 = 80$ dB; $r_2 = 2r_1$

Unknowns: β_2

Since you move twice the distance away and sound intensity falls off as $1/r^2$ the new intensity will be ¼ the original intensity. Thus $I_2 = I_1/4$. Then the difference in loudness is

$$\beta_1 - \beta_2 = 10\left[\log I_1/I_o - \log I_2/I_o\right] = 10 \log I_1/I_2 = 10 \log 4 = 6.02$$

$$\beta_2 = 80\,\mathrm{dB} - 6.02 = 73.98\,\mathrm{dB}$$

Here we have used the identity log A − log B = log A/B.

12.6

Knowns: $v_{\text{source}} = 45.0$ m/s; $v_{\text{observer}} = 25.0$ m/s; $f_{\text{true}} =$ 411 Hz; $T = 25$ °C

Unknowns: f_{observed}

The equation for the Doppler shift is Eq. 12.5

$$f_{\text{observed}} = \left[\frac{v_{\text{sound}} \pm v_{\text{observer}}}{v_{\text{sound}} \mp v_{\text{source}}}\right] f_{\text{true}}$$

In this problem, both the observer and the source are moving. The observer is moving towards the source. Thus, his motion will increase the frequency, so we must add his motion in the numerator of the equation. The source is moving away from the observer, which will decrease the frequency. Thus, we must add the speed of the source in the denominator of the equation. That gives us:

$$f_{\text{observed}} = \left[\frac{v_{\text{sound}} + v_{\text{observer}}}{v_{\text{sound}} + v_{\text{source}}}\right] f_{\text{true}}$$

To use this equation, however, we need to know the speed of sound which we get from Eq. 12.2

$$v = (331.5 + 0.606T)\text{m/s} = (331.5 + 0.606[25])\text{m/s} = 346.7\,\text{m/s}$$

Now we can calculate the observed frequency:

$$f_{\text{observed}} = \left[\frac{346.7\,\text{m/s} + 25.0\,\text{m/s}}{346.7\,\text{m/s} + 45.0\,\text{m/s}}\right](411\,\text{Hz}) = 3.90 \times 10^2\,\text{Hz}$$

Since the observer and source are moving away from each other, the net effect is to reduce the frequency of the horn. Note that if the train and car were moving at the same speed, the observed frequency would be the same as the true frequency, since the source and observer would not be moving relative to one another.

12.7

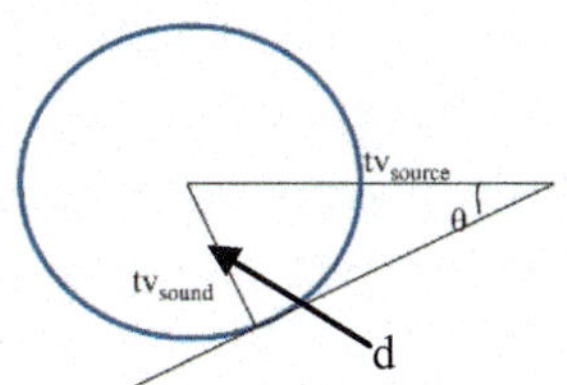

Knowns: v_{source} = Mach 1.5 = 514.4 m/s; t = 3 s

Unknowns: d

From Eqs. 12.6 and 12.7,

$$\theta = v_{\text{sound}}/v_{\text{source}} \text{ and Mach Number} = v_{\text{source}}/v_{\text{sound}} = 1.5$$

$$\theta = \sin^{-1}(1/1.5) = 41.81°$$

$$d = tv_{\text{source}} \sin\theta = (3\,\text{s})(514.5\,\text{m/s})\sin 41.81° = 1029\,\text{m}$$

12.8

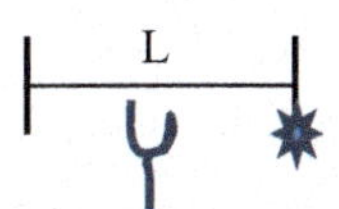

Knowns: f_2 = 392 Hz; μ = 0.0150 kg/m; T = 144 N

Unknowns: L

We know that the frequency of this string's second harmonic is 392 Hz, because that's the frequency at which sound waves are hitting the string, and that frequency resonates the second harmonic. Since we know the tension and linear mass density, we can get the speed of the waves in the string with Eq. 11.4

$$v = \sqrt{\frac{T}{\mu}} = \sqrt{\frac{144\,\text{N}}{0.0150\,\text{kg/m}}} = 98.0\,\text{m/s}$$

Now we can determine the wavelength:

$$f = v/\lambda$$

$$\lambda = v/f = (98.0\,\text{m/s})/(392\,\text{Hz}) = 0.250\,\text{m}$$

Now we can determine the length of the string. Since this is the second harmonic there are 3 nodes. The wavelength is

$$\lambda = 2L/(N-1) = 2L/2 = L = 0.250\,\text{m}$$

So the length of this string is 0.25 m.

12.9

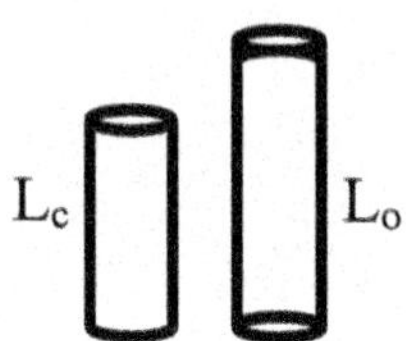

Knowns: $f_c = f_o$
Unknowns: L_c/L_o

For a tube that is closed on one end, the possible wavelengths are:

$$\lambda_c = 4L_c/(2N-1)$$

For a tube open on both ends, the possible wavelengths are:

$$\lambda_o = 2L_o/N$$

For both cases, n is the number of nodes.

Since most of what you hear is the first harmonic, we can set $N = 1$. Also, since the air is the same temperature in both tubes, the same pitch (frequency) also means the same wavelength.

Thus, if we divide these equations:

$$\lambda_c/\lambda_o = \frac{4L_c/(2n-1)}{2L_o/n} = 1$$

$$2L_c/L_o = 1$$

$$L_o = 2L_{cf}$$

The open tube, then, is twice as long as the closed tube.

12.10

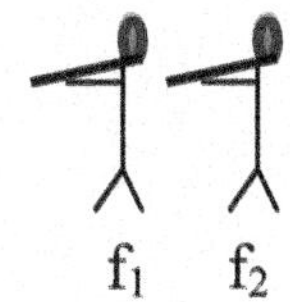

Knowns: $f_1 = 440$ Hz; $T_B = (1/3)$s

Unknowns: f_2

The period of the beat frequency is 0.33 s so the beat frequency is

$$f_B = 1/T_B = 1/(0.33\,\text{s}) = 3\,\text{Hz}.$$

The beat frequency from Eq. 12.8 is

$$f_{\text{beat}} = |f_1 - f_2| = 3\,\text{Hz}$$

Since f_1 is 440 Hz, f_2 must be either 443 Hz or 437 Hz.

Study Guide for This Chapter

1. Imagine a sound wave with a particular wavelength and frequency that is traveling through the air. If it strikes a brick wall, compare the frequency and wavelength of the sound wave that is reflected to that of the original wave. Compare the phase angle of the two waves.
2. Imagine the situation given in question #1, and now consider the portion of the wave that travels into the wall. Compare the frequency and wavelength of the sound wave in the wall to what it was when it was traveling through the air. Compare the phase angle of the two waves as well.
3. When you breathe in a lot of helium (by sucking on a helium balloon, for example), the pitch of your voice goes up substantially. What can you say about the speed of sound in helium as opposed to air?
4. Suppose a singer wants to break a glass using her voice. How could she determine the pitch at which she needs to sing in order to break the glass?
5. A young girl is running away from a mean boy that is chasing her with a frog. As is often the case, the young girl is screaming at a constant pitch and loudness. As they run towards a wall, the boy hears the loudness of the girl's scream vary at a regular frequency. However, the girl is not varying the loudness of her scream. What is happening?
6. Why will plastic strings and metal strings make different sounds on a guitar?
7. Why do many musicians in rock bands develop hearing problems?
8. How are RADAR and SONAR alike and how are they different?
9. In tuning your flute, you find you are slightly flat compared to the rest of the orchestra. Should you move your head joint in or out compared to the body of your flute?

10. A man hears music from a car that is about 100 m away from him. He estimates that he is hearing the music at a loudness of about 40 dB. If someone were unfortunate enough to be standing only 1 m from the car, how loud would the music be?
11. A bat uses sonar to navigate. Suppose the bat emits a "chirp" and 0.55 s later hears the echo from that chirp. How far away is the obstacle that caused the sound wave to bounce back? Assume the temperature is 20.0 °C.
12. A pipe that is open at both ends is 25.0 cm long. What is the fundamental frequency of the pipe when the air is at a temperature of 25.0 °C?
13. Suppose you took a 12.8 cm pipe that is closed on one end and put it next to the pipe in problem #3. If you blew on both so that the air (25.0 °C) began vibrating at the pipes' fundamental frequencies, you would hear beats. At what frequency would you hear the beats?
14. A policeman is in a school zone with a speed limit of 25 mph. A car is approaching him honking the horn which he hears at a frequency of 643 Hz. He knows that model car has a horn that emits at 600 Hz. How much is the car exceeding the speed limit? (The speed of sound is 345 m/s; there are 1609.34 m/mile.)
15. A person standing on a platform at a train station sees two trains approaching the station from opposite directions going different speeds. They both blow their whistles at exactly the same frequency of 325 Hz and the person hears a beat frequency 6 Hz. If one train is going 45 mph how fast is the other train going? (Sound travels at 750 mph.)
16. A boat operates a SONAR system at 38 kHz in an area where the speed of sound in air is 1100 ft/s and the speed of sound in water is 4200 ft/s. What are the frequencies and wavelengths of the SONAR signal in air and in water? If it gets an echo return in 0.5 s from the bottom of the ocean, how deep is the ocean at that point?
17. Two loudspeakers on a stage are 4 m apart. They are both emitting sound waves at 750 Hz. If the speed of sound is 350 m/s, find the positions of the first three intensity minima on a line going between the speakers. Consider the same two speakers. If one has an intensity of 0.6 W/m^2 and the other 0.3 W/m^2, what is the total intensity at the first intensity minimum you found if it is closest to the lower intensity speaker?
18. As you are standing on a corner, a car goes by you at a constant speed honking its horn. If you hear the pitch change from 280 to 260 Hz, how fast is the car going? (Speed of sound is 350 m/s.)

Next Level

19. You purchase an activities venue that is 20 m high, 100 m long, and 50 m wide all made of material that reflects twice the amount of sound that it absorbs. What is its reverberation time and is it better used for lectures or orchestra concerts?

20. If you re-model the venue in problem 19 so that the floor absorbs 0.8 as much sound as it reflects, the ceiling absorbs half as much sound as it reflects, and the walls absorb 0.75 as much sound as they reflect, what is the new reverberation time and is it better for lectures or concerts?

Properties of Matter I: Thermodynamics

13

Chapter Summary

All of Creation is made of matter. Therefore, to understand the creation we live in we must understand the physical characteristics of the matter of which it is made. This chapter and the following chapter describe some of the properties of matter from a macroscopic point of view while the final two chapters look at the microscopic properties of matter. There is still much we don't understand about matter but we are learning more every day. Some of the themes of the chapter are represented in Fig. 13.1.

Main Concepts in This Chapter

- Basic Thermodynamics
- Heat Flow
- Laws of Thermodynamics
- State Variables
- P–V Diagrams

13.1 Introduction

Matter is anything that occupies space and has mass. The four states of matter are solid, liquid, gas, and plasma. A sample of matter can change back and forth between these different states by being exposed to changes in temperature and pressure. All matter is made of many atoms. In the final two chapters of this book

R. C. Powell, *Physics*, Undergraduate Lecture Notes in Physics,
https://doi.org/10.1007/978-3-032-09361-5_13

we will study the properties of individual atoms, but in this chapter we are dealing with the properties of many atoms when they act together as a piece of matter.

The main difference between the states of matter is how tightly the atoms are bonded together. In a solid the atoms are so tightly bonded that it has a rigid shape no matter what container it is in. The atoms can form a lattice with a geometrical shape associated with it, in which case it is a crystal. On the other hand, the atoms may be distributed randomly in which case the solid is a glass. The atoms vibrate about their equilibrium positions but cannot move past their neighboring atoms.

Liquids and gases are called fluids because they flow. That is, they are bound loosely enough that an atom can move past a neighboring atom. Fluids take the shape of the container that holds them. Liquids maintain a constant volume when their shape changes. Gases expand to fill the volume of their container. The separation between the atoms (or molecules) that make up a gas is far greater than it is in a liquid or solid. A plasma is a special type of gas made up of atoms that have had their electrons stripped off to become ions, along with the free electrons. Thus plasmas are made up an equal number of positive and negative electrical charges. The electromagnetic properties of a plasma are much different than those of a normal gas, so it is considered to be a fourth state of matter. Examples of where plasmas are found are lightning strikes and fluorescent light bulbs.

These different configurations of matter allow solids, liquids, gases and plasmas to have very different thermal, mechanical, and electrical properties as discussed in the following sections. All of these types of properties are important for each of the states of matter. However, thermal properties have most significance in gases and liquids, mechanical properties are most significant for liquids and solids, and electrical properties are most significant for solids and plasmas. The properties of solids are so important they are treated later in a chapter of their own.

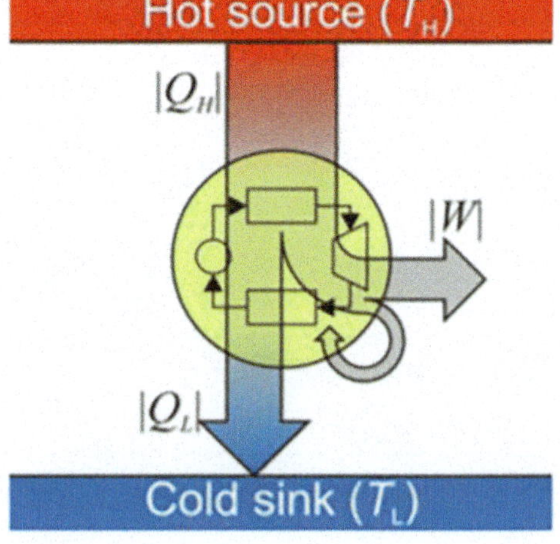

Fig. 13.1 Heat engine. Credit: Brundl 15, Creative Commons Attribution-Share Alike 4.0

13.2 Basic Thermodynamics

The physics of thermal properties of matter is called thermodynamics. Three important aspects of thermodynamics are thermal energy, temperature, and heat. All of these are associated with the motion of the atoms or molecules in the sample. Thus, they reflect an internal kinetic energy of the sample. Thermal energy is the total amount of this type of energy in the sample whereas temperature is the average kinetic energy of one of its atoms. Heat is the flow of this type of energy from one area to another. The characteristics of this type of energy depend on the state of matter of the sample. In gases the particles are free to move around and collide with each other and the walls of the container. In solids the atoms vibrate around equilibrium positions like masses on a set of springs. In a liquid the atoms or molecules can flow over one another without escaping from the surface. The vibrational and linear motion of the particles contributes to the thermal energy. In some molecular systems, the internal vibrations and rotations of the molecules can also contribute to the thermal energy (Fig. 13.2).

Since it is impossible to describe the individual motion of each of the millions of atoms or molecules in the sample, we can use two possible approaches. The first is to consider the properties of the total ensemble of particles while the second is to consider the average property of a particle making up the sample. These are quite different approaches. The first depends on the size of the sample while the second is independent of size. The total kinetic energy of all the particles in the sample is called its thermal energy. The average kinetic energy of an atom or molecule in the sample is called its temperature. Both of these are important concepts.

When you touch a piece of metal that is glowing red, it burns you. This is because the atoms in the metal are vibrating back and forth at a very fast rate. This motion is indicative of a large amount of kinetic energy in the metal which is its thermal energy. When you touch the metal, the atoms begin colliding with your skin. The atoms of your skin have much less thermal energy and are moving much slower. The collisions of the metal atoms with your skin atomstransferring a lot of

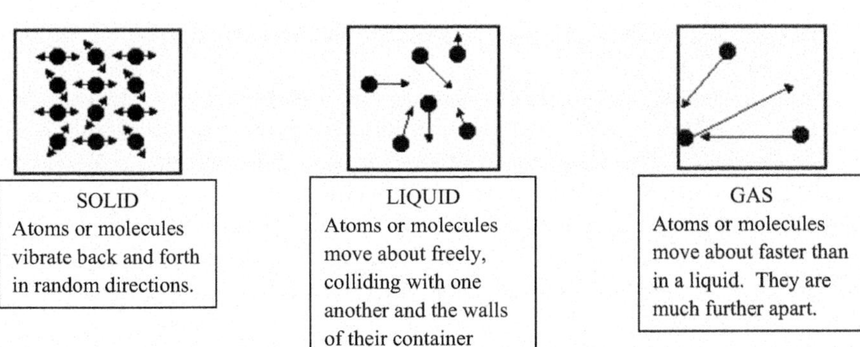

Fig. 13.2 Atomic motion in matter

thermal energy from the metal to your skin. Your cells cannot handle all of that energy, so they die. The result is a burn. The thermal energy in the glowing metal is not what burned you, however. What burned you was the fact that the energy was transferred to you from the metal. That's what we call heat. Heat is energy that is transferred as a consequence of temperature differences. When two objects with different thermal energies come into contact with each other, the collisions of the atoms of the two objects cause thermal energy to flow from the hot object to the cold object. The energy that flows from one object to another object is heat. Heat will continue to flow until the two objects reach an equilibrium condition where they have the same temperature. That means the average kinetic energy of the moving atoms is the same in both objects. In the example given above, the original temperature of the metal is very high (it is hot) while the original temperature of your skin is low (it is cold). Note that temperature is not a measure of heat or total internal energy.

A thermometer is used to measure temperature. It makes use of the fact that substances tend to expand as they get warmer. A typical thermometer consists of a substance (usually mercury or alcohol) that is contained in a column of glass. As the substance gets warmer, it begins to expand. Since the only way it can expand is in the upward direction, it begins to move up the glass column. The height that it moves up the column is directly related to the temperature.

There are three different temperature scales in common use. The temperature unit that you are probably most familiar with is Fahrenheit (abbreviated as °F). Although this is a very common temperature unit, it is not used by physicists. Instead, physicists usually use one of two temperature units: Celsius (°C) or Kelvin (K). These units all rely on their definition. We can define any temperature scale that we want. The only thing we know for sure is that as the temperature increases, the average kinetic energy of the moving atoms of the object increases.

The boiling and freezing of water are typically used as reference points to define a temperature scale because both processes are easy to observe in the course of normal activities. In the Fahrenheit scale, water freezes at 32.0 °F and boils (under normal atmospheric pressure) at 212.0 °F. In the Celsius scale, water freezes at 0.0 °C and boils (under normal atmospheric pressure) at 100.0 °C. Finally, in the Kelvin scale, water freezes at 273.15 K and boils (under normal atmospheric pressure) at 373.15 K.

In the Celsius and Fahrenheit scales the two reference points are exact and therefore contain an infinite number of significant figures. However, this is not the case for the Kelvin temperature scale. The Kelvin temperature scale is considered an absolute temperature scale, because we cannot reach a temperature of 0 K. We can get arbitrarily close to it, but we can never reach it. For example, in the laboratory, a temperature of 0.00000001 K has been obtained. Since the absolute temperature scale was defined based on experiment, it has experimental error associated with the measurement. Thus, the reference points given above for the Kelvin scale do not have an infinite number of significant figures.

The equations to use to convert between these temperature scales are:

$$°C = (5/9)(°F - 32.00) \tag{13.1}$$

$$K =° C + 273.25 \tag{13.2}$$

For example, the Fahrenheit equivalent of 0.00 K is

$$°C = K - 273.25 = -273.15$$

$$°F = (9/5)\,°C + 32.00 = -459.67\,°F$$

13.3 Heat Flow

Heat is thermal energy that is transferred from a hot area to a cold area. This can occur through conduction, convection or radiation. It is designated by Q and has units of Joules. When heat is flowing out of an area Q is negative. When heat is flowing into an area Q is positive.

The transfer of energy can have several effects on a physical system. For example, the temperatures of the different parts of the system may change. The study of the relationship between heat and temperature is often called calorimetry. This is because another unit for energy is the calorie (abbreviated as cal). This unit, common in chemistry, is defined as the amount of energy required to warm up one gram of water one degree Celsius. Please note that this unit is different from the food calorie (usually abbreviated as Cal). It takes 1000 cals to make one Cal.

If energy is transferred from one object to another object, one object must lose energy while the other object gains energy. Since temperature is a measure of the thermal energy in a substance, then it is possible that when energy is transferred, the temperature of both objects will change. If that happens, the amount of energy transferred can be related to the change in temperature by the following equation:

$$Q = mC\Delta T \tag{13.3}$$

In this equation, Q is the heat (the amount of energy transferred), m is the mass of the object that is gaining or losing energy, C is the specific heat capacity (sometimes referred to as the specific heat) of the object, and $\Delta T = (T_{\text{final}} - \text{T}_{\text{initial}})$ is the change in temperature of the substance. Notice, then, that this equation clearly shows that temperature is not a measure of heat. Since heat is energy that is being transferred, the change in temperature can be related to heat, but temperature is definitely not a measure of heat.

Table 13.1 Specific heat capacities of common substances

Substance	Specific heatcapacity (J/g °C)
Copper	0.382
Iron	0.452
Aluminum	0.900
Glass	0.837
Liquid water	4.19

The definition of specific heat capacity is:

Specific heat capacity: The amount of heat necessary to raise the temperature of 1.0 g of a substance by 1.0 °C.

In other words, specific heat capacity tells how easy it is to warm up a substance. For example, if we put a metal plate on the burner of a stove and we put a wooden plate on another burner of the stove, which one would get hot faster? As long as the wood did not catch on fire, the metal plate would get much hotter much faster. This is because the metal plate has a lower specific heat capacity than the wood. Since metals have low heat capacities, then even a small amount of heat changes the temperature significantly. On the other hand, wood has a relatively large specific heat capacity. This means that it takes a lot of heat to make even a small change in the temperature of the wood.

Each substance has its own, unique specific heat capacity. Table 13.1 lists some common substances and their heat capacities.

Note the units of specific heat capacity are energy divided by grams and by °C. If you look at Eq. 13.3, you will see that these units allow the equation to work out so that Q is in Joules. This makes sense, since the Joule is an energy unit. Also, since Eq. 13.3 deals with a temperature difference, and the size of a K is the same size as a °C, either K or °C could be used in the units of C.

Based on the numbers in the table, if we add the same amount of energy to equal masses of each of the metals listed, copper will experience the greatest change in temperature and aluminum will experience the least change in temperature.

The relationship between heat and temperature for a gas depends on the *way* that the gas is warmed. Later in this chapter we will learn that there are two different ways for a gas to change temperature. One of these occurs at a constant pressure and the other occurs at a constant volume. The amount of heat needed to change the temperature is different for these two processes. Therefore there are two values for the specific heat of a particular type of gas, C_{p} and C_{v}. The former refers to the temperature change at constant pressure and the latter at constant volume. These are related by the equation,

$$C_{\mathrm{p}} = C_{\mathrm{v}} + R \tag{13.4}$$

where R is a constant, $R = 8.31$ Pa m^3/mol K that will be discussed in a later section. Note that C_{p} is always greater than C_{v} because at constant pressure some of the added heat is used to expand the volume instead of raising the temperature.

If a gas is allowed to expand as it is warmed, some of the energy that it absorbs will be used to expand the gas. For the gas to expand, it must push outwards. That's work, and work requires energy. Thus, if the gas is allowed to expand while it is warmed, its temperature will not increase as rapidly as when it is not allowed to expand.

The number of particles in a sample gas N is a very large number. Because of this, a quantity called a mole has been defined. This is a term with which you should be familiar if you took chemistry. Note that it appears in the denominator of the units for R. Remember, a mole refers to a certain number of molecules, 6.02×10^{23}. This number is called **Avogadro's Number**, in honor of Amedeo Avogadro. If we have 1 mol of gas molecules, we know that we have 6.02×10^{23} gas molecules. If we have 2 mol of gas molecules, we have 12.04×10^{23} gas molecules. The term "mole," then, simply refers to a certain number of molecules. The number of moles of a substance is designated by $n = N/N_A$.

Now consider the following examples of heat flow.

Example 13.1

A piece of copper and a piece of glass have exactly the same mass (45.0 g) and are at the same temperature (25.0 °C). If both absorb the same amount of energy (450.0 J), what will be the final temperature of each?

Q

m_c, T_c m_g, T_g

Knowns: $m_c = 45.0\,\text{g}$; $m_g = 45.0\,\text{g}$; $T_c = 25.0\,°\text{C}$; $T_g = 25.0\,°\text{C}$; $Q = 450.0\,\text{J}$

Unknowns: T_{fc}; T_{fg}

Using Eq. 13.3 and the values of C from Table 13.1, we can solve for ΔT:

$$Q = mC\Delta T$$

For copper:

$$\Delta T_c = Q/mC_c = 450.0\,\text{J}/(45.0\,\text{g})(0.382\,\text{J/g}\,°\text{C}) = 26.2\,°\text{C}$$
$$26.2\,°\text{C} = T_{fc} - T_{ic} = T_{fc} - 25.0\,°\text{C}$$
$$T_{fc} = 51.2\,°\text{C}$$

For glass:

$$\Delta T_g = Q/mC_g = 450.0\,\text{J}/(45.0\,\text{g})(0.837\,\text{J/g}\,°\text{C}) = 11.9\,°\text{C}$$
$$11.9\,°\text{C} = T_{fg} - T_{ig} = T_{fg} - 25.0\,°\text{C}$$
$$T_{fg} = 36.9\,°\text{C}$$

Note that this answer is exactly what we would expect from the heat capacities.

Since glass has a higher specific heat capacity than copper, its temperature does not rise as high for the same amount of energy absorbed.

Example 13.2

A sample of gas that contains 2.00 mol at 45.0 °C is held at constant volume while 565 J are transferred to the gas. What is the final temperature of the gas? What would the final temperature of the gas be if the gas were held at constant pressure rather than constant volume? (C_v = 20.8 J/[molK])

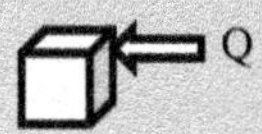

Knowns: $n = 2\,\text{mol}$; $T_i = 45\,°\text{C}$; $Q = 565\,\text{J}$; V or P constant

Unknowns: T_f

The gas is being held at a constant volume, so first, we use Eq. 13.3 with C_v

$$Q = nC_v\Delta T$$
$$565\,\text{J} = (2.00\,\text{mol})(20.8\,\text{J}/[\text{mol K}])\Delta T$$
$$\Delta T = 13.6\,\text{K}$$

(Note that we are given specific heat in units of J/[molK] so the mass must be given in number of moles to make the units work out.)

Converting to degrees Celsius, ΔT is also 13.6 °C.

Next, we need to determine the final temperature of the gas if its pressure were held constant. For that, we need C_p, which we get from Eq. 13.4.

$$C_p = C_v + R = 20.8\,\text{J}/[\text{mol K}] + 8.31\,\text{J}/[\text{mol K}] = 29.1\,\text{J}/[\text{mol K}]$$

Now we can use Eq. 13.3 with C_p

$$Q = nC_p\Delta T$$
$$565\,\text{J} = (2.00\,\text{mol})(29.1\,\text{J}/[\text{mol K}])\Delta T$$
$$\Delta T = 9.71\,\text{K}$$

Once again, that's the same as $\Delta T = 9.71$ °C, so the final temperature is 54.7 °C. Note that this final temperature is lower. That's because some of the energy was used for the work required to expand the gas. As a result, the temperature will not rise as high as it would if the volume were held constant.

Now try the following problem.

Student

13.1 A 4.50-mol gas absorbs 1000.0 J of energy while the pressure is held constant. As a result, its temperature increases by 21.5 °C. What is its molar heat capacity *at constant volume*?

Suppose you took a hot piece of metal and dropped it into some cold water. What would happen? The metal would begin to lose energy, lowering its temperature. The water would gain that energy, raising its temperature. This will end when the two objects have the same temperature. This is called thermal equilibrium. Since the objects have the same temperature, there will be no net energy transfer in thermal equilibrium. Energy will still be transferred from the metal to the water, but an equal amount of energy will be transferred from the water back to the metal. Thus, since there is no net energy transfer, there will be no temperature changes, and the metal and water will stay at the same temperature. Equation 13.3 allows us to determine what the temperature of thermal equilibrium will be as shown in the following example.

Example 13.3
A piece of iron (m = 50.0 g) at 100.0 °C is dropped into an insulated cup that contains 200.0 g of water at 24.0 °C. Once thermal equilibrium is reached, what will be the temperature of both the water and the metal?

Knowns: $m_{\text{i}} = 50.0\,\text{g}$; $T_{\text{i}} = 100.0\,°\text{C}$; $m_{\text{w}} = 200.0\,\text{g}$; $T_{\text{w}} = 24.0\,°\text{C}$

Unknowns: T_{f}

We know that the metal will lose energy and the water will gain energy. We also know that all of the energy lost by the metal will be gained by the water. Thus, the Q of the water will be equal to the Q of the metal. However, we have to be careful about this. The water *gains* energy, but the metal *loses* energy. If you look at Eqs. 13.3 and 13.4, when a substance loses energy, its ΔT is negative. That means its Q is negative as well. This makes sense, since a negative Q would imply a loss of energy. In this case, then, the metal'sQ is negative (it loses energy) and the water's Q is positive (it gains energy). In the end, then, the Q of the metal will equal the negative of the Q of the water:

$$Q_{\text{i}} = -Q_{\text{w}}$$

Since we have Eq. 13.3 for Q, we can plug that equation into the equation above. We can also insert the equation for ΔT, into Eq. 13.3:

$$m_{\mathrm{i}} \cdot C_{\mathrm{i}} \cdot (T_{\mathrm{f}} - T_{\mathrm{i}}) = -m_{\mathrm{w}} \cdot C_{\mathrm{w}} \cdot (T_{\mathrm{f}} - T_{\mathrm{w}})$$

Note that T_{f} is the same for both the water and the iron because thermal equilibrium is reached.

Now we can plug in our numbers, and the only thing we don't know is T_{f}:

$$(50.0\,\mathrm{g})(0.452\,\mathrm{J/g\,^\circ C})(T_{\mathrm{f}} - 100.0\,^\circ\mathrm{C})$$
$$= -(200.0\,\mathrm{g})(4.19\,\mathrm{J/g\,^\circ C})(T_{\mathrm{f}} - 24.0\,^\circ\mathrm{C})$$

$$T_{\mathrm{f}} = 22{,}400\,\mathrm{J}/(861\,\mathrm{J/^\circ C}) = 26.0\,^\circ\mathrm{C}$$

Notice, then, that even though the iron was 76.0 °C hotter than the water, the water's temperature increased only 2.0 °C. This is due to the mass of the water in this example and the large specific heat capacity of water.

Now try this problem.

Student

13.2 A sample of unknown metal ($m = 150.0$ g) at 95.0 °C is dropped into an insulated cup that contains 300.0 g of water at 25.0 °C. Thermal equilibrium is achieved at 27.4 °C. What is the specific heat capacity of the metal?

Energy flowing into or out of an object can do more than change the temperature of the object. We already mentioned in passing the possible change in volume of a gas that we will discuss later. Another possibility is that the change in thermal energy can be used to change the phase of the object. As we add energy, the solid phase gives way to the liquid phase (we call that melting), and the liquid phase eventually gives way to the gas phase (we call that evaporating). Alternatively, if we remove energy, the gas phase gives way to the liquid phase (we call that condensing), and the liquid phase eventually gives way to the solid phase (we call that freezing). Thus, when energy is released or absorbed, the result might be a change in phase instead of a temperature change. Please perform the following experiment to demonstrate this.

Experiment 13.1
The Energy Associated With a Phase Change

Supplies:

- Table salt
- Water
- Three Styrofoam coffee cups
- Thermometer (It needs to read temperatures from − 10 to 100 °C.)
- Freezer
- Stove
- Pot in which to boil water
- One-quarter cup measuring cup
- Empty ice cube tray

Introduction: A phase change requires energy, as illustrated in this experiment.

Procedure:

1. Fill one of the Styrofoam cups ¾ of the way with water and dissolve as much salt as possible into the water. If there is left-over salt at the bottom of the cup, don't worry about it.
2. Measure out ¼ of a cup of water, and pour it into as many ice cube depressions in the ice cube tray as are necessary to accept all of the water in the measuring cup.
3. Put both the ice cube tray and the cup of salt water into the freezer, and leave them there until the water in the ice cube tray freezes. The salt water in the cup should not freeze, because by adding salt, you have reduced the freezing point of the water. This is called **freezing point depression**, and you should have learned about it when you took chemistry.
4. Place the thermometer in the freezer as well, so that it can measure the temperature of the freezer.
5. When the water in the ice cube tray is nearly frozen, boil at least 2 cups of water in the pot.
6. Nest one coffee cup into the other. You now have a **calorimeter**, which is something else you should have learned about in chemistry.
7. Read the temperature on the thermometer and remove it from the freezer.
8. Take the salt water out of the freezer and quickly measure out ¼ of a cup of it. Pour that into the coffee cup calorimeter.
9. Place the thermometer in the cup so that the bulb is immersed in the salt water.

10. Quickly measure out ¼ cup of boiling water and pour the hot water into the coffee cup calorimeter.
11. Stir the solution with the thermometer and read the temperature occasionally. After a while the temperature should level out. Read that as the final temperature.
12. Dump out the solution in the coffee cup calorimeter and rinse it out.
13. If you took the pot of water off of the stove, put it back on, because you will need boiling water again right away.
14. Take the ice cube tray out of the freezer and remove the cubes that have been made from the ¼ cup of water.
15. Place the ice cubes into the coffee cup calorimeter and put the thermometer in again.
16. Once again, quickly measure out ¼ cup of boiling water and dump it in the calorimeter.
17. Once again, stir with your thermometer and continue to read the thermometer until the temperature levels out. Record the final temperature.
18. Clean everything up.

In the first part of the experiment, energy was transferred from the hot water to the cold water (heat). This warmed the cold water and cooled the warm water until thermal equilibrium was achieved. Since the boiling water was approximately 100 °C (depending on atmospheric pressure), and since the mass of cold water was about the same as the mass of warm water (because of the salt, the mass of the cold water was a bit higher), the final temperature should have been the midpoint between the temperature of the cold water and 100.0 °C. Because your experiment was not perfect (energy leaked into the surroundings), and because the cold water was more massive than the hot water, the final temperature was probably lower than the midpoint.

In the second part of your experiment, the final temperature should have been significantly lower than the final temperature in the first part of the experiment. In fact, your final temperature might have ended up being very near 0 °C. The ice was less massive than the cold salt water, since it had no salt in it. Nevertheless, more energy was transferred from the hot water to the ice than was transferred from the hot water to the cold salt water. The reason is that in order to melt ice, you need to supply more energy than what is required to raise its temperature. Once the ice's temperature was raised to the melting point, additional energy had to be added in order to actually melt the ice. If additional energy was not required to make the phase change, the final temperature in the second part of the experiment would have been nearly equal to the final temperature in the first part of the experiment.

This is an important point. Energy was being transferred from the hot water to the ice, but once the ice reached its melting point (0 °C), it did not warm up while the ice was melting. That's because the energy it was receiving was dedicated solely to melting the ice and thus could not be used to raise its temperature. This means:

During a phase change, the temperature of the substance does not increase while energy is added or decrease while energy is removed.

This happens with all phase changes. For example, suppose you take some water and begin heating it on the stove. The temperature of the water will increase, because energy is being transferred from the burner to the water. Eventually, the water will reach 100.0 °C and begin to boil. However, while the water is boiling, the temperature will not continue to increase! It will remain at a constant 100.0 °C.

Why does the temperature stay constant during a phase change? It takes energy to go from solid to liquid and from liquid to gas. In order to go from solid to liquid, the atoms that make up the substance must break away from a rigid arrangement in which they can only vibrate back and forth. Once they break these strong chemical bonds, they can move around more freely. It takes energy to break the bonds. In the same way, molecules in the liquid state stay relatively near each other because they are chemically attracted to one another. To get into the gas phase, those molecules must move far apart from one another. It takes energy to overcome the attraction that they have for one another, so it takes energy to go from the liquid phase to the gas phase.

Similarly, when a substance changes from gas to liquid, it must release energy. The molecules or atoms that make up the substance are traveling around freely while in the gas phase. When they come together to form a liquid, they must move more slowly, because they feel the effects of their mutual attraction. Thus, they must lose a lot of kinetic energy to go from the gas phase to the liquid phase. As a result, energy is released. When going from the liquid phase to the solid phase, the molecules must lose a lot of kinetic energy once again, because they must go from being able to move around to being arranged in a rigid structure which only allows for vibration. Thus, they must lose kinetic energy and, once again, energy is released.

When you are warming a substance up, once it begins to melt or evaporate, all of the energy you are supplying goes towards the phase change. As a result, there is no energy left to increase the temperature. In the same way, while you are cooling a substance down, during the time it is freezing or condensing, the temperature cannot change because any energyyou are taking away from the substance is being released as a result of the phase change. Once again, then, the temperature cannot decrease.

The amount of energy needed to produce a phase change depends on the substance involved and its mass. The equation for the heat required is

$$Q = mL \tag{13.5}$$

where Q is the heat required to make the transition, m is the mass of the substance, and L is called the latent heat of transformation. This is unique for every substance as well as for each phase change. The L for freezing (or melting) is called the latent heat of fusion, while the L for evaporating (or condensing) is called the latent heat of vaporization. The latent heat of fusion for water, for example, is 3.34×10^5 J/kg. This means that for every kilogram of ice at 0.0 °C, 3.34×10^5 J of heat must be added to turn it into a kilogram of liquid water at 0.0 °C. The latent heat of vaporization for water is 2.26×10^6 J/kg, which means it takes 2.26×10^6 J of heat to change one kilogram of liquid water at 100.0 °C to one kilogram of gaseous water (water vapor) at 100.0 °C.

In order to boil water, then, it is not enough to just warm it up to 100.0 °C. You must also supply enough energy to actually make the phase change occur. Study the following example to see what this means.

Example 13.4

How much energy must be added to 150.0 g of water at 25.0 °C in order to completely boil the water away? ($L_{vaporization}$ = 2260 J/g for water)

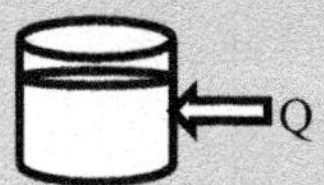

Knowns: $m = 150.0\,\text{g}$; $T = 25.0\,°\text{C}$; $L_{vaporization} = 2260$ J/g

Unknowns: Q

To get the water boiling to begin with, we must raise its temperature to 100.0 °C. To calculate the heat for that, we must refer to Eq. 13.3.

$$Q = mC\Delta T = (150.0\,\text{g}) \cdot (4.19\,\text{J/g}\,°\text{C}) \cdot (100.0\,°\text{C} - 25.0\,°\text{C}) = 47{,}100\,\text{J}$$

That's not enough, however. We must now supply the energy for the phase change from Eq. 13.4:

$$Q = mL = (150.0\,\text{g}) \cdot (2260\,\text{J/g}) = 339{,}000\,\text{J}$$

Notice that the energy required to cause the phase change is significantly higher than the energyrequired to warm the water. The total energy required, then, is the sum of these energies.

$$Q_{total} = Q_{heating} + Q_{vaporization} = 47{,}100\,\text{J} + 339{,}000\,\text{J} = 386{,}000\,\text{J}$$

Now try following problem.

Student

13.3 How much energy does it take to completely evaporate 50.0 g of water if the water starts out as ice at a temperature of − 11.0 °C? [The specific heat capacity of ice = 2.02 J/(g °C), $L_{vaporization}$ = 2260 J/g for water, L_{fusion} = 334 J/g for water.]

13.4 Laws of Thermodynamics

Thermodynamics is based on three fundamental laws and one postulate that is sometimes referred to as the Zeroth Law of Thermodynamics. This postulate states:

The Zeroth Law of Thermodynamics: If object A is in thermal equilibrium with object C, and if object B is in thermal equilibrium with object C, objects A and B are in thermal equilibrium with each other.

This says that when two objects are each in thermal equilibrium with a third object, they are also in thermal equilibrium with each other. This tells us something very important. When two objects of different temperature are placed in thermal contact, energy will flow from the hot object to the cold object. You already knew that, but it is a fundamental postulate of thermodynamics. Thermal equilibrium can only be reached by the objects coming to the same temperature. Much of what we have been discussing in the previous section is associated with the zeroth law of thermodynamics.

The first law of thermodynamics is essentially a restatement of the conservation of energy. It says:

First Law of Thermodynamics: Energy cannot be created or destroyed, it can only change forms.

This law applies to any type of physical process. In this area of physics, we are interested in turning heat energy into mechanical energy which is governed by the first law of thermodynamics written as

$$\Delta U = Q - W \tag{13.6}$$

Here the left side of the equation is the change in thermal energy and the right side is the heat flow in or out of the system minus the work done by or on the system. Each of the quantities has units of Joules. Heat is positive when it is absorbed and negative when it is emitted. The internal energy of a gas is proportional to its temperature. If the internal energy of a gas increases, its temperature increases. If the internal energy of a gas decreases, its temperature decreases.

Work is positive when it is done by the system and negative when it is done on the system. Suppose you took a balloon that was full of air and squished it. You would be working on the system, and W would be negative if we were analyzing the gas in the balloon. If, on the other hand, you warmed the balloon and it expanded, it would be doing work on the surroundings (pushing air out of the way), so W would be positive if you were analyzing the gas in the balloon.

A device that operates according to Eq. 13.6 is called a heat engine. It consists of three parts as shown schematically in Fig. 13.3. The net heat flow for the system is

$$Q = Q_{\mathrm{H}} - Q_{\mathrm{L}} \tag{13.7}$$

When the heat engine is operating at its equilibrium conditions, there is no change in the thermal energy of the system so the net heat exchange given by

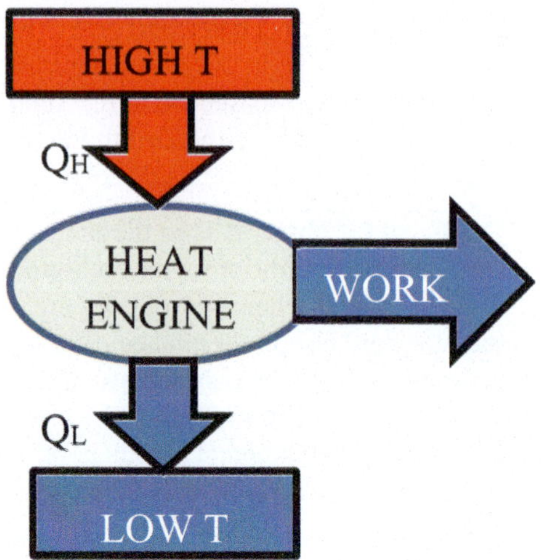

Fig. 13.3 Heat engine

Eq. 13.7 is equal to the work by or on the heat engine. There is always unused heat that is exhausted so the efficiency of the heatengine given by

$$\eta = W/Q_{\mathrm{H}} = 1 - Q_{\mathrm{L}}/Q_{\mathrm{H}}. \tag{13.8}$$

is never 100%. We will do more problems using the first law of thermodynamics later in this chapter.

The second law of thermodynamics applies only to special cases. One example is specific types of physical processes called spontaneous processes. These are processes where nothing outside the system is driving it to change. An isolated system is one in which no mass or energy crosses its boundary. A closed system is one in which a small amount of heat but no mass can cross the system boundary. In isolated systems, only spontaneous processes occur while in open or closed systems either spontaneous or nonspontaneous processes can occur. A process can occur in such a way that it is either reversible or nonreversible. A reversible process takes place slowly with small increments of change so that the system is always in thermal equilibrium with its surroundings. All of these distinctions are important in considering the second law of thermodynamics.

In the previous section, we discussed thermal energy as a property of a sample related to the motion of its atoms. A somewhat similar property is the entropy of a sample. Entropy measures the amount of disorder in a sample.

Entropy: A measure of the disorder that exists in any system.

The standard unit for entropy is J/K.

Disorder is one of the most fundamental things we can learn about a system. For example, suppose you have a priceless vase sitting on a shelf somewhere. Suppose further that something happens which causes the vase to fall onto the floor and shatter into a million pieces. Clearly, when the vase is shattered, it is more disordered than when it was sitting on the shelf in one piece. Thus, we would say that while the vase was sitting on the shelf, it was in a state of low entropy, because it was not very disordered. However, when the vase shatters, it moves into a state of high entropy, because now it is very disordered.

Although you can imagine a vase falling from a shelf and shattering into a million pieces, you can never imagine the reverse process happening where a shattered vase spontaneously rearranges itself into a whole vase again. This is an irreversible process. The reason that the reverse process cannot happen is because of the Second Law of Thermodynamics:

Second Law of Thermodynamics: Spontaneous processes always go in a direction that maintains or increases the total entropy of the universe.

This can be expressed mathematically as

$$\Delta S_{\text{universe}} = \Delta S_{\text{system}} + \Delta S_{\text{surroundings}} \geq 0. \tag{13.9}$$

where S is the symbol for entropy. We are sure we are dealing with a spontaneous process if the second law is applied to an isolated system. In this case, a process continues to evolve spontaneously until thermodynamic equilibrium is reached and the system has a maximum value of entropy.

Any process that proceeds as a series of slow, equilibrium steps is considered reversible. For example, suppose you allowed a gas to expand in a cylinder. As long as the gas expanded in a series of many tiny expansions so that the system was in equilibrium with its surroundings throughout the process, we would call that a reversible process. For a reversible processes in a closed system where a small amount of heat can cross the system boundary, the second law of thermodynamics can be written as

$$\Delta S_{\text{rev}} = Q/T, \tag{13.10}$$

where T is in units of K. When heat is added to an object, its disorder increases so S increases. When heat is removed from an object S decreases. Sometimes the heat is written as δQ to indicate this only applies to very small amounts of heat transfer. For processes in which the temperature changes, Eq. 13.10 must be modified to be a calculus expression. Since this is beyond the math we are using in this text, we can just use the average temperature in Eq. 13.10. An important result is that for a process with no heat transfer, the change in entropy is zero. All isolated or closed systems tend spontaneously to become more disordered with time. The first law of thermodynamics given by Eq. 13.6 still holds for spontaneous processes as well as Eq. 13.10, but for nonspontaneous and irreversible processes only Eq. 13.6 is applicable.

From the discussion above, it is obvious that in applying the second law of thermodynamics, it is important to consider the system of interest and its surroundings and to know whether the system is isolated, closed, or open. Sometimes it might seem like second law is violated and entropy is decreasing. This can occur for one part of a system. However, if you correctly consider the entire isolated or closed system and its surroundings, the total entropy will never decrease. Remember the statement of the second law of thermodynamics given above applies to the entropy of the entire universe, not one part of it.

The second law of thermodynamics is also responsible for the fact that a heat engine can never be 100% efficient as discussed previously. Since a heat engine converts heat to work, and heat is a more random form of energy than work, this process decreases the entropy of the system. However, if some of the heat is exhausted, this can cause an increase in the entropy of the surroundings. This must occur in a way that the total entropy of the system plus its surroundings increases or is constant. Therefore Q_L can never be zero. Entropy is sometimes thought of as the amount of internal energy not available to do work.

Consider the following example.

Example 13.5

A 50.0 g cube of ice is at − 5.00 °C. What is the ΔS of the ice cube if it is melted at 0.00 °C? What is the *minimum* ΔS of the surroundings? (C_{ice} = 2.02 J/(g °C), L_{fusion} = 334 J/g)

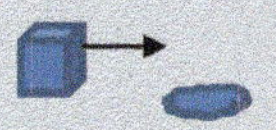

Knowns: $m = 50.0\,\text{g}$; $T_i = -5.00\,°\text{C}$; $C_{ice} = 2.02\,\text{J}/(\text{g}\,°\text{C})$; $L_{fusion} = 334\,\text{J}/\text{g}$

Unknowns: ΔS

Since ΔS depends on Q, we need to think about how heat is added to this ice cube in order to make it melt. The ice must first be warmed to 0.00 °C. The Q for that is:

$$Q = mC_{ice}\Delta T = (50.0\,\text{g})(2.02\,\text{J/g}\,°\text{C})(0.00\,°\text{C} - [-5.00\,°\text{C}]) = 505\,\text{J}$$

Now we can calculate ΔS using Eq. 13.10, just taking the average temperature of the process, which is − 2.50 °C, or 270.65 K.

$$\Delta S_{rev} = (505\,\text{J})/(270.65\,\text{K}) = 1.87\,\text{J/K}$$

That's not the end of the story, however. Now that the ice is at 0.00 °C, it must be melted. The heat required for that is:

$$Q = mL = (50.0\,\text{g}) \cdot (334\,\text{J/g}) = 16{,}700\,\text{J}$$

The ΔS associated with that heat is:

$$\Delta S_{rev} = (16{,}700\,\text{J})/(273.15\,\text{K}) = 61.1\,\text{J/K}$$

The total ΔS of the ice cube, then, is 63.0 J/K. Now, if this is to happen, the ΔS of the universe must be greater than or equal to 0. We have just calculated the ΔS of the system, so:

$$\Delta S_{\text{system}} + \Delta S_{\text{surroudings}} \geq 0$$
$$63.0\,\text{J/K} + \Delta S_{\text{surroudings}} \geq 0$$
$$\Delta S_{\text{surroudings}} \geq -63.0\,\text{J/K}$$

The entropy of the surroundings, then, can actually decrease, since the entropy of the system increases. This will, in fact, happen, since the system absorbs heat from the surroundings in order for the ice to melt. Thus, the Q of the surroundings is negative, so the ΔS of the surroundings will be negative as well.

The sign of the ΔS we calculated in the example is positive. That means the entropy of the ice cube increased. This should make sense. First, we increased the ice cube's temperature. That makes the water molecules in the ice cube move faster. The faster the molecules move, the harder it is to "keep track" of them, so the more disordered the system is. If we had cooled the ice cube rather than warmed it, the value of Q would be negative, and therefore the value of ΔS would be negative. When something absorbs heat, its entropy increases. When it releases heat, its entropy decreases. Second, we melted the ice cube. In a solid, the molecules are held in a tight, rigid structure and they are only allowed to vibrate. In a liquid, the molecules move around more freely. Thus, it is once again more difficult to "keep track" of them, so the entropy of a liquid is higher than that of a solid.

Notice also the numerical values we calculated. When the ice cube was warmed by 5.00 °C, the ΔS was 1.87 J/K. Thus, entropy increased a small amount. When the ice cube was melted, the ΔS was 61.1 J/K. In that step, then, the entropy increased significantly more than when the ice cube was heated. In general, the entropy changes associated with phase changes are much higher than those associated with simply warming or cooling a substance. This should make sense, as a phase change denotes a huge change in the behavior of the molecules, while warming or cooling produces only a change in the speed of the molecules.

It is important to emphasize that the ΔS of a system can be negative without violating the Second Law of Thermodynamics. As Eq. 13.9 tells us, the second law says nothing about the ΔS of a system by itself. It only says that when you add the ΔS of the system and the ΔS of the surroundings, you must come up with zero or a positive number. Thus, the ΔS of a system is free to be negative, as long as the ΔS of the surroundings is sufficiently positive so that the sum of both ΔS values is greater than or equal to 0.

The classic example of how this works is the freezing of water. When water freezes, it releases energy based on its mass and L_{fusion}. Since the water releases energy, its Q is negative and thus its ΔS is negative. That does not violate the second law, however, because when it releases its energy, that energy heats up the surroundings, increasing the entropy of the surroundings. The increase in entropy of the surroundings is exactly equal to the decrease in entropy of the system, so the sum of the two ΔS's is 0. Thus, the freezing of water is consistent with the second law and can therefore occur.

Now try the following problem.

Student

13.4 A gas expands at a constant temperature of 314 K so that the work done is 2450 J. What is the ΔS of the gas?

The third law of thermodynamics is applicable to closed systems in thermodynamic equilibrium. Simply stated, it says that:

Third Law of Thermodynamics: The entropy of a system approaches zero or a constant value as its temperature approaches absolute zero.

At a temperature of 0 K, the system must be in its ground state which is the state in which it has minimum energy. If this is a unique state its entropy is zero. If it is not a unique state there is some residual entropy of the system.

This law does not have as many practical applications as the Zeroth, First, and Second Laws because it deals with such extremely low temperatures that we don't normally experience a situation where it is relevant. However, recently it has been the topic of physics research into the fundamental properties of matter.

13.5 State Variables

In physics it is important to distinguish between two types of parameters, state functions and processes functions. The former are variables used to describe a specific state of the system while the latter describe something that changes the state of the system. For example, in our study of systems in mechanics, position and velocity were typical state functions. In thermodynamics state functions include temperature, pressure and volume as well as internal energy and entropy. Other parameters such as heat and work are processes functions. It turns out that there are some important relationships between temperature, pressure and volume that we need to learn about. We talked previously about things that can happen when heat is added to a system. A change in volume is another possibility. We will see how this happens in solids in the next chapter while in this chapter we will talk about what happens in a gas.

In a gas, the particles have momentum changing collisions with the surface of the container that holds it. The impulse of these collisions on the surface of the container is what we call pressure. We can define pressure as the force on a surface divided by the area of the surface

$$P = F/A. \tag{13.11}$$

The units of pressure are called Pascals, abbreviated Pa. One Pascal is one Newton per meter squared. There are other units used for pressure. The **atm** is a unit of pressure that references the normal pressure of earth's atmosphere. A pressure of 1.0 atm (101.3 kPa) is the pressure exerted by earth's atmosphere on an object at sea level. The unit **mmHg**, which is often called the **torr**, measures the height of a column of mercury in a barometer. There are 133 Pa in 1.00 mmHg.

The relationship between the pressure, temperature, and volume of a gas is determined by two laws. The first is called Boyle's law. This law states that if the temperature is constant, the volume of the gas and its pressure are inversely proportional to one another. Mathematically this is expressed as

$$PV = \text{constant (for constant } T). \tag{13.12}$$

Thus the parameters for a gas in state 1 and in state 2 are related by the equation

$$P_1V_1 = P_2V_2. \tag{13.13}$$

The second important relationships between state parameters in a gas is called Charles' law. This law states that if the pressure of the gas is constant, the volume of the gas varies directly with its temperature. Mathematically this is written as

$$V/T = \text{constant (for constant } P). \tag{13.14}$$

Thus the parameters for a gas in state 1 and in state 2 are related by the equation

$$V_1/T_1 = V_2/T_2. \tag{13.15}$$

Boyle's law and Charles' law can be combined and written as.

$$\frac{P_1V_1}{T_1} = \frac{P_2V_2}{T_2} = \text{constant.} \tag{13.16}$$

Note that if $T_1 = T_2$ Eq. 13.16 reduces to Eq. 13.13 while if $P_1 = P_2$ Eq. 13.16 reduces to Eq. 13.15. Equation 13.16 is called the ideal gas law.

The constant on the right side of Eq. 13.16 is the number of particles in the gas, N, multiplied by the proportionality constant between the average kinetic energy of the particles and the temperature of the gas. This is written as

$$PV/T = kN. \tag{13.17}$$

k is an important fundamental constant in thermodynamics. It is known as Boltzmann's constant, and has a value of $k = 1.38 \times 10^{-23}$ m^2 kg/s^2 K. Note that the units of k are energy over temperature. Equation 13.17 can be rewritten as

$$PV = nRT \tag{13.18}$$

where $R = 8.31$ Pa m^3/mol K and n is the number of moles of the gas. Equation 13.18 is the most familiar form of the ideal gas law. It works well for most situations that don't involve extremely low temperatures or extremely high pressures.

Note that the three constants we have defined above are all related,

$$k = R/N_{\mathrm{A}} \tag{13.19}$$

If we want to work with the mass of the sample in grams instead of moles, we can use what we learned in chemistry that the mass of atoms and molecules is expressed in terms of atomic mass units (amus). The amu of a molecule is the mass in grams it takes to make one mole of the gas. Thus if we know the amu of a molecule we can divide by Avogadro's number to get the mass of a single molecule in grams.

Note that Eq. 13.18 predicts that at a constant pressure, an increase in temperature is accompanied by an increase in volume. This is called "thermal expansion." Although Eq. 13.18 applies only to gases, thermal expansion occurs in all states of matter. As the substance expands, its atoms (or molecules) move away from each other and it becomes less dense. The gravitational attraction will be less on the less dense material than it is on the denser material. Therefore hot air (or water) will rise while cold air (or water) falls in a multi-temperature sample. These are called convection currents. Thermal expansions in solids will be discussed in the following chapter.

To show how the ideal gas law works, consider blowing up a balloon. Why does the size of the balloon increase when you blow it up? While you are standing in a room, the temperature stays pretty constant, as does the pressure. Thus, P and T are constants. R is also a constant. Thus, the only variables in Eq. 13.18 for this situation are V and n. When you blow into a balloon, you are adding gas to the balloon. Thus, you are increasing n. As Eq. 13.18 predicts, as n increases, V will have to increase; therefore, the balloon expands. Why will the balloon shrink if you put it in the freezer? In this case, the quantities n and pressure P are constant, as is the ideal gas constant R. As a result, the only two variables in Eq. 13.18 for that situation are V and T. As the equation predicts, when T goes down, V will go down. So, the balloon got smaller.

Consider the following example.

Example 13.6

Standard temperature and pressure (STP) is defined as $T = 273.15$ K and $P = 101.3$ kPa (normal atmospheric pressure). How much volume does 1.00 mol of gas occupy at STP?

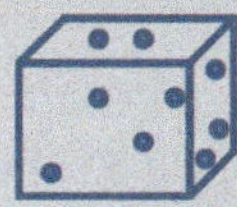

Knowns: $n = 1$ mol; $T = 273.15$ K; $P = 101.3$ kPa

Unknowns: V

In this problem, we know the pressure, temperature, and the number of moles. Since our units are all standard, we will use 8.31 J/(molK) as the value of R. The ideal gas law tells us

$$PV = nRT$$

So

$$V = nRT/P = (1.00\,\text{mol})(8.31\,\text{J/mol K})(273.15\,\text{K})/1.013 \times 10^5\,\text{Pa} = 0.0224\,\text{m}^3$$

Notice how the units work out. Since a Joule is a N m and a Pa is a N/m^2 the volume works out to the standard volume unit, m^3. When using the ideal gas law, the units work out as long as you use the units that are consistent with the value of R. If you prefer to have volume in liters, 1 m^3 = 1000 L.

This example shows that 1.00 mol of gas occupies 0.0224 m^3 at STP. Notice that the results of this calculation are independent of the type of gas. As long as the gas behaves as an ideal gas, 6.02×10^{23} molecules of the gas (1.00 mol), will occupy 0.0224 m^3 of volume. Thus, whether we have 6.02×10^{23} molecules of hydrogen gas, 6.02×10^{23} molecules of oxygen gas, or 6.02×10^{23} molecules of ammonia gas, they will all occupy 0.0224 m^3 at STP.

Now try the following problems.

Student

13.5 A gas is contained in a spherical balloon at atmospheric pressure (P = 101.3 kPa).

a. If the balloon has a radius of 10.0 cm, what is the average force exerted by the gas molecules as they strike the balloon? (The surface area of a sphere is $4\pi r^2$.)

b. If the temperature of the situation described in problem 13.8 is 25.00 °C, how many moles of gas are contained in the balloon? How many *molecules* of gas are contained in the balloon?

(The volume of a sphere is given by $4\pi r^3/3$.)

One standard type of problem used to demonstrate work done in a system of gas molecules is a cylinder with a piston. Suppose we put a gas in a cylinder whose top is a piston which can move up and down and then watch the gas as it expands as shown in Fig. 13.4. In thermodynamics, it is very important to identify the system as well as the surroundings. In general, the system is whatever you happen to be studying, and the surroundings are essentially everything else that can be affected by the system. In this example, the system is the gas itself. The surroundings are the cylinder, the piston, and the air surrounding the outside of the cylinder and piston.

In the situation we are analyzing the gas is doing work by moving the piston. The gas is exerting a force against the piston, and the force is being exerted over a distance as the piston moves. Thus, work is done. We learned in Chap. 3 that we express work as

$$W = Fx\cos\theta$$

where θ is the angle between the force and the displacement. In this case, $\theta = 0$ so

$$W = Fx$$

In order to move the piston, the gas must overcome whatever forces are pushing down on the piston. The force of gravity is one of these forces, but if the piston is not very massive we can ignore it. The other, larger force is the force that results from the air pressure which is pushing against the cylinder/piston arrangement. That pressure is given by:

$$P = F/A \tag{13.20}$$

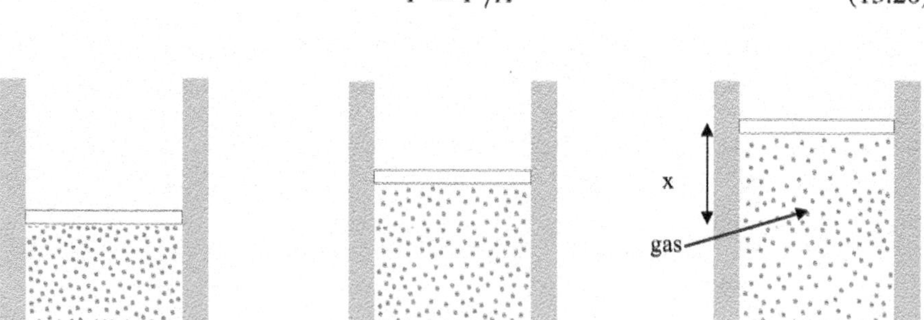

Fig. 13.4 Gas expanding in a cylinder/piston system

We can rearrange this equation to solve for force, and we can put the resulting expression into the equation for work given above to get:

$$W = PAx$$

Air pressure is being exerted over the entire cylinder/piston, but the only thing we are interested in is the piston, because that's what is moving. Thus, A is the surface area of the piston. The "x" in this equation represents the distance that the piston moves up. That means that Ax represents the change in volume that the gas experiences when the piston moves up. We can call the change in volume ΔV so that:

$$W = P \cdot \Delta V \tag{13.21}$$

This tells us that the work which a gas does as it expands under constant pressure is equal to the pressure times the change in volume. Notice how the units work out in this equation. Pressure has the units N/m^2, and volume has the units m^3. Multiplying these units together, gives N m, which is the same as a Joule.

The value of work in this equation can be positive or negative. If the gas expands, the final volume is greater than the initial volume, so ΔV is positive and work is positive. If the gas contracts, the final volume is smaller than the initial volume, so ΔV is negative and work negative. When the gas expands, the system does work on the surroundings. When the gas contracts, the surroundings are doing work on the system. This is consistent with how we defined positive and negative work previously. Consider the following example.

Example 13.7

A 1.50 mol sample of gas is contained in a cylinder/piston system such as the one shown in Fig. 13.4. It starts out at a pressure of 156 kPa and a volume of 0.0151 m^3 and then expands to a volume of 0.0251 m^3 while the pressure stays constant. What is the change in the internal energy of the gas? [The C_p of an ideal gas is 20.8 J/(molK).]

Use Fig. 13.4	Knowns: $n = 1.5$ mol; $P = 156$ kPa; $V_i = 0.0151$ m^3; $V_f = 0.0251$ m^3; $C_p = 20.8$ J/mol K
	Unknowns: ΔU

This problem depends on the First Law of Thermodynamics, given by Eq. 13.6

$$\Delta U = Q - W$$

To determine Q and W we can use Eqs. 13.3 and 13.21. Starting with the equation for work done,

$$W = P \cdot \Delta V = \left(1.56 \times 10^5\,\text{Pa}\right) \cdot \left(0.0251\,\text{m}^3 - 0.0151\,\text{m}^3\right) = 1560\,\text{J}$$

Notice that we first had to convert kPa to Pa to keep everything in standard units. In addition, notice that the work is positive. That means the system did the work.

Now we have to determine Q using Eq. 13.3 since pressure stays constant:

$$Q = n \cdot C_\text{p} \cdot \Delta T$$

So to get Q, we need ΔT but there are no temperatures given in the problem. However, we can calculate the temperatures using the ideal gas law. Notice that the problem gives us the pressure, the volume, and the number of moles. Thus, the ideal gas law can give us the temperature at each point.

At the beginning of the problem, $P = 1.56 \times 10^5$ Pa, $V = 0.0151$ m^3, and $n = 1.50$ mol.

$$PV = nRT$$

$$T_\text{i} = (PV)/(nR) = \left(1.56 \times 10^5\,\text{Pa}\right)\left(0.0151\,\text{m}^3\right)$$

$$/(1.50\,\text{mol})\left(8.31\,\text{Pa}\,\text{m}^3/\text{mol}\,K\right) = 189\,\text{K}$$

The final temperature is

$$T_\text{f} = (PV)/(nR) = \left(1.56 \times 10^5\,\text{Pa}\right)\left(0.0251\,\text{m}^3\right)$$

$$/(1.50\,\text{mol})\left(8.31\,\text{Pa}\,\text{m}^3/\text{mol}\,\text{K}\right) = 314\,\text{K}$$

Now we can finally calculate Q:

$$Q = nC_\text{p}\Delta T = (1.50\,\text{mol})(20.8\,\text{J/mol K})(314\,\text{K} - 189\,\text{K}) = 3.90 \times 10^3\,\text{J}$$

Notice that Q is positive which means that the system absorbed energy. That should make sense, since the temperature increased. Now that we have Q and W, we can determine ΔU

$$\Delta U = Q - W = 3.90 \times 10^3\,\text{J} - 1560\,\text{J} = 2340\,\text{J}$$

So the internal energy of the system increased by 2340 J.

The First Law of Thermodynamics is like having an "energy meter" in the gas, because it can be used to track how the energy of the gas changes. In order to change the temperature of the gas from 189 to 314 K, the gas needed to absorb 3900 J of energy. However, the gas did not keep all of that energy in the form of thermal energy. Because the gas expanded, some of that energy went into work. Thus, the change in internal energy of the gas was not 3900 J, it was 2340 J. The difference (1560 J) is the energy it "cost" for the gas to expand.

Try the following problems.

Student

13.6 A 2.50-mol sample of gas is compressed under a constant pressure. It begins with a volume of 0.0514 m^3 and ends with a volume of 0.0201 m^3. If the compression causes the internal energy of the gas to drop by 2570 J, what is the pressure? [$C_p = 20.8$ J/(molK)]

13.7 A gas is in a cylinder/piston system. The system is connected to a heater which allows a person to add heat gently in a very controlled way. The gas expands at a constant temperature. If the work done is 2020 J, how much heat was added to the gas?

We said earlier that the temperature of a substance represented the average kinetic energy of the particles making up the substance. The relationship between T and KE can be derived by considering the collisions of a single particle of a gas with the side of a container. If a particle having a speed v_x in the x direction has an elastic collision with a wall and rebounds with a speed $-v_x$ in the minus x direction, its change in momentum is

$$\Delta p = m(-v_x) - mv_x = -2mv_x.$$

Equation 3.12 defined this change of momentum as an impulse equal to the force causing the change in momentum multiplied by the time the force acts, $J = \Delta p = F\Delta t$. For a particle bouncing back and forth in the container it hits the wall once every round trip or $\Delta t = 2d/v_x$ where d is the width of the container. Then using Eq. 3.12 gives the force exerted on the wall as

$$F = 2mv_x/\Delta t = mv_x^2/d.$$

The pressure exerted by the gas on the wall is the force per unit area,

$$P = F/A = mv_x^2/(dA) = mv_x^2/V.$$

This is the pressure exerted by one particle of the gas. For N particles with an average squared velocity in the x-direction $\overline{v_x^2}$, the total pressure on the wall is

$$PV = Nm\left(\overline{v_x^2}\right).$$

Since the particle will have the same average speed squared in all three directions, its total average speed squared is simply three times the average speed squared in one direction. So,

$$PV = Nm\left(\overline{v^2}\right)/3.$$

From Eqs. 13.18 and 13.19, $PV = NkT$, so

$$NkT = Nm\overline{v^2}/3.$$

Multiplying both sides by ½ gives the usual expression for kinetic energy,

$$\text{KE} = m\overline{v^2}/2 = (3/2)kT. \tag{13.22}$$

This shows that the temperature of a substance is related to the average kinetic energy of its particles. The $3k/2$ is simply a proportionality constant.

Next Level

Maxwell–Boltzmann Distribution

The average speed of a particle in a gas is given by Eq. 13.22 to be $\overline{v} = (3kT/m)^{1/2}$. The particles of a sample of matter have speeds ranging from much less than this to much greater than this. The distribution of the number of particles having a specific speed is found from statistical mechanics to have a shape as shown in Fig. 13.5 at a specific temperature. This is called a Maxwell–Boltzmann distribution. As temperature increases, the velocity distribution curve shifts to having a maximum at higher speed, a broader width, and a lower maximum number.

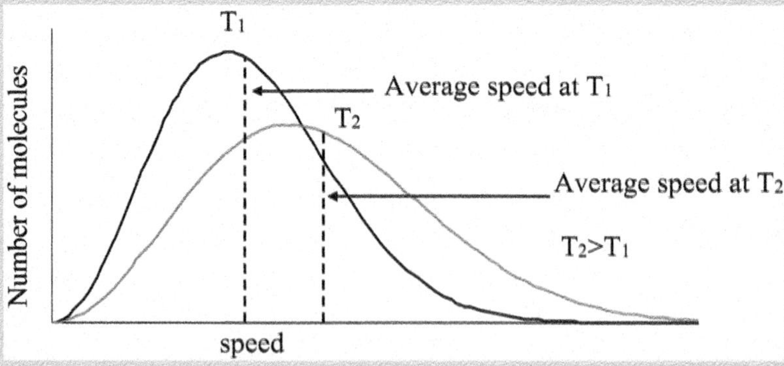

Fig. 13.5 Distribution of particle speeds

For Maxwell–Boltzmann statistics, the probability density distribution $f(v)$ is the probability per unit speed of finding a particle with a speed in a small range of values Δv near v. It is given by

$$f(v)\Delta v = \left(\frac{m}{2\pi kT}\right)^{3/2} 4\pi v^2 e^{-\frac{mv^2}{2kT}} \Delta v$$

The maximum of the probability density curve is a speed of $v_{max} = \sqrt{2kT/m}$ and the average particle speed is slightly larger, $v_{avg} = \sqrt{3/2}\, v_{max}$. The temperature is $T = \frac{mv_{avg}^2}{3k} = \frac{mv_p^2}{2k}$. This is consistent with Eq. 13.22.

Now try the following problem.

Student

13.8 The average kinetic energy of a mole of gas is 3123 J. If the average speed of the gasmolecules is 512 m/s, what is the temperature? What is the mass of one of the molecules? You can give your answer in kg or kg per mole, whichever you prefer.

13.6 P–V Diagrams

We have been discussing how a system changes its state under certain imposed conditions such as constant pressure or constant volume. These changes are called processes and there are several different types of processes that are important in thermodynamics. These are defined below.

- An isobaric process is one in which volume expansion occurs at a constant pressure. In this case the work done is given by $W = P\Delta V$. Thus, $\Delta U = Q - P\Delta V$.
- An isothermal process occurs at a constant temperature. The thermal energy of the system is also unchanged in an isothermal process, $\Delta U = 0$. Thus $Q = W$.
- An isometric process is one in which the volume is constant. In isometric processes, the work done is zero, $\Delta W = 0$. Thus $\Delta U = Q$.
- An adiabatic process is one in which the heat flow between the system and its surroundings is zero, $\Delta Q = 0$. In adiabatic processes the work done is equal to the change in thermal energy, $W = -\Delta U$.

As an example of adiabatic compression and expansion, do the following experiment.

Experiment 13.2
Adiabatic Compression and Expansion

Supplies:

- A large plastic bottle with a lid (You need to be able to see into the bottle, so remove any labels that might be on the bottle.)
- A thermometer that fits comfortably inside the bottle (The thermometer needs to read slightly above and below room temperature. It must also fit into the bottle easily, because you will be compressing the bottle.)
- Water
- A match

Introduction: In this experiment, you will be observing adiabatic compression and expansion in two different ways. You will see that an adiabatic process is not necessarily isothermal.

Procedure:

1. Open the plastic bottle and dump out any contents.
2. Place the thermometer into the bottle.
3. Screw the lid onto the bottle tightly so that the seal is airtight.
4. Watch the thermometer for a few moments. It might raise or lower slightly until it equilibrates.
5. When the thermometer is reading a constant temperature, squeeze the bottle hard so that the gas inside is compressed. Keep your eye on the thermometer and continue to squeeze.
6. As long as your system is airtight, you should notice a change in the thermometer as you squeeze. Once you have noted the change, release the bottle so that it expands back to it original size.
7. Note the change in the thermometer when you release the bottle.
8. Open the bottle and remove the thermometer.
9. Add some water to the bottle so that about one-tenth of the volume is water.
10. Light the match and then drop it into the bottle so that it hits the water and goes out.
11. Immediately put the lid back on the bottle and screw it tightly so that the seal is airtight.
12. Once again, squeeze the bottle hard and hold it for a second.
13. Release the bottle so that it expands to it initial size. Note what happens.

In this experiment, you were dealing with an adiabatic process. Since the temperature of the bottle contents was roughly the same as the temperature of the surroundings, and since plastic is an insulator, little if any heat was transferred from the gas inside the bottle to the surroundings, or vice-versa. Nevertheless, in the first part of the experiment, you saw that ΔT was *not* equal to zero.

When you compressed the air inside the bottle, you did work on the air. As a result, $W < 0$. Remember, when work is done on the gas, the work is negative. Also, since the process was adiabatic, $Q = 0$. Thus, $\Delta U = - W$ according to the First Law of Thermodynamics. If $W < 0$ and $Q = 0$, then $\Delta U > 0$. In other words, when you worked on the gas, the internal energy of the gas increased. That's why the temperature in the bottle increased when you squeezed the bottle. Now, of course, the increase was small (probably about 1–2 °C), because the amount of work you did was small. Had you been able to compress the gas to an even smaller volume, the increase in temperature would have been greater.

When you released the bottle and allowed the air inside it to expand, you should have seen the temperature drop. When the gas expands, $W > 0$. Since $Q = 0$, the first law says $\Delta U < 0$, which means the temperature must decrease. In other words, since heat was not being added, and since the air inside the bottle had to do work, the energy for that work had to come from the air's internal energy. As a result, the internal energy and thus the temperature decreased. Because the thermometer takes time to respond, however, that effect might not have been as noticeable. That's why we did the second part of the experiment.

In the second part of the experiment, you squeezed the bottle while it had some water in it. This heated up the water, causing it to evaporate. When you released the bottle, the air cooled back down. As the air cooled, it could not hold as much water vapor as it did when it was warmer. Thus, the water vapor had to condense out of the air, and that's why the bottle clouded up when you released it. The water vapor that had been put into thc air by the higher temperature had to condense back out, so it formed a cloud inside the bottle.

Why did you need to use the match? If you ever studied weather, you probably learned that in order for a cloud to form, there must be something onto which the water can condense. The match put small particles of smoke into the bottle. The water then could condense onto those smoke particles, forming a cloud. These particles are called "cloud condensation nuclei," because they form the center of water condensation, which forms clouds.

A good example of a practical use of an adiabatic process is a refrigerator. This is shown in Fig. 13.6. The "functional unit" of a refrigerator is composed of pipes, a compressor, and a gas. First, the gas is compressed by the compressor. This is mostly adiabatic, so the gas'stemperature increases substantially. The gas then travels through several twists and turns of the pipes where it exchanges most of its heat with the surroundings. This takes place outside of the refrigerator, typically behind it.

Once the gas has traveled through the twists and turns of pipes outside of the refrigerator, it encounters an expansion valve. In this area, the volume that the gas can occupy increases, so the gas expands. Once again, this is essentially an adiabatic process, so as the gas expands, it cools. The cold gas then travels through a series of twists and turns which are *inside* the refrigerator. This is what cools the inside of the refrigerator. When the gas gets back to the compressor, the process

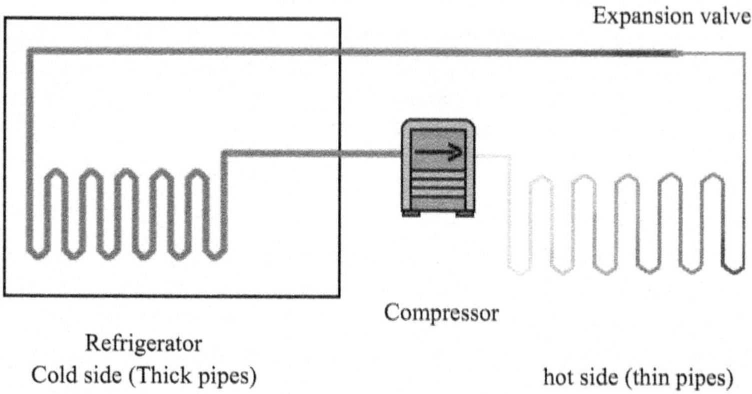

Fig. 13.6 Refrigerator

starts all over again. A refrigerator, then, just uses adiabatic expansionto produce a "cold side" and adiabatic compression to produce a "hot side." The outside of the refrigerator is the hot side, and the inside is the cool side.

The only time Eq. 13.21 is useful in calculating the work done in a thermodynamic process is for an isobaric process. If we are dealing with other types of processes, we can calculate the work done using pressurevolume diagrams such as the one shown in Fig. 13.7. The work done in a process is the area under the process line in a P–V diagram. In this figure, we plot the pressure and volume of a gas as it changes from its initial state to its final state. The gas starts out at a pressure of 1.0×10^5 Pa and a volume of 0.5 m^3 and ends up at a pressure of 2.0×10^5 Pa and a volume of 1.5 m^3. The gas expanded, but it did not expand under constant pressure, since the final pressure is different from the initial pressure. The arrow between the two points illustrates how the pressure and volume changed as the gas went from its initial to its final state.

The work done by the gas in this expansion can be calculated by finding the area underneath the arrow on this P–V diagram. This is a general result:

The area beneath a P–V curve is equal to the work done when a gas changes state.

We can calculate the area in the P–V graph in the same way we did when we were finding the area under velocity versus time diagrams in Chap. 2. We split the area under the curve up into simple geometrical shapes whose areas we know how to calculate. In Fig. 13.7, for example, we can divide the region under the arrow into two regions: a triangle (above the dashed, gray line) and a rectangle (below the dashed gray line). The rectangle goes from 0.5 to 1.5 m^3, so it has a base of 1.0 m^3. It also goes from 0 to 1.0×10^5 Pa, so it has a height of 1.0×10^5 Pa. Its area, then, is:

$$\text{Area of rectangle} = (b) \cdot (h) = \left(1.0\,\text{m}^3\right) \cdot \left(1.0 \times 10^5\,\text{Pa}\right)$$

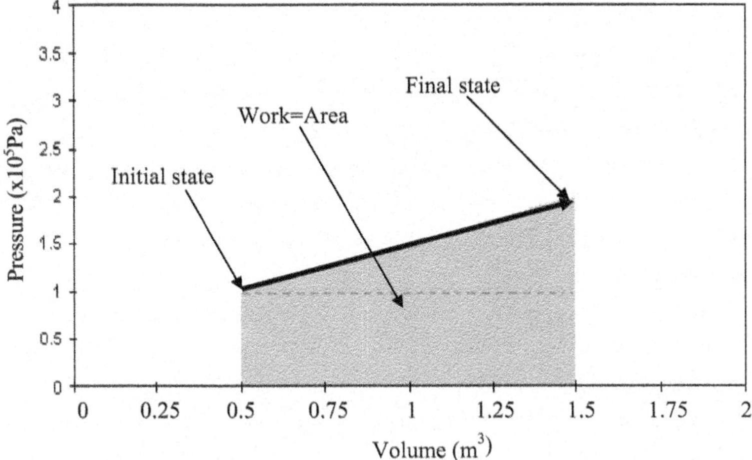

Fig. 13.7 P–V diagram

$$= 1.0 \times 10^5\,\mathrm{Pa\,m^3} = 1.0 \times 10^5\,\mathrm{J}$$

Notice that the units work out into energy units. That should make sense, since the area under a P–V curve is equal to work.

To complete our calculation of the area under the curve, we need to determine the area of the triangle. The triangle goes from 0.5 to 1.5 m^3, so it has a base of 1.0 m^3. It also goes from 1.0×10^5 to 2.0×10^5 Pa, so it has a height of 1.0×10^5 Pa. The area is:

$$\text{Area of triangle} = {}^1\!/_2 \cdot (b) \cdot (h) = {}^1\!/_2 \cdot \left(1.0\,\mathrm{m^3}\right) \cdot \left(1.0 \times 10^5\,\mathrm{Pa}\right) = 5.0 \times 10^4\,\mathrm{J}$$

The total area (which is also the work done) is simply the sum of the two:

$$\text{Work done} = 1.0 \times 10^5\,\mathrm{J} + 5.0 \times 10^4\,\mathrm{J} = 1.5 \times 10^5\,\mathrm{J}$$

This tells us that in making the transition shown in the P–V diagram of Fig. 13.7 the gas did 1.5×10^5 J of Work.

When using the area under the curve method, you must determine the sign of the work. In Fig. 13.7, the gas expanded, because it went from 0.5 to 1.5 m^3. Thus, work was done by the gas and it is therefore positive. Had the gas contracted, however, we would still have gotten a positive area for the work done. Thus, we would have had to add the negative sign in order to indicate that work was done on the gas. You will see how that works in the following problem.

Student

13.9 Two samples of gas are compressed indifferent ways. The first is compressed from 0.40 to 0.20 m^3 at a constant pressure of 1.50 × 10^5 Pa as shown by the dashed arrow in the figure to the right. The second is compressed from 0.40 to 0.20 m^3 while the pressure increases to 2.5 × 10^5 Pa. Then, the gas is cooled so that the pressure is decreased to 1.5 × 10^5 Pa while the volume does not change. This process is shown with the two solid arrows in the figure to the right. As a result, both gases have the same initial and final conditions. Calculate the work in each case.

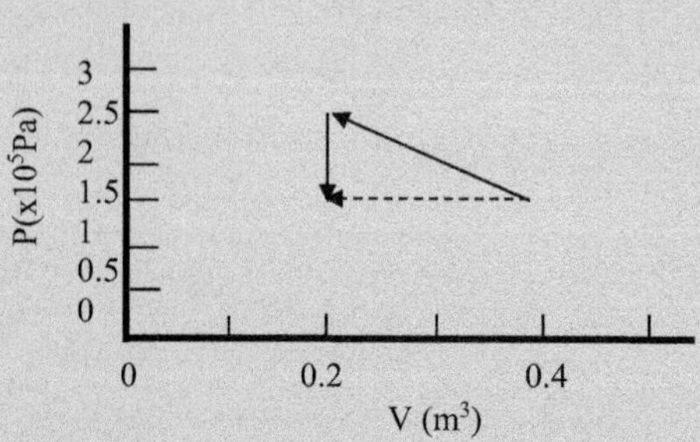

A thermodynamic cycle is a four-step process on a P–V diagram where the system ends up in the same state from which it started. This is illustrated in Fig. 13.8. There are many different kinds of cyclic processes, so the illustration is just an example. In this particular cyclic process, the gas is first expanded. We can see that, because in going from A to B, the gas's volume increases. The expansion is isobaric, because pressure stays constant at 2.0 × 10^5 Pa. In step 2, the gas's pressure decreases, but it does so in an isometric fashion, because the volume stays constant at 0.40 m^3. The gas is then compressed in step 3, which is an isobaric process, and the pressure is then increased in step 4, which is isometric.

Cyclic processes are quite common. As discussed in the previous section, for example, a refrigerator uses cyclic expansion and compression of a gas to produce a hot side and a cold side. One thing that is very important to remember about cyclic processes is that the gas returns to the same state at which it started. Since

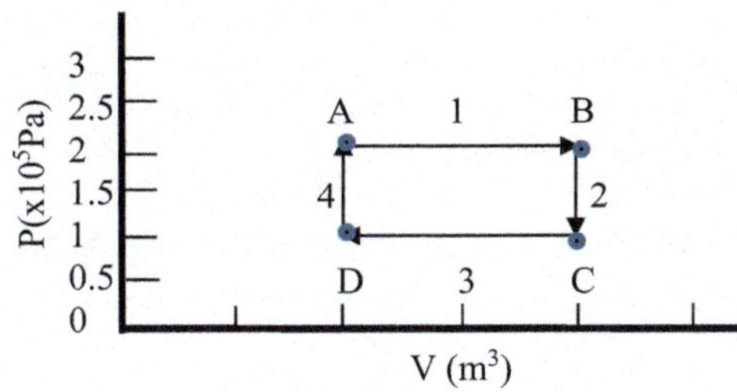

1. Isobaric expansion from A to B
2. Pressure decreased from B to C in an isometric process
3. Isobaric compression from C to D
4. Pressure increased from D to A in an isometric process

Fig. 13.8 P–V diagram of a cyclic process

that's the case, the internal energy of the gas must be the same at the end of the cyclic process as it was at the beginning.

In a cyclic process, $\Delta U = 0$ over a complete cycle.

This, of course, should make sense. The ideal gas law tells us that $PV = nRT$. In a cyclic process, P and V are the same at the end of the cycle as they were at the beginning. Assuming no gas was lost, n is also the same, and R is a constant. Well, if P, V, n, and R are all the same at the beginning and end, then T must be the same as well, which tells you that the internal energyis the same. Let's see how all of this works out in an example problem.

Special Topic

One of the most important contributions that thermodynamics gave to society was the invention of the steam engine. In the middle of the eighteenth century, James Watt invented a reciprocating engine where the piston motion was driven by steam working on a thermodynamic cycle. Mechanical work was done either directly by the moving piston or indirectly by connecting it to a rotating shaft. This invention was a major impetus for the industrial revolution. It found many applications in automating different types of factories and allowed factories to be built away from water sources which had been the major provider of power. It became the basis for steam locomotives which allowed the development of major railroads. Later steam turbines were developed that work on a thermodynamic cycle to generate electricity. Currently the great majority of electrical power on our grid is generated by steam turbines. We are definitely a society that depends on thermodynamic cycles.

Example 13.8

A 1.50-mol sample of ideal gas undergoes the cyclic process shown in Fig. 13.8. Calculate the work done and the heat exchanged in each step as well as one full cycle. (C_p = 20.8 J/molK)

Use Fig. 13.8 Knowns: $n = 1.50$ mol; $C_p = 20.8$ J/molK

Unknowns: W; Q

Reading from the graph, the gas starts out at a pressure of 2.0×10^5 Pa and a volume of 0.20 m^3. It then expands in an isobaric process to a volume of 0.40 m^3. Calculating the work done is easy, since the process is isobaric:

$$W = P \cdot \Delta V = \left(2.0 \times 10^5\,\text{Pa}\right) \cdot \left(0.40\,\text{m}^3 - 0.20\,\text{m}^3\right) = 4.0 \times 10^4\,\text{J}$$

Notice that the work is positive. That means the gas did work.

Now we have to determine Q for this step. We can use Eq. 13.3 with C_p since pressure stays constant:

$$Q = n \cdot C_p \cdot \Delta T$$

To get ΔT, we must use the ideal gas law. At the beginning of the process, $P = 2.0 \times 10^5$ Pa, $V = 0.20$ m^3, and $n = 1.50$ mol.

$$PV = nRT$$

$$T = PV/nR = \left(2.0 \times 10^5\,\text{Pa}\right)\left(0.20\,\text{m}^3\right) /\left[(1.5\,\text{mol})\left(8.31\,\text{Pa}\,\text{m}^3/\text{mol}\,\text{K}\right)\right] = 3200\,\text{K}$$

At the end of the process

$$T = \left(2.0 \times 10^5\,\text{Pa}\right)\left(0.40\,\text{m}^3\right)/\left[(1.5\,\text{mol})\left(8.31\,\text{Pa}\,\text{m}^3/\text{mol}\,\text{K}\right)\right] = 6400\,\text{K}$$

Now we can calculate Q

$$Q = (1.5\,\text{mol})(20.8\,\text{J/mol}\,\text{K})(6400\,\text{K} - 3200\,\text{K}) = 1.0 \times 10^5\,\text{J}$$

For step 2, we can immediately conclude that $W = 0$. This is a vertical line, and there is no area under a vertical line. Since the work done is the area under the P–V curve, the work done in step 2 is zero. The value for Q, however, is not zero. We cannot use C_p here, because pressure is changing. However, we can use C_V, but we will have to determine it first:

$$C_p = C_V + R$$

$$C_V = C_p - R = 20.8\,\text{J/mol K} - 8.31\,\text{J/mol K} = 12.5\,\text{J/mol K}$$

Using the ideal gas law calculation as we did for the first process, gives the beginning temperature for step 2 as $T = 6400$ K, and temperature at the end of step 2 as, $T = 3200$ K. Thus:

$$Q = nC_V\Delta T = (1.50\,\text{mol})(12.5\,\text{J/mol K}) \cdot (3200 - 6400) = -6.0 \times 10^4\,\text{J}$$

Notice that Q is negative. That tells us that the gas cooled down. This should make sense, since the pressure of the gas decreased but the volume did not. The only way a gas's pressure can decrease when the volume stays constant is if the gas cools down.

We can calculate the work and heat of steps 3 and 4 in the same way. The work in step 3 is calculated with Eq. 13.3 again, and it turns out to be -2.0×10^4 J. The work is negative because the gas was compressed so work was done on the gas. The heat is calculated by determining the initial and final temperatures with the ideal gas law and then calculating Q using C_p, because the process is isobaric. The result is -5.0×10^4 J. Once again, Q is negative because the gas cools down.

In the last step, no work is done, because the line is vertical and thus there is no area under it. Once again, if we calculate the temperature at the beginning and end of the process, we can use C_V to determine Q. The value for Q works out to be 3.0×10^4 J. It is positive because the gas warmed up. Now let's look at all of the steps:

1. $W = 4.0 \times 10^4\,\text{J},\ Q = 1.0 \times 10^5\,\text{J}$
2. $W = 0,\ Q = -6.0 \times 10^4\,\text{J}$
3. $W = -2.0 \times 10^4\,\text{J},\ Q = -5.0 \times 10^4\,\text{J}$
4. $W = 0,\ Q = 3.0 \times 10^4\,\text{J}.$

Notice what happens when we calculate the *total* work done and heat exchanged. The total work done is 2.0×10^4 J, and the total heat exchanged is 2.0×10^4 J. In this thermodynamic process cycle, $Q = W$ for one complete cycle. This makes sense since $\Delta U = Q - W$, and we know that for one full cycle, $\Delta U = 0$. Thus, Q must equal W in a cyclic process. Interestingly enough, if you calculate the area enclosed by the square, you will get 2.0×10^4 J. This is a general result.

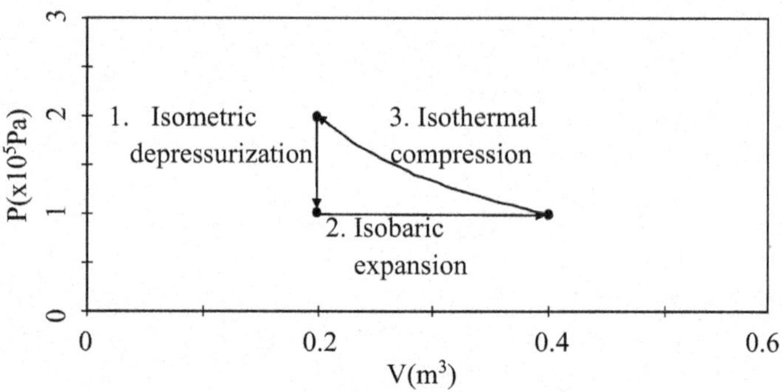

Fig. 13.9 Cyclic process with an isothermal step

An important result from this example is that:

The work done in one full cycle is the area enclosed by that cycle on a P–V diagram.

Even though the system started and ended in the same state so that $\Delta U = 0$, heat was converted to work. The process steps in P–V diagrams are not always straight lines as shown in Fig. 13.9. This cyclic process is composed of three steps: an isometric process that lowers the pressure, an isobaric expansion, and then an isothermal compression. It turns out that all isothermal processes have a curved shape, because in an isothermal process, T is constant. Since n and R are also constant, the ideal gas law becomes:

$$PV = \text{constant}$$

So

$$P = \text{constant}/V$$

Therefore

On a P–V diagram, an isothermal process will have the shape of a $y = 1/x$.

Since this is a cyclic process, we know that ΔU equals 0. This means that for the entire process, $Q = W$. This is important, because it would be nice to know Q and W for each step as well as the entire cycle. However, we cannot calculate W for the isothermal compression by splitting the area underneath the curve into nice, recognizable geometric shapes because of the curved nature of the line. However, using the First Law of Thermodynamics and the fact that $\Delta U = 0$ for a cyclic process, we can determine the work done in the isothermal compression if we know something about the heat transferred. This is shown in the following example.

Example 13.9

A 1.0 mol sample of an ideal gas is taken through the cyclic process shown in Fig. 13.9. If $Q = -9.0 \times 10^3$ J for one complete cycle, what is the work done during the isothermal compression?

Use Fig. 13.9 Knowns: $n = 1$ mol; $Q = -9.0 \times 10^3$ J

Unknowns: W

We know that through one cycle, $\Delta U = 0$. This means that $Q = W$ for the entire cycle. Thus, if $Q = -9.0 \times 10^3$ J for the entire cycle, then $W = -9.0 \times 10^3$ J for the entire cycle. However, that's not quite what the question asked for. It asked for the work of the isothermal compression. We can calculate that by taking the total work done and subtracting out the work done in the other steps. In the isometric depressurization, the work done is zero. Remember, there is no area under a vertical line, so there is no work done in an isometric process. Thus, $W = 0$ for that step.

In the isobaric expansion, we can calculate the work with Eq. 13.21:

$$W = P \cdot \Delta V = \left(1.0 \times 10^5 \,\text{Pa}\right) \cdot \left(0.40\,\text{m}^3 - 0.20\,\text{m}^3\right) = 2.0 \times 10^4 \,\text{J}$$

We know that the total work is -9.0×10^3 J, so:

$$W_{\text{total}} = W_{\text{isometric}} + W_{\text{isobaric}} + W_{\text{isothermal}}$$

$$-9.0 \times 10^3 \,\text{J} = 0 + 2.0 \times 10^4 \,\text{J} + W_{\text{isothermal}}$$

$$W_{\text{isothermal}} = -29{,}000 \,\text{J}$$

The work in the isothermal step, then, is $-29{,}000$ J. The negative sign means that the surroundings worked on the gas, which makes sense, since the gas was compressed.

Try this problem.

Student

13.10 Given the P–V diagram below:

a. calculate the total work done and total heat exchanged in one cycle.
b. what would happen to the values of W and Q if the cycle were done in reverse? In other words, what would happen if the arrows were pointing opposite of the way they are pointing now?

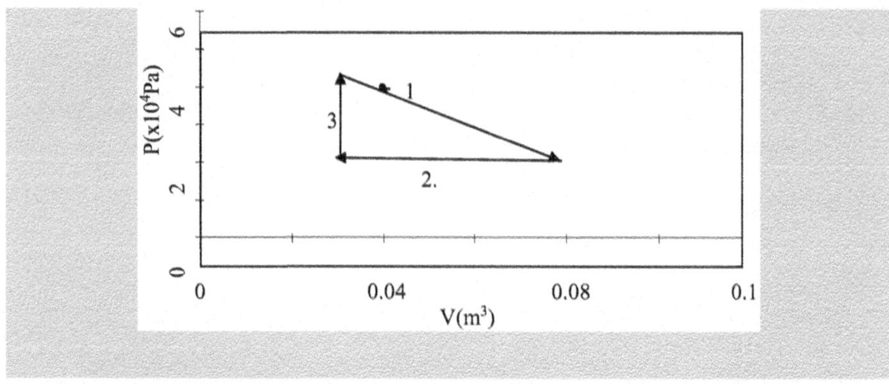

P–V diagrams are useful in describing the operation of heat engines that we discussed in an earlier section of this chapter. As we found, no cyclic heat engine can convert all of its heat input into useful work because this would violate the second law of thermodynamics. They must exhaust some heat and operate at less than 100% efficiency.

We can calculate the efficiency of a heat engine. It takes in a certain amount of energy (Q_H) and then loses some of that energy (Q_L). The work it creates is given by $W = Q_H - Q_L$. The efficiency (e), then, is then the work done divided by the total energy input,

$$e = W/Q_H = 1 - Q_L/Q_H \tag{13.23}$$

Since the Second Law of Thermodynamics says that Q_L cannot be zero (some of the heat must be lost to the surroundings), then the Second Law of Thermodynamics tells us that the efficiency of a heat engine can never be equal to 1.

It would be nice to determine what the most efficient design is for a heat engine. Equation 13.23 tells us the maximum efficiency for a generic heat engine, but some heat engines must be more efficient than others. What is the most efficient heat engine possible? Sadi Carnot (kar' noh) answered that question in the early 1800s. He demonstrated that according to the Second Law of Thermodynamics, the most efficient heat engine possible would work according to the following cycle, which we now call the Carnot cycle. This is shown in Fig. 13.10.

The Carnot cycle is composed of four basic steps. The area enclosed by these steps is the work done. First, a gas undergoes an **isothermal compression**. This is the step in which energy is lost. Since $\Delta U = 0$, that means $Q = W$. Thus, the work it takes to compress the gas is lost. Physically, this happens because as energy is lost to the cold reservoir, the gas cools down. However, since this is isothermal, the gas's temperature does not decrease. It stays the same as the cold reservoir, T_L. Thus, the volume decreases instead.

The second step of the Carnot cycle is an **adiabatic compression**. No heat is added or lost in adiabatic processes, so $Q = 0$. As a result, $\Delta U = -W$. Thus, as work is done on the gas, the internal energy and thus the temperature of the

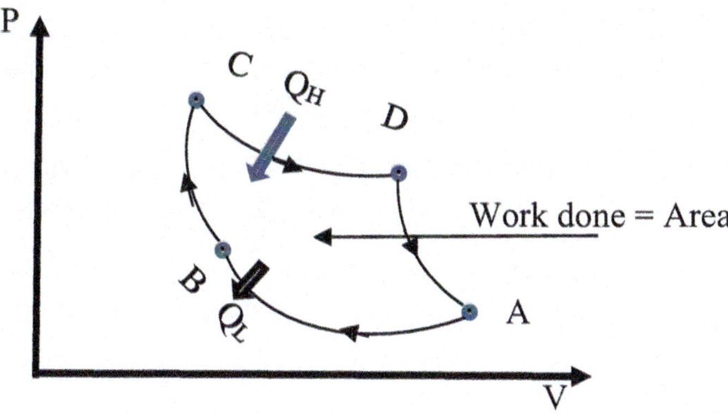

Fig. 13.10 Carnot cycle

gas increases. Now remember, during the previous step, the temperature of the gas stayed steady at T_L. Now that work is being done on the gas, however, the temperature begins to increase. The temperature continues to increase until it reaches T_H, the temperature of the hot reservoir.

Once the gas has reached the high temperature, it is ready to start the third step, which is **isothermal expansion**. The gas is at the same temperature as the heat reservoir, so heat will not flow from the reservoir to the gas unless the gas begins to expand. Since it expands isothermally, $\Delta U = 0$ and thus $Q = W$. Any energy absorbed from the heat reservoir, then, gets converted directly to work. This is the step in which the heat engine produces work. The problem is, the first step takes some of that work away, so the net work produced is the work produced in this step *minus* the work used in the first step (the isothermal compression).

Once the isothermal expansion is completed, an **adiabatic expansion** occurs in order to get the gas back to its original state so the process can start all over again. Since $Q = 0$, the gascools, because $\Delta U = -W$. Thus, the gas does work by expanding, and that work lowers the internal energy of the gas back to T_L.

As is the case with any cyclic process, the work done in the Carnot cycle is equal to the difference in the heat transferred from the heat reservoir and that taken away by the cold reservoir. This is also equal to the area enclosed in the P–V diagram.

Why is this such an efficient heat engine? Since the step that takes in heat (C–D) is isothermal, all heat taken in is converted to work. None is used to warm up the gas. In the same way, when heat is removed, that step (A–B) is isothermal as well. Thus, the gas loses no internal energy. Also, the two adiabatic steps which heat (B–C) and cool (D–A) the gasinvolve no heat transfer. They are adiabatic. Thus, the only time heat is transferred, that heatdoes work.

After Carnot proposed the Carnot cycle, Lord Kelvin demonstrated that the ratio of Q_L to Q_H is equal to the ratio of the two temperatures. Mathematically, this is expressed as

For the Carnot cycle ONLY: $Q_L/Q_H = T_L/T_H$

This allows us to simplify Eq. 13.13 for the efficiency of a Carnot cycle

$$\eta_{Carnot} = 1 - T_L/T_H \tag{13.24}$$

Once again, in this equation, temperature must be in Kelvin. Equation 13.24 tells us two things. First, it tells us that even the most efficient heat engine designed is not 100% efficient. Its efficiency is reduced from one by the ratio of the temperature of the cold reservoir to the temperature of the hot reservoir. Second, it tells us that the hotter the hot reservoir and the cooler the cold reservoir, the more efficient the engine.

Summing Up

In this chapter you learned about the phases of matter and how matter can change from one phase to another obeying the laws of thermodynamics. The most important laws state that energy is conserved and that the disorder of the universe always increases or stays the same for natural processes. You learned about the differences between internal energy, temperature, and heat and how to get work from a heat engine. For gasses, you learned about the relationship between temperature, pressure and volume and how to diagram processes showing the changes in these variables and the work done during the process. All of this has important applications in the world we live in. In the next chapter we extend this discussion to solids and to other properties of matter.

Answers to the Problems

13.1

Knowns: $n = 4.5$ mol; $Q = 1000.0$ J; $P =$ const.; $\Delta T = 21.5$ °C

Unknowns: C_v

We can get the molar heat capacity at constant pressure rather easily:

$$Q = nC_p\Delta T$$

$$C_p = Q/n\Delta T = 1000.0\,\mathrm{J}/(4.50\,\mathrm{mol})(21.5\,\mathrm{K}) = 10.3\,\mathrm{J/mol\,K}$$

Remember, when dealing with ΔT, Kelvin and Celsius are the same.

Now we can find the molar heat capacity at constant *volume*:

$$C_v = C_p - R = 10.3\,\mathrm{J/mol\,K} - 8.31\,\mathrm{J/mol\,K} = 2.0\,\mathrm{J/mol\,K}$$

13.2

Knowns: $m_m = 150.0$ g; $T_m = 95$ °C; $m_w = 300.0$ g; $T_w = 25$ °C; $T_f = 27.4$ °C

Unknowns: C_m

In this problem, the metal loses energy while the water gains energy. Thus:

$$q_{metal} = -q_{water}$$

$$(150\,\text{g})C_m(27.4\,°\text{C} - 95.0\,°\text{C}) = -(300.0)(4.19\,\text{J/g}\,°\text{C})(27.4\,°\text{C} - 25.0\,°\text{C})$$

$$C_m = 0.30\,\text{J/g}\,°\text{C}$$

13.3

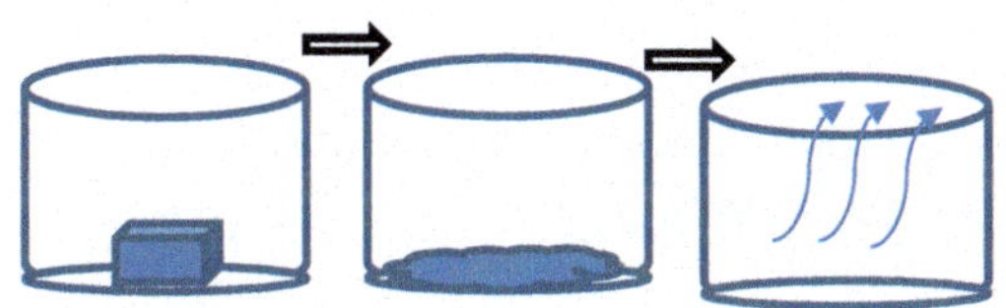

Knowns: $m = 50.0$ g;
$T_i = -11.0$ °C;
$C_{ice} = 2.02$ J/g °C;
$L_{vap} = 2260$ J/g; $L_{fus} = 334$ J/g

Unknowns: Q

To evaporate all the mass in the ice cube, we must first warm the ice cube to 0.0 °C:

$$Q = mC\Delta T = (50.0\,\text{g}) \cdot (2.02\,\text{J/g}\,°\text{C}) \cdot (0.0° - 11.0\,°\text{C}) = 1110\,\text{J}$$

Now that the ice is at 0.0 °C, it can melt.

$$Q = mL = (50.0\,\text{g}) \cdot (334\,\text{J/g}) = 16{,}700\,\text{J}$$

Now we have to take the liquid water and warm it up to the boiling point:

$$Q = mC\Delta T = (50.0\,\text{g}) \cdot (4.19/°\text{C}) \cdot (100.0\,°\text{C} - 0.0\,°\text{C}) = 2.10 \times 10^4\,\text{J}$$

Now that the water can boil, we just have to add the energy required to evaporate it:

$$Q = mL = (50.0\,\text{g}) \cdot (2260\,\text{J/g}) = 113{,}000\,\text{J}$$

The sum of all of the energies calculated above represents the total amount of energy needed:

$$Q_{tot} = 1110\,\text{J} + 16{,}700\,\text{J} + 2.10 \times 10^4\,\text{J} + 113{,}000\,\text{J} = 152{,}000\,\text{J}$$

13.4

Knowns: $T = 314\,\text{K}$; $W = 2450\,\text{J}$

Unknowns: ΔS

In an isothermal process $\Delta U = 0$ so $W = Q$. Thus, $Q = 2450$ J. Then Eq. 13.10 is

$$\Delta S_{rev} = Q/T = 2450\,\text{J}/314\,\text{K} = 7.80\,\text{J/K}$$

13.5

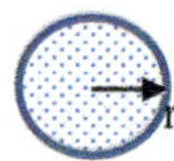

Knowns: $P = 101.3\,\text{kPa}$; $r = 10\,\text{cm}$

Unknowns: F

a. To determine the average force, we can use Eq. 13.20 if we know the area over which the force is exerted. This can be calculated from the surface area equation:

$$A = 4\pi r^2 = 4\pi(0.100\,\text{m})^2 = 0.126\,\text{m}^2$$

To get the units to work out, we must use standard units, so $P = 1.013 \times 10^5$ Pa:

$$P = F/A$$
$$F = PA = \left(1.013 \times 10^5\right) \cdot \left(0.126\,\text{m}^2\right) = 12{,}800\,\text{N}$$

That's a LOT of force!

b. To determine the number of moles, we will need the ideal gas law. To use that equation, however, we will need to calculate the volume:

$$V = 4\pi r^3/3 = 4\pi(0.100\,\text{m})^3/3 = 0.00419\,\text{m}^3$$

Now we need to get temperature into K:

$$K = ^\circ\text{C} + 273.15 = 25.00 + 273.15 = 298.15\,\text{K}$$

Now we are all set:

$$PV = nRT$$

$$n = (PV)/(RT) = \left(1.013 \times 10^5\,\text{Pa}\right)\left(0.00419\,\text{m}^3\right)$$
$$/(8.31\,\text{J/mol K})(298.15\,\text{K}) = 0.171\,\text{mol}$$

Since 1 mol contains 6.02×10^{23} molecules.

$$\text{No. Molecules} = (0.171\,\text{mol})\left(6.02 \times 10^{23}\,\text{molecules/mol}\right)$$
$$= 1.03 \times 10^{23}\,\text{molecules}$$

13.6

Knowns: $n = 2.50\,\text{mol}$; $P = \text{constant}$; $V_i = 0.0514\,\text{m}^3$; $V_f = 0.0201\,\text{m}^3$; $\Delta U = -2570\,\text{J}$; $C_p = 20.8\,\text{J/(molK)}$
Unknowns: P

Combining Eqs. 13.3, 13.6, and 13.21 gives

$$\Delta U = Q - W = mC_p\Delta T - P\Delta V$$

$$-2570\,\text{J} = (2.50\,\text{mol})(20.8\,\text{J/(mol K)})\Delta T - P\left(0.0201\,\text{m}^3 - 0.0514\,\text{m}^3\right)$$

Note that ΔU is negative because the internal energy *drops*.
We can get both temperatures in terms of P through the ideal gas law, Eq. 13.18

$$T_f = PV_f/nR = P\left(0.0201\,\text{m}^3\right)/[(2.50\,\text{mol})(8.31\,\text{J/mol K})]$$
$$= 9.67 \times 10^{-4} P\,\text{m}^3\text{K/J}$$

$$T_i = PV_i/nR = P\left(0.0514\,\text{m}^3\right)/[(2.50\,\text{mol})(8.31\,\text{J/mol K})]$$
$$= 24.71 \times 10^{-4} P\,\text{m}^3\text{K/J}$$

$$\Delta T = \left(-15.04 \times 10^{-4} P\right)\text{K}$$

Substituting into the equation above:

$$-2570\,\mathrm{J} = (2.50\,\mathrm{mol})(20.8\,\mathrm{J/(mol\,K)})\left(-15.04 \times 10^{-4}P\right)\mathrm{K} \\ - P\left(0.0201\,\mathrm{m}^3 - 0.0514\,\mathrm{m}^3\right)$$

$$-2570\,\mathrm{J} = -(0.0782P)\,\mathrm{J} + (0.0313P)\,\mathrm{J} = -(0.0469P)\,\mathrm{J}$$

$$P = 54{,}7007\,\mathrm{Pa}$$

13.7

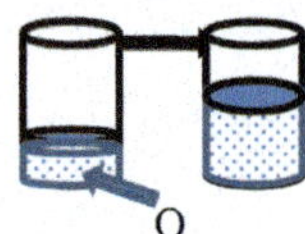

Knowns: T = constant; $W = 2020$ J

Unknowns: Q

Since the *temperature* did not change, that tells you the *internal energy* did not change, so

$$\Delta U = 0.$$

This means:

$$\Delta U = Q - W = 0$$

Therefore,

$$Q = W = 2020\,\mathrm{J}$$

Therefore, 2020 J of heat were added to the gas. The heat powered the expansion only so there was no energy left over to increase the temperature of the gas.

13.8

Knowns: $\mathrm{KE}_{\mathrm{avg}} = 3123\,\mathrm{J}$; $v_{\mathrm{avg}} = 512\,\mathrm{m/s}$; $n = 1\,\mathrm{mol}$

Unknowns: T; mass per molecule

The first part of the problem is a straightforward application of Eq. 13.22. However, we must understand that this equation uses the kinetic energy *per molecule.* This sample has one mole, which has 6.02×10^{23} molecules. Thus, the kinetic energy per molecule is:

$$\begin{aligned} \mathrm{KE_{avg}} &= 3kT/2 = 3123\,\mathrm{J}/6.02 \times 10^{23}\,\mathrm{molecules/mol} \\ &= 5.19 \times 10^{-21}\,\mathrm{J} \end{aligned}$$

We can relate *that* to the temperature:

$$T = 2\mathrm{KE_{avg}}/(3k) = 2\left(5.19 \times 10^{-21}\,\mathrm{J}\right)/3\left(1.38 \times 10^{-23}\,\mathrm{J/K}\right) = 251\,\mathrm{K}$$

Now we need to calculate the mass. You could either use the definition of kinetic energy (KE = ½ mv^2) with Eq. 13.22:

$$m = 3kT/v^2 = 3\left(1.38 \times 10^{-23}\,\mathrm{J/K}\right)(251\,\mathrm{K})/(512\,\mathrm{m/s})^2 = 3.96 \times 10^{-26}\,\mathrm{kg}$$

13.9

Use the figure that is given	Knowns: Gas 1: $V_i = 0.40\ \mathrm{m}^3$; $V_f = 0.20\ \mathrm{m}^3$; $P = 1.50 \times 10^5$ Pa Gas 2: $P2 = 2.5 \times 10^5$ Pa Unknowns: W_1; W_2

The work done on the first gas is easy to calculate, because pressure is constant:

$$W = P \cdot \Delta V = \left(1.5 \times 10^5\,\mathrm{Pa}\right) \cdot \left(0.20\,\mathrm{m}^3 - 0.40\,\mathrm{m}^3\right) = -3.0 \times 10^4\,\mathrm{J}$$

The work done on the second gas is the area under the line in the first step in the processes. In the second step, the work is zero, because volume did not change. You can also see that work is zero because the line in the P–V diagram is vertical. There is no area under a vertical line, so no work is done. The area under the process line for the first step can be divided into a rectangle and a triangle. Area of the triangle:

$$\mathrm{Area} = (bh)/2 = \left(0.20\,\mathrm{m}^3\right)\left(1.0 \times 10^5\,\mathrm{Pa}\right)/2 = 1.0 \times 10^4\,\mathrm{J}$$

Area of the rectangle:

$$\mathrm{Area} = w \cdot h = \left(0.20\,\mathrm{m}^3\right) \cdot \left(1.5 \times 10^5\,\mathrm{Pa}\right) = 3.0 \times 10^4\,\mathrm{J}$$

The total work for the first step, then, is -1.0×10^4 J + -3.0×10^4 J = -4.0×10^4 J. Note that the work is negative. We had to figure that out on our own, since the sign is not taken care of by any equation. We know that work is negative because something had to work *on* the gas to get it to compress. Work done by the gas is positive; work done *on* the gas is negative.

Notice that the work is *not the same* for the two gases. Although both gases started and ended in the same state, the work done is different. That's because work can vary depending on the *path* through which the gas must go. The second gas was taken through a path that required more energy. The only quantity whose value is independent of path is U. Thus, the ΔU of both gases in this problem is the same, but both W and Q are not.

13.10

Use the given figure	Knowns: see figure
	Unknowns: W; ΔQ

a. The work is the area of the triangle enclosed by the cycle.

$$W = bh/2 = \left(0.50\,\text{m}^3\right)\left(2.0 \times 10^4\,\text{Pa}\right)/2 = 5.0 \times 10^2\,\text{J}$$

We have to figure out the positive or negative sign for the work. In step 1, work is positive, since the gas expands. In step 2, the work is negative, since the gas contracts. In step 3, work is 0, since the volume is constant. The area under the step 1 line is greater than the area under the step 2 line, meaning there is more positive work than negative work, so the total work is positive. Thus, the work is 5.0×10^2 J.

In a cyclic process, $\Delta U = 0$ for a complete cycle, which means $Q = W$. Thus, the heat is also 5.0×10^2 J.

b. If the direction of the process is reversed, the area enclosed in the cycle is the same but the gas is compressed so the work is negative and the heat transfer is negative.

Study Guide for This Chapter

1. When a gas is held at a constant volume, the pressure it exerts increases with increasing temperature. Why?
2. Consider a sample of liquid water. Suppose you cooled it down to 0.0 °C but kept it from freezing by gently stirring it. Then, suppose you stopped stirring and carefully measured the temperature of the water's surroundings as the water froze. Would the temperature of the surroundings increase, decrease, or remain the same as the water froze?
3. A gas exerts a given amount of pressure on its cubic container. Suppose the length of each side of the cube was decreased by a factor of 2. If the temperature of the gas and the amount of gas does not change, what is the change in pressure?

4. You have a certain amount of energy that you can transfer to a sample of gas. If you want the temperature of the gas to change as little as possible, should you keep the gas at constant volume or constant pressure?
5. A gas is stored in an insulated container so that it cannot exchange heat with its surroundings. If the gas is compressed, what will happen to its temperature?
6. Is it possible for a gas to expand both isothermally and adiabatically? Why or why not?
7. Is it possible for a gas to expand both isobarically and adiabatically? Why or why not?
8. A gas experiences a decrease in entropy and an increase in temperature. Is W positive or negative for this process? What happened to the entropy of the surroundings?
9. What is special about a Carnot engine? State the four steps by which such an engine operates.
10. A 50.0 g chunk of metal initially at 75.0 °F absorbs 350.0 J of heat. If the specific heat capacity of the metal is 0.250 J/(g °C), what is the final temperature of the metal in °F?
11. A 75.0 g ice cube ($T = -$ 15.0 °C) is placed in 300.0 g of water ($T = 95.0$ °C). What is the final temperature of the resulting mixture? ($C_{ice} = 2.02$ J/(g °C), $L_{fusion} = 334$ J/g for water)
12. A total of 5.0 g of water condenses on the surface of a glass ($m = 200.0$ g, $C = 0.837$ Jg °C). Assuming that all heat released by the water is transferred to the glass, by how many degrees Celsius does the temperature of the glass increase? ($L_{vap} = 2260$ J/g)
13. A 1.00 mol sample of a gas is held at atmospheric pressure (101.3 kPa) and heated from 25.0 to 50.0 °C. What was the volume of the gas at 25.0 °C? What is the volume of the gas at 50.0 °C?
14. A 4.50 mol sample of gas has a mass of 319.5 g. If it is held at a temperature of 50.0 °C, what is the average speed of the molecules?
15. A balloon is inflated to a volume of 0.0050 m^3 at a pressure of 1.01×10^5 Pa and a temperature of 25 °C. It is then placed in a freezer which is at the same pressure but has a temperature of $-$ 5 °C. What is the change in internal energy of the gas? ($C_p = 20.8$ J/[moleK])
16. A gas expands both adiabatically and isobarically. It starts out at $P = 1.0 \times 10^5$ Pa, $V = 0.020$ m^3. If the gas loses 1560 J of internal energy, what is the volume after expansion?
17. Consider a 3-step cyclic process in which a sample of gas started at $P = 1.5 \times 10^5$ Pa, $V = 0.020$ m^3, first expanded to $P = 1.5 \times 10^5$ Pa, $V = 0.040$ m^3, then was compressed to $P = 2.5 \times 10^5$ Pa, $V = 0.030$ m^3, and finally compressed again to $P = 1.5 \times 10^5$ Pa, $V = 0.020$ m^3.
    a. What are Q and W for this cyclic process?
    b. Does this cycle best represent a heat engine or a refrigerator?

18. A gas contracts isothermally at a temperature of 315 K. When the contraction is over, ΔS of the gas is -1.5 J/K.
    a. What is the work associated with this process?
    b. What is the minimum change in entropy of the surroundings?
19. Suppose you want to create a Carnot engine whose maximum efficiency is 0.800. If the heat source you use has a temperature of 565 °C, what must the cold temperature of the engine be?

Next Level

20. A volume of gas of atoms of mass 3×10^{-27} kg has a temperature of 300 K. What is the maximum velocity of an atom in the gas and what is the average value?
21. If the temperature of a gas is doubled, how much does the average velocity of an atom of the gas increase?

Properties of Matter II: Mechanical, Electrical, Optical, and Thermal Properties

14

Chapter Summary

In the previous chapter, we focused on the thermal properties of matter in the gas and liquid phases. In this chapter these discussions will be extended to other types of properties and focus on matter in the solid state. Solid state physics has become so important in practical applications that it is treated as a separate subarea of physics. We will see how the knowledge of the electrical and optical properties of solids has led to some of the greatest developments of the twentieth century, microelectronics and lasers. Some of the themes of the chapter are represented in Fig. 14.1. Figure 14.2 shows a typical crystal shape.

Main Concepts in This Chapter

- Thermal Properties of Solids
- Mechanical Properties of Matter
- Electrical Properties of Matter
- Optical Properties of Matter

R. C. Powell, *Physics*, Undergraduate Lecture Notes in Physics,
https://doi.org/10.1007/978-3-032-09361-5_14

14.1 Introduction

The thermodynamics we discussed in the previous chapter focused on matter in the gas phase, although some areas like calorimetry involved solid and liquid matter also. Now we want to continue our discussion of the properties of matter by focusing on matter in the liquid and solid states.

So far we have talked about the properties of solids in relation to the physics topic we are studying such as electricity, magnetism or optics. Now we will reconsider these topics as properties of the solid material we are studying. Specifically, we will look at the thermal properties, mechanical properties, electrical properties and optical properties of solids. When we treat mechanical properties we will include the consideration of fluids as well as solids.

Solids can be divided into two general types depending on their structure: crystals and glasses. The atoms or molecules making up a crystal are organized in geometric shapes called lattices This gives the crystal long range order with a symmetry that is critical in determining its physical properties. There are 14 different shapes of crystal lattices. Some physical properties depend on the long-range symmetry of the crystal lattice while others depend on the symmetry an atom experiences at a specific point in the crystal. Glasses have an amorphous structure with no long-range order to the placement of the atoms or molecules. This structural difference leads to significant differences in the properties of crystals and glasses.

Fig. 14.1 Octahedral fluorite crystal. *Credit* W. Carter, Creative Commons Attribution-Share Alike 4.0

14.2 Thermal Properties of Solids

In the previous chapter, we saw that the ideal gas law predicted that at a constant pressure, an increase in temperature is accompanied by an increase in volume. This is called "thermal expansion." Although Eq. 13.18 applies only to gases, thermal expansion occurs in all states of matter, solids, liquids and gases. As the substance expands, its atoms (or molecules) move away from each other and it becomes less dense.

In solids thermal expansion occurs because as temperature rises the kinetic energy of the vibrating particles increases. As their vibrational amplitude increases, the average distance between the particles increases. The amount of expansion is proportional to the increase in temperature. For expansion in one dimension,

$$\Delta L = \alpha L_1 \Delta T \tag{14.1}$$

where α is called the "coefficient of linear expansion." It has units of K^{-1}. L_1 is the length of the sample. In three dimension the expression for thermal expansion is,

$$\Delta V = \beta V_1 \Delta T \tag{14.2}$$

where β is called the "coefficient of volume expansion." It has units of K^{-1}. V_1 is the volume of the sample.

For a typical material like aluminum, $\alpha = 2.5 \times 10^{-5}\ K^{-1}$ and $\beta = 7.5 \times 10^{-5}\ K^{-1}$. It is usual for the coefficient of volume expansion to be three times the coefficient of linear expansion as is the case for aluminum.

Example 14.1

If a 0.32 m³ piece of aluminum is heated from 280 to 300 K, what is its final volume?

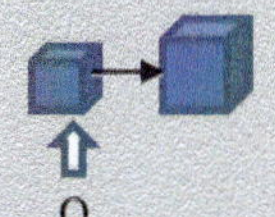

Knowns: $V = 0.32\ m^3$; $T_1 = 220$ K; $T_2 = 320$ K; $\beta = 7.5 \times 10^{-5}\ K^{-1}$

Unknowns: V_f

Using Eq. 14.2 and the coefficient of volume expansion for aluminum gives the change in volume:

$$\Delta V = \beta V_1 \Delta T = \left(7.5 \times 10^{-5}\,K^{-1}\right)(0.32\,m^3)(100\,K) = 2.40 \times 10^{-3}\,m^3$$

$$V_f = V + \Delta V = (0.320\,m^3) + \left(2.40 \times 10^{-3}\,m^3\right) = 0.3224\,m^3$$

Note that the volume change for thermal expansion is solids is much smaller than it was for gases.

Now try the following problem.

Student

14.1 A 10 m length copper wire has a coefficient of linear expansion of 1.7×10^{-5}/K. If it is currently at 300 K and constrained to expand in one dimension, what temperature do you have to heat it to for it to be 10.1 m long?

14.3 Mechanical Properties of Matter

Mechanical properties are the properties of matter that are the reaction to a force. The forces can be internal to the matter itself or external applied forces. Both are important in describing mechanical properties. The particles in gases are so far apart and weakly interacting that their mechanical properties are essentially described by their response to changes in pressure as discussed in the previous chapter. In liquids, the application of a force at some point is felt throughout the entire sample. In some cases, it can cause the liquid to flow. The application of a force that causes a solid sample to move was discussed in Chap. 2. Forces can also cause the properties of a solid sample to change. This section will focus on forces in liquids and solids.

Let us begin by considering the effects of forces on a liquid. The particles (atoms or molecules) that make up a liquid are held together by an electromagnetic force of attraction. This is weak enough to allow the particles to move over each other so the liquid flows, but strong enough that the particles do not fly away from each other like a gas. The particles are always in motion and this kinetic energy determines the temperature of the liquid. These internal forces holding the liquid together are called cohesive forces. These cohesive forces and the collisions between particles as they move provide an internal friction that inhibits the flow of the liquid. A liquid's resistance to flowing is called its viscosity. Viscosity is a measure of the internal friction of a liquid. The stronger the cohesive forces in a liquid, the higher its viscosity. As temperature increases, the viscosity of a liquid decreases because the particles of the substance move farther apart thus lowering their cohesive forces so the liquid flows more easily.

In the middle of a liquid, a particle experiences cohesive forces in all directions so the net force is zero. However, there are no particles above a particle on the surface of the liquid so there is no cohesive force in the upward direction. The net cohesive force of a particle on the surface of a liquid is in the downward direction producing a compressed surface layer. This is called surface tension and it prevents the particles on the surface of a liquid from escaping. This is shown in Fig. 14.2. The stronger the cohesive forces of a liquid, the stronger its surface

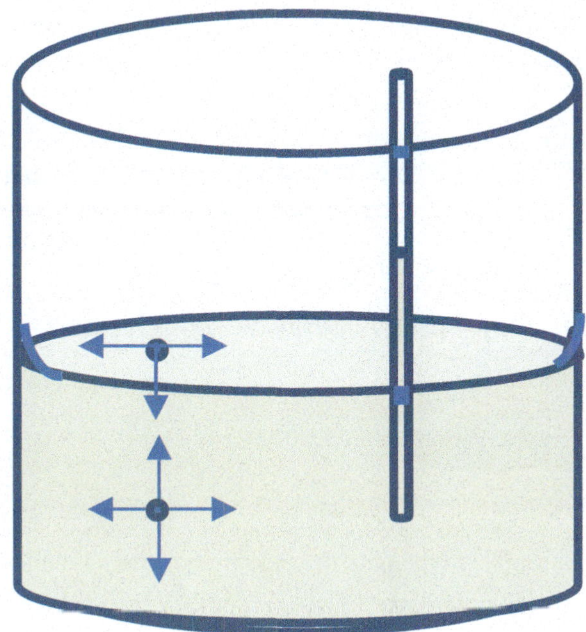

Fig. 14.2 Surface tension and capillary action

tension. Surface tension causes a liquid to contract to its smallest possible shape. This is why small samples of a free liquid form spherical droplets. Surface tension is a property of the liquid and represented by γ with units of N/m.

Evaporation of a liquid occurs when its most energetic particles can break away from the surface tension holding them in the liquid. Since these are the particles with the highest amount of kinetic energy, the average kinetic energy of the sample is decreased and this lowers the temperature. This effect is known as "evaporative cooling." The ability of a liquid to undergo evaporation is known as its volatility. The weaker the cohesive forces of a material are, the more volatile it is. The reverse of evaporation is called condensation. During condensation, the cohesive forces of the particles in a sample of matter pull the nearby particles of a gas into the liquid.

The electromagnetic forces between the particles of different types of substances are called adhesive forces. These can be especially important if the forces are between a liquid and its container. A liquid like water moves higher on the edges of a glass container than in the center of the glass because the adhesive forces between the molecules on the surface of the liquid balances the cohesive forces of surface tension. If a small diameter tube is inserted in the liquid as shown in Fig. 14.2, the adhesive forces cause the liquid to move a significant distance up the inside of the tube. This is called capillary action and is demonstrated in the following example. Capillary action is especially important for blood flow in your body. The top of the liquid in a capillary tube can have a curved shape called a meniscus because of the adhesive forces on the edges and the surface tension on the top. The angle the meniscus curve makes with the edges is called the contact angle *c*.

Example 14.2

A glass capillary tube of internal radius 0.20 mm is dipped into water that has a surface tension 0.070 N/m. How high does the water rise in the tube? (Note that for pure glass and pure water the contact angle of contact for the meniscus is zero.)

Knowns: $r = 20 \times 10^{-5}$ m; $\gamma = 0.070$ N/m; $\rho = 1000$ kg/m^3

Unknowns: y

The adhesive forces acting with the surface tension to pull the water in the capillary tube up a distance y compared to the level of water in the beaker. This force acts around the circumference of the capillary tube to give

$$F_{\text{up}} = 2\pi r\gamma$$

If there were a contact angle from the meniscus that was not zero, this expression would need an additional factor of cos θ to account for the component of force in the upward direction.

The downward force on the water in the tube is its weight. This is given by

$$F_{\text{down}} = (\rho g)\left(\pi r^2 y\right)$$

where ρ is the density of water. In equilibrium the upward and downward forces must be equal,

$$2\pi r\gamma = (\rho g)\left(\pi r^2 y\right)$$

Solving for y gives

$$\begin{aligned} y &= 2\gamma/(\rho g r) = 2(0.070\,\text{N/m})/\left(1000\,\text{kg/m}^3\right)\left(9.81\,\text{m/s}^2\right)\left(20 \times 10^{-5}\,\text{m}\right) \\ &= 0.0714\,\text{m} \end{aligned}$$

Try the following problem.

Student

14.2 What is the surface tension on a column of water in a capillary tube of radius 0.08 mm if it rises 6.0 mm above the surface of the water in the beaker it is in? (For water, $\rho = 1000\ \text{kg/m}^3$.)

When external pressure is applied at any point on a fluid, it is transmitted throughout the entire fluid regardless of the shape of the container holding the fluid. This is called Pascal's Principle. It forms the basis for a hydraulic lift shown schematically if Fig. 14.3. In the left side of the figure, a force of F_1 is applied to depress a piston with an area A_1. This is instantaneously transmitted to the right side of the figure as a force F_2 pushing up on a piston with an area A_2. The pressure on the left side and the right side must be equal so,

$$P_1 = F_1/A_1;\ P_2 = F_2/A_2;\ P_1 = P_2.$$

Therefore,

$$F_2 = F_1(A_2/A_1). \tag{14.3}$$

Thus, with a favorable ratio of areas, the effective force can be larger than the force that was originally applied to the system. This is like the mechanical advantage of levers and compound pullies we studied in Chap. 4 as demonstrated in the following example.

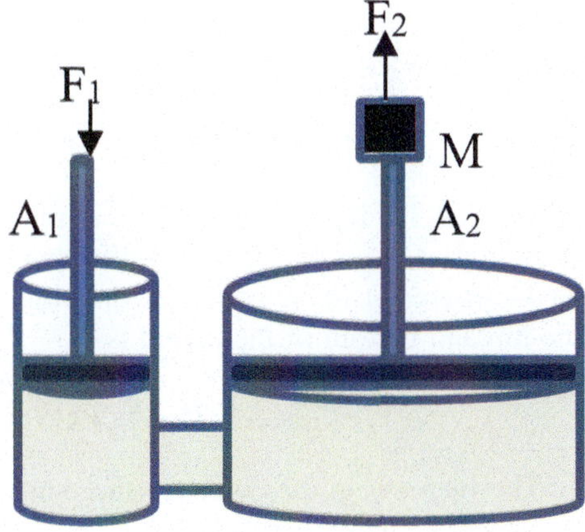

Fig. 14.3 Hydraulic lift

Example 14.3

Consider a hydraulic lift such as the one shown in Fig. 14.3. If $A_2 = 10A_1$, how much force must be exerted on the left side of the apparatus to lift a mass of 10 kg on the right side?

Use Fig. 14.3 Knowns: $F_2 = mg$; $g = 9.81$ m/s^2; $m = 10$ kg; $A_2 = 10A_1$

Unknowns: F_1

Using Eq. 14.3 with these values gives

$$F_1 = F_2/(10) = \left(9.81\,\mathrm{m/s^2} \times 10\,\mathrm{kg}\right)/(10) = 9.81\,\mathrm{N}.$$

Thus 9.81 N of force are needed to lift a weight of 98.1 N. So this gives you a mechanical advantage of 10 since you can lift something that is 10 times heavier than the force you apply.

Now you can do the problem.

Student

14.3 A hydraulic press has a small piston that can use 20 N to lift a 400 N load with the big piston. If the large piston has a diameter of 1.5 m, what is the diameter of the small piston and what is the mechanical advantage of the press?

When an object is immersed in a liquid, it experiences a pressure due to the weight of the column of liquid above it. This can be expressed as

$$P = F_g/A = mg/A = \rho Vg/A = \rho Ahg/A = \rho gh \tag{14.4}$$

where h is the depth of the object in the liquid and ρ is the density of the liquid. If the top of the object is at a depth h and the bottom of the object is at a depth $h + \Delta h$ where Δh is the height of the object, then the difference in the pressure on the top and bottom of the object is

$$P_{\mathrm{bottom}} - P_{\mathrm{top}} = \rho g(h + \Delta h) - \rho gh = \rho g \Delta h.$$

The increase in the force on the bottom of the object compared to the top of the object is than given by

$$F_{\mathrm{buoyancy}} = \left(P_{\mathrm{bottom}} - P_{\mathrm{top}}\right)A = \rho g \Delta V, \tag{14.5}$$

where ΔV is the volume of the liquid the object displaces and ρ is its density. This is called the force of buoyancy and its direction is upward. Equation 14.5 tells us that the force of buoyancy on an object immersed (or partially immersed) in a liquid is equal to weight of the liquid it displaces. This is known as Archimedes' Principle.

Archimedes' Principle—An object will float if its weight is less than the buoyant force exerted on it and it will sink if its weight is greater than the buoyant force exerted on it.

Example 14.4

Will a canoe with a mass of 45 kg that displaces a 5 m^3 volume of water float?

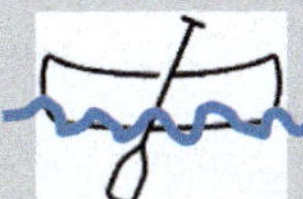

Knowns: $m_{canoe} = 45$ kg; $V = 5$ m^3; $\rho_{water} = 997$ kg/m^3; $g =$ 9.81 m/s^2

Unknowns: W_{canoe}; $F_{buoyancy}$

The weight of the canoe is

$$W_{canoe} = m_{canoe}g = (45\,\text{kg})\left(9.81\,\text{m/s}^2\right) = 441.5\,\text{N}.$$

Equation 14.5 can be used to find the force of buoyancy.

$$F_{buoyancy} = \rho g V = \left(997\,\text{kg/m}^3\right)\left(9.81\,\text{m/s}^2\right)\left(5\,\text{m}^3\right) = 48{,}903\,\text{N}.$$

Therefore:

$$F_{buoyancy} > W_{canoe}$$

Thus, the canoe will float even if it is heavily loaded. Note that to work this problem you have to look up the density of water.

Do the following experiment and then try the problem.

Experiment 14.1

Archimedes Principle

Supplies:

- Empty aluminum foil container
- Sink, bathtub, or Pot large enough to hold the foil container in water
- Water
- Weight scale
- Millimeter ruler and marker

- Objects you can weigh that will fit into the container.

Introduction: This experiment shows that the buoyancy force is equal to the water it displaces.

Procedure:

1. Measure the area of the bottom of the container and a scale in millimeters on one side.
2. Float the container in water so that the scale if visible.
3. Weigh one of your objects and place it in the floating container.
4. Measure on the scale of the container how much it sinks into the water.
5. Calculate the volume of water that was displaced and determine how much it weighed.
6. Compare your result with how much you measured the object to weigh.
7. Repeat steps 3–6 for your other objects.
8. Write down your results and explain any differences between an object's weight and the weight of the water it displaced.
9. Put everything away and clean up the area.

Student

14.4 A temporary floating bridge is used for a truck to cross a river. The bridge is 5 m wide and 50 m long. It sinks by 0.5 m when the truck is on it. What is the weight of the truck? (The density of water is 10^3 kg/m^3.)

So far we have been dealing with the mechanical properties of matter at rest. Now we look at how forces change due to the motion of matter through fluids. These changes apply to all fluids, both liquids and gases.

The effect of having a relative velocity between an object and a fluid leads to what is called Bernoulli's Principle.

Bernoulli's Principle—As the velocity of an object increases, the pressure exerted on it by a fluid decreases.

This is essentially a statement of conservation of energy.

The Bernoulli effect of a decrease in pressure associated with an increase in fluid velocity has many important applications. One of these is shown in Fig. 14.4a. When a liquid like water goes through a hose to a restricted region like a nozzle, its velocity increases. The velocity increase must occur if the same mass of liquid entering the large end of the hose per unit time exits the small end per unit time. This results in a decrease in pressure in the nozzle compared to the pressure in the

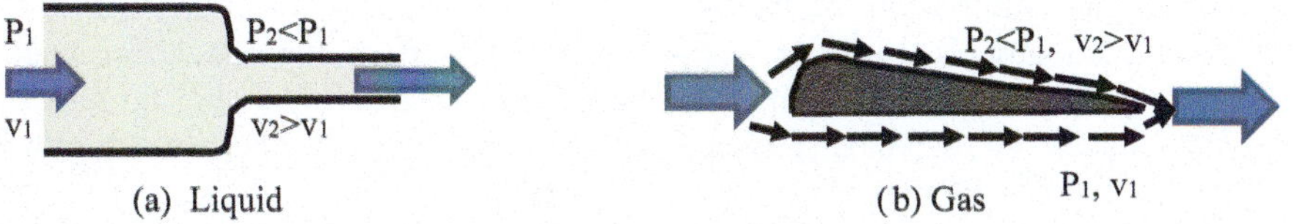

Fig. 14.4 Bernoulli effect

hose itself. We can understand this as a consequence of conservation of energy. As we learned in Chap. 3, energy is related to the work done by the system. The velocity is related to the kinetic energy so as v increases the kinetic energy increases. This means that the potential energy must decrease to keep the total energy constant. The change in potential energy can be expressed as the difference in the work done in two parts of the system. For example, in the hose-nozzle system shown in Fig. 14.4a, the difference in the work done to force the liquid through the nozzle and the work done forcing the liquid through the entrance to the hose represents the change in potential energy in the system. Since work is force times distance, this can be expressed as

$$\Delta \mathrm{PE} = F_2 d_2 - F_1 d_1 = P_2 A_2 d_2 - P_1 A_1 d_1 = P_2 V_2 - P_1 V_1 .$$

Here we have used the definition of pressure as force per unit area. The values of d_i are the distances moved through the hose and nozzle at the two different speeds and the A_i are the areas of the hose and nozzle. For the same amount of water mass moving through the hose and nozzle per unit time, the volumes V_2 and V_1 must be equal. Thus the only way for ΔPE to be negative in order to offset the increase in KE is for P_1 to be greater than P_2.

An example of the Bernoulli effect in a gas is shown in Fig. 14.4b. The wing of an airplane has a shape that forces the air going on the underside to travel a shorter distance than the air on the top side. Thus the air on top must travel faster than the air on the bottom to reach the back of the wing at the same time. Since the velocity of the air on top is greater than the velocity of the air on the bottom, the upward pressure the air exerts on the bottom of the wing is greater than the downward pressure the air exerts on the top of the wing. This difference in pressure is called lift. If the force of lift on an object in the air is greater than the force of gravity pulling it down, it can fly.

This discussion about moving fluids can be expressed mathematically starting with the conservation of energy,

$$\begin{aligned} \Delta \mathrm{PE} &= -\Delta \mathrm{KE} \\ P_2 V - P_1 V &= \rho V v_1^2/2 - \rho V v_2^2/2 \end{aligned}$$

where we have assumed the volume of fluid entering and leaving the system is the same and mass is given by $m = \rho V$. The final result is

$$P_1 + \rho v_1^2/2 = P_2 + \rho v_2^2/2 \tag{14.6}$$

This is called Bernoulli's equation.

There are many other everyday applications of the Bernoulli effect. For example, when a pitcher throws a curve ball in a baseball game, the spin he puts on the ball makes the air velocity greater on one side than the other resulting in a difference in pressure that causes the ball to curve.

Consider the following example.

Example 14.5

Water enters a hose with a pressure of 5.0 x 10² Pa at a speed of 20 cm/s. Its pressure drops to 1.0 x 10² Pa in a nozzle. What is its exit speed? (The density of water is 10³ kg/m³.)

Knowns: $P_1 = 5.0 \times 10^2$ Pa; $v_1 = 0.20$ m/s; $P_2 = 1.0 \times 10^2$ Pa

Unknowns: v_2

This is a straightforward application of Bernoulli's equation,

$$P_1 + \rho v_1^2/2 = P_2 + \rho v_2^2/2$$

$$5.0 \times 10^2 (\mathrm{kg/ms^2}) + (10^3\,\mathrm{kg/m^3})(0.20\,\mathrm{m/s})^2/2$$
$$= 1.0 \times 10^2 (\mathrm{kg/ms^2}) + (10^3\,\mathrm{kg/m^3}) v_2^2/2$$
$$520(\mathrm{kg/ms^2}) - 100(\mathrm{kg/ms^2}) = 500(\mathrm{kg/m^3}) v_2^2$$

$$v_2 = \left[420(\mathrm{kg/ms^2})/500(\mathrm{kg/m^3})\right]^{1/2} = 0.92\,\mathrm{m/s}$$

Thus the exit speed is 4.6 times faster than the entrance speed.

Now do the following experiment and then try the following problem.

Experiment 14.2

Bernoulli's Principle

Supplies:

- Two empty aluminum cans of identical size with a hole punched in the bottom so they can be hung by a string
- Two strings about 40 cm long
- Hair dryer
- Ruler
- Table

Introduction: This experiment shows the relationship between velocity and pressure of a fluid.

Procedure:

1. Tie the aluminum cans to one end of each of the strings.
2. Tape the other ends of the strings to the top of the table so the cans hang freely over the table.
3. The cans should hang at the same height (about 30 cm from the table top) and about 6 cm apart.
4. The blower end of the hair dryer should just fit between the cans. It would be best if the hair dryer had a rectangular attachment that could be oriented in the vertical direction. This would allow the cans to be as close together as possible.
5. Turn the hair dryer on and blow the air between the two cans without blowing on the cans themselves.
6. Measure how much the cans move toward each other.
7. If the hair dryer has different operating speeds, repeat steps 5–6 for each speed.
8. Record your results.
9. Use Bernoulli's equation to derive an expression for the change in pressure created by the air velocity between the cans.
10. Derive an expression to show how you can use the angle between the string and the vertical with the hair dryer on to determine the weight of the can.
11. Put everything away and clean up.

Student

14.5 The wing of an airplane has air flowing on the bottom at 200 m/s and across the top at 250 m/s. What lift does it provide? (The density of air is 1.225 kg/m^3.)

Now let's turn to forces in solids. Solids have internal cohesive forces like liquids that hold their atoms or molecules together. However, in solids these electromagnetic forces are strong enough so a solid maintains its shape instead of flowing or expanding to take the shape of its container. The particles of a solid vibrate around their equilibrium positions providing it with a kinetic energy and thus a temperature. At temperatures where the kinetic energy is less than the binding energy of the cohesive forces, the particles freeze into fixed positions to form a solid. When this happens, the positions of the particles can either be random making a glass or form a lattice with a geometric shape making a crystal. The same chemical substance can form either a crystal or a glass depending on the thermodynamic conditions under which it is formed. For example, silicon dioxide (SiO_2) can be either a crystal of quartz or a glass called fused silica.

The mechanical properties of a solid are related to how it responds to an applied force. In general, the response to an applied force is deformation. The shape of the solid is changed. The property known as the elasticity of a solid describes how well it returns to its original shape when the force is removed. If the force distorting the solid is too great, it exceeds the elastic limit. In this case the shape of the solid does not return to its original form so the deformation is permanent. The magnitude of the elastic limit depends on the strength of the electromagnetic forces among the particles of the material. Two important types of properties of a solid are how ductile it is and how malleable it is. The former refers to the ability to pull the material into long thin wires. Ductility is important for applications such as making copper wires for transmitting electricity or making optical fibers for optical communication. The latter refers to the ability to roll the material into thin flat sheets. There are many uses for thin sheets of material such as forming them into aluminum or tin cans.

A deformation force can either be compressive (pushing in) or tensile (pulling out). When it is applied in one dimension, the distortion it causes follows a simple linear relationship known as Hooke's Law,

$$F = -kx \tag{14.7}$$

where k is the restoring force constant related to the internal electromagnetic forces between the particles of the material. It has units of N/m. A typical graph showing Hooke's Law is shown in Fig. 14.5. This is the same relationship we used in Chap. 6 for treating a mass on a spring where k was the spring restoring constant.

When forces greater than the elastic limit are applied to the material, one of three things can happen. The first is, it can immediately fracture. A material that does this is said to be brittle. Second, it can continue to deform before fracturing

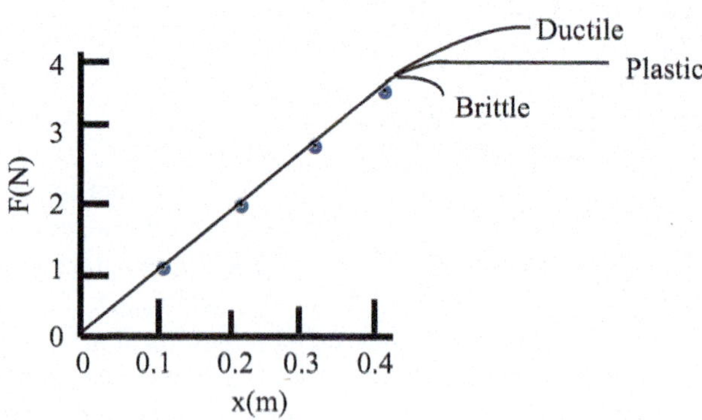

Fig. 14.5 Hooke's Law

at some higher force, but it no longer follows Hooke's Law. This is called plastic deformation. Third, it can undergo much greater levels of plastic deformation. This type of material is called ductile. These three cases are shown schematically in Fig. 14.5.

The response of a material to an external force depends on how that force is applied. For example, when a force is applied continually over a long period of time, deformation occurs slowly. This is known as creep. When a force is applied repeatedly many times for short periods of time, the material weakens. This is called fatigue.

Along with its ductility, malleability, and brittleness as defined above, a material has several other properties related to how it reacts to a force of deformation. One of these is called the strength of a material. Strength is defined as the maximum stress (force per unit area) a material can withstand and still be within its elastic limit where the deformation is reversible. It has units of N/m^2. The toughness of a material is the amount of energy it can absorb before fracturing. The hardness of a material is the amount of force that causes permanent deformation to its surface. This is measured in terms of a scale of relative hardness where the hardest material is diamond at a value of 10 and the least hard is talc at a value of 1. This is called the Mohs scale. The stiffness of a material defines how flexible it is. This is given by its Young's Modulus which is the ratio of how much it deforms to the load that is applied, or stress divided by strain,

$$E = \frac{F/A}{\Delta L/L} = (FL)/(A\Delta L). \tag{14.8}$$

Example 14.6

Calculate how much force is required to change the length of a piece of aluminum by 0.001 m if it is 1 m long and has a cross sectional area of 0.02 m^2. The Young's modulus for aluminum is 69 GPa.

L ΔL

F

Knowns: $E_{Al} = 69 \times 10^9$ Pa; $L = 1$ m; $A = 0.02$ m^2; $\Delta L = 0.001$ m

Unknowns: F

Then Eq. 14.8 gives,

$$F = EA\Delta L/L = (69 \times 10^9\,\text{Pa})(0.02\,\text{m}^2)(0.001\,\text{m})/1\,\text{m}$$
$$= 138 \times 10^4\,\text{N}.$$

Thus it takes a large force to create a small deformation in a piece of aluminum this size.

Now try the following problem.

Student

14.6 A metal block of volume 600 cm^3 is suspended from a steel wire 150 cm long that has a cross-sectional area of 2×10^{-3} cm^2 and a Young's modulus of 190 GPa. If the block is immersed in water, how much does the length of the wire change? (The density of water is 10^3 kg/m^3.)

In crystals an applied external force is called stress and the resulting deformation is called strain. For crystals with a different lattice structure in different directions the relationship between stress and strain is different in different directions. To account for this, the properties are represented by tensors which are multi-dimensional vectors. The mathematics of tensor analysis is beyond the scope of this book but is an important technique for dealing with the directional properties of crystals.

These mechanical properties of a material and the cause for its fracturing are associated with structural defects in the material that decrease the cohesive forces holding the material together. These can include point defects such as chemical impurities or vacancies where an atom is missing. In addition, linear defects where the atomic structure changes abruptly affect the material's properties. These are called dislocations. Grain boundaries which define structural changes in three dimensions are also important. In order to strengthen a material by minimizing the effects of these defects, the material can be heated and slowly cooled. This is called annealing. The relationships between the structure of a material and its mechanical properties are treated in detail in advanced courses on solid-state physics and material science.

14.4 Electrical Properties of Matter

In Chaps. 7 and 8 we learned about the general properties of electricity and electrical circuits. In this chapter we extend this discussion to the electrical properties of matter. The discussion will focus on the solid state of matter because of its great importance in practical applications. In general, electrical properties are not as important for the gas state of matter. However, when electrical charges are ripped from atoms to form a plasma of electrons and charged ions, electrical properties are present. The plasma state of matter is not very common in our daily life, although plasmas are present in the center of every star, so it is a common state of matter in the universe. As far as liquids are concerned, the most important electrical properties are associated with batteries which is more of a discussion for chemistry than physics.

The most important type of solids for electrical applications are crystals. Therefore our discussion will focus on crystalline solids and not amorphous solids (glass). The major topics are how solids conduct electricity and how we can control their properties of electrical conduction. We learned previously that materials can be divided into three classes based on their ability to conduct electricity. The first is conductors that conduct electricity very well. These include metals like copper or aluminum that are useful in making wires to transmit electrical current in circuits or power lines. The second class is insulators that do not conduct electricity well like rubber or plastic. These are useful in wrapping around conductors to keep the electrical conduction from harming people or the environment. The third class is called semiconductors. These are materials like silicon and germanium that have intermediate electrical conduction properties. Semiconductors are useful because their electrical conduction properties can be controlled. This makes them especially useful in electrical devices as discussed below.

Electrical conductivity occurs when particles with an electrical charge are free to move. The most important type of mobile particle for producing an electrical current is an electron, which carries a negative charge. A crystalline solid is made up of a very large number of atoms bound together in a geometric lattice structure. Each atom has a nucleus with a positive electrical charge and is surrounded by electrons so that it is electrically neutral. As discussed in Chap. 16, each electron is in a state with a specific discrete energy. When these atoms come close to each other to form a solid, they interact with each other in a way that causes the energy of their electron states to be slightly different. These closely spaced states form energy bands for the electrons in a crystalline solid. The lowest band of energy states is called the valance band. Since electrons always go into the lowest possible energy state, every state in this band has an electron in it. There is an energy gap between this band and the next lowest band. No electron can exist at energies within this gap. The next band is called the conduction band. It may have some electrons in it, but it also has many unfilled states. This band structure is shown schematically in Fig. 14.6.

For an electron to be mobile, it must be able to move from its current state to an empty energy state. This cannot happen in the valance band since all the states

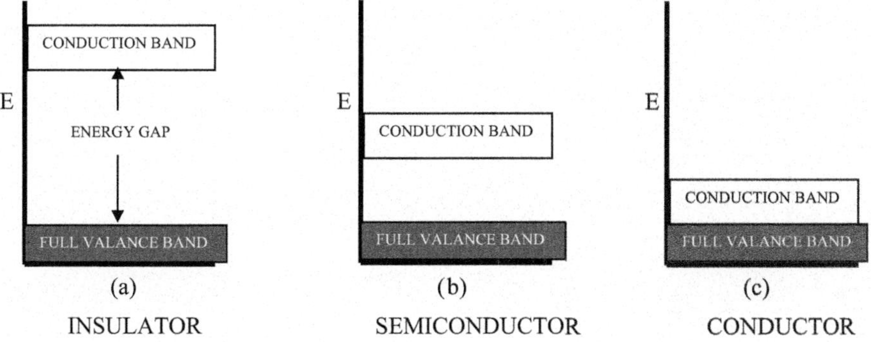

Fig. 14.6 Electron energy bands in a crystal

are full. However, it can happen in the conduction band, so the mobile electrons in this band are the ones that are responsible for the electrical current in the crystal. The electrons in the valance band are important in providing the cohesive force for bonding the atoms of the crystal together. In addition, their vibrations about their equilibrium positions provide a resistance to the movement of mobile electrons. As shown in Fig. 14.6, an insulator material has a large energy gap (around 5 eV) between the conduction band and the valance band. Conductors have a very small energy gap, so thermal agitation or the application of an external electric field can give some of the electrons enough energy to be in conduction band states and carry electrical current. Semiconductors have an intermediate energy gap (around 1 eV). Raising the temperature of a semiconductor causes more electrons to be in the conduction band and thus increases its conductivity. However, for conductors raising the temperature causes an increase in the thermal vibrations of the valance state electrons thus increasing resistance and decreasing conductivity.

The discussion above shows that the number of electrons in the conduction band is a critical factor in determining the electrical properties of a crystal. This property is controlled best in semiconductors by the processes of doping. Doping refers to the process of replacing a small number of atoms of the material by atoms of a different chemical substance. These dopant atoms must be close to the same size of the host crystal atoms, so the presence of impurities does not negatively affect the mechanical properties of the crystal as discussed in the previous section.

Depending on the type of ion used for doping, this can either add or subtract electrons to the system. For example, silicon (Si) is a semiconductor with a band gap of 1.744×10^{-19} J. It is part of Group IV in the periodic table of elements and has 4 valance electrons that are used for bonding the Si atoms together to form a silicon crystal. One common type of dopant for a silicon crystal is arsenic (As). Arsenic is part of Group V in the periodic and has 5 valance electrons. The extra electrons from the dopant ions go into impurity states that exist in the band gap of the Si crystal. As shown in Fig. 14.7a, these extra states are called donor states and they are just below the conduction band. It takes very little thermal or electrical energy to excite an electron from a donor state into the conduction band. Semiconductors doped with atoms having extra electrons are called n-type semiconductors.

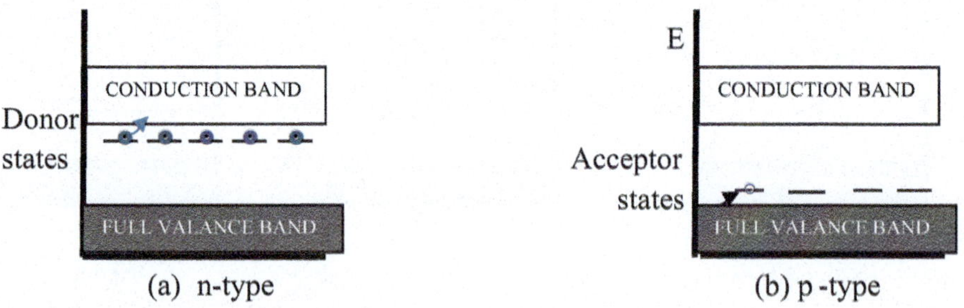

Fig. 14.7 Donors and acceptors in semiconductors

It is also possible to dope a semiconductor like silicon with atoms that have fewer electrons. For example, boron (B) atoms are in Group III in the periodic table and have 3 valance electrons. In this case they create empty impurity states just above the valance band in the Si crystal band gap as shown in Fig. 14.7b. These are called acceptor states. It takes very little thermal energy for an electron to be excited out of the valance band into an acceptor state. When this happens, it leaves an empty state in the valance band. Since neighboring electrons can move into this empty state this creates some mobility for electrons in the valance band. The empty state is called a hole. Because of the missing electron, it has an effective positive charge. The movement of an electron into a neighboring hole causes the hole to act like a positively charged particle moving in the opposite direction. Thus, the presence of acceptor states allows for an electrical current to be carried by holes in the valance band. This is called a p-type semiconductor.

Semiconductors with no doping ions rely on thermal energy for getting electrons from the valance band to the conduction band and are called intrinsic semiconductors. Those relying on dopant ions are called extrinsic semiconductors.

Once the electrons are in the conduction band (or holes are in the valance band) they are free to move and carry electrical current. There are two reasons that mobile charge carriers move, diffusion and drift. Diffusion occurs whenever a system of particles has an uneven distribution where there is one region with a higher density of particles than another region. We know from Chap. 13 that the second law of thermodynamics tells us that the system will naturally evolve to maximum entropy where the particles are spread out with a uniformly random distribution. This process is called diffusion. In semiconductors or metals charge carriers injected into one area will diffuse to fill up the material. Their mobility and diffusion length (how far they will move before combining with a charge of opposite sign) is determined by amount of doping, defects, and interaction with lattice vibrations (temperature). The drift movement of charge carriers takes place when an electric field is applied. We learned in Chap. 7 that a charged particle in an electric field feels a force that will cause it to accelerate. This is what drives an electric current in the conduction band of a metal wire. We will see later in this section that both the diffusion and drift of mobile charges play an important role in the electrical properties of solids.

The development of n-type and p-type semiconductors led to the creation of a new field called microelectronics. Electrical devices such as radios are made out of components with electrical circuits that amplify and control electric currents. Before microelectronics, these functions were provided by large vacuum or gas filled tubes. Once it was proven that semiconductors could perform the same functions, small solid-state devices replaced tubes. The ability to control electrical circuits with small, rugged devices stimulated the development of solid-state electronics for many applications.

The key element in electronic devices made from semiconductors is called a p-n junction. In this case, a semiconductor material is made with one part p-type and the other part n-type. The p-n junction is where these two parts meet. When the p-n junction is first made, the mobile electrons in the n-type region will attempt to diffuse into the p-type region and the mobile holes in the p-type region will attempt to diffuse into the n-type region. These electrons will be trapped in empty acceptor states in the p-type region while these holes will be filled with electrons in the n-type region. This creates a depletion region spanning the junction where there are no mobile charges. On the n side of the depletion region, there is a layer with a positive charge compared to the bulk of the n-type region because of the depletion of mobile electrons there. Similarly, on the p side of the depletion region there is a layer with a negative charge compared to the bulk of the p-type region because of the depletion of mobile holes there. This creates an internal electric field across the depletion layer as shown in Fig. 14.8a. This acts as a potential barrier which stops further diffusion of charge carriers into the depletion region. An electron in the middle of the depletion region will feel a force pushing it to drift into the n-type region and a hole in the middle of the depletion region will feel a force pushing it to drift into the p-type region.

The key to the properties of a p-n junction is the depletion layer which controls the flow of current through the semiconductor device. When it is connected in an electrical circuit with a forward bias voltage as shown in Fig. 14.8b, the external voltage provides enough energy to the mobile charges to overcome the potential barrier of the depletion region. The electrons in the n-type region move toward the p-type region and the holes in the p-type region move toward the n-type region. This causes the depletion region to become smaller which lowers the potential barrier and allows current to flow. On the other hand, if the external battery is connected in the opposite direction, this reverse bias pulls the donor electrons and acceptor holes apart thus making the depletion region wider shown in Fig. 14.8c. This makes the potential difference across the depletion layer greater and more effective in stopping current flow in the device. These directional properties cause the device to act as a diode which allows current to flow in one direction but not the opposite direction. The symbol for a diode in a circuit is ▶|. Figure 14.8d shows a simple circuit with a forward biased diode. The current-voltage characteristics of a diode are shown in Fig. 14.8e. For a reverse bias voltage, a small current flows in the negative direction but when the voltage becomes positive a large current flows in the positive direction.

Active control of the depletion layer in a p-n junction acts as the basis for important microelectronic devices like transistors which amplify or switch electric signals. As an example, two different types of transistor designs are shown in Fig. 14.9. Figure 14.9a of the figure shows a bipolar transistor device that has two p-n junctions. The device shown has two parts that are n-type separated by one part that is p-type. Bipolar transistors can also be made as two p-type regions separated by one n-type region. This type of transistor has two p-n junctions forming two back-to-back diodes in a circuit. The two important aspects of the n-p-n configuration in Fig. 14.9a are first, the n-type region on the left (emitter) is much more heavily doped than the n-type region on the right (collector), and second, the p-type region (base) is thinner than diffusion length of the donor electrons. The voltage V_{CB} keeps the collector more positive than the emitter and provides a reverse bias to the to the base-collector junction. This makes the depletion region

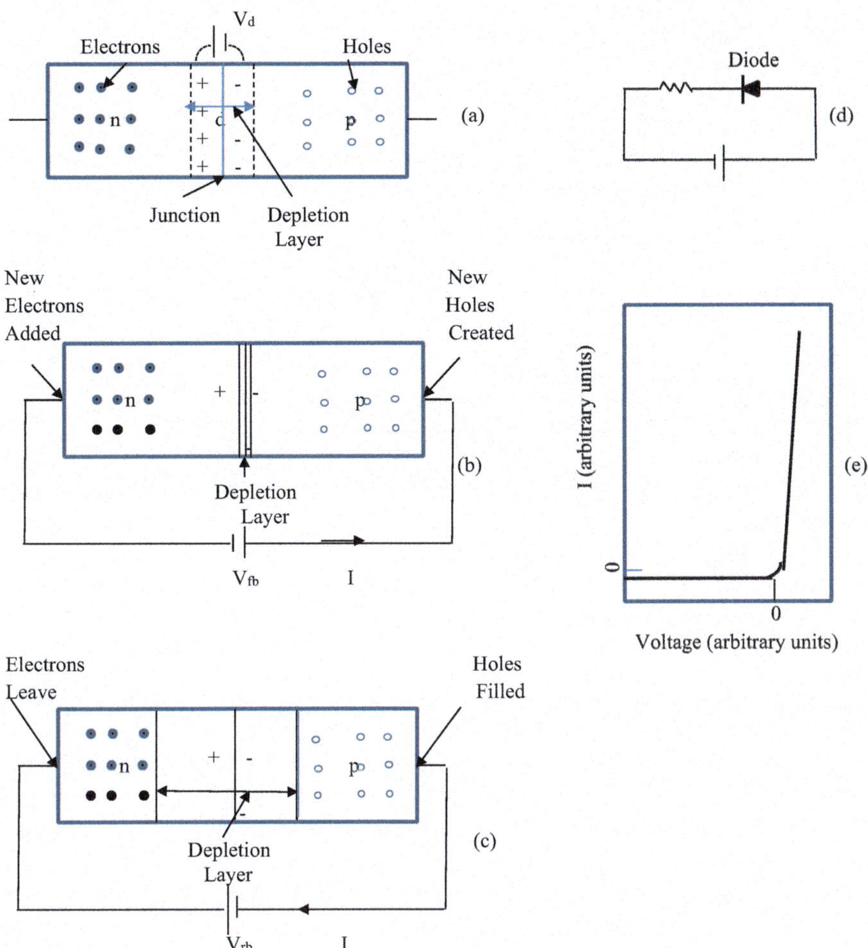

Fig. 14.8 **a** p-n junction diode, **b** forward bias, **c** reverse bias, **d** diode circuit, **e** I–V characteristics of a diode

of this junction (the dotted area) wide enough to stop current flow from the collector to the base. The V_{EB} voltage makes the base more positive than the emitter and provides a forward bias to the base-emitter diode. This makes the depletion region for this junction (the dotted area) very small. An input device is attached on the emitter side of the transistor. This can be something like a solar cell, a thermocouple, or a microphone. The small current from this device flows into the base and causes the depletion regions to become even smaller. The excess electrons in the emitter region have a strong diffusion potential pushing them toward the base and collector. When this exceeds the drift potential in the opposite direction provided by the p-n junction, the electrons flow through the thin base into the collector. This is like closing a switch that lets current flow in the external

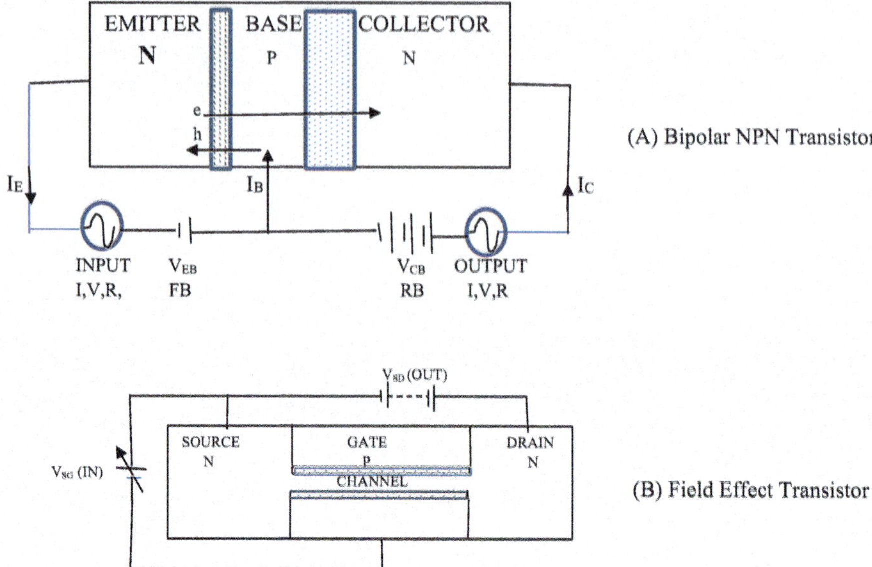

Fig. 14.9 Two types of transistors

circuit from the emitter to the collector. Since V_{CB} is much larger than V_{EB} the current I_C is much larger than I_B. To satisfy Kirchhoff's laws at the junction, $I_E = I_B + I_C$. The magnitude of I_C is proportional to I_B. This current flows through an output device like a lamp and provides it with the current it needs to operate. Thus, the transistor acts as a switch, a current regulator, and a power amplifier. The current gain is the ratio of I_C to I_B. This current gain can be between 50 and 100 times. The power gain involves the current gain and also the ratio of the load resistance to the input resistance. Note that there is a small hole current from the base to the emitter which is why it is called a "bipolar" transistor. The following problem demonstrates the use of this type of transistor as an amplifier.

Student

14.7 A bipolar transistor is used as a power amplifier between a microphone and a speaker. The microphone is connected to the base circuit so it produces a base current of 0.01 A and an emitter-base voltage of 0.3 V. The collector current is 60 times greater than the base current and the collector-base voltage is 3 V. What is the power amplification between the microphone and the speaker?

Next Level

Field Effect Transistor

In a piece of equipment like a computer, millions of transistors are needed. Making them work by controlling current generates too much heat. Transistors that work by controlling voltage instead of current are called field effect transistors (FETs). A schematic diagram of a typical field effect transistor is shown in Fig. 14.9b. This is constructed by having two n-type regions (source and drain), separated by a p-type region (gate) and connected by a thin n-type channel. The channel has a p-n junction with a depletion layer in it (shown by dotted regions). When no potential difference V_{GS} is supplied between the p-type region (gate) and the n-type region on the left (source), electrons can flow freely through the channel from source to drain (n-type region on the right). This completes the circuit from source to drain so current also flows on the circuit outside the device providing an output voltage V_{SD}. However, when V_{GS} is supplied as a reverse bias, the depletion layer formed around the p-n junction in the channel gets wider making the channel smaller which increases the resistance to the flow of electrons. Thus the V_{GS} input's effect on the channel acts as a variable resistor in the source to drain circuit. The size of the depletion layer and thus the resistance to the flow of electrons is proportional to the magnitude of the reverse bias that is applied to the junction. This internal flow of electrons determines the external current flow and voltage between the source and the drain. In other words, the negative input voltage from the gate to the source controls the external current flow from the source to the drain. The output current between the drain and source is controlled by the V_{GS} input voltage. The voltage V_{DS} is shown since it changes along the length of the gate because the width of the depletion layer changes creating a change in resistance. In this case the output current is controlled by the input voltage instead of current as it is in a bipolar transistor. This can act as a voltage amplifier or as a variable resistor in an electrical circuit.

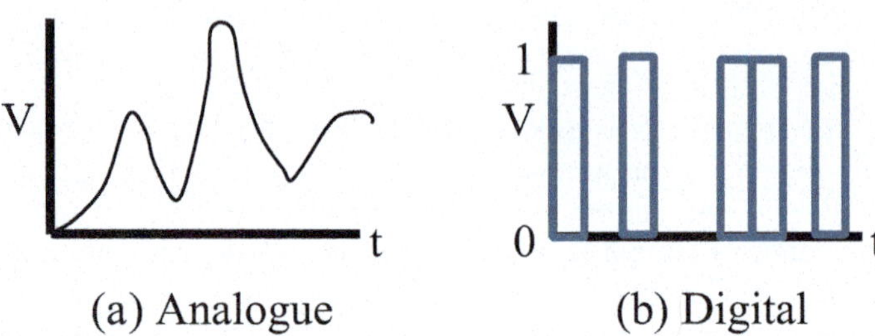

Fig. 14.10 Analogue versus digital signals

The transistor was invented in 1947. It stimulated the development of solid-state microelectronics in the second half of the twentieth century. Because of its importance in the development of computers, television, smart phones, and other electronic devices we use in our daily lives, the transistor has been heralded as the most important invention of the twentieth century. One important type of diode emits light when the electrons and holes recombine. Light emitting diodes (LEDs) are important in lighting applications as discussed in the following section.

Using microelectronic switching devices has led the way for the development of digital electronics. One way to encode information into an electrical signal is by modifying its amplitude or frequency. This is called an analogue signal and an example is shown in Fig. 14.10a. Another way to encode information is called a digital signal and an example is shown in Fig. 14.10b. In this case the signal has only two states, zero and one. These are called *bits* and they can be used to encode text, pictures, music and any other type of information using binary mathematics. Information can be transmitted and stored digitally with much less interference and distortion than analogue signals having the same information. The switching operation of transistors make them ideal for converting information to digital signals.

14.5 Optical Properties of Solids

In Chap. 11 we learned a little about the optical properties of transparent solids. Remember, an optical beam is a wave of oscillating electric and magnetic fields with its polarization in the direction of the oscillating electric field. The beam travels through a crystal by inducing the electrons on the atoms of the crystal to oscillate in synchronization with its electric field. If the electromagnetic wave exits the other side of the crystal with the same frequency and polarization as the incident wave, this is referred to as linear optics. In some materials, the interaction between neighboring atoms makes it easier for the electrons on the atoms to oscillate in one direction compared to another direction. Therefore the crystal structure causes the light to travel differently in different directions and this results in nonlinear optics. Usually nonlinear optical effects are important only for high intensity light beams. The use of lasers has made these useful for practical applications in controlling the properties of optical beams. Several types of nonlinear optics effects are described below.

In some crystals, the lattice structure results in the index of refraction being different in different directions. This property is called birefringence. Since $n = c/v$ and c is constant, if n is different in different directions the velocity of the optical wave will be different for different directions of travel and polarization. For example, in a quartz crystal the refractive index in one direction is $n = 1.5534$ and in another direction it is $n = 1.5443$. Because of this, it is possible that the direction of polarization of the transmitted beam can be rotated with respect to that of the incident beam. Along with natural birefringence, the spatial dependence of the refractive index can also be induced in a crystal by applying an external electric or magnetic field. Inducing a polarization change by use of an external electric field or magnetic field is called an electrooptic effect or a magnetooptic effect. Similar effects can be caused by applying external stress to a crystal. This is called the photoelastic effect. These effects are used in devices for modulating laser beams and for sensors. The multi-dimensional directionality of these effects requires that they be treated using tensor algebra which is beyond the math level of this book.

As an example, Fig. 14.11 shows the rotation of the polarization of a light beam in a crystal with an applied magnetic field. This is called the Faraday Effect and is described in the following example and the problem. The amount of rotation depends on the length of the material and the strength of the applied magnetic field as well as the type of material that is used. The material properties describing its usefulness in the Faraday Effect are called Verdet Constants designated by V with units of rad/T m. The amount of rotation in radians is given by

$$\theta = VBd. \tag{14.9}$$

The magnetic field is generally created by an electric current in a coil like the solenoids we studied in Chap. 10. The direction of the rotation of the polarization angle is in the same sense as the direction of the current in the coil if the Verdet constant is positive. If V is negative, the rotation direction is in the opposite direction of the current in the coil.

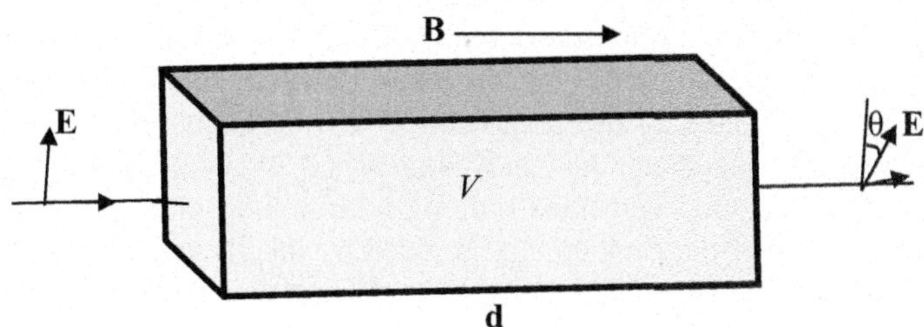

Fig. 14.11 Faraday effect

Example 14.7

Terbium gallium garnet is a crystal with a Verdet constant of − 134 rad/T m for a laser beam of wavelength of 632.8 nm. What is the angle of rotation of a vertically polarized beam in a magnetic field of 0.06 T going through a 10 cm long crystal as shown in Fig. 14.11?

Use the figure that is given. Knowns: $V = -134$ rad/T m; $B = 0.06$ T; $d = 0.1$ m.

Unknowns: θ

This is a direct application of Eq. 14.9,

$$\theta = VBd = (-134\,\text{rad/T m})(0.06\,\text{T})(0.1\,\text{m}) = -0.804\,\text{rad} = -46.07^\circ$$

Figure 14.11 shows a positive angle of rotation. Since our answer came out negative, the rotation will be in the opposite direction.

Now try the problems.

Student

14.8 At a wavelength of 1064 nm, the Verdet constant for terbium gallium garnet is − 38 rad/T m. For a Faraday rotator with a magnetic field of 0.08 T, how long does a crystal have to be to rotate the polarization angle by − 20°?

14.9 You need to decrease the intensity of your 1064 nm laser beam to 0.8 its original intensity. If you have a terbium gallium garnet crystal that is 12 cm long, how high a magnetic field must you apply to your Faraday rotator? (Assume no intensity loss in the rotator.)

Lasers usually emit light at one specific wavelength and many applications require light at some other specific wavelength. Therefore, one of the most useful applications of nonlinear optics is changing the wavelength of a laser beam. An example of doing this using nonlinear optics is second harmonic generation. This is based on the fact that the index of refraction has its fundamental term but also higher order terms related to the polarizability of the atoms and the strength of the electric field driving them. Just like we learned in Chap. 12 that a vibrating string emits a fundamental wavelength along with higher order harmonics whose wavelengths are that of the fundamental divided by integer numbers, at high electric fields the vibrating atoms produce a fundamental optical wave and related harmonic waves. The second harmonic wave has one half the wavelength of the fundamental. The transfer of energy from the fundamental wave to the second harmonic wave is maximized if the two waves are traveling in phase with each other. This means that they have to be moving in a direction and with a polarization direction so that their refractive indices are equal so their speeds are equal. A common example of second harmonic generation is a green neodymium YAG (yttrium aluminum garnet) laser. The fundamental emission of Nd-YAG is in the infrared at $\lambda = 1064$ nm. After going through a KTP (potassium titanyl phosphate) crystal in exactly the right direction, the wavelength is shifted to $\lambda = 532$ nm which is green. The green wavelength is preferred for some medical procedures.

Another optical property of solids that was not covered in Chap. 8 is the absorption and emission of light. This is characterized by spectroscopy. When you look at a light beam with a range of frequencies hitting a sample, the intensity of the light will decrease at frequencies where it is absorbed by the material. This is the sample's absorption spectrum. Conversely, if you look at light with a range of frequencies coming from a sample, it will have maximum intensities at frequencies where the sample is emitting light. This is a sample's emission spectrum.

Absorption happens when some of the energy of the light wave is transferred to some of the electrons in the solid and they move from one electronic state to another. The band theory of solids discussed in Sect. 14.4 shows that there will be an absorption transition from the valence band to the conduction band at a specific frequency. For light at higher frequencies the material will be opaque since the light is absorbed. For light at lower frequencies the material will be transparent since it is not absorbed. Thus small band gap materials like metals are opaque at most frequencies of light while insulators with large band gaps are transparent for much of the visible light spectrum. For doped semiconductors, there will be other specific absorption frequencies associated with the electrons or holes from the dopant ions.

When an electron in a solid is in an excited electronic state it can return to its ground state either by emitting heat or be giving off light. The latter case is called fluorescence. In microelectronics, diodes have been designed specifically so the electrons and holes recombine to cause fluorescence. These LEDs (light emitting diodes) have become important in lighting applications as well as television screens.

We found previously that doping semiconductors was important in controlling their properties for microelectronic applications. Doping of insulators is equally important in controlling their light absorption and emission properties. The application that makes this so important is the development of solid-state lasers. Lasers are discussed in Chap. 16 but the properties of solids relevant to lasers are described here.

A good example of a doped insulator is ruby. The first laser was made from a ruby crystal. The host crystal for ruby is a material called sapphire made of aluminum oxide (Al_2O_3). When a few of the Al^{3+} ions (less than a tenth of a percent) are replaced by Cr^{3+} ions the crystal turns red and is called ruby. The change in how ruby absorbs and emits light compared to sapphire is due to the chromium dopant ions. The electrons on the Cr^{3+} ions have a number of possible energy states they can move to and from by absorbing and emitting light energy. The chromium ion in the aluminum oxide lattice is surrounded by oxygen ions.

The electrostatic field of these ions affects the energy of the Cr^{3+} electrons as we learned in Chap. 7. Since these ions undergo thermal vibrations, the electrostatic field at the site of the Cr^{3+} electrons is modulated in magnitude. This modulation results in some of the electron transitions being broad spectral bands and others being sharp spectral lines.

The optical absorption and fluorescence spectra for ruby are shown in Fig. 14.12. The data shown in the figure were obtained at room temperature for a specific direction of light polarization. They change with temperature and polarization changes. The intensity scales are in arbitrary units. For ruby, there are two broad bands in the visible light absorption spectrum that allow it to absorb a significant amount of optical energy. The electrons excited in these transitions to high energy electronic states relax down to the lowest electronic excited state by giving off heat. Figure 14.12a shows these absorption bands. The small line near 7000 Å is the transition to the lowest excited state. The transition from this lowest excited level to the ground state is a pair of sharp lines that can be used in laser emission. These are shown in the fluorescence spectrum of ruby in Fig. 14.12b. The ability to absorb so much optical energy and emit it all in the sharp R_1 spectral line makes ruby ideal for a solid-state laser material.

Special Topic

Stimulated emission of microwave radiation was demonstrated by Charles Townes and his research group at Columbia University and shown to be a useful way to amplify microwave signals. Townes and Arthur Schawlow at Bell Labs then published a theoretical paper describing how this concept could be useful in amplifying visible light using a ruby crystal. Their first attempt to demonstrate a ruby laser was not successful, but Theodore

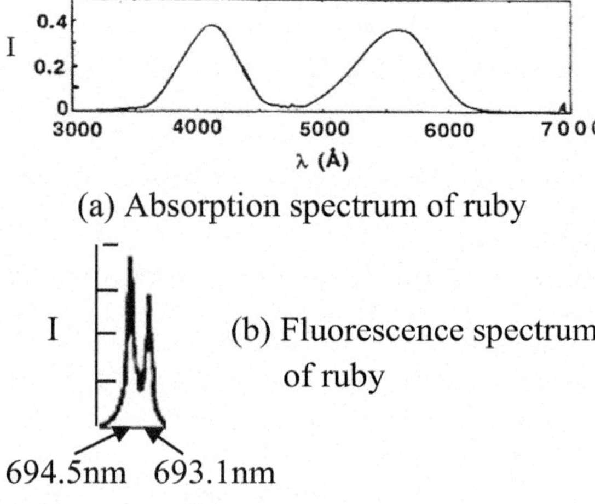

Fig. 14.12 Optical spectroscopy of ruby

Maiman at Hughes Research Laboratories used their original ideas to demonstrate the first ruby laser in 1960. The key to getting the laser to work had to do with using a ruby crystal with the right amount of chromium ions. This was a clear demonstration of how important it was to understand the fundamental physics of doping crystals. It is also important to think about the fact that the development of the laser came about through curiosity driven research and was not aimed at solving a specific problem. In fact, it was initially referred to as a solution in search of a problem. Of course, since its invention the laser has solved many different problems. It is a great example of the importance of curiosity driven research.

Typical solid-state laser materials are made of oxide or fluoride crystals or glass hosts doped with transition metal ions or rare earth ions. These types of dopant ions are used because they have unfilled energy levels that can be used for optical absorption and emission transitions. We will return to our discussion of lasers in Chap. 16.

Now do this final problem.

Student

14.10 You have an application for laser light at 347.25 nm. What is a good system to use for this?

Summing Up

You now know a lot about the matter that makes up the creation in which we live. For example, you know why some matter conducts electricity and some does not. You know why different types of matter absorb and emit light of different colors. You know why some matter floats and other matter sinks. You know why some matter flies and other matter does not. Mankind has used this knowledge to make things like microelectronic devices, lasers, ships and airplanes. In the final two chapter we will learn about what matter is made of.

Solutions to the Problems

14.1

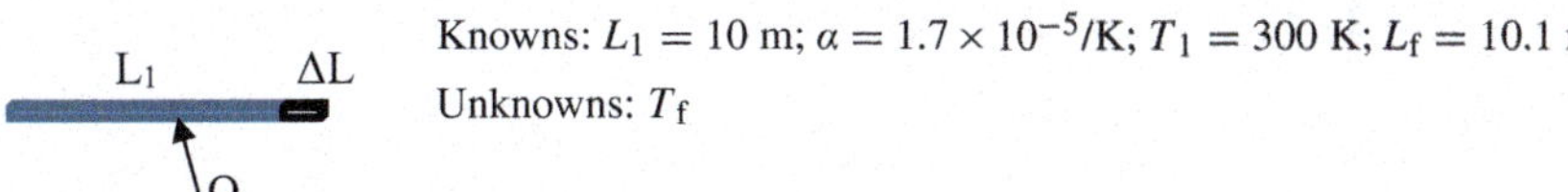

Knowns: $L_1 = 10$ m; $\alpha = 1.7 \times 10^{-5}$/K; $T_1 = 300$ K; $L_f = 10.1$ m

Unknowns: T_f

From Eq. 14.1,

$$\Delta L = \alpha L_1 \Delta T = L_f - L_1 = \alpha L_1 (T_f - T_1)$$
$$0.1\,\text{m} = \left(1.7 \times 10^{-5}/\text{K}\right)(10\,\text{m})(T_f - 300\,\text{K})$$
$$T_f = 300\,\text{K} + (0.01)/\left(1.7 \times 10^{-5}/\text{K}\right) = 888\,\text{K}$$

14.2

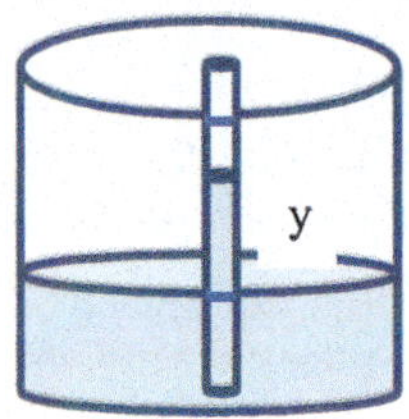

Knowns: $y = 6.0 \times 10^{-3}$; $\rho = 1000$ kg/m^3; $r = 8.0 \times 10^{-5}$ m
Unknowns: γMm

From the example,

$$y = 2\gamma \cos\theta/(\rho g r)$$
$$\gamma = y\rho g r/2 = (6.0 \times 10^{-3}\,\text{m})(1000\,\text{kg/m}^3)\left(8.0 \times 10^{-5}\,\text{m}\right)(9.81\,\text{m/s}^2)/2$$
$$= 0.00235\,\text{kg/s}^2 = 0.00235\,\text{N/m}$$

14.3

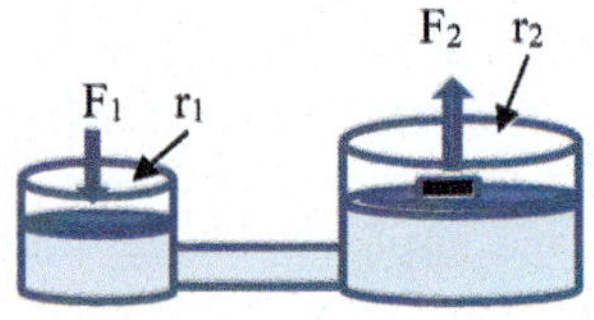

Knowns: $F_1 = 20$ N; $F_2 = 400$ N; $r_2 = 0.75$ m
Unknowns: d_1; MA

The pressures in the two tanks are equal. Therefore,

$$F_1/A_1 = F_2/A_2 \text{ so } 20\,\text{N}/\left(\pi r_1^2\right) = 400\,\text{N}/(\pi)(0.75\,\text{m})^2$$

$$r_1 = \left[(20\,\text{N})(0.75\,\text{m})^2/400\,\text{N}\right]^{1/2} = 0.15\,\text{m}$$

$$d_1 = 2r_1 = 0.3\,\text{m}$$

Also, the mechanical advantage is

$$\mathrm{MA} = F_2/F_1 = 400/20 = 20$$

14.4

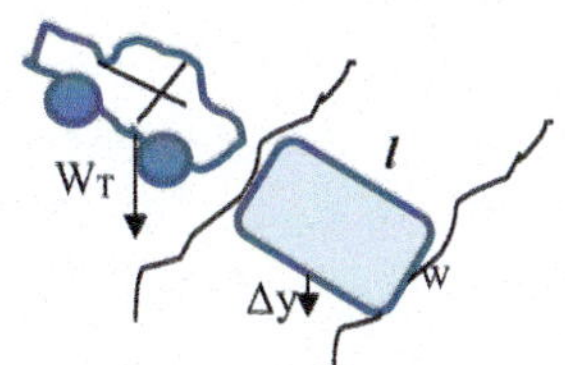

Knowns: $w = 5$ m; $l = 50$ m; $\Delta y = 0.50$ m; $\rho = 1000$ kg/m^3

Unknowns: W_T

The volume of water displaced is

$$V = wl\Delta y = (5\,\mathrm{m})(50\,\mathrm{m})(0.50\,\mathrm{m}) = 125\,\mathrm{m}^3$$

The weight of the displaced water is,

$$W = \rho g V = \left(1000\,\mathrm{kg/m}^3\right)\left(9.81\,\mathrm{m/s}^2\right)\left(125\,\mathrm{m}^3\right) = 12.26 \times 10^5\,\mathrm{N}$$

Since the truck caused this much water to be displaced, its weight is

$$W_T = 12.26 \times 10^5\,\mathrm{N}$$

14.5

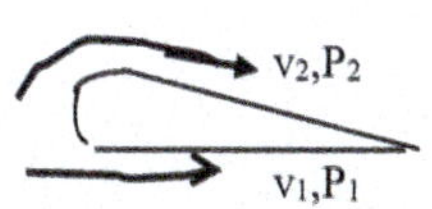

Knowns: $v_1 = 200$ m/s; $v_2 = 250$ m/s; $\rho = 1.225$ kg/m^3

Unknowns: Lift

According to Bernoulli's equation (Eq. 14.6),

$$P_1 + \rho v_1^2/2 = P_2 + \rho v_2^2/2$$

The lift is the difference between the two pressures,

$$\begin{aligned}\text{Lift} &= P_1 - P_2 = \rho v_2^2/2 - \rho v_1^2/2 \\ &= 0.6125\,\text{kg/m}^3\left[(250\,\text{m/s})^2 - (200\,\text{m/s})^2\right] \\ &= 13{,}781\,\text{kg/m s}^2 \\ &= 13.781\,\text{kPa}\end{aligned}$$

14.6

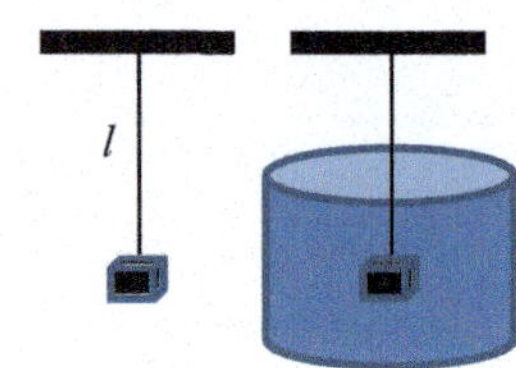

Knowns: $V = 600\ \text{cm}^3 = 60 \times 10^{-6}\ \text{m}^3$; $l = 150\ \text{cm} = 1.50\ \text{m}$; $A = 2 \times 10^{-3}\ \text{cm}^2 = 2 \times 10^{-4}\ \text{m}^2$; $E = 190 \times 10^9\ \text{Pa}$; $\rho = 10^3\ \text{kg/m}^3$

Unknowns: Δl

From Eq. 14.5,

$$E = FL/(A\Delta l)$$

So,

$$\Delta l = Fl/(EA)$$

The force causing the change in length of the wire is the force of buoyancy of the displaced water.

$$F = \rho g V = \left(10^3\,\text{kg/m}^3\right)\left(9.81\,\text{m/s}^2\right)\left(60 \times 10^{-6}\,\text{m}^3\right) = 0.589\,\text{kg m/s}^2$$

Substituting this into the equation for the change in length gives,

$$\Delta l = \left(0.589\,\text{kg m/s}^2\right)(1.50\,\text{m})/\left(190 \times 10^9\,\text{Pa}\right)\left(2 \times 10^{-4}\,\text{m}^2\right) = 2.33 \times 10^{-8}\,\text{m}$$

14.7

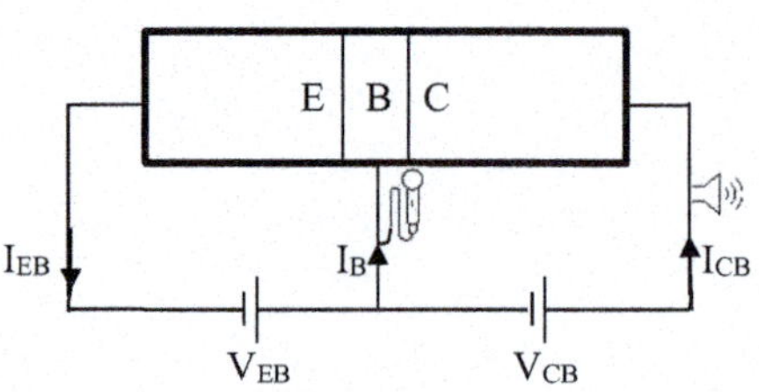

Knowns: $I_{CB} = 60 I_B$; $I_B = 0.01$ A; $V_{CB} = 3$ V; $V_{EB} = 0.3$ V

Unknowns: Power amplification

The power from the microphone is

$$P_{\text{mic}} = I_{\text{B}} V_{\text{EB}} = (0.01\,\text{A})(0.3\,\text{V}) = 3 \times 10^{-3}\,\text{W}$$

The power to the speaker is

$$P_{\text{spe}} = I_{\text{CB}} V_{\text{CB}} = (0.6\,\text{A})(3\,\text{V}) = 1.8\,\text{W}$$

The amplification factor from 3×10^{-3} to 1.8 W is a factor of 600.

14.8

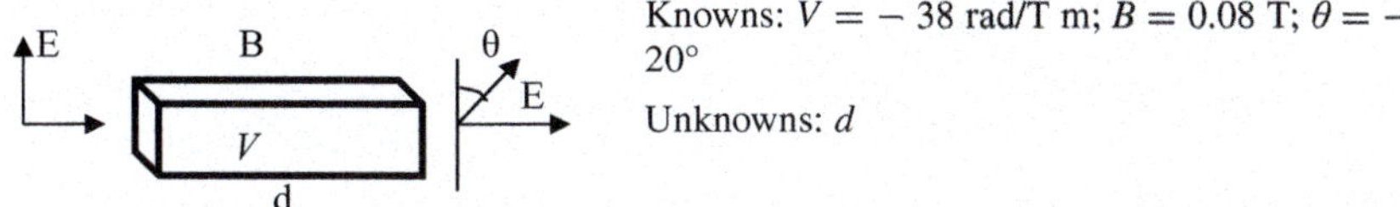

Knowns: $V = -38$ rad/T m; $B = 0.08$ T; $\theta = -20°$

Unknowns: d

The equation for the rotation angle in the Faraday effect is Eq. 14.9

$$\theta = VBd$$

The angle must be in radians so $\theta = -20°(\pi\,\text{rad}/180°) = -0.349\,\text{rad}$. So,

$$d = \theta/VB = -0.349\,\text{rad}/(-38\,\text{rad}/\text{T m})(0.08\,\text{T}) = 0.115\,\text{m}$$

14.9

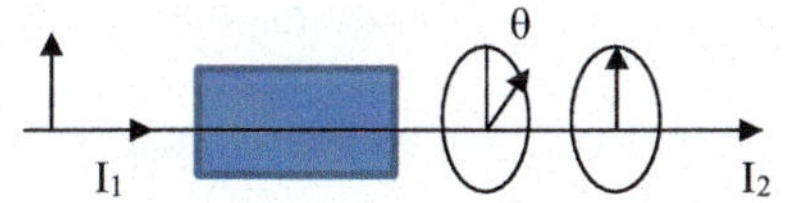

Knowns: $I_2/I_1 = 0.8$; $V = -38$ rad/T m; $d = 0.12$ m

Unknowns: B

The intensity decrease between two polarizers can be found from Malus's equation given by Eq. 11.17

$$I_2 = I_1 \cos^2 \theta$$

Solving this for θ gives

$$\theta = \cos^{-1}(I_2/I_1)^{1/2L}$$

Now this angle can be used in Eq. 14.9 for Faraday rotation,

$$\theta = VBd = \cos^{-1}(I_2/I_1)^{1/2}$$

Solving for B gives,

$$B = \cos^{-1}(I_2/I_1)^{1/2}/Vd = 0.465\,\text{rad}/[(38\,\text{rad/T m})(0.12\,\text{m})] = 0.10\,\text{T}$$

14.10

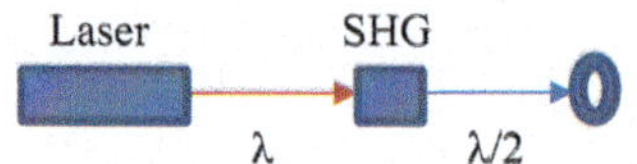

Knowns: $\lambda_{ruby} = 694.5$ nm

Unknowns: λ final

After going though a second harmonic generation crystal, the final wavelength is half the initial wavelength,

$$\lambda\ \text{final} = \lambda_{ruby}/2 = 694.5\,\text{nm}/2 = 347.25\,\text{nm}$$

Study Guide for This Chapter

1. What is the difference between a glass and a crystal?
2. What causes capillary action?
3. Why do some objects float in a liquid and others sink?
4. What is Pascal's Principle?
5. What is the Bernoulli Effect?
6. Explain the differences between materials that are elastic, plastic, ductile, and brittle.
7. What is the most important distinguishing feature between insulators, conductors, and semiconductors?
8. What is the difference between intrinsic and extrinsic semiconductors?
9. What is a p-n junction?
10. Explain the difference between linear and nonlinear optics.
11. An aluminum wire of length 0.5 m is hanging from the ceiling support a weight. If the room temperature goes up from 300 to 500 K, how much does the weight sag? ($\alpha_{aluminum} = 2.5 \times 10^{-5}\,\text{K}^{-1}$)
12. Two capillary tubes of radii 0.5 and 0.1 mm are immersed in water with a surface tension of 0.07 N/m. How much higher does the water raise in one tube versus the other tube?
13. The effort side of a hydraulic press has a radius of 0.5 m. The load side has a radius three times larger. How much effort does it take to lift a 0.51 kg mass?
14. A ship is said to weigh 20 tons. How much water does it displace?

15. Water enters a hose at 1.0 m/s. If the pressure drops by 100 Pa between the hose and the nozzle, what is the speed of the water in the nozzle?
16. A wire is 0.5 m long with a cross sectional area of 0.05 m^2 and a Young's modulus of 80×10^9 Pa. If you pull on it with a force of 2×10^6 N, how much longer does it get?
17. You want to decrease the intensity of your laser by a factor of 0.6. You have a polarizing filter and a Faraday rotator that has a magnetic field of 0.20 T and can hold a crystal that is 10 cm long. What is the Verdet constant of the crystal you need to use in your resonator?
18. A bipolar transistor power amplifier has an emitter-base voltage of 0.1 V and a collector-base voltage of 5.0 V. If the emitter-base current is 2 A for a power amplification factor of 500, what is the base current?

Next Level

19. A Field Effect Transistor acts as a $\times$ 10 voltage amplifier. At a gate to source input voltage of $-$ 5 mV a source to drain current of 0.4 mA flows. What is the resistance of the source to drain circuit?
20. What is the voltage amplification factor of a Field Effect Transistor if an input gate to source voltage of $-$ 2 mV gives a resistance between source and drain of 200 Ω with a source to drain current of 0.5 mA?

Modern Physics I: Quantum Mechanics and Relativity

15

Chapter Summary

The principles of classical mechanics that we studied in the early chapters of this text give us an appropriate way to describe the motion of objects with a normal size moving at normal speeds. Also, the basic principles of electromagnetism and electromagnetic waves were summarized in Maxwell's Equations and they did a good job of explaining this area of classical physics. However, in the early part of the twentieth century physicists began studying atomic size particles and objects traveling at speeds close to the speed of light. At these extreme conditions they found that normal classical physics could not explain all of their observations. This spawned the birth of Modern Physics based on quantum mechanics and relativity. After discussing the basic ideas of these theories in this chapter, we will see how they are applied to atomic and nuclear physics in the final chapter. Some of the themes of the chapter are represented in Fig. 15.1.

Main Concepts in This Chapter

- Quantum Mechanics I: Duality
- Quantum Mechanics II: Exclusion, Probability, and Uncertainty
- Relativity I: Special Relativity
- Relativity II: General Relativity

R. C. Powell, *Physics*, Undergraduate Lecture Notes in Physics,
https://doi.org/10.1007/978-3-032-09361-5_15

15.1 Introduction

Two different theories form the fundamental basis for modern physics: quantum mechanics and relativity. Both theories are discussed in this chapter. Although both of these had their beginnings at about the same time, quantum mechanics was developed by several different physicists trying to explain observations that couldn't be explained by classical physics while the development of relativity was due primarily to the ingenious ideas of one individual, Albert Einstein.

Studying the properties of electrons played an important role in the early development of quantum mechanics. The ancient Greeks understood some matter possessed electrical properties, but it wasn't until 1897 that J. J. Thompson identified the electron as a particle with an electrical charge. After that, there were important experiments in the early 1900s to establish the basic properties of electrons. The first was conducted by Robert Millikan who determined the electrical charge on an electron by observing the Coulomb force on charged oil droplets in an electric field. He found $q_e = 1.602 \times 10^{-19}$ C. Then J. J. Thompson used the properties of a beam of electrons traveling in electric and magnetic force fields to determine the mass of an electron. From our discussions of the motion of charged particles in an electric field (Chap. 7) and in a magnetic field (Chap. 9), a beam of electrons will experience a force $F_E = Eq$ in the former and a force $F_B = Bqv$ in the latter. In Thompson's experiment, the electric and magnetic fields were perpendicular to each other, so their forces were in the opposite directions. By balancing **B** and **E**, so the electrons traveled in a straight line,

$$Eq = Bqv \quad \text{or} \quad v = E/B.$$

When the electric field is turned off, the magnetic field causes the particle to undergo circular motion with the centripetal acceleration given by (see Chap. 5)

$$Bqv = mv^2/r$$

or

$$q/m = v/(Br).$$

Knowing B and v from the first part of the experiment and measuring r from the second part of the experiment, Thompson determined the ratio of the charge and mass of an electron. Using the charge Milliken had measured, he determined the mass of an electron to be 9.107×10^{-31} kg. He used the same technique to measure the mass of a proton. As more information was obtained about these extremely small particles, important questions arose which are discussed in the next section.

Another important experiment in that era was done by Michaelson and Morley to accurately measure the speed of light. Their work helped establish that light traveled in a vacuum at a constant speed of 2.998×10^8 m/s. This result was important both for the physicists thinking about quantum mechanics and for Einstein's ideas about relativity.

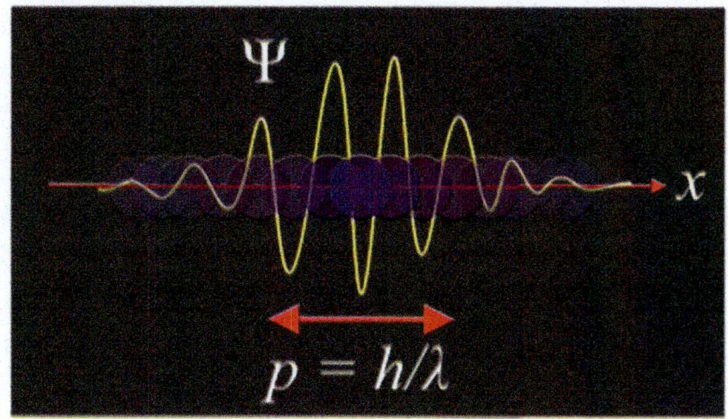

Fig. 15.1 Quantum wave packet. Credit: Maschem. Creative Commons Attribution-Share Alike 0. Note that the astronomy pictures are by the author

15.2 Quantum Mechanics I: Duality

Quantum mechanics involves some strange new concepts in physics including duality, exclusion, probability, and uncertainty. We will begin with the concept of wave-particle duality as shown in Fig. 15.1.

In Chap. 11 we described the physical properties of light from the perspective of light waves. However, in some cases, light acts like a beam of particles instead of a wave. A good example of this is the photoelectric effect. When light is shined on the surface of a metal, electrons can be emitted from the surface. If light acts like a wave, we would expect that the kinetic energy of the emitted electrons would depend on the lights' intensity but not on its frequency. The higher the light intensity the higher the maximum kinetic energy of the emitted electron. Instead, in the early 1900s the scientists measuring the photoelectric effect in different metals found that the maximum kinetic energy of the emitted electrons depended on the frequency of the light and not its intensity. For each type of metal, there was a threshold frequency below which no free electrons were observed. Above this threshold frequency, the maximum kinetic energy of the emitted electrons depended on the frequency of the light and not its intensity. These experiments and their results are shown in Fig. 15.2.

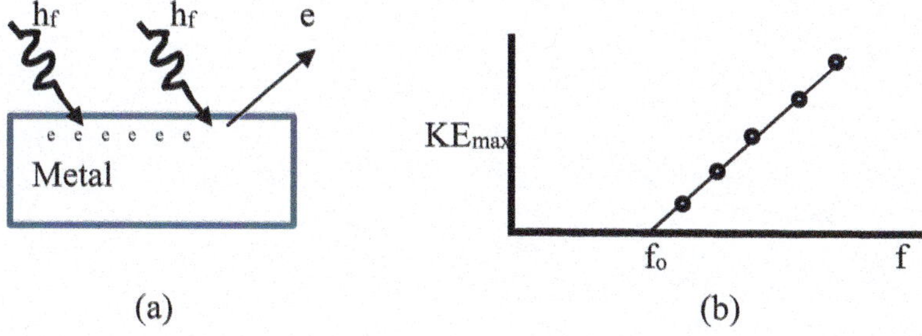

Fig. 15.2 Photoelectric effect

As shown in Fig. 15.2a, the maximum kinetic energy of the emitted electron in the photoelectric effect is proportional to the frequency of the light hitting the surface of the metal. It does not depend on the intensity of the light waves. In addition, electrons are emitted immediately when the incident light hits the surface of the metal. There is no time delay. Below a certain threshold frequency of light, no electrons are emitted. This threshold is different for different types of metals.

The explanation of the photoelectric effect was provided by Albert Einstein who suggested treating the light as a beam of particles instead of treating it as light waves. The particles are called photons and they each have a discrete amount of energy. The discrete energy of each photon is proportional to its frequency. It is given by

$$E = hf. \tag{15.1}$$

In this expression, h is a fundamental constant in physics known as Planck's constant. Its value is $h = 6.63 \times 10^{-34}$ J s. Photons of all frequencies have no mass and travel with the speed of light. When a photon hits an electron, all of its energy is transferred to the electron.

The threshold energy required to overcome the cohesive forces holding the electrons in the metal is called the work function. It is designated by ϕ and is different for each type of metal. The equation explaining the data shown in Fig. 15.2b is

$$\mathrm{KE}_{\max} = hf - \phi = h(f - f_0). \tag{15.2}$$

The maximum kinetic energy of the emitted electron is the energy of the photon that hit it minus the amount of energy used to overcome the work function of the material. By extrapolating the straight line in Fig. 15.2b back to the y-axis where $f = 0$, the work function frequency can be determined. Note that increasing the intensity of the incident light does not change its frequency and thus does not change the kinetic energy of the emitted electrons. The increased intensity means an increase in number of photons and thus an increase in the number of emitted electrons but not an increase in their energy.

This experiment and Einstein's explanation confirmed the particle nature of light. Because the light particles have discrete amounts of energy, they are said to be "quantized." This is demonstrated in the following example and problems.

Example 15.1

A sheet of aluminum has a work function of 6.9 x 10^{-19} J. If it is exposed to light of wavelength 250 nm what is the maximum kinetic energy of the emitted electrons?

Use Fig. 15.2 Knowns: $\phi = 6.9 \times 10^{-19}$ J; $\lambda = 250$ nm.

Unknowns: $\mathrm{KE}_{\max}$.

Using Eq. 15.2,

$$\begin{aligned} KE_{max} &= hf - \phi = hc/\lambda - \phi \\ &= \left(6.63 \times 10^{-34}\,\mathrm{J\,s}\right)\left(2.998 \times 10^{8}\,\mathrm{m/s}\right)/\left(2.5 \times 10^{-7}\,\mathrm{m}\right) - 6.9 \times 10^{-19}\,\mathrm{J} \\ &= 1.05 \times 10^{-19}\,\mathrm{J}. \end{aligned}$$

So the photons hitting the metal each have an energy of 7.95×10^{-19} J. This is transferred to an electron in a photon-electron collision. The electron uses at lease 6.9×10^{-19} J of this energy to overcome the work function and escape from the surface of the metal. This leaves the electron with 1.05×10^{-19} J or less for its kinetic energy. For photons with wavelengths greater than 288 nm no photoelectrons will be emitted.

Now try the following problems.

Student

15.1 What is the work function of a metal if light of wavelength 300 nm emits photoelectrons with a maximum kinetic energy of 1.5×10^{-19} J?

15.2 Silver emits electrons when illuminated with light. However, if the light wavelength is below 264 nm, no electrons are emitted. What is the work function of silver? ($1\ \mathrm{nm} = 10^{-9}\ \mathrm{m}$)

The photoelectric effect is not the only evidence for light being a beam of particles with quantized energy. Nor are photons the only physical quantity that is quantized. Another example of both of these concepts is the frequency distribution of the radiation emitted by a hot body. This can not be adequately explained by classical physics which predicts that the radiated power per unit frequency should be proportional to the frequency squared. Therefore, the radiated power is predicted to go to infinity at very high frequencies. This prediction is called the ultraviolet catastrophe and it really bothered the physicists of the early 1900s. Max Planck solved this problem by assuming that a system of atoms or molecules was quantized and could only absorb or emit radiation in discrete quantized amounts. We will discuss this further in the next chapter. However, treating the absorption and emission of radiation as quantized photons being absorbed and emitted by quantized energy states of atoms and molecules predicted the observed results for blackbody radiation and solved the problem of the ultraviolet catastrophe. The detailed mathematical description of this problem is beyond the scope of this textbook. However, it is important to note that solving this ultraviolet catastrophe problem was further proof of the particle nature of light. At this point it is clear that light has a dual nature.

Next Level
Blackbody Radiation

A blackbody is a hot cavity in thermal equilibrium that absorbs and emits all wavelengths of electromagnetic radiation from the infrared through the visible and ultraviolet regions. The observed wavelength distribution of the radiation per unit volume is shown in the figure at the left. In classical physics, the theoretical prediction for this distribution was given by the Rayleigh-Jeans law for the energy per unit wavelength per unit volume

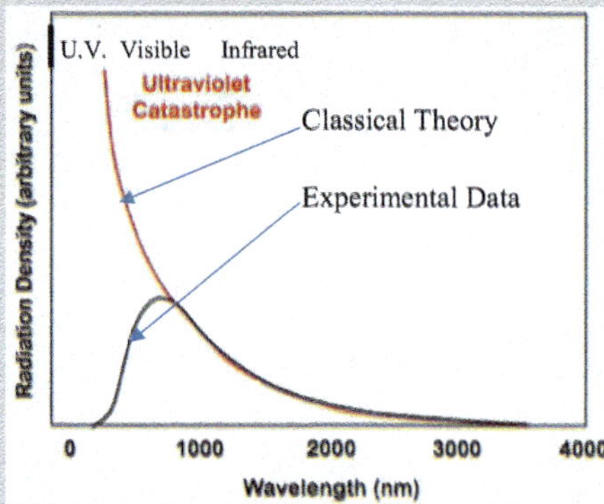

$E_{\lambda V} = 8\pi kT/\lambda^4$
which assumed that all states had an equal energy of kT. Each state was treated as a resonant mode of the cavity where the wavelength had to fit the cavity dimensions like a wave on a string treated in Sect. 12.4. This theory fit the long wavelength data as shown in the figure but diverged at small wavelengths due to the λ^4 in the denominator. This was referred to at the "ultraviolet catastrophe."

Using the Planck–Einstein quantum theory of electromagnetic radiation, the cavity modes are photons with energy $E = hc/\lambda$. Bose and Einstein derived an expression for the statistical distribution of occupied photon states in thermal equilibrium. Multiplying this distribution function by the photon energy gives a quantum mechanical expression for the energy density of a blackbody,

$$E_{\lambda V} = \frac{8\pi hc}{\lambda^5\left(e^{\frac{hc}{\lambda kT}} - 1\right)}.$$

This predicts the observed experimental fairly accurately. At long wavelengths the factor in parenthesis in the denominator becomes $hc/\lambda kT$ and the quantum expression is the same as the Rayleigh-Jeans expression. At short wavelengths, the factor in parenthesis in the denominator overrides the λ^5

factor and causes the curve to go to zero. This was another success of quantum theory in explaining experimental results that could not be explained by classical physics.

The photoelectric effect assumes there are collisions between photons and electrons in which energy and momentum are transferred from one to the other. This implies that photons have momentum like a regular particle even though they don't have mass. To prove that photons really do have momentum, in 1923 A. H. Compton fired a beam of X-ray photons into a graphite target. He found that the wavelength of the scattered photons had a higher wavelength than the wavelength of the incident photons. Photons were presumed to have energy and momentum given by $E = hc/\lambda$ and $p = h/\lambda$. The increase in wavelength shows that the scattered photons had smaller energy and momentum than the incident photons. Compton also observed that electrons were ejected from the graphite in his experiments. He measured their energy and momentum and found they matched the energy and momentum lost by the scattered photons. He concluded from this that the photons underwent scattering collisions with the electrons in which both energy and momentum were conserved. This demonstrated that photons have momentum like particles.

The conclusion of these experiments is that light sometimes acts like waves that undergo diffraction and interference and it sometimes acts like particles when it interacts with matter. The obvious question is "does matter also have a dual nature and sometimes act like a wave?" Louis de Broglie suggested that the answer to this question is "yes" and the wavelength of a matter wave is called its de Broglie wavelength. An expression for this is derived below.

In Einstein's description of photons, they have an energy given by Eq. 15.1 and travel at the speed of light, c. This can be written as

$$E = hc/\lambda \tag{15.3}$$

where c is the constant speed of light and λ is the wavelength of light. Einstein also proposed that mass and energy are related by the expression

$$E = mc^2. \tag{15.4}$$

For normal matter, its momentum p is equal to its mass times its velocity. Although photons have no mass per se, the equivalent expression for their momentum can be derived from Eq. 15.4 as

$$p = E/c. \tag{15.5}$$

Combining Eqs. 15.3 and 15.5 gives an expression relating momentum and wavelength,

$$p - h/\lambda.$$

Table 15.1 Properties of photons

• $E = hf = hc/\lambda$
• $v = c$
• $p = E/c$
• $\lambda = h/p$

Inverting this gives the de Broglie expression for the wavelength of a matter wave,

$$\lambda = h/p. \tag{15.6}$$

This holds for both photons with their momentum given by Eq. 15.5 and for particles of mass m with a momentum $p = mv$. The properties of photons are listed in Table 15.1.

Now go through the following example and try the problem.

Example 15.2

Compare the de Broglie wavelength of an electron that is moving at a speed of 50 m/s with that of a baseball moving at the same speed.

(The mass of an electron is 9.11 x 10^{-31} kg; the mass of a baseball is 0.15 kg.)

m,λ v

Knowns: $v = 50$ m/s; $m = 9.11 \times 10^{-31}$ kg

Unknowns: λ

Using Eq. 15.6 for the de Broglie wavelength of an electron,

$$\lambda = h/mv = 6.63 \times 10^{-34}\,\text{J s}/\left(9.11 \times 10^{-31}\,\text{kg}\right)(50\,\text{m/s}) = 0.0146\,\text{mm}.$$

On the other hand, the de Broglie wavelength of a baseball of mass 0.15 kg moving at a speed of 50 m/s is,

$$\lambda = h/mv = 6.63 \times 10^{-34}\,\text{J s}/(0.15\,\text{kg})(50\,\text{m/s}) = 1.1 \times 10^{-34}\,\text{m}.$$

These examples show that for objects of normal size, their de Broglie wavelength is far too small to measure while for very small size particles the de Broglie wavelength is long enough to be measured. This is one reason that quantum effects are important on the atomic scale but not in everyday life.

Student

15.3 Calculate the de Broglie wavelength of an electron moving at 1510 m/s. How does this compare to the wavelength of visible light (roughly 400–700 nm)? ($m_e = 9.11 \times 10^{-31}$ kg)

The fact that electrons exhibit wavelike properties was demonstrated conclusively by several experiments in the mid 1920s. Davisson and Germer looked at the distribution of electrons scattered off the surface of a crystal of nickel. Instead of the results having a smooth angular distribution of electrons, they exhibited regions of high concentrations and regions of low concentrations of electrons. This is consistent with the interference pattern of waves scattered from a geometric distribution of closely spaced atoms in the crystal. A similar result was obtained by Thompson who sent a beam of electrons through thin sheets of crystal line metals and observed that the transmitted electrons had high concentrations and low concentration regions similar to the diffraction pattern described in Chap. 11 for light waves going through small slits. The observation of these diffraction patterns for light waves was explained by the interference phenomena for waves. Since they were also observed for electrons, it was interpreted as being a consequence of the wave-like nature of particles.

We still do not understand why our creation has a dual wave-particle nature. However, duality is an important, well-established part of quantum mechanics.

15.3 Quantum Mechanics II: Exclusion, Probability, and Uncertainty

Duality isn't the only strange phenomenon in quantum mechanics. Another one is exclusion. When a system is quantized, some of the parameters describing the state of the system can only have certain discrete values and not any arbitrary value. The following chapter gives a specific example of this for atoms. The electrons bound to an atom can only be in specific states. These states are designated by a set of quantum numbers that designate the energy, angular momentum, and spatial orientation of the electron's motion. Elements with atomic numbers greater than one have more than one electron. One of the fundamental rules of quantum mechanics is called the Pauli exclusion principle. This says that no two electrons can have the same set of quantum numbers at the same time. To generalize this specific example, no two particles in a quantized system can exist in the same space at the same time. We will discuss the specific example of electrons on atoms again in the next chapter, but you should remember that the Pauli Exclusion Principle applies to all quantum mechanical systems.

Two other strange phenomena of quantum mechanics we need to discuss are probability and uncertainty. In classical physics we can measure the exact values of all of the parameters that describe a system and use this information to say that the system is in a specific physical state. This is not true in quantum mechanics. The mathematical formalism of quantum mechanics only gives us the ability to know the probabilities of a system being in each of its possible states at any given time. If we make a measurement on the system and find it to be in a specific state at a specific time, we can only determine the probability of the system evolving to one of its other possible states at some future time. To describe the evolution of a quantum mechanical system we need to determine the probabilities of it going to each of its possible states.

One of the important aspects of quantum mechanics is the special relationship between the variables defining the state of a system, such as position and momentum or energy and time. In classical physics, these are all independent variables and can be measured to any degree of accuracy. However, this is not true in quantum mechanics. In quantum mechanics the accuracy of the measurement of one variable is determined by our knowledge of the measurement of another variable. This is called the Heisenberg uncertainty principle. For related variables, it says that the more accurately we know the value of one variable the less accurately we know the value of the other variable. For position and momentum this is written as

$$\Delta x \Delta p \geq h/(4\pi) \tag{15.7}$$

where $h = 6.63 \times 10^{-34}$ J s is Planck's constant. Here, Δx is the uncertainty in the value of position and Δp is the uncertainty in the value of momentum. As one decreases the other must increase. Remembering that $p = mv$ where the mass m is a constant and v is velocity, the term $\Delta \mathrm{p} = \mathrm{m}\Delta \mathrm{v}$. Moving m to the denominator on the right side of Eq. 15.7 gives

$$\Delta x \Delta v \geq h/(4\pi m).$$

Since h is a very small number of the order of 10^{-34} J s, if the particle of interest has a mass of macroscopic size, the uncertainty in the values of the variables is too small to make a difference. However, for atomic size particles such as an electron with a mass of the order of 9.11×10^{-31} kg the uncertainty in the values of the variables is very relevant. Thus, the uncertainty principle is important in quantum mechanics but not in classical physics.

Energy and time are another set of related variables, so the uncertainty principle requires

$$\Delta E \Delta t \geq h/(4\pi). \tag{15.8}$$

The more accurately we know the energy of a system the less accurately we know the time it will be in that energy state. In the next chapter, we will see how important this is when considering an electron in a high energy state of an atom.

Like duality, probability and exclusion are important concepts in quantum mechanics that we don't need to use in classical physics. Even if we don't fully understand why they are present, it is important to use these concepts when we deal with quantum mechanical systems. Before going on to the next section, try the following problems.

Student

15.4 You measure the velocity of an electron $(m = 9.11 \times 10^{-31}\,\text{kg})$ with an accuracy of plus or minus 1mm/s. How accurately can you simultaneously measure its position?

15.5 If you measure the energy of a system to be $2 \pm 0.1\,\text{J}$, how long can you reasonably expect it to have that energy?

15.4 Relativity I: Special Relativity

The other important area of modern physics is Einstein's work on relativity theory. A detailed treatment of relativity is far beyond the scope of this physics text. However, a brief overview is included for the sake of completeness. Einstein developed two different types of theories, one called special relativity and one called general relativity.

Special relativity deals with describing physics events involving objects moving at speeds near to the speed of light. Einstein based his work on two important postulates. The first is that the laws of physics are the same in all reference frames that are not undergoing acceleration. The second is that the speed of light in a vacuum is the same for all observers regardless of the motion of the observer or the light source. For events not involving objects moving at speeds near the speed of light, the results are the same as predicted by normal classical physics. However, at relativistic speeds, things can change such as length measurements can become smaller, time measurements can become slower, mass and energy can be exchanged. Einstein's formulation of special relativity has been successful in explaining these changes. It also shows that the speed of light in a vacuum c is not only constant, but also the maximum speed to which any object can be accelerated.

We begin our discussion of special relativity with a reminder of what we learned in Chap. 2 about inertial reference frames. We learned that in physics, we must define motion *relative* to something else. Although we can't say for sure whether or not we are moving, we can certainly tell whether or not we are moving relative to a specific object. Thus, motion must be stated relative to something else. Using this idea, an inertial reference frame is a reference frame that moves at a constant velocity relative to another inertial reference frame. Although it is not absolutely true, you learned in Chap. 2 that we treat the earth as an inertial reference frame. Thus, anything moving at a constant velocity relative to the earth is an inertial reference frame. An elevator in free fall, for example, is not an inertial reference frame, because it is accelerating relative to the earth. If the elevator is falling at a constant velocity, however, it is an inertial reference frame, since it is moving at a constant velocity relative to the earth. A race car traveling at a constant speed around a circular track is not an inertial reference frame, because its direction (and therefore its velocity) is changing. Thus, it is not moving at a constant velocity relative to the earth. A race car traveling along a straight-away at a constant speed is an inertial reference frame, however, because its speed and direction (and thus its velocity) is not changing with respect to earth.

Now remember what is special about an inertial reference frame. In an inertial reference frame, Newton's Laws of Motion work. In a non-inertial reference frame, they do not necessarily work. When a car is traveling at a constant velocity, for example, boxes on the seat which are sitting still will not begin moving until they are acted on by a force. That's Newton's First Law of Motion. On the other hand, if the car goes around a curve, the boxes might begin moving seemingly of their own accord. That's because once the car goes around the curve, it is no longer an inertial reference frame, and as a result, Newton's First Law is no longer valid in the reference frame of the car. We said in Chap. 2 that Newton's Laws work in *any* inertial reference frame. What about the laws of electromagnetism? Will they work in *any* inertial reference frame? If Newton's Laws work in any inertial reference frame, the laws of electromagnetism work in any inertial reference frame. That may sound like a simple statement, but the consequences of that statement are profound. The laws of electromagnetism (Maxwell's equations) state that the speed of light in a vacuum is 2.998×10^8 m/s, which we usually abbreviate as "c." Since that is a consequence of the laws of electromagnetism, and since we want to believe that the laws of electromagnetism work in all inertial reference frames, we must therefore agree that the speed of light is equal to c in all inertial reference frames.

To illustrate just how profound all of this is, let's suppose you could run at the speed of light. As long as you are running in a straight line and your speed does not change relative to the earth, you are in a inertial reference frame. Thus, Newton's Laws work, and the laws of electromagnetism work. Suppose you are holding a mirror while you run. If you look in the mirror, what will you see? In order to see yourself in the mirror, light must reflect off your face, hit the mirror, and return to your eye. But light can't reach the mirror. You are running with the mirror at the speed of light. The light reflects off of your face at the speed of light. Thus, the light will never hit the mirror, and it will therefore never have a chance to reflect off of the mirror and reach your eyes, right? *WRONG!*

Remember, the laws of electromagnetism work in *all* inertial reference frames. Thus, even in your inertial reference frame (running at the speed of light), light travels at a speed of c. In your inertial reference frame, the mirror is not moving. It is staying at the same position relative to you, so it is at rest in your inertial reference frame. Since it is at rest in your inertial reference frame, and since light moves at a speed of c in your inertial reference frame, light will reflect off of your face, hit the mirror, and reach your eyes just as if you are not moving. Whether you stand still relative to the earth, run at 100 m/s, run at 1,000,000 m/s, or run at c, light will always reflect off of your face, hit the mirror, and reflect back into your eyes, all with a speed of c. It must, if the laws of electromagnetism are the same in all reference frames.

The situation discussed above was originally discussed by Einstein in an attempt to explain the reasoning behind his special theory of relativity. It can be summed up in one statement:

Einstein's Special Theory of Relativity—The laws of physics work the same in all inertial reference frames.

Regardless of whether we are stationary relative to the earth or moving at a constant velocity relative to the earth, the laws of physics work the same. If we do experiments while we are stationary, or on a spaceship that is moving at a constant velocity relative to earth, the experiments will have exactly the same results, because the laws of physics work in all inertial reference frames. If the laws of physics are the same in all inertial reference frames, and if the laws of electromagnetism state that the speed of light in a vacuum is c, then "common sense" will not always describe physical situations.

For example, suppose you are standing on a street and far, far away, your friend turns on a flashlight. The light approaches you at a speed of c. If you had some device that you could use to measure the speed of light, you would determine that the light is approaching you with a speed of 2.998×10^8 m/s. Now, suppose you start running towards your friend, at a constant speed of one-half the speed of light, 1.499×10^8 m/s. If you then used the same device to measure the speed at which the light is approaching you, what would you get? Common sense would say that you would measure that the light is approaching you at 4.497×10^8 m/s, which is the sum of your speed and the speed of light. However, that is *WRONG!* Remember, the laws of physics must be the same in all inertial reference frames. Thus, the speed of light must be 2.998×10^8 m/s in all inertial reference frames. This means that whether you are standing still relative to your friend, or whether you are running at your friend at a constant speed of one-half the speed of light, you will measure that the light is approaching you at exactly 2.998×10^8 m/s.

If we drive a car at 55 miles per hour east, and another person drives a car at 55 miles per hour west, how quickly are the cars approaching one another? They are approaching one another at 110 miles per hour. However, if we are traveling at the speed of light east and light is heading towards us from the east, the light is approaching us at exactly the speed of light. Whether we stand still or move towards the light source, the light will *always* be approaching at exactly 2.998×10^8 m/s—nothing more and nothing less! That's a direct consequence of the idea that the laws of physics work the same in all inertial reference frames. Since the laws of physics state that light moves at a speed of 2.998×10^8 m/s in a vacuum, then no matter what inertial reference frame we are in, we will always measure the speed of light to be 2.998×108 m/s.

All of this might sound really odd to you, because it goes against your common sense. Remember, however, that common sense is built on your experiences and these are limited to moving at normal speeds. If we want to believe that the laws of physics work the same in all inertial reference frames, then all of the situations we discussed above must be true. This is a basic premise of Einstein's theory of special relativity and there is strong experimental evidence that this theory is correct.

In special relativity, a physical event is described as taking place in a 4-dimensional space–time where x, y, z and t are variables in a specific reference frame. Someone using a different reference frame will measure different properties of the same physical event. For example, the motion of an object is not an absolute quantity, it is motion relative to a specific point of reference. The main question in special relativity is deriving the relationships between the properties of an event given by observers having two different inertial reference frames. There may be a relative velocity between the two reference frames but there cannot be any acceleration. In general, we can assume one reference frame is fixed to the surface of the earth and the other reference is moving with respect to it at a constant velocity v. The moving reference frame might be attached to a train or a car on a straightaway, but not on a curved track since that would involve acceleration. For events not involving objects moving at speeds near the speed of light, the results of measurements are the same as predicted by normal classical physics. However, at relativistic speeds, things can change.

The equations relating the measurements of position and time in the two reference frames are called the Lorentz transformation equations. For a reference frame moving with a velocity v in the x direction with respect to a fixed reference frame, these are,

$$\begin{aligned} x' &= (x - vt)/\sqrt{1 - v^2/c^2}, \\ y' &= y, \\ z' &= z, \\ t' &= \left(t - vx/c^2\right)/\sqrt{1 - v^2/c^2} \end{aligned} \tag{15.9}$$

where (x', y', z', t') are parameters of an event in the moving reference frame measured by an observer in the fixed reference frame, while (x, y, z, t) are the same parameters measured by an observer in the moving reference frame. Note that when v is small compared to c these equations are the classical mechanics equations from Chap. 1.

If we deal with a time interval in the moving reference frame measured by an observer in that frame, $\Delta t = t_2 - t_1$. The same time interval in the moving reference frame measured from the fixed reference frame is $\Delta t' = t_2' - t_1'$ and the clock is in the same position for both time measurements. The second terms of the time equation in Eq. 15.9 cancel out and leave the time interval equation

$$\Delta t' = \Delta t/\sqrt{1 - v^2/c^2} \tag{15.10}$$

So the time an earth observer measures for an event on a rocket ship is longer than the time measured by an astronaut on the rocket ship.

There is also an effect on the measurement of length in the direction of motion. The length between two points along the direction of motion is the velocity times the time it takes to move between the two points. This is $\Delta L = v\Delta t$ for an observer in the moving reference frame and $\Delta L' = v\Delta t'$ for an observer in the fixed reference frame. Note that the velocity is the same in both these equations. This follows from the fact that both the length and the time change,

$$\begin{aligned} v' &= \Delta L'/\Delta t' \\ &= \left[(x_2 - vt)/\sqrt{\mathbf{1} - \mathbf{v}^2/\mathbf{c}^2} - (x_1 - vt)/\sqrt{\mathbf{1} - \mathbf{v}^2/\mathbf{c}^2}\right]/\Delta t/\sqrt{1 - v^2/c^2} \\ &= (x_2 - x_1)/\Delta t = \Delta L/\Delta t = v \end{aligned}$$

Therefore the length of travel between two points in the x direction in a moving reference frame measured by an observer in a fixed reference frame is

$$\Delta L' = v\Delta t' = v\left(t - vL/c^2\right)/\sqrt{1 - v^2/c^2} = \Delta L\left(1 - v^2/c^2\right)\left(1 - v^2/c^2\right)^{-1/2}$$

$$\Delta L' = \Delta L\sqrt{1 - v^2/c^2} \tag{15.11}$$

This shows that the length of a moving object measured from a frame at rest with respect to the motion $\Delta L'$, will decrease from its length measured in its rest frame ΔL as its speed increases near to the speed of light.

According to Eq. 15.11, a measurement of distance in the x-direction in the moving reference frame by an observer in the fixed reference frame will be shorter than the same measurement made by an observer in the moving reference frame. In addition, Eq. 15.10 shows that the time of an event in the moving reference frame measured by an observer in the fixed reference frame will be longer than the time measured by an observer in the moving reference frame. These two profound consequences of Einstein's Special Theory of Relativity are called length contraction and time dilation. Despite these effects, the fundamental laws of mechanics, electromagnetic theory, and other areas of physics will be the same in both reference frames. Let's look at some examples of this.

Suppose you wanted to measure the speed of light. You could do this by setting up a laser that shoots out a pulse of light and, at the same instant, start a clock. The light travels to a mirror, reflects straight back, and travels back to a detector at the position of the laser. At the instant the light hits the detector, the clock is stopped. The speed of light, then, is twice the distance between the laser and the mirror divided by the measured round trip time $2x/t$. This is shown in Fig. 15.3a.

If you were to do this experiment in a lab that is stationary relative to earth, you would find that the speed of light is 2.998×10^8 m/s, just as the laws of physics demand. Now suppose you set up this experiment in a rocket ship in which you are traveling at a constant velocity very near the speed of light. Once again, you would measure the speed of light to be 2.998×10^8 m/s, because the speed of light is the same in all inertial reference frames. Let's suppose the rocket in which

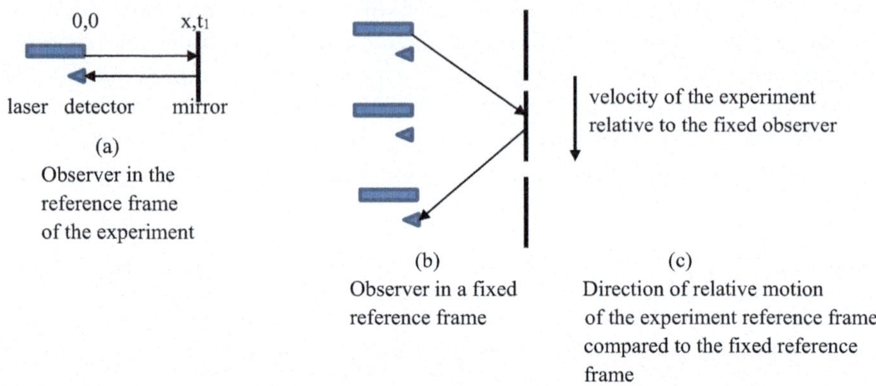

Fig. 15.3 Measuring the speed of light with a laser

you are doing your experiment is transparent, and let's have an observer at rest on earth watch your experiment through a telescope as you pass by. Suppose you are passing by so that the setup is perpendicular to your velocity. What will this observer see? Figure 15.3b illustrates this.

In the reference frame of the experiment, the experiment is stationary. Thus, you see the light shoot straight towards the mirror, hit the mirror, and come straight back. However, that's not what the observer at rest relative to the moving reference frame sees. Suppose the observer decided to time the light in order to determine its speed. Thus, he starts a clock the instant he sees the laser fire, and he stops the clock the instant he sees the light hit the detector. Since the experiment is moving relative to him, he sees the whole apparatus move. As a result, the light does not travel straight for him. Instead, it travels at a diagonal, because it hits the mirror, which is moving, with respect to him. Since the light travels diagonally relative to the observer, he sees the light traveling a longer distance than "$2x$." Then, he takes the distance that he saw the light travel and divides by the time. What does he get for the speed of light? If the laws of physics work the same in all inertial reference frames, he will also get a speed of 2.998×10^8 m/s. Remember, the speed of light is a consequence of the laws of physics. Thus, if the laws of physics are the same in *all* inertial reference frames, then he must get the speed of light demanded by those laws, which is 2.99×10^8 m/s. The only way this can happen is if the time he measures is longer than the time measured in the reference frame of the experiment. In other words, from his point of view, ***time runs slower in the moving reference frame*** compared to what an observer in the reference frame measures.

There is a substantial amount of evidence that time does, indeed, vary from one reference frame to another. Let's clarify *exactly* what special relativity tells us with regard to time. When you are in a reference frame that is moving relative to someone else, *you do not notice anything unusual in your own reference frame.* Time does not move slowly as far as you are concerned. Indeed, everything is perfectly normal for you. However, if someone who is at rest relative to you looks at your reference frame, he or she observes that time runs slowly in your reference frame compared to what you measure. If this person were to see a clock in your spaceship, they would see the clock run slower than the clocks in their reference frame. The clocks in your ship are keeping perfect time. However, someone at rest relative to you sees a much different picture. They sees the clocks running slowly, because, according to them, time is dilated in your reference frame. Of course, we know of one thing that they see the same in your reference frame as in their's—the speed of light. The speed of light is the same in *all* reference frames.

Consider the following example:

Example 15.3

A person travels to a distant star on a spaceship that moves at a constant velocity of 0.900*c* relative to earth. If the star is 20.0 light years from earth, how much time will pass on earth before the ship reaches its destination? How much time will pass in the spaceship?

A light year is the distance over which light travels in a year. Thus, if the rocket ship could travel at the speed of light (c), 20 years would pass on earth before the spaceship reached the star. However, the rocket ship can only travel at $0.900c$, so it will take a bit longer. If light takes 20 years to travel that far, we can say that the distance is (20 years) (c). As long as c is expressed in meters per year, then the units work out. We know that the spaceship is traveling at $0.900c$. Thus, we have the distance and the time:

$$\text{time} = \text{distance/speed} = (20.0\,\text{years}) \cdot c/(0.900c) = 22.2\,\text{years}$$

That means 22.2 years will pass on earth before the spaceship gets to the star.

How much time will pass on the ship? Since the ship moves relative to the earth, it experiences time differently than those on earth. To find out how differently, we use Eq. 15.10. Now remember, $\Delta t'$ is the time that is measured of an event in the moving reference frame by an observer on earth, and Δt is the time of the event measured by someone in the moving reference. In this case, the earth is stationary relative to the spaceship, and the spaceship is moving. Thus, we need to solve for Δt, because we want to know how much time passes in the moving reference frame. This means that $\Delta t' =$ 22.2 years, because that's the time measured from earth.

$$\Delta t = \Delta t' \sqrt{1 - v^2/c^2} = (22\,\text{years}) \sqrt{1 - (0.900c)^2/c^2} = 9.68\,\text{years}$$

This means that on the spaceship, only 9.68 years pass.

What does this mean? Don't 22.2 years "really" pass on the spaceship? *NO!* Only 9.68 years pass. Suppose, for example, a man leaves on the spaceship, but his twin stays behind on earth. The man travels to the star and then turns around and travels back. When he gets back to earth, a total of 44.4 years (22.2 years there, 22.2 years back) have passed. Thus, his twin is 44.4 years older. However, on the spaceship, only 19.4 years have elapsed. As a result, the man is only 19.4 years older. This means that *his twin is now 25 years older than he is.* That's the way it works!

This gets even a bit more strange. For the person in the space ship, only 9.68 years pass as he travels to the star. Of course, he knows the speed at which he was traveling relative to earth, so he could calculate the distance that he traveled over those 9.68 years. What would he calculate? Well, he knows that he was traveling at $0.900c$ relative to earth, so he knows that his speed is $(0.900)(2.998 \times 10^8 \text{ m/s})$, or 2.698×10^8 m/s relative to earth. He also knows that he has been traveling for 9.68 years. Thus, he traveled:

$$(2.698 \times 10^8 \text{ m/s})(9.68 \text{ years})(31{,}557{,}600 \text{ s/year}) = 8.24 \times 10^{16} \text{ m}$$

His friends on earth, however, know that the distance from the earth to the star is 20.0 light years, which is:

$$(2.998 \times 10^8 \text{ m/s})(20 \text{ years})(31{,}557{,}600 \text{ s/year}) = 1.89 \times 10^{17} \text{ m}$$

That's *longer* than the distance that the man in the spaceship calculates. Which one is right? Both are right!

Just as time is not a constant from one reference frame to another, distance is not, either. In the spaceship's reference frame, the spaceship traveled only 8.24×10^{16} m, while those on earth watched the spaceship travel 1.89×10^{17} m. Thus, just as time dilates in a reference frame that is moving relative to you, length contracts! Suppose, for example, there was a meter stick on that spaceship. To the person on the spaceship, it would be exactly 1.000 m long. However, if the spaceship were transparent and someone on earth were to look at the meter stick, it would not be 1.000 m long. According to Eq. 15.11, it would be shorter than that! For this example,

$$\Delta L' = \Delta L\sqrt{1 - v^2/c^2} = (1.89 \times 10^{17} \text{ m})\sqrt{1 - (0.900c)^2/c^2} = 8.24 \times 10^{-16} \text{ m}$$

This explains the differences in the lengths measured in our example.

Length contraction is a bit more complicated than time dilation because it occurs only in the direction of travel. If the meter stick was perpendicular to the velocity of the spaceship, the meter stick would look the same in both reference frames. Nevertheless, if a stationary person were to observe a moving reference frame whose speed is comparable to the speed of light, he would see that time moves slowly and length is contracted in the direction of the velocity! Distance and time, then, are not absolutes. They vary from reference frame to reference frame.

These results may seem strange but there is plenty of evidence to support them. Some of this comes from observations in astronomy. First let's look at the result that the speed of light is the same in all reference frames. Consider, for example, the fact that we can see stars which are far from us. Because we are orbiting our sun, we are moving relative to the star. Sometimes we are moving away from the star, and sometimes we are moving towards the star. In addition, some stars orbit other stars in what is called a **binary star system**. Sometimes, those stars are traveling towards us, and sometimes they are traveling away from us. Nevertheless, the light from those stars is always approaching us at a speed of c, regardless of which way we are moving and regardless of which way the star is moving.

How do we know that? Well, suppose the light approached us more quickly when we moved towards the star or when the star moved towards us. During those times, the light would reach us in a shorter time. However, if this were to happen, when we moved away from the star or the star moved away from us, the light would take longer to reach us. If this were the case, the path of the star would look erratic to us. It would seem to spend less time in the part of its orbit during which it is moving towards us and more time in its orbit when it is moving away from us. However, we do not see that. The orbit is smooth, indicating that light travels towards us at the same speed regardless of our motion relative to the star or the star's motion relative to us.

Back in the 1950s scientists studied a group of particles called **muons** (myou' ons), which constantly rain down on the earth at a speed of $0.994c$. These muons decay at a fixed rate, and we know that rate for stationary muons. (Note, we will discuss particle decay in the next chapter.) Thus, we know how long a muon will "live" before it decays. However, the muons that are raining down on us at a speed of $0.994c$ "live" a lot longer than stationary muons. How long do they live? They live for exactly the length of time predicted by Eq. 15.10. Thus, we know that in their reference frame, time actually does move more slowly than it does for earth, and the dilation is given exactly by Eq. 15.10.

Lots of studies like this one have put the concept of time dilation on firm experimental footing. For example, in another classic experiment, two very precise clocks were synchronized and observed over a long period of time. During the entire time that they were at rest relative to one another, they kept exactly the same time. Then, one clock was placed in a very high-speed jet and flown around the world a few times. When that clock was compared to the clock that remained on earth, it was behind by exactly the amount of time predicted by relativity! This gives us good evidence that time dilation is real, and therefore time is not constant. It varies based on the speed of the reference frame.

Time dilation and length contraction are real effects that must be taken into account in high-speed situations. For example, in a television, the electrons that make the image on the CRT screen (discussed in Chap. 7) move at about $0.3c$. If the engineers who design televisions did not take length contraction into account, televisions would not have focused properly. Now try the following problem.

Student

15.6 Consider the experiment described in this section. You are on a spaceship traveling at $0.80c$ east relative to the earth, and you measure the speed of light as described in Fig. 15.3. The distance from the laser to the mirror is 10,000.0 m. (It's a *big* spaceship!) If an observer on the earth watches your experiment and times how long it takes for the light to travel from the laser to the mirror, what time will he measure?

One famous puzzle resulting from time dilation is called *The Twin Paradox*. Remember, there is no way of telling whether or not a reference frame is really moving or fixed. We can only say that a reference frame is moving relative to some object, but not that a reference frame is moving absolutely. Thus, if we are driving a car at 55 miles per hour towards a tree, we could say that we are moving towards the tree and the tree is stationary, but we could just as easily say that we are stationary and the tree is moving towards us. From a physics point of view, both statements are equally correct.

This means that there was something that was not "quite right" about the twins in the spaceship and on earth that we talked about. When the spaceship gets back to earth, we said that the man in the spaceship would be younger than the twin on earth. But that seems to indicate that we know which reference frame was moving. After all, if the ship is moving away from the earth at $0.900c$, isn't that equivalent to the spaceship standing still and the earth moving away from it at $0.900c$? According to everything we know about physics, it should be. If we look at it that way, the twin that is on the earth is the one that is moving, so time runs slowly in *his* reference frame compared to that of the twin in the spaceship, so why isn't the twin on earth younger than the twin on the spaceship?

This is called the **twin paradox**, because it seems to break the rule that you cannot determine which twin was moving. If you treat the earth as stationary and the spaceship as moving, the twin on the spaceship will be younger than the twin on earth when he returns. On the other hand, if you treat the spaceship as stationary and the earth as moving away from it, the twin on earth will be younger than the twin on the spaceship when the earth returns to the space ship. However, that's not how it works. The twin on the spaceship will be the younger one every time. That means we know which twin was moving.

Even though we know which twin was moving, that does not really break any of the rules. In order to get back to earth, the twin on the spaceship must turn around. Thus, for a while, he is not in an inertial reference frame. The twin on earth never turns around. Thus, he is always in an inertial reference frame. As a result, we know that the spaceship was the moving reference frame. In inertial reference frames, there is no way to tell which is moving. However, since Newton's Laws of Motion only work in inertial reference frames, as soon as the twin in the spaceship turns around, his motion is defined relative to the twin on the earth.

Until the spaceship turns around, there is no way of knowing which twin is moving. If the twin in the spaceship were to observe the twin on the earth, the twin on the spaceship would think that time is running slowly on earth. If the twin on earth were to observe the twin on the spaceship, he would think that time on the spaceship was running slowly. Both would think that time was slow in the other's reference frame. However, once the spaceship turns around, its motion is defined relative to the other twin. On the way back, both twins would still think time was running slowly in the other's reference frame, but when they came together, they would see that compared to the earth, time ran slowly in the spaceship, because the twin in the spaceship would be younger.

In order to understand why c is the maximum velocity to which any object can be accelerated, we turn to the turn to the most famous equation in physics, Eq. 15.4. Einstein published his thoughts on the relationship between mass and energy a couple of years after he published paper on special relativity.

$$E = mc^2$$

This equation tells us that energy and mass are really just different forms of the same thing. The speed of light is the "conversion factor" between the two. If energy and mass are really just different measures of the same thing, it must be possible to transform one into the other. We will demonstrate this in the nuclear physics section of Chap. 16. For now, consider the following example and work the problem.

Example 15.4

The mass of an electron (and that of a positron) is 9.11 x 10^{-31} kg. What is the energy of a gamma ray that can produce an electron and a positron?

If a gamma ray produces an electron and a positron, its energy is being converted into mass. The total mass produced is:

$$\begin{aligned} \text{total mass} &= \text{mass of electron} + \text{mass of positron} \\ &= 9.11 \times 10^{-31}\ \text{kg} + 9.11 \times 10^{-31}\ \text{kg} = 1.822 \times 10^{-30}\ \text{kg} \end{aligned}$$

To determine how much energy is required to produce this mass, we just use Eq. 15.4

$$\begin{aligned} E = mc^2 &= \left(1.822 \times 10^{-30}\ \text{kg}\right)\left(2.998 \times 10^{8}\ \text{m/s}\right)^2 \\ &= 1.638 \times 10^{-13}\ \text{kg}\,\text{m}^2/\text{s}^2 = 1.638 \times 10^{-13}\ \text{J} \end{aligned}$$

That must be the energy of the gamma ray. This doesn't sound like a lot of energy, but not much matter is being created, either. Note how the units work out here. As long as mass is in kg and the speed of light is in m/s, the energy unit works out to be in Joules.

Student

15.7 On the star ship Enterprise, the crew has access to "replicators," which take energy and covert it into any kind of matter. If Captain Jean-Luc Picard walks up to a replicator and asks it to make Earl Grey Tea, how much energy must the replicator use? The total mass of the tea and the cup from which to drink the tea is 600.0 g.

In order to understand why c is the maximum velocity to which any object can be accelerated, we need one more transformation equation. This is for mass,

$$m = m_0/\sqrt{1 - v^2/c^2}. \tag{15.12}$$

Here m_0 is the mass of the object when it is at rest, called the rest mass, and m is the mass of the object when it is moving at a velocity v. Remember from Newtonian mechanics that mass is a measure of inertia, the resistance to a change in motion. According to Eq. 15.12, as the velocity of an object increases to near the speed of light, its mass increases so its resistance to further acceleration increases. Remember that mass and energy are related through the expression in Eq. 15.4, $E = mc^2$. In special relativity where we are dealing with velocities near to the speed of light, this expression becomes

$$E = m_0c^2/\sqrt{1 - v^2/c^2}. \tag{15.13}$$

As v approaches c, the relativistic mass approaches infinity so it would take an infinite amount of energy to accelerate an object to a velocity equaling the speed of light in a vacuum.

15.5 Relativity II: General Relativity

What Einstein really wanted to do with his general theory of relativity was to generalize the idea that the laws of physics are the same in all inertial reference frames to an idea in which the laws of physics are the same in *all* reference frames, *regardless* of whether or not the reference frame is inertial. Einstein had the idea that the laws of classical physics weren't exactly correct. They were really only special cases of more general laws that worked in all reference frames. Thus, what Einstein really wanted to do was formulate general laws of physics which work in all reference frames, regardless of the acceleration of the frame. He started with the principle of equivalence.

One of his first questions concerned the nature of mass. He pointed out that classical mechanics involves mass in two different ways. The first way is providing inertia as a resistance to motion,

$$m_{\text{inertia}} = F_{\text{net}}/a.$$

The second way is in gravitational attraction,

$$m_{\text{gravity}} = r^2 F_{\text{gravity}}/(G m_{\text{source}}).$$

Einstein claimed that these two concepts of mass were the same. This is called the principle of equivalence.

Consider, for example, an astronaut on the space shuttle. They can float in space because they are weightless. However, being weightless does not come from the lack of gravity. They are weightless because they are in free fall. That is, they are continually falling toward the earth due to gravitational attraction, but they don't feel the effects of gravity because their reference frame is falling with them. Because of the shuttle's velocity, which is perpendicular to the acceleration, the shuttle keeps "missing" the earth. As a result, it is in orbit.

Einstein said to consider a reference frame that is freely falling. Objects *inside* the reference frame do not experience any force due to gravity. The reference frame itself does, but the objects inside do not. This reference frame is equivalent in every way to a reference frame in deep space that is so far from any star, galaxy, or planet that there is no gravity whatsoever. Thus, we can produce the absence of gravity by simply allowing our reference frame to accelerate at whatever acceleration is produced by gravity. Under those conditions, the people inside the reference frame would have no way of knowing whether or not gravity is really there. Einstein also said that you could reverse the situation. Suppose, for example, the space shuttle was sitting on the earth. The people in the space shuttle would no longer be weightless. They would "feel" the gravity produced by the earth. However, we could fool the occupants of the space shuttle by taking them into deep space (so far away from any matter so that there really is no gravity) and then have the space shuttle accelerate directly upwards at 9.81 m/s^2. If we did that, the occupants of the space shuttle would think that there was gravity, because they would weigh what they weighed on earth, and everything would work just as if they were at rest on earth's surface. However, there really is no gravity at all. The shuttle simulates the effect of gravity by accelerating upwards.

Have you ever watched a science fiction movie in which a space station is rotating, and the people on the inside of the space station are walking around as if they are under the influence of gravity? There is no gravity, however. The rotation *simulates* the effect of gravity. If the space station is rotating, the floor of the space station must push against a person inside the space station in order to keep that person rotating with the space station. The force with which the floor must push is the centripetal force, which depends on the speed of the rotation, the radius of the rotation, and the mass of the person. The person, however, does not notice this as long as the radius of rotation is large. As a result, the person just feels themselves being pushed upwards by the floor. This is exactly what the person experiences when he is on earth, under earth's gravitational influence. This situation is equivalent in every way to being in a gravitational field, but there is no gravity anywhere.

The point is, we can construct reference frames in which gravity is really there but the objects inside the reference frames behave as if there is no gravity (freely falling reference frames like the space shuttle). We can also construct reference frames in which there is no gravity, but the objects behave as if there is gravity (the space shuttle accelerating upwards or the rotating space station). These reference frames are equivalent in every way. That's the principle of equivalence since it says the inertial mass and gravitational mass are the same.

The principle of equivalence led Einstein to conclude that gravity is not a force. He said that a real force could not be "taken away" by constructing the proper reference frame. The electrical force that attracts positive and negative charges, for example, exists in all frames of reference. You cannot construct a frame of reference in which the objects inside behave as if there is no electrical force. The same can be said for the magnetic force. There is no way to construct a frame of reference in which the objects behave as if there is no magnetic force. However, you can construct a frame of reference in which objects behave as if there is no gravitational force. Therefore, gravitational force is not a real force. It must be an artifact—something that we interpret as a force but is really not a force.

Although this sounds strange, we actually apply this logic to other situations. Consider the car that rounds a curve. When the car rounds a curve, the packages fly off of the seat, seemingly of their own accord. Some people say that this is because the packages experience a "centrifugal force." However, there is really no such thing as a centrifugal force. The fact that the packages move off of the seat is the result of Newton's First Law of Motion as observed by someone at rest relative to the car. However, people in the car cannot tell that their velocity is changing, so they assume that there is a "centrifugal force" causing the packages to move. The "centrifugal force" is really an artifact, however. It is an artifact of the acceleration of the reference frame.

Einstein said that gravity is another artifact of our reference frame and perhaps other reference frames as well. To understand this, consider another comparison between two reference frames. One is in deep space where there is no gravity. Light enters a hole in the box and hits a target on the other side. The other reference frame is freely falling in a gravitational field. Light enters a hole in the box and hits a target on the other side. However, since the target is in free fall, the light must bend to hit it. This must happen the same in both reference frames since they are equivalent so experiments must have the same results in both cases. Thus, gravity causes light waves to bend.

Newton's Law of Universal Gravitation does not predict this. According to Newton, the gravitational force that exists between objects depends on the mass of the objects. Light has no mass. Thus, according to Newton, gravity should not affect light. However, it does. If the principle of equivalence is correct, in order for the experiments above to have the same result in both reference frames, light must bend when it is in a gravitational field.

This is where we can refer back to special relativity. In special relativity, we learned that neither time nor space were absolute. They change depending on the reference frame. However, Einstein constructed a four-dimensional substance called **spacetime** which is absolute. The four dimensions of spacetime are the three dimensions of space plus a fourth dimension, which is time. This four-dimensional quantity, spacetime, is the same from reference frame to reference frame.

According to Einstein, the laws of motion could really be boiled down to one statement:

In the absence of external forces, objects travel the straightest possible path in spacetime.

Since light bends in the places we think of as having a gravitational field, Einstein decided that:

Mass and energy cause spacetime to curve.

Thus, what we observe as a "gravitational force" is no force at all. It is simply a consequence that objects want to travel along a straight line in spacetime but mass curves spacetime. Instead of moving in a straight line, then, the moon moves in an orbit around earth not because it is really attracted to earth, but because it is moving along a straight path in spacetime, but spacetime is curved by the masses of the earth and the moon. As a result, the straight line in spacetime is in the shape of an orbit. To better understand this concept, try the following experiment.

Experiment 15.1
Simulating Curved Spacetime

Supplies:

- A soft seat cushion from a couch (A soft bed will work as well.)
- A bowling ball (A heavy rock will work as well.)
- A marble

Introduction: Einstein's General Theory of Relativity concludes that the gravitational "force" is not really a force at all. It is actually a result of the fact that mass curves spacetime. This experiment will help illustrate such a strange concept.

Procedure:

1. Lay the seat cushion on the floor. If you are using a bed, just stand next to the bed.

2. Find a spot on the cushion which is away from the center of the cushion but relatively flat. Put the marble on that point so that it stays there without rolling.
3. Now lay the bowling ball at the very center of the cushion. Note what happens to the marble.
4. Next, take the bowling ball off of the cushion and smooth it out so that it is reasonably flat again.
5. Roll the marble (slowly) straight across the cushion, but not near the center. Note that it rolls reasonably straight.
6. Put the bowling ball back in the center of the cushion and roll the marble along the same path that you rolled it before, with the same slow speed. Note the path that the marble takes.

Einstein's General Theory of Relativity states that space is not always the way it appears to us. Suppose, for example, that you did not know the world was round. Would you think that it was? Probably not. After all, the earth looks pretty flat all around you. Thus, you would probably think that the earth is flat. We know that this is not the case, though. Despite what it looks like from our vantage point, we know that the earth is round. In the same way, Einstein postulated that although it does not appear to change at all, space actually bends in the presence of an object with mass.

In the first part of your experiment, the seat cushion represents spacetime. With nothing on the seat cushion, it stayed relatively flat. However, when a massive object (the bowling ball) was placed there, the entire seat cushion bent. The bend was greatest in the middle and least around the edges, but nevertheless, the entire seat cushion bent. In response, the marble rolled towards the bowling ball. This is Einstein's picture of gravity. Spacetime (the seat cushion) bends in the presence of mass (the bowling ball). As a result, all objects (the marble) accelerate towards the mass. This makes it look like a force is being applied to the object.

In the second part of your experiment, you watched the marble roll straight across the flat seat cushion. When the bowling ball was once again placed on the seat cushion, the marble did not roll straight. Instead, it rolled in a curved path. According to Einstein, this is why planets orbit the sun. In his theory, the planets are all actually moving in a straight line in spacetime. Because spacetime is so strongly bent by the mass of the sun, however, that straight line is deformed until it becomes a circle. Thus, as far as the planets are concerned, they are traveling in a straight line. Curved spacetime, however, causes that straight line to become a circle.

However, the experiment is not a true analogy. Spacetime is four-dimensional, while the surface of your cushion is two-dimensional. Also, gravity plays a role in this experiment, even though there is no gravity in general relativity. Thus, the experiment was just an attempt to get you to visualize a part of what general relativity states. Another visualization is shown in Fig. 15.4. In the presence of mass, spacetime bends so much that a "well" is created. A planet will appear to move in a straight line as far as anyone on the planet can tell. Nevertheless, since spacetime is curved so much, that apparent straight line is actually a circle.

In Einstein's General Theory of Relativity, then, the laws of physics are the same in all reference frames, regardless of whether or not those reference frames are in uniform motion or being accelerated. However, Newton's Laws of Motion are not really correct. They are approximations of the true laws of motion, which are based on the idea that, in the absence of all forces, objects move in a straight line in spacetime. If we accept this view, then gravity is not a force at all. It is simply a consequence of the fact that mass curves spacetime.

Because of the importance of gravity in the university, it is not surprising that much of the evidence for general relativity come from astronomical measurements. Einstein produced one bit of evidence himself for his theory. Long before Einstein, astronomers had noticed that Mercury did not obey Newton's Law of Universal Gravitation exactly. Newton's Law of Universal Gravitation predicts that the planets will orbit the sun in ellipses that never change. However, it became apparent that Mercury's orbit does change a bit. The ellipse which defines the orbit actually tilts relative to the other planets, and that tilt changes slightly from year-to-year. This was a puzzle in physics, but since the effect was very small, it was regarded as more of a curiosity than anything else.

Einstein realized that Mercury was a planet on which he could test his theory. The planet Mercury is the planet closest to the sun. Thus, it experiences the sun's gravity more than any other planet. Einstein noted that his theory (which is full of a lot of incredibly complicated math) was essentially equal to Newton's Laws when the gravitational force between objects is weak. That's why Newton's Laws work so well for us. However, as gravity gets stronger, Einstein's equations of general relativity differ from Newton's Laws. When he applied his equations to Mercury, he found out that his equations predict Mercury's orbit's tilt. In fact, they predict exactly the behavior that Mercury has exhibited over the years.

As we have been able to observe situations in space in which large gravitational forces play a role, we have seen more and more experimental confirmations of

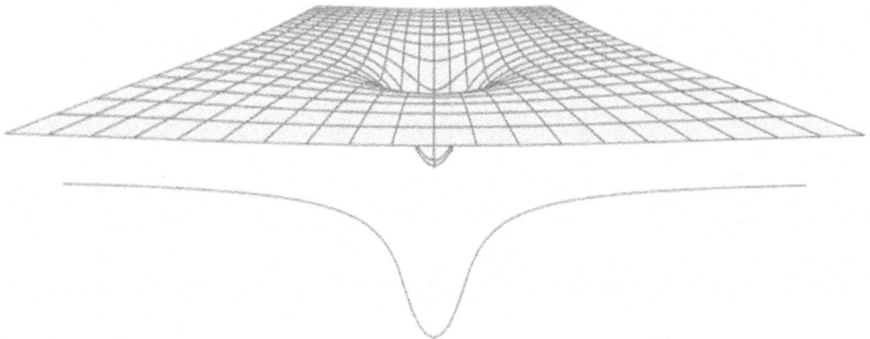

Fig. 15.4 The orbit of a planet, according to general relativity

Einstein's Theory of General Relativity. For example, we have actually observed the bending of light predicted by the equivalence principle. This was first seen by Sir Arthur Eddington on May 29, 1919. Eddington was observing stars during a solar eclipse, and he noticed that the positions of the stars were different when the sun was near the line of sight of the star. In other words, when the star's light had to pass near the sun, it bent, making the observer think that the star was in a different place. When the sun was not in the way, the star's light was not bent, and the star seemed to be in its "normal" position on the horizon. This is called "gravitational lensing," and it is now a phenomenon that has been seen time and time again by astronomers.

Einstein's Theory of General Relativity predicts many other things that we are now observing in space, one of which is a black hole where the gravity is so high that nothing including light can escape. The escape velocity is defined as the velocity needed for an object to overcome the gravitational attraction of a larger object. This can be derived by using conservation of energy. The kinetic energy due to the velocity of the object is converted into the gravitational potential energy which increases as the two objects get further apart. If the kinetic energy is all used up just when the gravitational potential energy is zero, the initial velocity of the object is the minimum velocity needed to escape the gravitational attraction.

Since the total energy is the sum of the kinetic and potential energies, and the final total energy is zero, the initial total energy must also be zero,

$$m_2 v^2/2 - G m_1 m_2 / r = 0$$

where m_2 is the mass of the body trying to escape the gravitational attraction of the object with mass m_1. Therefore, the escape velocity is

$$v_e = \sqrt{\frac{2Gm_1}{r}}. \tag{15.14}$$

For a photon trying to escape a black hole, $v_e = c$. Using this in Eq. 15.14 shows that in order to stop photons from escaping, a black hole must have a mass to radius ratio of at least

$$\begin{aligned} m_h/r_h &= c^2/2G = \left(3 \times 10^8\,\mathrm{m/s}\right)^2/2\left(6.67 \times 10^{-11}\,\mathrm{Nm^2/kg^2}\right) \\ &= 6.747 \times 10^{26}\,\mathrm{kg/m}. \end{aligned}$$

Our sun has a mass to radius ratio of $1.982 \times 10^{30}/6.696 \times 10^8 = 2.96 \times 10^{21}$ kg/m which is more than four orders of magnitude too small to keep a photon of light from escaping. This demonstrates the high density of mass there must be in a black hole. This is consistent with what physicists think would happen at the event horizon of a black hole. The event horizon is the point at which light is close enough to the black hole to be caught in its gravitational pull and thus is unable to escape. As a result, you cannot see light coming from any point inside the event horizon. Also, there are several observations of galaxies in which the behavior of the stars close to the center of the galaxy is consistent with the existence of a black hole at the galaxy's center. Thus, there is indirect evidence for the existence of such structures.

We now know that most galaxies have black holes at their center. Astronomers have even taken a "picture" of a black hole. The Hubble space telescope has captured events in which light emitted by highly excited matter seems to be disappearing at a particular point in space.

Special Topic

If Einstein's theory of general relativity is correct, any disturbance of spacetime will permeate the entire universe. This can be triggered by an event that causes a massive change in gravity at its location which produces ripples in gravity that travel out from the event like ripples in water when a pebble hits a pond. Of course, the universe is so large that these changes in gravity will be quite small after traveling billions of miles to get to earth. One of the great successes of modern physics in the twenty-first century was building the Laser Interferometer Gravitational-wave Observatory (LIGO). There are three systems working together in the northwest and southeast United States and in Italy. They consist of detectors that can measure a change in distance less than the size of a proton. In 2016 they detected the first gravitational wave. It was attributed to the merger of two black holes over 1.3 billion light years from earth. Between then and the end of 2019 fifty gravitational wave events were detected. This helps confirm the spacetime understanding of gravity.

Now try a couple of problems.

Student

15.8 Calculations indicate that a reasonable radius for a black hole is about 10.0 km. What is the mass necessary for a sphere of radius 10.0 km to be a black hole?

15.9 How much kinetic energy must a 5.0 kg mass have to escape a black hole with a mass to radius ratio of 7.0×10^{26} kg/m?

There are other predictions of general relativity which have been confirmed. For example, general relativity predicts that time moves more slowly in areas where gravity is more intense (the greater that spacetime is curved). This has been demonstrated experimentally.

Student

15.10 On the surface of the moon, the gravitational acceleration is 1.62 m/ s^2. Suppose we wanted to construct a lab on earth in which all of our experiments would work as if we were on the surface of the moon. Neglecting the absence of an atmosphere on the moon, temperature, etc., what would we have to do to the lab to make this happen? Think about the principle of equivalence.

It is important for us to understand gravity because it is so important in the formation and operation of creation. Astrophysicists call the current model for the origin of the universe the Big Bang Theory. It is based on the experimental observations that the universe is about 13.8 billion years old and has been expanding since its inception. In this model, all the matter in the universe today originated as elementary particles coming out of one tiny point in space. Physics provides no insights as to how the particles came into existence or what caused the big bang to occur. These are questions for theology to answer.

Gravity existed from the very beginning of creation and caused the particles to come together to form matter, including the nuclei of the lightest element, hydrogen. When these nuclei captured an electron to become a hydrogen atom, photons could be absorbed and emitted so the universe had light. Due to gravity, giant clouds of hydrogen gas called nebula began to coalesce into stars. Gravity pulled the stars together to form galaxies. Galaxies aligned themselves in groups and clusters which formed structures in the universe which continues to expand. Examples of a nebula and a galaxy are shown in Fig. 15.5.

(A) Orion Nebula.

(B) Bode's galaxy.

Fig. 15.5 Diffuse nebula and galaxy

In the center of stars, nuclear fusion processes take place. These are discussed in detail in Chap. 16. These processes make heavier elements up to and including iron. This process of creating new elements is called nucleosynthesis. There is a delicate balance between gravity trying to make the star collapse and nuclear fission processes trying to make the star explode. However, stars do not live forever. When their nuclear fuel burns itself out, stars the size of our sun collapse into dense white dwarfs. This creates a shock wave that ejects all of the elements the star created into a gas cloud that is called a planetary nebula. Over the course of many years, the gas in the planetary nebula expands throughout the universe and spreads the elements that were created onto everything in the universe, including our planet earth.

When giant stars, much bigger than our sun die, they explode as a supernova. The extremely high temperatures in a supernova explosion allow for nucleosyntheses of heavier elements with a greater atomic mass number than iron. The supernova remnants are gas clouds containing these heavy elements. Over the course of many years, these clouds spread out over the entire universe leaving heavy elements on all the bodies in the universe including earth. Thus, all the elements of matter we have on earth were produced in the center of some distant star and came to earth when the star died.

Figure 15.5a shows a picture of a diffuse nebula called the Orion Nebula. By analyzing the spectral line of the red emission, it has been identified as a cloud of hydrogen gas. The bright white central region is where new stars are being created. This is the first deep space object studied by Galileo when he first used a telescope for astronomy. Figure 15.5b shows one of the two trillion galaxies in the universe. This is called Bode's galaxy. It is a spiral galaxy held together by a gigantic black hole about 12 million light years away from earth.

Figure 15.6a, b show a planetary nebula and a supernova remnant. The planetary nebula is called the Helix Nebula. It was created by the death of a star very much like our sun. The colors show a variety of light gases including hydrogen, helium, nitrogen and oxygen. The supernova remnant is called the Crab Nebula. It was created by a giant star that died in 1054. The supernova appeared as a second sun in the sky for 23 days. It has a pulsar at the center giving off high energy radiation at 33 beats per second. Both this planetary nebula and this supernova remnant are expanding and sending their element filled gases throughout the universe.

The red shift in the positions of spectral lines of the light from distant galaxies shows that the galaxies are moving away from us as the universe expands. Surprisingly, the velocity of the receding galaxies gets larger the further they are away from us. This implies that there is some energy in the university causing the distant galaxies to speed up. This energy is called dark energy. Currently, we do not understand what dark energy is.

Another mystery of the universe is dark matter. There are several experimental observations demonstrating that far more matter exists in the universe than the type of matter we are familiar with. These include the bending of light rays traveling by a galaxy and the modeling of stars spiraling in the arms of a galaxy. It appears

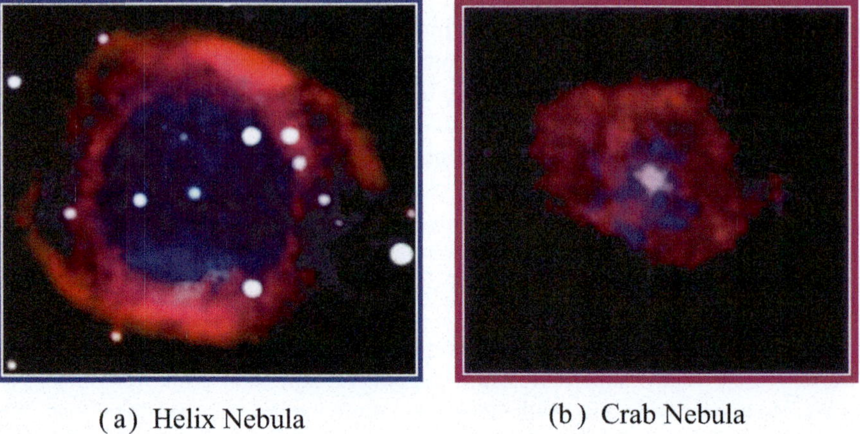

Fig. 15.6 The death of stars

that dark matter makes up the great majority of matter in the universe, but we have no idea what it is.

In conclusion, general relativity gives us a much better understanding of gravity and thus the workings of the universe. This is important for our understanding of the creation in which we live. However, there are still many unanswered questions about our creation.

Summing Up

You now have learned that physics is quite different on the atomic scale. We must account for the dual wave-particle nature of matter as well as dealing with probabilities in our measurements. We have new laws such as the Heisenberg uncertainty principle and the Pauli exclusion principle that are important. You also have learned about what happens to anything that moves at a speed near to the speed of light and the importance of choosing a reference fame for observation. Finally, you learned how to look at gravity as a perturbation in space–time instead of a normal force. You were left with two of the biggest questions in science today: What is dark energy? and What is dark matter? In the final chapter we will extend these ideas to understand the basics of matter: atoms, nuclei, and elementary particles.

Answers to the Problems

15.1

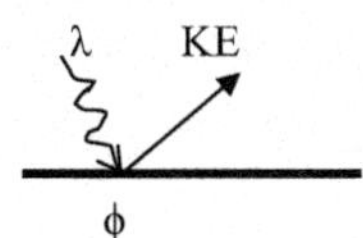

Knowns: $\lambda = 300$ nm; $KE_{max} = 1.5 \times 10^{-19}$ J
Unknowns: ϕ

Using Eq. 15.2,

$$E = hc/\lambda - \phi$$

So,

$$\phi = hc/\lambda - E = (6.63 \times 10^{-34}\,\text{J s})(2.998 \times 10^{8}\,\text{m/s})/3.00 \times 10^{-7}\,\text{m} - 1.5 \times 10^{-19}\,\text{J} = 5.13 \times 10^{-19}\,\text{J}$$

15.2

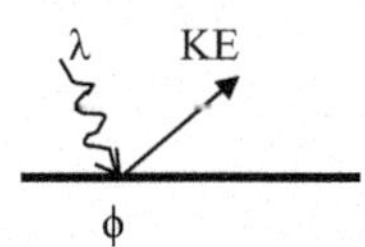

Knowns: $\lambda_{min} = 264$ nm
Unknowns: ϕ

If electrons cannot be emitted once the wavelength of light falls below 264 nm, we know that the energy of the light below 264 nm is less than the work function. Thus, the energy of light at 264 nm is equal to the work function. To get the energy, we start with the frequency:

$$f = v/\lambda = (2.998 \times 10^{8}\,\text{m/s})/(2.64 \times 10^{-7}\,\text{m}) = 1.14 \times 10^{15}\,\text{s}^{-1}$$

Then the energy is

$$E = h \cdot f = \left(4.14 \times 10^{-15}\,\text{eV s}\right) \cdot \left(1.14 \times 10^{15}\,\text{s}^{-1}\right) = 4.72\,\text{eV}$$

The work function, then, is 4.72 eV. If you did it in Joules, the answer is 7.56×10^{-19} J.

15.3

Knowns: $m = 9.11 \times 10^{-31}$ kg; $v = 1510$ m/s

Unknowns: λ

The de Broglie wavelength comes from the momentum Eq. 15.6:

$$\lambda = h/p = 6.63 \times 10^{-34}\,\mathrm{J\,s}/\left(9.11 \times 10^{-31}\,\mathrm{kg}\right)(1510\,\mathrm{m/s}) = 4.82 \times 10^{-7}\,\mathrm{m}$$

The wavelength of the electron, then, is 482 nm, which is right in the range of visible light. Thus, the wave characteristics of this electron cannot be ignored, since the wavelength is not small compared to other waves with which we are familiar.

15.4

Knowns: $m = 9.11 \times 10^{-31}$ kg; $\Delta v = 1$ mm/s

Unknowns: Δx

Using Eq. 15.7, the Heisenberg uncertainty principle for position and momentum gives us

$$\Delta x \Delta v \geq h/(4\pi m).$$

So,

$$\Delta x \geq h/(4\pi m \Delta v) = 6.63 \times 10^{-34}\,\mathrm{J\,s}/\left(4\pi 9.11 \times 10^{-31}\,\mathrm{kg}\right)\left(2 \times 10^{-3}\,\mathrm{m/s}\right)$$
$$\geq 29\,\mathrm{mm}$$

Therefore we can only measure the position at this time to an accuracy of plus or minus 14.5 mm.

15.5

Knowns: $\Delta E = 0.2$ J

Unknowns: Δt

Using Eq. 15.8, the Heisenberg uncertainty principle for energy and time gives us

$$\Delta E \Delta t \geq h/(4\pi)$$

So

$$\begin{aligned}\Delta t &\geq h/(4\pi)(\Delta E) = 6.63 \times 10^{-34}\,\text{J s}/(4\pi 0.2\,\text{J}) \\ &\geq 2.64\,\text{s}\end{aligned}$$

The expected lifetime of this state of the system is 2.64 s.

15.6

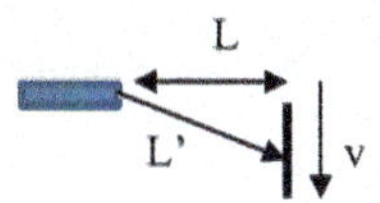

Knowns: $v = 0.80\,c$; $L = 10^4\,\text{m}$

Unknowns: $\Delta t'$

First, we have to determine the time that *you* measure. If the distance is 10,000.0 m, and the speed of light is $2.998 \times 10^8\,\text{m/s}$, you measure:

$$L = c\Delta t$$

So

$$\Delta t = L/c = 10{,}000.0\,\text{m}/2.998 \times 10^8\,\text{m/s} = 3.336 \times 10^{-5}\,\text{s}$$

That's t (the time measured by an observer in the reference frame). According to Eq. 15.10, your friend measures t':

$$\begin{aligned}\Delta t' &= \Delta t/\sqrt{1-(v/c)^2} = 3.336 \times 10^{-5}\,\text{s}/\sqrt{1-(0.80)^2} \\ &= 5.6 \times 10^{-5}\,\text{s}\end{aligned}$$

While only 3.336×10^{-5} s pass measured in your reference frame, 5.6×10^{-5} s pass measured in your friend's reference frame.

15.7

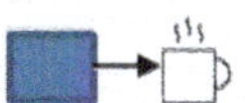

Knowns: $m = 600.0\,\text{g}$

Unknowns: E

This is a straightforward application of Eq. 15.4. You just have to make the units work by putting mass in kg.

$$\begin{aligned} E &= mc^2 \\ &= (0.6000\,\text{kg}) \cdot \left(2.998 \times 10^8\right)^2 = 5.393 \times 10^{16}\,\text{J s} \end{aligned}$$

That's a *lot* of energy. It is the same amount of energy that a 100 W light bulb uses if the light bulb burns for 17 *million* years!

15.8

Knowns: $r = 10.0\,\text{km}$

Unknowns: m

In order for something to be a black hole its escape velocity must be greater than the speed of light. From Eq. 15.14 this gives us

$$\begin{aligned} m_{\text{h}}/r_{\text{h}} &= c^2/2G = \left(3 \times 10^8\,\text{m/s}\right)^2/2\left(6.67 \times 10^{-11}\,\text{Nm}^2/\text{kg}^2\right) \\ &= 6.74 \times 10^{26}\,\text{kg/m}. \end{aligned}$$

Thus

$$\begin{aligned} m_{\text{h}} &= \left(6.74 \times 10^{26}\,\text{kg/m}\right) r_{\text{h}} = \left(6.74 \times 10^{26}\,\text{kg/m}\right)\left(10^4\,\text{m}\right) \\ &= 6.74 \times 10^{30}\,\text{kg} \end{aligned}$$

Note that this is just slightly more than the mass than that of our sun. Thus if our sun were shrunk from its current radius (6.96×10^8 m) to just under 10.0 km, it would become a black hole.

15.9

Knowns: $m_{\text{h}}/r_{\text{h}} = 7.0 \times 10^{26}\,\text{kg/m}$; $m = 5.0\,\text{kg}$;
$G = 6.67 \times 10^{-11}\,\text{Nm}^2/\text{kg}^2$

Unknowns: $\text{KE}_{\text{escape}}$

Begin by getting the escape velocity from Eq. 15.14,

$$\begin{aligned} v_{\text{e}}^2 &= 2Gm_{\text{h}}/r_{\text{h}} = 2\left(6.67 \times 10^{-11}\,\text{Nm}^2/\text{kg}^2\right)\left(7.0 \times 10^{26}\,\text{kg/m}\right) \\ &= 93.38 \times 10^{15}\,\text{m}^2/\text{s}^2 \end{aligned}$$

Then the kinetic energy of the object at the escape velocity is

$$\text{KE}_{\text{escape}} = mv_{\text{e}}^2/2 = (5.0\,\text{kg})\left(93.38 \times 10^{15}\,\text{m}^2/\text{s}^2\right)/2 = 233.5 \times 10^{15}\,\text{J}$$

15.10

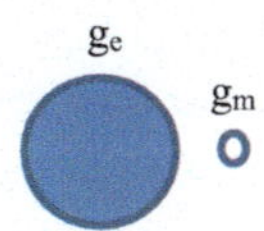

Knowns: $g_e = 9.81\,\text{m/s}^2$; $g_m = 1.62\,\text{m/s}^2$

Unknowns: need to make $g_e = g_m$

The principle of equivalence says that we have to do construct a reference frame in which things accelerate downwards at 1.62 m/s^2. As you learned in Chap. 5, when a reference frame is falling, the acceleration of the reference frame subtracts from the acceleration due to gravity. Near the surface of the earth, the acceleration due to gravity is 9.81 m/s^2. Thus, if we do experiments in a lab that is falling with an acceleration of 8.19 m/s^2, the net acceleration for objects falling inside the lab is 9.81 m/s^2 – 8.19 m/s^2, which is 1.62 m/s^2. That will be just like the surface of the moon (neglecting the lack of an atmosphere, the lower temperature, etc.).

Study Guide for This Chapter

($h = 6.63 \times 10^{-34}$ J s $= 4.14 \times 10^{-15}$ eV s; $G = 6.67 \times 10^{-11}$ Nm^2/kg^2; $c = 2.998 \times 10^8$ m/s)

1. Suppose you are in a spaceship traveling at 0.8*c*. Another spaceship approaches yours, also traveling at 0.8*c*. Are the spaceships approaching each other at 1.6*c*? Why or why not?
2. Consider the two spaceships discussed above. If the first were to turn on a light and shine it at the second, at what speed would the light be approaching the second spaceship? At what speed would the light be traveling away from the first?
3. If you ask a physicist how many different types of force there are in Creation, he or she would say three: the electroweak force, the strong nuclear force, and the gravitational force. If Einstein's General Theory of Relativity is true, how many different types of force are there in Creation?
4. Which of the following is the more correct statement:
   a. Physicists currently think that light has both wavelike and particle-like properties.
   b. Physicists currently think that light as well as matter has both wavelike and particle-like properties.
5. State whether or not classical physics could account for each of the following descriptions involving the photoelectric effect.
   a. When light is shined on a metal, electrons are emitted from the metal.
   b. The maximum kinetic energy of the electrons is independent of the light's intensity.
   c. There is a frequency of light under which electrons will not be emitted by the metal.

d. The maximum kinetic energy of the electrons is proportional to the light's frequency.
e. The electrons are emitted almost immediately once the light is turned on.

6. What was Einstein's major assumption in his explanation of the photoelectric effect?
7. If de Broglie is right, why doesn't a baseball thrown during a baseball game exhibit wavelike properties?
8. In quantum mechanics, if you measure the exact position of an object, with what accuracy can you measure its momentum?
9. Why can't you accelerate an object to a speed greater than the speed of light?
10. Is the principle of conservation of energy violated by the equation $E = mc^2$?
11. What is the maximum kinetic energy of electrons that are emitted by platinum when it is illuminated with light whose wavelength is 154 nm? (Platinum's work function is 6.35 eV.)
12. Aluminum is illuminated with light of unknown wavelength. The work function of aluminum is 4.08 eV, and the maximum kinetic energy of the electrons being emitted is 1.02 eV. What is the wavelength of the light being used?
13. A physicist wants to investigate the wavelike properties of protons and electrons. He accelerates a beam of electrons to a speed of "ve." If he then wants to accelerate the protons so that they have the same wavelength as the electrons, what must the proton's speed be in terms of "ve?" (mass of a proton = 1.67×10^{-27} kg, mass of an electron = 9.11×10^{-31} kg)
14. An astronomer observes indirect evidence for the existence of a black hole. Based on the behavior of the substances around this black hole, the astronomer estimates that the mass of the black hole is 5.67×10^{35} kg. What is the radius of the black hole?
15. A space ship leaves earth traveling at $0.999c$. It is headed to a location that is 677 light years from earth, as reckoned by observers on earth. How much time will pass on the ship before it reaches the location? How far will the people in the ship have traveled?
16. A 1000.0 mW coal-burning power plant burns about 4 million tons of coal each year. A nuclear power plant, however, uses the conversion of matter into energy in order to produce electricity. Assuming that the nuclear power plant is 100% efficient, how much mass would it consume each year? (Remember, Watt is a power unit, and P = Work/time).
17. Suppose you had a spherical space station whose radius is 3500.0 m. To provide the same gravitational effects as the surface of the earth, at what speed (in m/s) must the space station rotate?
18. Your softball team recruits a girl who can pitch the 0.18 kg ball at a speed of $0.58c$. What momentum will the catcher feels when they catch the ball?

Next Level

19. Compare the predicted black body energy density ($E_{\lambda v}$) for photons with $\lambda =$ 100 nm at $T = 10{,}000$ K in the classical and quantum models.
20. Compare the predicted black body energy density ($E_{\lambda v}$) for photons with $\lambda =$ 2000 nm at $T = 10{,}000$ K in the classical and quantum models.

Modern Physics II: Atomic and Nuclear Physics

16

Chapter Summary

The theories of modern physics discussed in the previous chapter are critical to our understanding of matter. In this final chapter we will discuss what matter is made of. Physicists have now been able to identify the elementary particles that make up all the known matter in creation but no one knows how they were made or where they came from. In addition there is evidence for "dark matter" that is a completely unknown substance. Some of the themes of the chapter are represented in Fig. 16.1. Note that atoms can be identified by the wavelengths that they absorb and emit light.

Main Concepts in This Chapter

- Atomic Physics I: The Bohr Model
- Atomic Physics II: Beyond the Bohr Model
- Absorption and Emission of Photons
- Nuclear Physics I: Structure of the Nucleus
- Nuclear Physics II: Nuclear Processes
- Beyond the Nucleus

16.1 Introduction

Although the concept of all matter being made of atoms dates back to the ancient Greeks, it was only in the early 1900s that we began to understand what an atom is. To support these theoretical advances, major experimental facilities were constructed. These included particle accelerators that smash atoms into pieces.

R. C. Powell, *Physics*, Undergraduate Lecture Notes in Physics,
https://doi.org/10.1007/978-3-032-09361-5_16

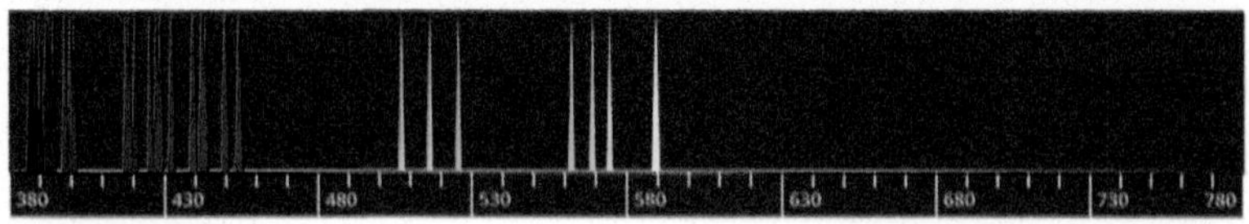

Fig. 16.1 Atomic spectrum of sodium. Credit: Nucleus hydro elemon, Creative Commons Attribution-Share Alike 0

Advances in cryogenics have allowed us to determine how atoms act near absolute zero temperature. Enhancing our understanding of light-matter interaction on the atomic scale led to the development of the laser which is used in so many applications that it has revolutionized our lives.

The discussion in the previous chapter makes it clear that the concepts of normal, classical physics cannot be used to describe events on the atomic scale. One of the main motivations of the development of quantum mechanics was to have a better understanding of an atom. This was followed by the development of a better understanding of the nucleus of an atom which has created a whole new field of elementary particle physics. This chapter provides an overview of the basic concepts of atomic, nuclear, and elementary particle physics.

16.2 Atomic Physics I: The Bohr Model

The lightest element is hydrogen. Since it has the least complicated structure, it was the most useful element in the development of models for an atom in the early 1900s. Physicists knew that atoms were comprised of electrons and protons, but the discovery of the neutron didn't come until later. Prior to 1911, scientists thought that the atom was made of a positively charged "gel" or "pudding" with negatively-charged electrons sprinkled throughout the pudding. This view of the atom was commonly called the plum pudding model of the atom. Between 1909 and 1911, Ernest Rutherford performed experiments which showed that the atom is composed of a dense, positively charged center, called the nucleus, with electrons orbiting the nucleus much like the planets in our solar system orbit the sun. Rutherford's view of the atom was called the planetary model of the atom.

This model did not solve all of the problems of understanding the behavior of atoms. Two important examples were radiating electrons and atomic spectra. We mentioned in Chap. 11 that accelerating electrons emit radiation. When they do this, they lose energy. This implies that in a planetary model of the atom, the electrons will lose energy as they orbit the nucleus, causing them to spiral into the center.

In addition, since about 1860, scientists knew that gaseous atoms, when excited by electricity, emitted light. This was not a problem for classical physics, because physicists rightly concluded that the electricity provided energy to the atoms, and the atoms later released that energy in the form of light. However, the atoms would not emit just any light when excited by electricity. Each atom emitted specific wavelengths of light, and those specific wavelengths varied from atom to atom. They were unique for each element. For example, hydrogen atoms emitted light having four discrete frequencies of visible light and not light at any other visible frequencies. These patterns of specific wavelengths emitted by excited atoms are called **atomic emission spectra**, and they are completely unexplainable by classical physics.

In 1915 physicist Niels Bohr applied the principles of quantum mechanics to resolve these problems. To do this he used the concepts of discrete, quantized states of energy developed by Planck and Einstein in explaining the photoelectric effect and blackbody radiation. Bohr postulated that the electrons orbiting the nucleus of an atom obeyed these quantum mechanics concepts and did not obey the classical concepts of electromagnetic theory. In other words, the electrons were in stationary states having a specific amount of energy and not radiating. The electrons then had only *specific allowed orbits* in which they could orbit the nucleus and not any arbitrary orbit. Bohr had no rationale for making these assumptions. He did know that the assumption of discrete energy states had been useful in solving other problems in physics. Adding these concepts to our understanding of atoms solved both the problems of radiating electrons and discrete atomic emission spectra. Bohr's model of the atom was a major breakthrough in our understanding of atomic physics and a major success for our use of quantum mechanics.

Figure 16.2a shows the lowest three electron orbits of the Bohr model of the atom. The energy of each orbit is depicted in the energy level diagram shown in Fig. 16.2b. The lowest energy orbit is called the ground state and the higher energy orbits are called excited states. The electrons in the atom can only exist in one of these discrete states of energy. They cannot have any energy in between these energy states. The energy of the electron in an orbit is related to the radius of the orbit.

Since physical systems like to be in their lowest energy states, the electron on a hydrogen atom is usually found in the ground state orbit. The Bohr model explains the characteristics of absorption and emission of electromagnetic energy by electrons moving from one orbit to another. The absorption of a photon of light

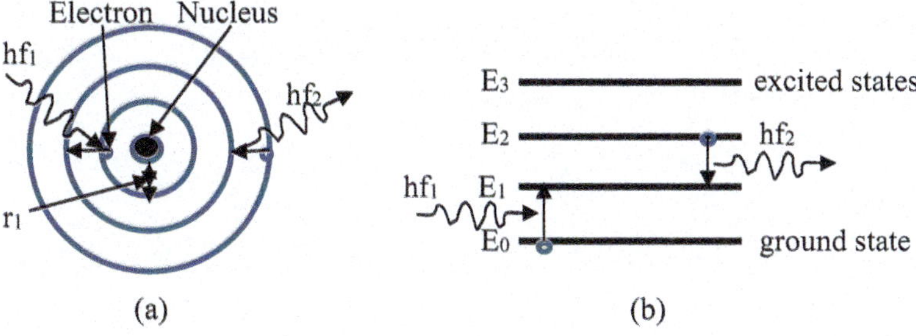

Fig. 16.2 Absorption and emission of light in the Bohr model of atoms

by an electron in the ground state of the atom can raise it to one of the excited states of the atom. When an electron in one of the excited states of an atom decays to one of the lower states, it emits a photon of light. Since the energy of photons are quantized and the energy levels of the atom are quantized, the photon energies in these transitions must equal the energy separation between the two states of the transition. This simply states the concept of the conservation of energy. Therefore, in Fig. 16.2,

$$hf_1 = E_1 - E_0 \text{ and } hf_2 = E_2 - E_1.$$

Photons with other energies that do not match transition energies between two states will not be absorbed or emitted. This is a result of quantum mechanics dealing with systems in discrete states instead of having a continuum of allowed possible states. It explains why the spectrum of light absorption and emission by hydrogen (and other atoms) appears as a series of discrete lines.

Consider the following problem.

Student

16.1 Using the Bohr model, suppose an atom absorbed enough energy so that an electron could jump from the first orbit to the fourth orbit. How many specific wavelengths of light could the atom emit as it released that energy?

As we discuss atoms in the remainder of this section, it would be good for you to remember the Periodic Table of Elements you learned about in your chemistry class. It is shown in Appendix D of this book. The atomic number of an element indicates the number of protons in an atom and for a neutral atom this is the same as the number of electrons it has. This electronic structure is responsible for most of the properties of a specific type of atom. Remember that an ion is an atom with a greater or lesser number of electrons than protons, so it has an electrical charge.

At this point we can put some mathematics to the Bohr model. In order to derive an expression for the radius of a circular electron orbit in the Bohr model of an atom, he used the classical mechanics equation for centripetal force, Eq. 5.8, with the force supplied by the Coulomb force between the nucleus with charge $+q$ and the electron with charge $-q$ given by Eq. 7.1. This resulted in

$$mv^2/r = Kq^2/r^2 \tag{16.1}$$

where K is the Coulomb constant. Solving this for r gives the radius of an orbiting electron as

$$r = Kq^2/(mv^2). \tag{16.2}$$

Equation 16.2 shows that the radius of an electron orbit is related to its velocity but it does not limit the value of r to discrete amounts. In Chap. 5 we learned that a particle of mass m in a circular orbit of radius r has an angular momentum of mvr where v is its linear velocity. Since the orbital radii in Bohr's model are quantized, Bohr's next assumption was that the angular momentum of an atomic state is quantized. He assumed the rule that the allowed values of angular momentum were integer values of $h/(2\pi)$. Using this assumption,

$$mvr = nh/(2\pi)$$

where h is Planck's constant and n is an integer. Thus, an electron on an atom is restricted to be in an orbit with a discrete amount of angular momentum. Solving this expression for velocity gives

$$v = \frac{nh}{2\pi mr} \tag{16.3}$$

Now we can substitute this expression for the velocity into Eq. 16.2 which gives an equation for the radius of an orbiting electron in a hydrogen atom as,

$$r_n = \frac{n^2h^2}{4\pi^2Kmq^2}. \tag{16.4}$$

Note that all of the quantities on the right side of this equation have known, constant values except n which is any integer.

The lowest electronic state of the hydrogen atom has $n = 1$. The radius of the electron orbit in this state is known as the *Bohr radius.* Using Eq. 16.4, the Bohr radius is

$$\begin{aligned} r_1 &= \frac{1^2\left(6.626 \times 10^{-34}\,\mathrm{Js}\right)^2}{4\pi^2\left(8.99 \times 10^9\,\mathrm{Nm^2C^{-2}}\right)\left(9.11 \times 10^{-31}\,\mathrm{m}\right)\left(1.6 \times 10^{-19}\,\mathrm{C}\right)^2} \\ &= 5.29 \times 10^{-11}\,\mathrm{m}. \end{aligned} \tag{16.5}$$

The next two higher orbits have radii of $r_2 = 4r_1$ and $r_3 = 9r_1$.

We have talked about atoms being small but now we can quantify this statement. Since there is only one electron in a hydrogen atom, in its ground state, the first Bohr orbit is the only orbit that is occupied. Thus, this tells us that the hydrogen atom itself has a radius of 0.529 Å in its ground state. (Å indicates angstroms. See Appendix B.) To understand this number look at a metric ruler and look at the markings that indicate millimeters. Based on the size we just calculated, approximately 19,000,000 hydrogen atoms in their ground state could fit in between two of those millimeter marks. This is really small. Equation 16.4 is specifically for a hydrogen atom. However, it can be used to estimate values for other single electron ions of light elements by including their atomic number Z in the denominator since the charge of their nucleus is Zq.

The requirement of angular momentum being quantized shows why the orbits only have specific values for their radii. Next we want to look at the energy of the electron in these orbits.

$$\mathrm{KE} = mv^2/2 = Kq^2/2r.$$

The potential energy of a charged particle in an electric field was discussed in Chap. 7. Using Eqs. 7.6 and 7.7, this can be expressed as

$$\mathrm{PE} = -Kq^2/r.$$

Remember the zero point of potential energy is arbitrary and for a bound particle it is chosen to be negative. Then

$$E_n = \mathrm{KE} + \mathrm{PE} = Kq^2/2r_n - Kq^2/r_n = -Kq^2/2r_n.$$

The minus sign indicates that the electron bound to the nucleus has lower energy than a free electron in its surroundings. Using Eq. 16.4 for the radius gives,

$$E_n = -\frac{2\pi^2 K^2 mq^4}{h^2}\frac{1}{n^2}. \tag{16.6}$$

The ground state energy for hydrogen is E_1 is

$$E_1 = -2\pi^2\left(8.99\times10^9\,\mathrm{Nm^2C^{-2}}\right)^2\left(9.11\times10^{-31}\,\mathrm{kg}\right)$$
$$\left(1.6\times10^{-19}\,\mathrm{C}\right)^4/\left(6.626\times10^{-34}\,\mathrm{Js}\right)^2 = -2.18\times10^{-18}\,\mathrm{J}.$$

This number is called the Rydberg energy, $R = 2.18\times10^{-18}$ J or 13.6 eV. Including a factor of Z^2 in the numerator of Eq. 16.6 allows for its use with other single electron ions. The first two excited states have energies of $E_2 = E_1/4$ and $E_3 = E_1/9$.

Equations 16.4 and 16.6 do a good job in describing the properties of the hydrogen atom. The latter can be simplified by using the Rydberg energy,

$$E_n = -RZ^2(1/n)^2. \tag{16.7}$$

Remember that this equation only is valid for ions with one electron.

Now consider the following examples and problems.

Example 16.1

For a Hydrogen atom, calculate the change in orbital radius when an electron moves from the second excited state to the first excited state and what is the energy of the photon that is emitted?

The picture for this problem is given in Fig. 16.2. The knowns and unknowns are:

Knowns: $r_3 = 9r_1$; $r_2 = 4r_1$; $r_1 = 5.3 \times 10^{-11}$ m; $E_3 = E_1/9$; $E_2 = E_1/4$; $E_1 = -2.17 \times 10^{-18}$ J.
Unknowns: Δr_{32}; E_p.

Using the known factors,

$$\Delta r_{32} = r_2 - r_3 = (4 - 9)r_1 = -5r_1 = -26.5 \times 10^{-11}\ \text{m};$$

$$\begin{aligned} E_p &= E_3 - E_2 = (1/9 - 1/4)E_1 \\ &= (-0.139)\left(-2.17 \times 10^{-18}\ \text{J}\right) = 3.01 \times 10^{-19}\ \text{J}. \end{aligned}$$

So by emitting a photon with 3.01×10^{-19} J of energy, the electron's orbital radius decreases by 26.5×10^{-11} m.

Example 16.2
What is the energy of an electron in the third Bohr orbit of a He^+ ion?

The picture for this problem is given in Fig. 16.2. The knowns and unknowns are:

Knowns: $Z = 2$; $n = 3$; $R = 13.6$ eV
Unknowns: E_3.

Since helium has two electrons, a He^+ ion has only one electron so Eq. 16.7 works. The atomic number of helium is 2, so $Z = 2$. Since the electron is in the third Bohr orbit, $n = 3$.

$$E_n = -RZ^2(1/n)^2 = -(13.6\,\text{eV})(2)^2/(1/3)^2 = -6.04\,\text{eV}$$

So the electron in this orbit has an energy of − 6.04 eV. Notice that the answer has three significant figures. This is because the Rydberg energy has three significant figures, while "Z" and "n" are perfect integers. As a result, they have infinite precision. Thus, the number of significant figures is determined only by the Rydberg energy.

Student

16.2 For which of the following atoms or ions can we use Eq. 15.13?

H, Li^+, Be, Li^{2+}, H^+, Be^{3+}, He

16.3 An electron in a hydrogen atom has an energy of -5.45×10^{-19} J.
 a. Which Bohr orbit does it occupy?
 b. How far from the nucleus is it orbiting?

16.4 What it the radius of a Li^{2+} ion in its ground state?

16.3 Atomic Physics II: Beyond the Bohr Model

Today our model for an atom is significantly more complex than the simple Bohr model. As important as the Bohr model was in advancing our understanding of the atom, as we learned more about the details of quantum mechanics, our view of the atom had to evolve to account for these details. However, this may never have happened without Bohr's initial assumption of the quantized states of electrons in atoms.

The current view of the atom still involves quantized energy states of the electrons bound to a nucleus. However, instead of circular orbits like those of the Bohr model, the quantum mechanical model of the atom says that electrons, can orbit the nucleus in patterns that can have quite interesting shapes. These shapes represent clouds of probabilities of finding the electron at any position in the orbital. The sizes and shapes of these orbitals are governed by a set of numbers called **quantum numbers**. The orbital clouds represent the probability nature of quantum mechanics and are associated with the wave nature of the electron. Also, we will see how the quantum numbers obey the Pauli Exclusion Principle. In addition, the transitions between orbitals are governed by the Heisenberg Uncertainty Principle. So all of the major features of quantum mechanics described in the previous chapter are included in this modern view of the atom.

The quantum mechanical model of the atom is governed by a very complicated equation called the Schrodinger equation. The mathematics of deriving, solving, and using this equation are well beyond the scope of this textbook. The solution of the Schrodinger equation for electrons bound to atoms is given in terms of the wavefunctions for the electron that can be used to determine the probability of finding the electron at specific positions. The results show that electrons orbiting a nucleus can have only certain, distinct energies and positions within the atom. These energies and positions are characterized by a series of four quantum numbers. Once you determine those four quantum numbers for an electron, you know how much energy it has and the probability of it being at different locations in the atom.

One way to think about why only certain electron orbits are allowed is to visualize a simple circular orbit with an electron in it that has a de Broglie wavelength. If the circumference of the orbit is an integral number of the de Broglie wavelength the wave will undergo constructive interference and act as a stable orbit. However if this is not the case, the wave will undergo destructive interference and the orbit will disappear. Using the expression for the de Broglie wavelength from the previous chapter, the requirement for a stable orbit is

$$2\pi r = n\lambda_n = nh/mv \qquad (16.8)$$

where n is an integer designating the specific orbit. In this view, the angular momentum of the electron in the orbit is also quantized. Using the expression for angular momentum from Chap. 4 along with Eq. 16.8 gives

$$L_n = mvr = lh/2\pi \qquad (16.9)$$

where we have used l instead of n for the integer number designating the angular momentum state since there is no reason that the same number should be used to designate the radius of the orbit and the angular momentum of the electron in the orbit. This is a simplified view of a complicated situation, but it provides some insight as to the quantization of an electron's orbit and angular momentum.

The quantum number for the size of an orbit is called the **principal quantum number**, abbreviated as n. This principal quantum number indicates how far the electron is from the nucleus. Just as is the case with the Bohr model, the larger the value for n, the farther away the electron is (on average) from the nucleus. In addition, the value for n has a major effect on the energy of the electron in the orbit. When two electrons have two different values for n, the electron with the larger value for n has more energy than the other electron. Once again, as is the case for the Bohr atom, the principal quantum number can have any whole number value other than zero.

Principal Quantum Number: $n = 1, 2, 3, 4 \ldots$

The second quantum number that results from solving the Schrodinger equation is called the azimuthal quantum number and it is usually abbreviated as l. This quantum number tells us what shape orbital the electron is moving in which is related to its angular momentum as discussed above. It also affects the energy, but not nearly as significantly as does the principal quantum number. The azimuthal quantum number can have any integer value, including zero, but the range of integer values is restricted by the value of n. For example, when n equals one, the only possible value of l is zero. When n equals 2, l can be either zero or one. In general, then, l can have any integer value from zero to one less than the principal quantum number.

Azimuthal Quantum Number: $l = 0, 1, \ldots (n - 1)$

An orbital with a specific azimuthal quantum number has a specific shape. Letters are associated with these orbital shapes. Figure 16.3 illustrates the basic shapes of electron orbitals for the first 6 values of the azimuthal quantum number, along with the letter associated with that shape. The figure shows that s-orbitals are round, p-orbitals have two lobes in a figure 8 shape, and higher orbitals have complicated shapes with additional lobes. The lobe patterns indicate that the electron goes through the center of the atom where the nucleus is. However, these orbitals are

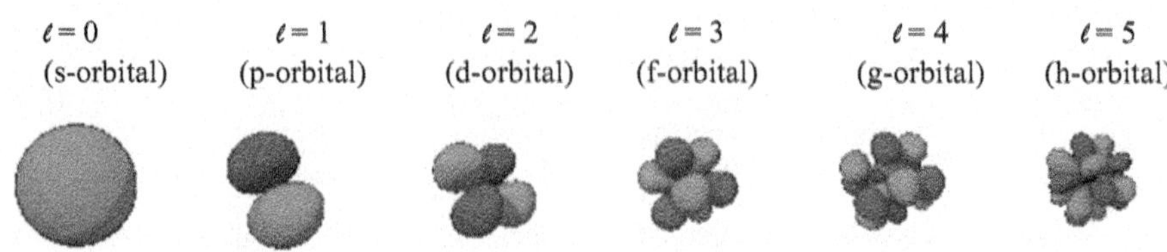

Fig. 16.3 Electron orbitals

probability clouds of where to find the electron and that probability must go to zero in the center.

Why does the value of "*l*" depend on the value of "*n*?" *n* tells us the average distance that the electron is from the nucleus. As the value for *n* gets larger, the larger that distance becomes. When $n = 1$, the electron is rather close to the nucleus; thus, there is not much room. As a result, only one orbital can "fit" within that space. Only the s-orbital ($l = 0$) is present when $n = 1$. When $n = 2$, the electron is farther away from the nucleus. As a result, there is more room, so more orbitals can "fit" within that space. There are actually three p-orbitals, so when $n = 2$, there is room for an s-orbital and three p-orbitals, so "*l*" can equal either 0 or 1. Likewise, when $n = 3$, the electron is even farther out, so now there is room for an s-orbital, three p-orbitals, and five d-orbitals. As a result, when $n = 3$, "*l*" can equal 0, 1, or 2. So the reason the value of "*l*" depends on the value of "*n*" relates to space. There are spatial limitations that slowly go away as "*n*" increases. Thus, for low values of "*n*," there are only a few possible orbitals. For large values of "*n*," however, many more orbitals are available.

The next quantum number is called the magnetic quantum number, and it is usually designated as "*m*." This number affects certain details of an orbital's shape, as well as its orientation in space. Just as the value of "*l*" depends on the value of "*n*," the value of "*m*" depends on the value of "*l*." The magnetic quantum number can have all integer values from $-l$ to $+l$.

Magnetic Quantum Number: *m* Ranges from $-l$ to $+l$ in Integer Steps

Orbitals with the same value of *l* but different values of *m* have the same energy but different spatial orientations. *m* essentially quantizes the vector direction of the angular momentum. If the atom is put in a magnetic field pointed in a specific direction, orbits with different orientations will have different energies because the magnetic field that interacts differently with the electron when it is pointed in different directions with respect to the motion of the electron. This is why *m* is called the magnetic quantum number.

When $l = 0$, the only possible value for "*m*" is 0. This tells us that there is only one possible s orbital. If we think of the atom in three-dimensional space, then, this looks like the s-orbital in Fig. 16.4. When $l = 1$, three values for "*m*" are possible, -1, 0, or 1. Since $l = 1$ means we are dealing with p-orbitals, then this tells us that there are 3 different kinds of p-orbitals. There is a p-orbital for which $m = 1$, a p-orbital for which $m = 0$, and a p-orbital for which $m = -1$. The shapes of those three orbitals are shown in Fig. 16.4. Notice that the basic shapes of these three orbitals are the same. That's because, for the most part, orbital shape is determined by "*l*." The difference between the three is their orientations. If we imagine the nucleus of the atom at the origin of our three-dimensional axis, you can see that all three of these orbitals can exist simultaneously around the atom. Now realize that the electron can be *anywhere* within the orbital, but it cannot be anywhere outside of the orbital. Thus, you can see that the electron cannot be anywhere right near the nucleus, but it can be anywhere within the two lobes of the orbital.

To carry the illustration one step further, when $l = 2$, there are five possible values of m: -2, -1, 0, 1, and 2. This means there are five different types of d-orbitals, as shown in Fig. 16.4. Notice that the value of m mostly affects the orientation of the orbital, except in the case where $m = 0$. There, the value of m does affect the shape. If you are good at three-dimensional visualization, you can see that all of these orbitals can exist simultaneously with the nucleus at the origin of the three-dimensional axis.

The point should now be clear. Because of the relationship between "*l*" and "*m*," the number of possible orbitals increases as the value of "*l*" increases. In general, the value of "*m*" determines the orientation of the orbital in space, and it has an effect on the shape of the orbitalin some cases. When you see drawings of orbitals such as those shown in these figures, remember what they mean. The electron can be found *anywhere* within the orbital, but nowhere outside of the orbital. These orbitals, then, represent regions around the nucleus in which the electrons can be found.

The last quantum number to learn about is called the **spin quantum number**, and it is usually designated by an "*s*." In a simple planetary model, the electron rotates on its axis as it orbits the nucleus. It has only one value for the magnitude

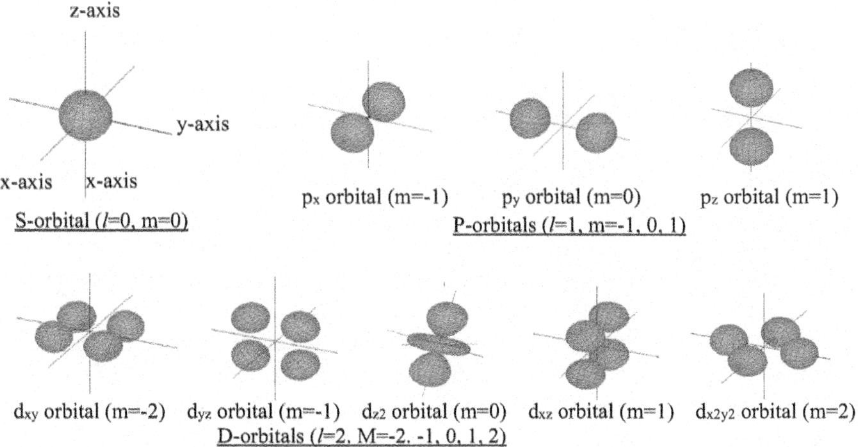

Fig. 16.4 S, P, and D orbital orientations

of spin but it can have two possible directions, clockwise and counterclockwise. These are generally called spin up and spin down and designated by a quantum number *s*. This is the simplest quantum number since it can have one of only two values, + 1/2 or − 1/2 for spin up and spin down. These values are independent of any of the other quantum numbers.

To review, then, there are four quantum numbers that designate the state of an electron on an atom:

Principal Quantum Number: $n = 1, 2, 3, 4 \ldots$	– Orbital energy
Azimuthal Quantum Number: $l = 0, 1, \ldots (n - 1)$	– Orbital shape
Magnetic Quantum Number: *m* **ranges from** $-l$ **to** $+l$ **in integer steps**	– Orbitalorientation
Spin Quantum Number: $s = +1/2$ **or** $-½$	– Spin direction

When you put together a list of four quantum numbers that meet these criteria, you have uniquely determined the average distance from the nucleus, general orbital shape, orbital orientation, spin direction and energy of an electron in an atom. This defines a specific quantum state of the atom.

Remember from Sect. 15.3 that the Pauli Exclusion Principle states that no two electrons can be in the same state at the same time. This means that no two electrons can have the same set of quantum numbers at the same time. Most of the time, we are dealing with atoms that have more than one electron. They will go onto an atom in the lowest unfilled orbital. Any spatial orbital can hold two electrons, one with spin up and the other with spin down. Thus, if we have an atom with ten electrons, the first two will fill the s-orbital and the next six will fill the three p-orbitals. The remaining two will go into two of the ten possible d-orbitals. The other d-orbitals will remain empty until some external perturbation causes the outer electrons to change orbitals. These electron configurations and especially the electrons in the highest energy orbits are critical in determining the properties of the atom as it interacts with neighboring atoms and its environment. Two examples are chemical bonding and the absorption and emission of light.

16.4 Absorption and Emission of Photons

The experimental measurement of the absorption and emission of photons by matter is called spectroscopy. We talked briefly about spectroscopy in Sect. 14.5 of this book dealing with the optical properties of matter. However, in order to really understand spectroscopy, we had to learn about how atoms absorb and emit photons. The central piece of equipment in a spectroscopy experiment is a spectrometer which uses a prism or diffraction grating to separate a light beam into its component frequencies. The absorption spectrum of a sample measures the amount of light absorbed at each wavelength or frequency of an incident beam of light. The emission spectrum of a sample measures the intensity of each wavelength or frequency of light emitted by the sample after its atoms have been excited to their higher energy levels. The spectral lines of a sample act as a fingerprint for identifying the type of atoms present in the sample.

One of the successes of Bohr's model of the atom was its ability to explain atomic emission spectra. Electrons in the Bohr model have a certain amount of energy when they sit in a given Bohr orbit. Based on Eq. 16.7, we can say that when "*n*" is small, the energy is large and negative. On the other hand, when "*n*" is large, the energy is small and negative. The smaller the negative number the larger the value, so in the end, orbits with large values of "*n*" are considered high energy orbits while orbits with low values of "*n*" are considered low energy orbits.

Suppose an electron is sitting in a low energy orbit. If it can absorb exactly the right amount of energy, it will jump up to a high energy orbit. This energy can be absorbed from the heat of the surroundings, or it can be absorbed by capturing a photon of light. In nature, everything wishes to end up in its lowest energy state, which is called the ground state. Thus, the electron will not stay in the high energy orbit to which it jumped. Instead, it will want to go back to the lowest energy orbit that is not completely occupied by other electrons. The only way it can do this is to release energy. It can do this by emitting heat or a photon of light.

So, in order to jump up to a higher energy orbit, electrons need to absorb energy from heat, from light, or from other source such as an electrical discharge. In order to jump back down into a lower energy orbit, the electron must release energy, usually in the form of light or heat. The amount of energy an electron needs to absorb or release depends on what orbit it is in and what orbit it is going to. For example, suppose an electron is in the third Bohr orbit and wants to jump back down to the first Bohr orbit. According to Eq. 16.7, the electron has a certain amount of energy in the third Bohr orbit, and it needs to have a certain amount of energy to be in the first Bohr orbit. The difference between these two energies is what the electron must release. Mathematically,

$$\Delta E = E_{\text{initial}} - E_{\text{final}} = -RZ^2(1/n_{\text{initial}})^2 - \left[-RZ^2(1/n_{\text{final}})^2\right]$$

or

$$\Delta E = RZ^2\left[(1/n_{\text{final}})^2 - (1/n_{\text{initial}})^2\right] \tag{16.10}$$

Equation 16.10 allows us to calculate the energy that an electron must either absorb or release in order to change orbits in the Bohr model of the atom. This is demonstrated in the following example.

Example 16.3
What color of light is emitted when the electron of a hydrogen atom moves from the fourth Bohr orbit to the second Bohr orbit?

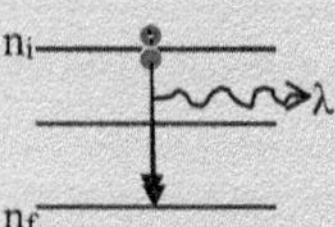

Knowns: $n_i = 4$; $n_f = 2$; $Z = 1$

Unknowns: λ

The electron starts out in the fourth Bohr orbit, so $n_{initial} = 4$. It ends up in the second Bohr orbit, so $n_{final} = 2$. According to Eq. 16.10

$$\Delta E = RZ^2\left[(1/n_{final})^2 - (1/n_{initial})^2\right]$$
$$= (13.6\,\text{eV})\left[(1/2)^2 - (1/4)^2\right] = 2.55\,\text{eV}$$

This tells us that the electron must lose 2.55 eV to make the transition, so that is the energy of the light that it emits. To get the wavelength, we first determine the frequency:

$$E = hf$$

So

$$f = E/h = (2.55\,\text{eV})/\left(4.14 \times 10^{-15}\,\text{eVs}\right) = 6.16 \times 10^{14}\,\text{s}^{-1}$$

Then, we use the speed of light to go from frequency to wavelength,

$$\lambda = c/f = \left(2.998 \times 10^8\,\text{m/s}\right)/\left(6.16 \times 10^{14}\,\text{s}^{-1}\right)$$
$$= 4.87 \times 10^{-7}\,\text{m} = 487\,\text{nm}$$

The electron must emit light with a wavelength of 487 nm, which is blue light.

When an electron is provided energy to absorb, it can jump up to any orbit that is not filled with other electrons, provided there is enough energy at its disposal. When the electron goes down to a lower energy orbit, it need not jump directly down to the lowest orbit available. Instead, it can make several jumps on its way down. Suppose, for example, that an electron jumps to the $n = 5$ orbit. In order to jump back down to the $n = 1$ orbit, it can do so in one jump, or it can go from $n = 5$ to $n = 3$ and then from $n = 3$ to $n = 1$. In that case, it would release light twice. First, it would release light that has the same energy as the difference between the energy in the fifth Bohr orbit and the third Bohr orbit, and then it would emit light that has energy equal to the difference in energy between the third Bohr orbit and the first Bohr orbit. Alternatively, it could jump down one orbit at a time, releasing light of four different energies. Because there are so many paths that an electron can take when jumping from a high energy orbit back to its ground state, there are many different energies of light that electrons will emit once they have absorbed energy. As a result, when atoms are exposed to a large amount of energy, they emit many different wavelengths of light. The sum total of all of those different wavelengths is the emission spectrum of the atom. Remember much of the electromagnetic spectrum is outside our visible range so the photon emission may be in the infrared or ultraviolet regions that we cannot see.

As an example of a spectroscopic fingerprint of an atom, we can consider the emission spectrum of hydrogen gas in the visible region of the electromagnetic spectrum. The results show four emission lines ranging from blue to red with wavelengths $\lambda = 410$, 434, 486, and 656 nm. The energy level diagram for hydrogen atoms is shown in Fig. 16.5. We used the Bohr model to calculate the ground state energy of a hydrogen atom as $-\ 2.17 \times 10^{-18}$ J and the excited states with energies given by Eq. 16.6. However, since it is the energy difference between levels that is relevant for photon emission, Fig. 16.5 starts with the ground state being zero and lists the energy above the ground state for each of the excited states. Using the expression for the energy of a photon given by Eq. 15.3, the observed wavelength of the light emission can be related to a transition between energy levels separated by an amount ΔE as

$$\lambda = hc/\Delta E = 1.99 \times 10^{-25}\,\mathrm{Jm}/\Delta E. \quad (16.11)$$

Using the energies of the levels shown in Fig. 16.5 indicates that the four observed spectral lines all terminate on level E_2 and originate on the levels E_3, E_4, E_5, and E_6. For example, the energy difference between E_3 and E_2 is 3.03×10^{-19} J. Substituting this into Eq. 16.11 gives a photon wavelength of $\lambda = 6.56 \times 10^{-7}$ m which is the red emission line of hydrogen. It should be mentioned that

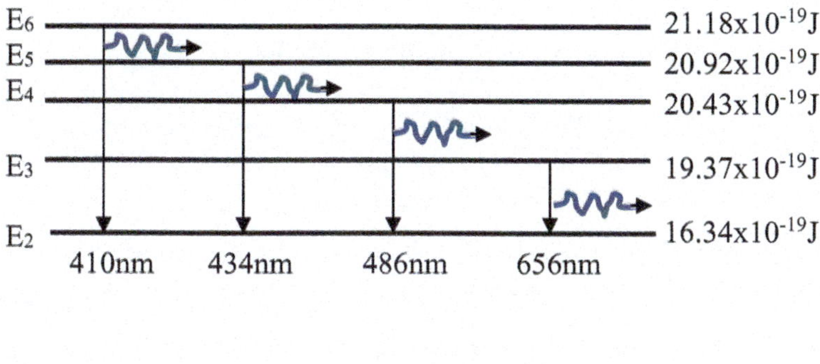

Fig. 16.5 Emission spectral lines for hydrogen in the visible region

hydrogen has many more spectral emission lines, but they occur either at higher energies in the ultraviolet region of the spectrum or at lower energies in the infrared region of the spectrum. These four lines are a fingerprint for hydrogen in the visible spectral region.

In normal absorption spectroscopy experiments, the transitions originate on the ground state and terminate on one of the excited states. As an example, calculate the wavelength of the photons absorbed in the lowest energy transition for hydrogen. According to the energy level diagram in Fig. 16.5, the energy of the photon in this transition is 16.34×10^{-19} J. Substituting this into Eq. 16.11 gives the wavelength of the absorbed light to be $\lambda = 1.22 \times 10^{-7}$ m.

Now try the following problems.

Student

16.5 An electron jumps from the $n = 3$ orbit of a hydrogen atom to the $n = 2$ orbit. What is the wavelength of the light emitted? Is it visible? If it is visible, what color is it?

16.6 An electron jumps from the fifth Bohr orbit to a lower orbit in a Li^{2+} ion. If the light emitted has a frequency of 6.67×10^{14} Hz, what orbit did the electron end up in?

The light emitted and absorbed by any element will be unique because each element has its own atomic number. Each molecule also has its own electronic structure and, as a result, the light that it can absorb or emit is unique to that molecule. Thus, if we excite an element or compound (typically by heat, light or electricity) and look at the pattern of wavelengths (visible and not visible) of light that it emits, we can unambiguously determine that atom or compound. This technique is called emission spectroscopy, and it is used to examine the elemental makeup of stars and to analyze chemicals. Alternatively, we could shine light through a compound or element and see what wavelengths of light are absorbed. This is called absorption spectroscopy.

There is one more aspect of quantum mechanics that shows up in the emission spectrum of atoms. An emission spectrum can be measured in units of energy, frequency, or wavelength. For photons, these are all connected through

$$E = hf = hc/\lambda$$

Emission spectra can be plotted in terms of energy, frequency, or wavelength. If we apply the Heisenberg Uncertainty Principle from Eq. 15.8,

$$\Delta E \Delta t \geq h/4\pi$$

Therefore the spread in energies, frequencies, or wavelengths we measure in a spectral transition is inversely proportional to the time increment these measurements are valid. For the initial state of an atomic transition, this time is called its lifetime and designated by $\Delta t = \tau$. Note that the ground state for an absorption transition has an infinite lifetime since we know its energy exactly. The shorter the lifetime of an excited state the broader the spectral features to or from that state in an absorption or emission spectrum. States with short lifetimes appear as broad bands in the spectra while states with long lifetimes appear as sharp lines in the spectra. Of course, Eq. 15.8 is an inequality so the width of an energy level can always be broader than the minimum required by the lifetime of the level. This can be due to additional effects such as experimental resolution or thermal motion. The lifetimes of electronic states of atoms range from nanoseconds to milliseconds. The states with long lifetimes are called metastable states.

The shape of spectral features is a further demonstration of the importance of quantum mechanics in our understanding of atoms. The enhanced understanding of atomic physics in the early part of the 1900s led to one of the most important inventions of all time in the middle part of the century—the laser. When Einstein was reformulating electromagnetic theory, he pointed out that photon emission by an atom could happen in two different ways. First, when an electron absorbs a photon and makes a transition to an excited energy level, it stays in that level for the lifetime of the level and then spontaneously decays to a lower level. This is what normally happens. However, if a photon with an energy equal to one of the emission transitions from that level to a lower level hits the atom with an electron in the excited state, it can stimulate the electron to undergo that transition to a lower level. When it does this, the photon it emits has the same energy and direction as the photon that hits the atom and the waves of the two photons are in phase with each other. This is shown schematically in Fig. 16.6. Stimulated emission is essentially the inverse of a stimulated absorption transition.

In 1960 scientists demonstrated how stimulated emission could be used to amplify light. The device they invented is called a laser which is an acronym for light amplification by stimulated emission of radiation. Figure 16.7 shows the difference between a normal light bulb emitting photons by spontaneous emission and a laser emitting photon by stimulated emission. The photons emitted spontaneously have many wavelengths, go in all directions, are emitted at different times, and have light waves with uncorrelated phases. This is called incoherent light. On the other hand, the photons from stimulated emission all have the same

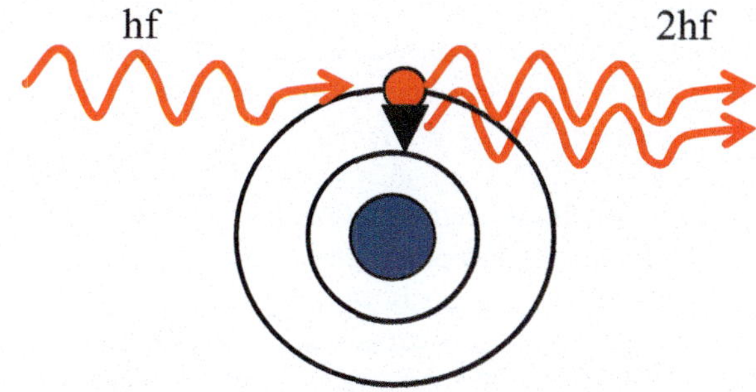

Fig. 16.6 Stimulated emission of radiation

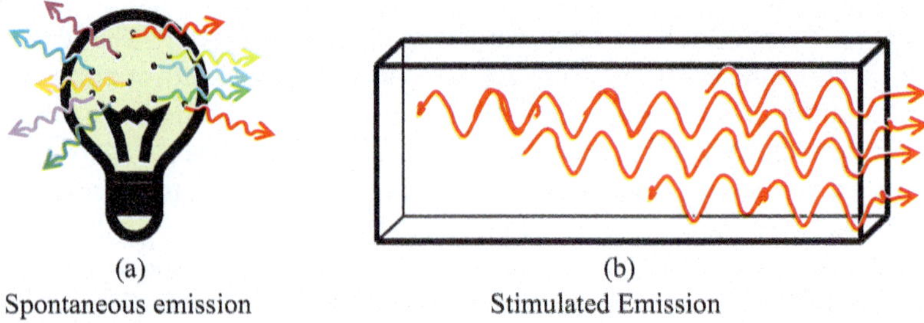

Fig. 16.7 Spontaneous versus stimulated emission

wavelength, go in the same direction at the same time and have light waves with correlated phases. This is called coherent light. The advantages of coherent light include going long distances without dispersion, generating ultrafast pulses, generating ultrahigh power, focusing to a small spot, and being monochromatic (having one wavelength).

There are many different types of lasers that produce coherent light. They range in size from as small as the head of a pin to as large as a whole building. The first laser was made with a ruby crystal. This was discussed in Chap. 14 in the section on the optical properties of solids. Figure 16.8a shows the components of a solid-state laser like ruby. The heart of the device is the ruby crystal. Ruby is aluminum oxide where a very small percentage of the aluminum ions have been replaced by chromium ions. Cr^{3+} ions. These ions have 21 electrons with a configuration of $1s^2 2s^2 2p^6 3s^2 3p^6 3d^3$ where the first number is the quantum number for the energy level, the letter is the designation of the orbital in that level, and the superscript is the number of electrons in the orbital. Note that the s-orbitals are filled with 2 electrons and the p-orbitals are filled with 6 electrons. There are 3 electrons in the outer most d-orbital which has the ability to hold 10 electrons. This means that there are close by empty energy levels that electrons can transition to and from. It also means that the six nearest neighbor oxygen ions produce an electric field at the site of the chromium ion that adds or subtracts energy from its free ion energy levels. When chromium ions are doped in aluminum oxide, their absorption spectrum has broad transition bands in the visible region that efficiently absorb white light. These energy bands are associated with transitions from the ground state to an excited state whose energy is modulated by the thermal vibrations of the oxygen ions. The energy then undergoes thermal decay transitions down to the lowest excited state. The lowest excited state is a metastable state with a long lifetime of about 3 ms at room temperature. Photon emission transitions to the ground state then occurs resulting in a sharp spectral line in the red region. The absorption and emission spectrum of ruby are shown in Fig. 14.11.

When a ruby crystal is working as a laser, it is exposed to a flashlamp like a camera flashbulb. This excites the chromium ions into the excited states and they decay quickly to the long lived metastable state. When more atoms have electrons in the metastable state than the ground state, it is called a population inversion. As

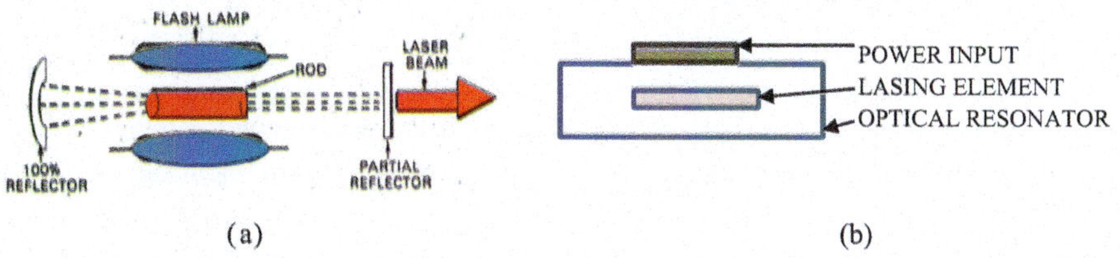

Fig. 16.8 Components of a laser

the ions decay to the ground state, they emit photons that can stimulate neighboring excited ions to emit similar photons. When the crystalabsorbing and emitting photons is placed in an optical resonator consisting of two mirrors, the photons travel back and forth between the two mirrors stimulating more transitions as they go. An optical resonator has one mirror that is 100% reflecting and the other mirror that has a reflectivity slightly less than 100%. This partial reflector transmits some of the circulating photons in the resonator as a beam of coherent radiation.

Although there are many different types and sizes of lasers, every laser has three basic components as shown in Fig. 16.8b. The key component is the lasing element. Many different types of crystals doped with optically active ions have been made into lasers. Transition metal ions like chromium and rare earth ions like neodymium doped into garnet crystals have been especially successful. Doped semiconductors described in Chap. 15 can be configured as p-n junctions where the recombination of electrons and holes emit light. These light-emitting diodes are an important type of laser. Many gases of optically active ions have also been made into ions. Argon, helium-neon, and carbon dioxide lasers are three important examples.

The second component of all lasers is a source of input power. Lasers do not create power. They convert some type of input power to a powerful beam of light. The optical pumping described above is a common type of power input for lasers made of doped crystals. In semiconductor diode lasers, an external current source provides the power. Gas lasers are usually excited by the electrical discharge in a gas-filled tube with a high voltage across it.

The third component for all lasers is the optical resonator. The output properties of the laser beam can be controlled by the focal lengths and reflectivity's of the resonator mirrors.

The reason that so many different types of lasers have been developed is that each application for a laser requires specific properties, especially the wavelength. The choice of a wide variety of lasers provides a wide variety of properties with wavelengths ranging from the ultraviolet through the visible into the infrared spectral region. Because of this, lasers have found applications in every area of our lives from barcode readers at store checkout counters to medical surgery to optical communications. This plethora of applications makes the laser rival the transistor in its importance as a scientific invention of the twentieth century.

16.5 Nuclear Physics I: Structure of the Nucleus

Early particle scattering experiments by Rutherford and others, and radioactivity experiments by the Curies and others, showed that an atom had a heavy, dense, positively charged nucleus at its center, surrounded by light orbiting electrons. Most of the atom was determined to be empty space. They found that atoms could emit radiation and change from one element to another. In the previous sections we learned a lot about electrons in atoms. In this section we will discuss the properties of the nucleus of atoms.

The nucleus of an atom is made of particles called nucleons. There are two main types of nucleons, protons and neutrons. The masses of these two types of particles are similar and about 1800 times the mass of an electron. This mass is designated as one atomic mass unit, $u = 1.66 \times 10^{-27}$ kg. The neutron has no electrical charge, but the proton has a positive electrical charge of the same magnitude as the negative charge on an electron.

Special Topic

It is hard to envision the size of the nucleus in an atom. One way is to imagine that a hydrogen atom is expanded until its average radius is as big as the walls of a major league baseball stadium. In this case the nucleus of the atom could be represented by a tiny marble located at the very center of the stadium. Most of an atom is empty space. Most of its mass is in the marble at the center.

The number of protons in a nucleus is called its atomic number designated as Z. The total number of protons plus neutrons of a nucleus is called its atomic mass number designated as A. The electrical charge of a nucleus is eZ and its mass is uA. The diameter of a nucleus is on the order of 10^{-14} m. The average density of nuclei is on the order of 10^{17} kg/m^3. Thus, an atom with a radius of about 10,000 time the size of the nucleus has most of it mass concentrated at its center.

The atom of a specific element is designated by the symbol for the element with a pre-subscript of its atomic number and a pre-superscript of its atomic mass number, ${}^{A}_{Z}E$. Nuclei of the same element have the same atomic number but can have a different number of neutrons and therefore a different atomic mass number. These are called isotopes of the element. An example is neon which has two isotopes ${}^{20}_{10}\mathrm{Ne}$ and ${}^{22}_{10}\mathrm{Ne}$ where the nucleus of the latter isotope has two more neutrons than that of the former isotope. Elements are listed in the Periodic Table of the Elements in Appendix D.

There must be some type of force that holds the nucleons together in a nucleus. This can not be the same electromagnetic force that holds a negatively charged electron in an orbit around a positively charged nucleus. This force would cause the positively charged protons to repel each other and would not act at all on neutrons that have no electrical charge. What we know about the force that holds nucleons together in the nucleus is that it is very short range, about 1.4×10^{-15} m, and very strong so it can overcome the electromagnetic force of repulsion between protons. It is an attractive force that is the same for two protons, two neutrons, or a proton and a neutron. This is simply referred to as the strong nuclear force. We don't really understand what this force is but we know it must exist or atomic nuclei wouldn't exist.

One clue about the nature of the strong nuclear force was found when it was discovered that a group of nucleons that are bound together in a nucleus have a total energy that is lower than the sum of the energies of the individual nucleons when they are unbound. This negative energy is called the binding energy of the nucleus. Since Einstein has shown that energy and mass are related by $E = mc^2$, the difference in energy between bound and unbound nucleons must reflect a difference in mass. In fact, the mass of a nucleus is less than the sum of the masses of the nucleons in the nucleus. This is called the mass defect. For example, the mass of a helium nucleus ${}^4_2He$ is 4.002603u. The sum of the masses of two protons ($m_p = 1.007276u$) and two neutrons ($m_n = 1.008665u$) is 4.03182u. So the mass defect is 0.029217u. According to the Einstein relationship this mass is equivalent to 4.36×10^{-12} J of energy. This is illustrated again in the following example.

Example 16.4

The mass of a ${}^7_3Li$ nucleus is 7.0160*u*. What is the binding energy of the nucleus? (The mass of a proton is 1.0073*u*, and the mass of a neutron is 1.0087*u*. The speed of light is 2.998 x 10⁸ m/s and 1*u* = 1.6605 x 10⁻²⁷ kg.)

Knowns: $m_p = 1.0073u$; $m_n = 1.0087u$; 3 protons; 4 neutrons; $m_L = 7.0160u$

Unknowns: E_B

Since lithium's atomic number is 3, all lithium atoms have 3 protons. The mass number, which is the sum of the protons and neutrons in a nucleus, therefore indicates that a ${}^7Li$ nucleus has 4 neutrons. The sum of the masses of 3 protons and 4 neutrons is:

$$3 \times (1.0073u) + 4 \times (1.0087u) = 7.0567u$$

Since the mass of a ${}^7_3Li$ nucleus is only 7.0160u, there is a mass deficit of $7.0567u - 7.0160u = 0.0407$ amu. This mass deficit is converted to energy but we must first convert the mass deficit to kg.

$$(0.0407u)\left(1.6605 \times 10^{-27}\,\text{kg}/u\right) = 6.76 \times 10^{-29}\,\text{kg}$$

Now we can convert mass defect to binding energy,

$$E_B = mc^2 = \left(6.76 \times 10^{-29}\,\text{kg}\right)\left(2.998 \times 10^8\,\text{m/s}\right)^2 = 6.08 \times 10^{-12}\,\text{J}$$

Although this doesn't sound like a lot of energy, remember that this is for a single atom. In a single gram of ${}^{7}_{3}\mathrm{Li}$ atoms, the total binding energy is 5.21×10^{11} J, which is quite a bit of energy!

As its name implies, binding energy tells us how tightly bound the nucleons are in the nucleus. The larger the binding energy, the stronger the nucleus holds its nucleons together. If we take the binding energy of a nucleus and divide it by the total number of nucleons in the nucleus, we get the binding energy per nucleon for that nucleus. This quantity gives us an idea of how strongly each nucleon is bound within the nucleus. The binding energy per nucleon tells us how stable a nucleus is. If the binding energy per nucleon is high in a nucleus, the nucleus holds tightly to each of its nucleons. If the binding energy per nucleon is low, the nucleus' hold on its nucleons is weak. If you calculate the binding energy per nucleon for several nuclei, you will find that this important quantity changes from nucleus to nucleus. In other words, some nuclei are more stable than others. Figure 16.9 illustrates a plot of binding energy per nucleon as a function of the mass number of a nucleus.

As you can see from the figure, the binding energy per nucleon rises with increasing mass number until the mass number reaches 56, which is the maximum binding energy per nucleon and then decreases for nuclei with higher mass numbers. This tells us that the most stable nuclei are those with mass numbers around 56. In fact, the most stable nucleus is ${}^{56}_{26}\mathrm{Fe}$ because it has the maximum binding energy per nucleon.

Since binding energy is negative and nature likes to be in the lowest possible energy state, naturally occurring nuclear processes that change one element into another element will tend to move toward iron. For nuclei with lighter atomic masses, nuclear reactions that increase A can occur naturally. For nuclei heavier than iron, nuclear reactions occur naturally that decrease A to form more stable isotopes. For example, uranium ${}^{238}_{92}\mathrm{U}$ can transform into thorium ${}^{234}_{90}\mathrm{Th}$ where the

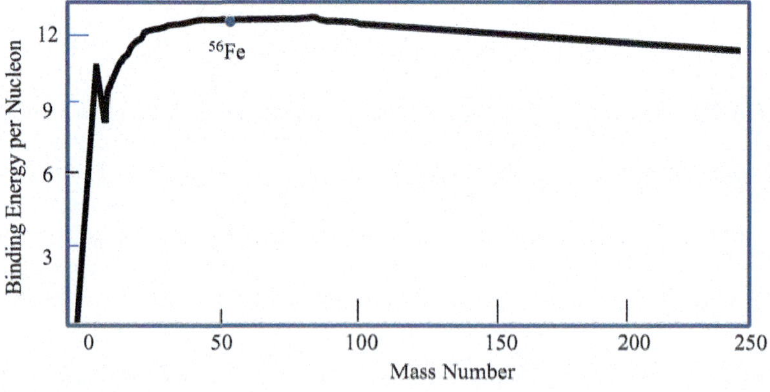

Fig. 16.9 Binding energy of nuclei

mass defect is contained in the mass and energy of the particles given off in the process. This is discussed further in the next section.

Now try the following problems.

Student

16.7 Calculate the binding energy per nucleon for $^{56}_{26}Fe$. (Use the data given in the example as well as the fact that an $^{56}_{26}Fe$ nucleus has a mass of $55.9349u$).

A more detailed theory of how the strong nuclear force provides the binding energy for the nucleus was proposed by Yukawa in 1937. He postulated that nucleons stayed together because, at short distances, they exchanged tiny particles called pions. Yukawa thought that the binding energy was used to give these pions kinetic energy, allowing them to travel from one nucleon to another. In other words, Yukawa believed that nucleons actually gave up a portion of their mass to form a small particle called a pion. Some of the mass that the nucleons gave up would go towards making the pion, and the rest would be converted to kinetic energy, allowing the pion to travel. Based on the properties of nuclei that were already known, Yukawa actually predicted what the mass of a pion should be.

Yukawa further believed that these pions can only exist for a very short time. As a result, he classified them as short-lived particles. Thus, Yukawa believed that a nucleon would form a pion, and the pion would begin to travel away from the nucleon. The pion, however, would not be able to live for very long. Thus, it would quickly encounter another nucleon and be absorbed by that nucleon. Since Yukawa believed that it is beneficial for nucleons to make, exchange, and absorb pions, he believed that nucleons crammed together into the nucleus in order to be able to do those things. All of this was just a hypothesis until 1947, when nuclear physicists discovered pions and found that they had almost exactly the mass that Yukawa predicted.

As a result of Yukawa's theorizing and the discovery of the pion, nuclear scientists now view the nucleus as a place full of busy activity. Nucleons in the nucleus are continually making, exchanging, absorbing, and re-making pions. The desire for nucleons to do this is so overwhelming that it overcomes the electromagnetic repulsion between protons, allowing protons to stay very close to one another. Because pions are short lived, nucleons can only exchange these particles when the nucleons are quite close. Thus, pion exchange exists only in the nucleus.

The binding energy of a nucleus, then, is mostly used to facilitate the exchange of pions. The attraction that nucleons feel as a result of this exchange is called the strong nuclear force. The strong nuclear force exists only between nucleons (because only they can exchange pions). It is also a very short-range force, because the pions that are exchanged can only exist for a brief period of time. Thus, a pion must travel from one nucleon to another before its lifetime is up. Finally, for very short distances, the nuclear force is incredibly strong, because the desire for nucleons to exchange pions is strong. As a result, for distances on the order of 10^{-15} m, the strong nuclear force is significantly stronger than the electromagnetic force. We will discuss further elementary particles like pions in Sect. 16.7 of this book.

Even with Yukawa's understanding of the strong nuclear force, we don't know why some combinations of nucleons form stable isotopes and some do not. Figure 16.10 shows plot of stable nuclei in terms of their numbers of neutrons and protons. The figure shows that for small nuclei (those that have only a few nucleons), the dots lie right along a line that corresponds to the number of protons equal to the number of neutrons. This means that for small nuclei, an equal number of neutrons and protons leads to a stable nucleus. Indeed, nuclei like ${}^{4}_{2}\mathrm{He}$, ${}^{16}_{8}\mathrm{O}$, and ${}^{40}_{20}\mathrm{Ca}$ are all stable. Notice, however, that as the number of nucleons in the nucleus increases, the dots begin to rise farther and farther above the line that indicates an equal number of protons and neutrons. This tells us that as a nucleus gets larger, it can only be stable if it has more neutrons than protons. Unstable nuclei are radioactive as discussed in the next section.

16.6 Nuclear Physics II: Nuclear Processes

There are three important types of nuclear processes, radioactivity, fission, and fusion. The first of these deals with a set of radioactive elements that spontaneously emit particles and so their nuclei decay into more stable nuclei accompanied by a release of excess energy. Radioactivity is classified in terms of the type of particles that are emitted as α-decay, β-decay, and γ-decay.

An alpha particle is essentially a helium nucleus, ${}^{4}_{2}\mathrm{He}$. When α-decay occurs, the nucleus changes to a new nucleus with the atomic number reduced by two and the atomic mass number reduced by four. One example would be uranium transforming into thorium. This is represented by the equation

$$ {}^{238}_{92}\mathrm{U} \rightarrow {}^{234}_{90}\mathrm{Th} + \alpha $$

with the energy emitted as the kinetic energy of the α-particle. This isotope of thorium is still not stable so radioactive decay will continue until a stable a stable nucleus is created. Note that the superscripts on the right side of the equation add up to those on the left side of the equation and the same is true of the subscripts. These conditions lead to a balanced equation. The radioactive isotope is called the parent isotope and the result of the radioactive decay is the daughter isotope.

A beta particle is an electron emitted by the nucleus. Although electrons do not exist in the nucleus, a neutron can undergo a conversion to a proton plus an electron. This is expressed as

$$ {}^{1}_{0}n \rightarrow {}^{1}_{1}\mathrm{H} + {}^{0}_{-1}e $$

Note that the proton is represented as a hydrogen nucleus since they are equivalent. The electromagnetic charge balances to zero in this process while the atomic

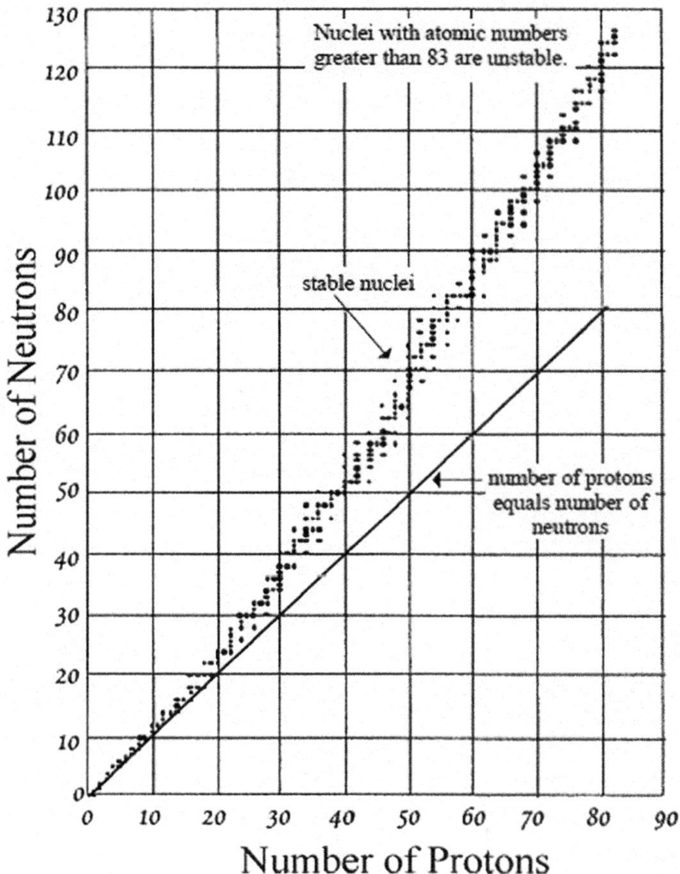

Fig. 16.10 Stable isotopes

number increases by 1 and the atomic mass number remains the same. The difference in mass between a neutron ($1.0087u$) and a proton ($1.0073u$) goes into the mass (9.11×10^{-31} kg) and kinetic energy of the emitted β-particle. This type of β-decay is a spontaneous reaction that all free neutrons undergo. However, if a neutron is part of a stable nucleus the pion exchange with other nucleons stops this decay from occurring. A typical β-decay process for an unstable nucleus is

$$^{90}_{38}\text{Sr} \rightarrow {}^{90}_{39}\text{Y} + {}^{0}_{-1}e$$

Note that this equation is balanced.

A gamma particle is a high energy photon. Gamma radiation occurs when there is a redistribution of energy in the nucleus. The γ-ray is emitted with no change in the atomic number or the atomic mass number of the nucleus. Since light has no protons and no neutrons in it, a gamma particle is symbolized as $^{0}_{0}\gamma$. The emission of a gamma particle does not affect the identity of the nucleus. However, it does remove energy from the nucleus. Thus, if a nucleus is stable but has too much energy, it will rid itself of the extra energy by emitting a gamma particle. For example, if a $^{90}_{39}Y$ nucleus has too much energy, it will emit a gamma particle.

$$^{90}_{39}Y \rightarrow ^{90}_{39}Y + ^{0}_{0}\gamma$$

This process simply rids the $^{90}_{39}Y$ nucleus of its excess energy. For historical reasons, gamma rays are also called **X-rays**. When you get an X-ray in order for a doctor to diagnose a condition, gamma rays are being shot at you.

Now that you have been introduced to the three forms of natural radioactivity, study the examples and solve the problems that follow to be sure you understand how to deal with nuclear equations.

Example 16.5

$^{14}_{6}C$ is a radioactive isotope that goes through beta decay. What is the daughter product of this decay? Write a balanced equation for the decay process.

According to the periodic chart, carbon has an atomic number of 6. This tells us that a $^{14}_{6}C$ atom has 6 protons and 8 neutrons in its nucleus. When a **radioactive** isotope undergoes beta decay, one of its neutrons turns into a proton. Thus, it will end up with one more proton and one less neutron. The daughter product then, will have 7 protons and 7 neutrons. According to the chart, all atoms with 7 protons are symbolized with an "*N*." The mass number of this nitrogen atom will be $7 + 7 = 14$. Thus, the daughter product is $^{14}_{7}N$.

Now that we know the daughter product, the balanced equation is rather simple. Using

$$^{14}_{6}C \rightarrow ^{14}_{7}N + ^{0}_{-1}e$$

Note that this equation is balanced, because the subscripts on one side add up to the subscripts on the other, as do the superscripts.

$^{232}_{90}Th$ is a radioactive isotope that goes through alpha decay. What is the resulting daughter product and balanced equation?

According to the chart, thorium (Th) atoms have 90 protons. Thus, this particular atom has 90 protons and 142 neutrons in it. When it goes through alpha decay, it actually spits out 2 protons and 2 neutrons in the form of a $^{4}_{2}He$ nucleus. The result will be only 88 protons and 140 neutrons in the daughter product. The chart tells us that Ra is the symbol for all atoms with 88 protons. The mass number of the resulting nucleus will be $88 + 140 = 228$. Thus, the daughter product is $^{228}_{88}Ra$.

Now that we know the daughter product, the balanced equation is rather simple. Using the notation for an alpha particle that was discussed above, the equation is:

$$^{232}_{90}\text{Th} \rightarrow {}^{228}_{88}\text{Ra} + {}^{4}_{2}\text{He}$$

Note that this equation is balanced, because the subscripts on one side add up to the subscripts on the other, as do the superscripts.

Write a balanced reaction for the gamma decay of $^{22}_{11}$Na.

Gamma decay simply takes energy away from the nucleus in the form of light. It does not change the identity of the nucleus. Thus, the daughter product is still ^{22}Na. The equation, then, is particularly easy to produce:

$$^{22}_{11}\text{Na} \rightarrow {}^{22}_{11}\text{Na} + {}^{0}_{0}\gamma$$

A radioactive decay process starts with a $^{234}_{90}$Th nucleus and produces a $^{234}_{91}$Pa nucleus. What kind of radioactive decay is this?

In $^{234}_{90}$Th, there are 90 protons and 144 neutrons. In ^{234}Pa, there are 91 protons and 143 neutrons. Thus, this must be beta decay, because the daughter product has one more proton than the radioactive isotope and one less neutron. This can only happen if a neutron turns into a proton.

The three kinds of radioactivity discussed above are phenomenon known as "natural radioactivity." The reason for this term is simple. These three types of radioactive decay are the only ones that occur naturally here on earth. However, scientists have artificially produced nuclei that decay via other mechanisms. The two "artificial" forms of radioactive decay are electron capture and positron emission.

In electron capture, a proton in a nucleus captures an electron (typically from the electrons that surround the nucleus). Electron capture is the reverse of β-decay. Thus, when a proton captures an electron, the result is a neutron. The following is an example of an electron capture reaction:

$$^{14}_{8}\text{O} + {}^{0}_{-1}e \rightarrow {}^{14}_{7}\text{N}$$

In this reaction, $^{14}_{8}$O has too many protons and not enough neutrons. To fix this problem, one of the protons captures an electron from the electron orbitals and the result is a stable $^{14}_{7}$N nucleus.

In positron emission, a proton emits a positron to become a neutron. A positron is a form of antimatter. It is an anti-electron. A positron has the same mass as an electron but a positive charge. It is symbolized as ${}^{0}_{+1}e$. When a positron and an electron collide with one another, they destroy each other, leaving nothing behind but a gamma ray (high energy light).

$$ {}^{0}_{+1}e + {}^{0}_{-1}e \rightarrow {}^{0}_{0}\gamma $$

This process, called annihilation, is what makes a positron antimatter. Matter and antimatter destroy each other. Thus, a positron is an anti-electron because, when it encounters an electron, the two particles destroy each other, leaving only energy (no mass) behind. This reaction, then, is an example of matter being converted into energy.

In the radioactive process known as positron emission, a proton emits a positron to turn into a neutron:

$$ {}^{8}_{5}\mathrm{B} \rightarrow {}^{8}_{4}\mathrm{Be} + {}^{0}_{+1}e $$

Notice that this process has the same effect as electron capture, because it transforms a proton into a neutron.

Spontaneous radioactive decay occurs on a specific time scale for each isotope. The activity of a radioactive isotope is characterized by its half-life. The half-life is defined as the time it takes for the number of isotopes to decay to half its initial value, N_{i}. As the decay continues, the number of radioactive isotopes N is given by

$$ N = N_{\mathrm{i}}\left(\frac{1}{2}\right)^{n} \qquad (16.12) $$

where n is the number of half-lives. The decay is exponential, so

$$ N = N_{\mathrm{i}}e^{-kt} \qquad (16.13) $$

where t is the time and k is a constant characteristic of the radioactivity of that particular isotope. The half-life can be derived from Eq. 16.13 to be

$$ t_{1/2} = 0.693/k. \qquad (16.14) $$

Half-lives of radioactive materials range from tenths of seconds to millions of years. Some radioactive decay reactions proceed quickly, and some do not. For example, the alpha decay of ${}^{214}\mathrm{Po}$ into ${}^{210}\mathrm{Pb}$ has a half-life of 0.00016 s. On the other hand, ${}^{238}\mathrm{U}$ alpha decays via a reaction whose half-life is 4.41×10^{9} years.

The N in Eqs. 16.12 and 16.13 can represent either the mass of the sample or the number of nuclei. This is demonstrated in the following example and problem.

Example 16.6

The alpha decay of ${}^{210}_{83}\mathrm{Bi}$ proceeds with a half-life of 5.00 days. A chemist makes 100.0 g of the isotope.

a. **How many grams will be left in 15.00 days?**

b. **How many grams will be left in 18.00 days?**

a. The amount of time elapsed is an integral multiple of the half-life. Thus, after 5.00 days, there are only 50.00 g left, after the next 5.00 days there are only 25.00 g left, and after a total of 15.00 days, there are only 12.50 g.
b. This one is not so easy, because the elapsed time is not an integral multiple of the half-life. Thus, we need to use Eq. 16.13. To use that equation, however, we need to know *k*. This comes from Eq. 16.14:

$$k = 0.693/t_{1/2} = 0.693/5.00\,\text{days} = 0.139\,\text{day}^{-1}$$

Now

$$N = N_{\text{i}}e^{-kt} = (100.0\,\text{g})e^{-(0.139/\text{days})(18.00\,\text{days})} - 8.19\,\text{g}$$

Consider the following problem.

Student

16.8 The daughter product of an alpha decay process is $^{220}_{86}\text{Rn}$. What was the radioactive isotope that went through alpha decay?
16.9 The half-life of $^{131}_{53}\text{I}$ is 8 days. How much of a 10.0 g sample will be left after 10 days?

Radiation can be harmful or helpful. When alpha, beta, or gamma particles collide with atoms or molecules in their way, the energy of the collision can ionize the atom or molecule with which the particle collides. Thus, alpha, beta, and gamma particles are referred to as ionizing radiation, because they ionize matter as they pass through it. When an alpha, beta, or gamma particle collides with a human cell, it results in the cell's death, because it ionizes chemicals in the cells that should not be ionized. If the particle hits the cell just right, the resulting ionization might mutate the cell's DNA rather than kill the cell. This can also be dangerous. However, your body expects cells to die so it is designed to reproduce cells. Thus, as long as your cells do not die faster than they can be replaced by your body, there is no real problem. If you are exposed to too much radiation too quickly, then your cells will be killed faster than your body can replace them, leading to radiation burns or organ damage.

The important thing to remember about the dangers of radioactivity is that it depends on the level of radioactivity to which you are exposed. A small amount of radioactivity is reasonably safe, a large amount is not. Nuclear scientists have come up with ways of measuring how much ionizing radiation people are exposed to. They refer to this as the dose of radiation. There are two units nuclear scientists use to measure radiation dosage. They are the rad (radiation absorbed dose) and the rem (roentgen equivalent in man). The rad is the amount of radiation that will deposit 100 J of energy into a kilogram of living tissue. This is a fine measure of radiation exposure, but it neglects the fact that certain types of radiation are more damaging to biological systems than others. Alpha particles, for example, do more damage to tissue per Joule they deposit than beta or gamma radiation because of the details of how alpha particles ionize matter. As a result, alpha particles are considered "more effective" at destroying living tissue. To take this into account, nuclear scientists have come up with the RBE (relative biological effectiveness) factor. This factor is different for each type of ionizing radiation. Alpha particles, for example have an RBE factor of 4 while gamma and beta particles have an RBE factor of 1. When the number of rads are multiplied by the RBE factor, the result is the dosage in rems.

$$\text{rems} = (\text{RBE}) \times \text{rads} \tag{16.15}$$

Thus, if you are exposed to 0.010 rads of beta particles, your radiation dose is $1 \times 0.010 = 0.010$ rems. If you are exposed to 0.010 rads of alpha particles, your radiation dose $4 \times 0.010 = 0.040$ rems.

If you add up all of the radiation you are exposed to from normal sources in the environment, your average radiation dose each year would be about 0.2 rems. Since studies by radiation biologists indicate that a lethal dose of radiation is about 470 rems, the dose of radiation you get as a result of everyday activity is simply too minimal to be worried about. Even if you are in a position in which you are exposed to large amounts of radioactivity, there are ways you can protect yourself. For example, alpha particles are extremely weak in terms of how much matter they can travel through. If you put a piece of paper between you and the radioactive source emitting the alpha particles, the vast majority of those alpha particles will stop in the paper. As a result, they will never hit you. Beta particles can travel through obstacles a bit better. It typically takes a thin sheet of metal to stop most of the beta particles coming from a radioactive isotope that emits them. Finally, gamma rays are the strongest type of radiation, requiring several inches of lead to stop them. Thus, one way you can protect yourself is to block the radiation before it hits you. This method is called "shielding." The other way you can protect yourself from an intense radioactive source is to simply move away from it. The farther you move away, the fewer particles can hit you.

Along with the dangers of radiation, it has many important uses. One very interesting use is radioactive dating. ${}^{14}_{6}\text{C}$ is probably the best known radioactive dating technique. ${}^{14}_{6}\text{C}$ decays by beta decay with a half-life of 5730 years. It turns out that all living organisms contain a certain amount of ${}^{14}_{6}\text{C}$, which is part of the reason that all living organisms are radioactive.

Living organisms continually exchange ${}^{14}_{6}\text{C}$ with their surroundings. Human beings, for example, exhale carbon dioxide, some of which contains ${}^{14}_{6}\text{C}$. In addition, human beings eat other organisms (plants and animals), which contain ${}^{14}_{6}\text{C}$

as well. Finally, part of the air that we inhale is made up of carbon dioxide, some of which contains ${}^{14}_{6}C$. Thus, organisms are continually exchanging ${}^{14}_{6}C$ with their environment. The practical result of all of this exchange is that, at any time when an organism is alive, it contains the same amount of ${}^{14}_{6}C$ as does the atmosphere around the organism.

This changes when the organism dies. At that point, the ${}^{14}_{6}C$ exchange ceases. Thus, the organism cannot replenish its supply of ${}^{14}_{6}C$, and the amount of ${}^{14}_{6}C$ in the organism begins to decrease because of the beta decay. Since the half-life of this process is 5730 years, the decay happens slowly. Nevertheless, it is a measurable effect. In general, then, organisms that have been dead a long time tend to have less ${}^{14}_{6}C$ in them as compared to those that have been dead for only a short time. With the known information, Eq. 16.13 will tell us how long the organism has been dead.

To use this technique, we need to make an assumption as to how much ${}^{14}_{6}C$ was in the organism when it died. We assume that the amount of ${}^{14}_{6}C$ it had when it died was the same as the amount of ${}^{14}_{6}C$ that was in the atmosphere at that time. We can measure the amount of ${}^{14}_{6}C$ that is in the organism now but we need to know how much was in it when it died to determine how long the organism has been dead. Luckily, the science of dendrochronology (study of tree rings) gives us a way we can measure the amount of ${}^{14}_{6}C$ in the atmosphere in years past. When you cut down a tree, you can count the rings in the tree's trunk to determine how old it is. You can actually measure the amount of ${}^{14}_{6}C$ in a tree ring and use it to determine how much ${}^{14}_{6}C$ was in the atmosphere during the year in which the tree ring was grown. As a result, scientists have determined the amount of ${}^{14}_{6}C$ in the atmosphere throughout a portion of the earth's past. Thanks to large depositories of tree ring sample at the University of Arizona and other laboratories, dating has been done of objects more than 12,000 years old. Using tree ring data to calibrate ${}^{14}_{6}C$ radioactive decay provides an accurate way to date archeological artifacts.

Another use for radiation is in medicine. Radiation has revolutionized the field of medicine. When a doctor wants to look at your bones, the doctor takes an X-ray picture of you. This is accomplished by placing the portion of your body that needs to be examined between a sheet of film (or a radiation detector) and a high-intensity gamma ray source. As you are exposed to the gamma rays, some pass through your body and hit the film, while others collide with cells in your body and stop. Gamma rays collide more with the dense portions of your body (the bones) than with the fleshy parts of your body. As a result, the film gets hit by gamma rays more frequently when bone is not between the gamma ray source and the film. When the film is developed, this will result in the parts of the film behind your bones being much whiter than those parts of the film behind the rest of your body. As a result, the gamma rays form an image of your bones on the film.

The dose of radiation required to make a clear X-ray is too small to hurt you. However, the person giving you the X-ray would be exposed to gamma rays all day if they were not shielded from them. That's why the person taking your X-ray stands behind thick shielding during the X-ray process.

Radioactive isotopes are also used to track things like blood flow inside the body. If a gamma emitting isotope is injected into your bloodstream, doctors can analyze how the blood flows to different parts of your body by detecting where the gamma rays are coming from inside your body.

Ionizing radiation is even used to kill cancerous cells in tumors. People with thyroid cancer often are given radioactive iodine ($^{131}_{53}\mathrm{I}$) to drink. Since iodine collects in your thyroid, drinking radioactive iodine will concentrate radiation in your thyroid, killing cancerous cells. The healthy cells will die as well, but your body is more likely to replace the healthy cells and not the cancerous cells, so this is a very common treatment for thyroid cancer.

These are only a few of the many uses of radiation, but they demonstrate how useful radiation can be.

Now we can move on to the other two types of nuclear processes, fission and fusion. A nuclear fission processes consists of splitting a large nucleus into two smaller nuclei accompanied by the emission of a significant amount of energy. One way to do this is to bombard a heavy nucleus with protons. A typical nuclear fission processes involves hitting uranium atoms with fast neutrons,

$$^1_0n + ^{235}_{92}\mathrm{U} \rightarrow ^{92}_{36}\mathrm{Kr} + ^{141}_{56}\mathrm{Ba} + 3^1_0n + 2.67 \times 10^{-11}\,\mathrm{J\ energy}.$$

On the left-hand side of this reaction, the neutron combines with the uranium nucleus to make it an unstable $^{236}_{92}\mathrm{U}$ nucleus. This isotope has a mass of $236.0456u$. It undergoes fission into an isotope of $^{92}_{36}\mathrm{Kr}$ with a mass of $91.9262u$ and an isotope of $^{141}_{56}\mathrm{Ba}$ with a mass of $140.9144u$ plus three neutrons each with mass $1.0087u$. The total mass on the right side of the equation is $235.8667u$. This means the mass defect is $0.1789u$ or about 2.67×10^{-11} J of kinetic energy in the fission fragments. Because this reaction produces more neutrons, these neutrons can hit other uranium atoms and create a chain reaction. There must be a minimum amount of $^{235}_{92}\mathrm{U}$ present to have a chain reaction occur. The amount of uranium needed to sustain a chain reaction is called its critical mass. Note that the equation is balanced.

When this chain reaction is controlled in a nuclear reactor, nuclear fission can be a source of clean energy. Under uncontrolled conditions it is an atom bomb. Normally, about 99% of the isotopes in a sample of uranium have an atomic mass number of 238 and do not have efficient fission reactions. Thus, a sample of uranium must be enriched to have a higher percentage of isotopes with atomic mass number of 235 to make it an efficient fuel for fission. It is important to note that the fission reaction that is shown is just one of many types of reactions that can occur when $^{235}_{92}\mathrm{U}$ is bombarded by a neutron. For example, the uranium nucleus might split into $^{112}_{45}\mathrm{Rh} + ^{121}_{47}\mathrm{Ag} + 3^1_0n$ or $2^{117}_{46}\mathrm{Pd} + 2^1_0n$ with various amounts of energy in the fission fragments.

The final type of nuclear processes to consider is fusion. In fusion processes two light nuclei bind together to form a heavier nucleus. An example is two isotopes of hydrogen forming a helium atom,

$$^2_1\mathrm{H} + ^2_1\mathrm{H} \rightarrow ^3_2\mathrm{He} + ^1_0n + 5.19 \times 10^{-13}\,\mathrm{J\ energy}.$$

These processes take place at such high temperatures that containing the fusion process is a problem. In nature, fusion processes take place in the center of stars like our sun. The energy created by fusion in the sun makes it glow and warm the earth. The process of making new elements through fusion is called nucleosynthesis. This is how most of the elements in the universe that are lighter than iron were created.

Notice the contrast between nuclear fusion and nuclear fission. Nuclear fusion has small nuclei as the reactants and nuclear fission has large nuclei as reactants. This makes sense in terms of nuclear binding energy. Remember the most stable nucleus $^{56}_{26}Fe$, because it has the most binding energy per nucleon. This means that as long as nuclei are smaller than $^{56}_{26}Fe$, they want to fuse with other nuclei so as to become more like $^{56}_{26}Fe$. Nuclei heavier than $^{56}_{26}Fe$, however, want to undergo fission so they lose nucleons and become more like $^{56}_{26}Fe$. Thus, nuclear fusion reactions between nuclei lighter than $^{56}_{26}Fe$ are spontaneous, whereas nuclear fusion reactions between ^{56}Fe nuclei and those that are heavier are not spontaneous. In the same way, nuclear fission reactions can be spontaneous for nuclei heavier than $^{56}_{26}Fe$, but not for $^{56}_{26}Fe$ and those nuclei that are lighter.

Although fusion reactions are spontaneous for light nuclei, they are non-existent unless the nuclei can be forced close to one another. This is tough because the nuclei are positively charged and therefore repel each other. Thus, unless there is enough activation energy to push the nuclei very close to one another, the nuclei will never fuse at any kind of appreciable rate. There are two places where this occurs. One is in instruments called particle accelerators that accelerate nuclei to such high speeds that they have enough energy to get close to and fuse with other nuclei. The second is in the center of stars, where the gravitational force is so strong and the temperature is so high that nuclei have enough energy to get close enough to fuse. In fact, most of the sun's energy comes from the fusion of light nuclei into heavier nuclei.

Study the following examples and Problem so that you are sure you understand how to deal with fission and fusion reactions.

Example 16.7

Fill in the blank for the nuclear fusion reaction below:

$$2\,^{3}_{1}\boldsymbol{H} \rightarrow {}^{4}_{2}\boldsymbol{He} + _\,^{1}_{0}\boldsymbol{n}$$

In order for this to be a valid reaction, it must balance. This means that the sum of the superscripts on both sides of the equation must equal each other as well as the sum of the subscripts. This is already the case for the subscripts. However, in order for the superscripts on the right side to equal 6, there must be 2 neutrons. Thus, the answer is 2.

What is the missing reactant in the following equation? Is this fusion or fission?

$$^{1}_{0}\boldsymbol{n} + _ \rightarrow {}^{112}_{45}\boldsymbol{Rh} + {}^{123}_{49}\boldsymbol{In} + 3^{1}_{0}\boldsymbol{n}$$

In order to get the subscripts to balance, the subscript in the blank must be 94. The chart tells us that the symbol for this atomic number is Pu. To get the superscripts to balance, the superscript in the blank must be 237. That way the sum of the mass numbers on the left side (1 + 237) equals the sum of the mass numbers on the right side (112 + 123 + 3 × 1). Thus, the missing reactant is $^{237}_{94}Pu$. This is fission, because a large nucleus is splitting into smaller nuclei.

Now consider the following problem.

Student

16.10 What is the missing reactant in the following nuclear equation? Is it a fusion or fission reaction?

$$^{27}_{13}\mathrm{Al} + {}^{4}_{2}\mathrm{He} \rightarrow _ + {}^{1}_{0}n$$

16.7 Beyond the Nucleus

For many years it was thought that the model of an atom consisting of electrons orbiting around a nucleus made of protons and neutrons was sufficient for understanding matter. However, in the second half of the twentieth century, a number of different giant particle accelerators were built. These were used to accelerate beams of atoms to extremely high speeds and study what happens when collisions occur at these high energies. This led to the discovery that protons and neutrons are themselves made of smaller particles. In fact, these particle accelerator experiments have resulted in the identification of many different types of elementary particles which give us a greater understanding of matter. Each of these particles has its own set of specific properties including mass, charge, spin, and other characteristics.

The Standard Model classifies the building blocks of matter into three classes of elementary particles. The first class is made up of particles called quarks. There are six types of quarks with different characteristics. Combinations of these quarks make up protons, neutrons, pions and many other types of particles that have been identified in particle accelerator experiments. The second class has particles are called leptons. There are six types of leptons with different characteristics. These include electrons, positrons (electrons with a positive charge), and neutrinos which are particles with no charge and extremely small mass. Quarks and leptons are the elementary particles that come together to form matter.

One other special type of elementary particle used in building matter is in a class by itself. It is called the Higgs boson. This was proposed as the particle that gives mass to leptons and quarks, but it was very difficult to detect its existence. However, in 2012 researchers at the Large Hadron Collider made the first observation of a Higgs boson.

The third class of elementary particles contains particles that are called force carriers. These are particles that transmit forces between particles of matter. The most common force carrier is the photon. The electromagnetic force between electrically charged particles is transmitted by the absorption and emission of photons. The strong force that holds quarks together in leptons is transmitted by gluons. Eight different types of gluons have been identified. Although pions are particles containing quarks held together by gluons, it is thought their exchange is involved in transmitting the strong force between nucleons to hold the nucleus together. There is also a weak force in the nucleus of an atom that acts between all quarks and leptons. This force is transmitted by three types of particles called gauge bosons. The fourth force in nature is gravity and the standard model would have this force transmitted between objects having mass by gravitons. However, the Standard Model is not as satisfactory for explaining gravity as Einstein's spacetime theory for special relativity discussed in the previous chapter.

The study of elementary particles has resulted in several interesting facts about matter. One is that anti-matter exists. For every type of particle there is an anti-particle. It has the same properties as the particle but with the opposite charge. For example, an electron has an anti-particle called a positron which has the same mass but a positive charge. When a particle collides with its anti-particle, they annihilate each other and create photons and other particles. We don't know what happened to all the antimatter particles at the time of creation.

Next Level
Standard Model

The elementary particles that make up the Standard Model of particle physics are shown in the table below. Each particle has its own box listing its name, symbol, charge (related to the standard charge, q), spin, and mass in Mega- or Giga-eV/c^2. The particles are separated into two types, those that produce matter and those that produce forces. The former are called "fermions" because they have half-integer spin and obey statistical rules developed by Fermi and Dirac. The latter are called "bosons" because they have integer spin and obey statistical rules developed by Bose and Einstein.

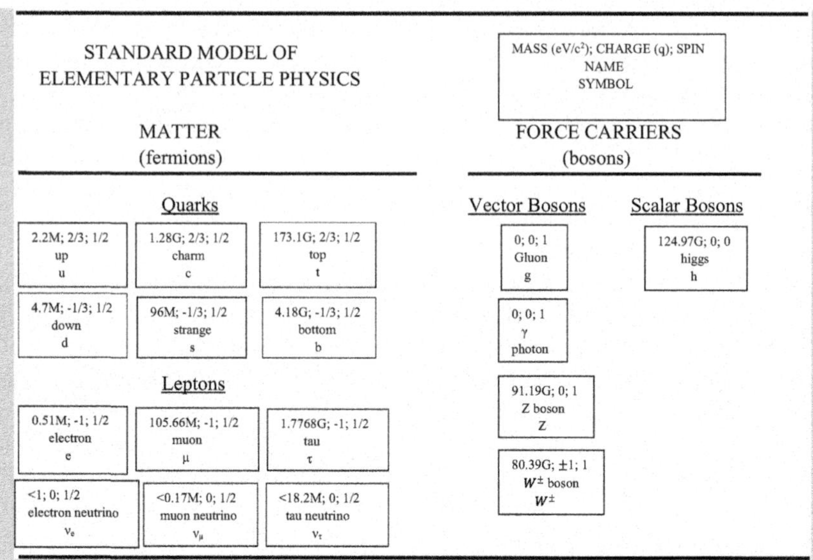

We learned before that photons transmit the electromagnetic force between charged particles. Similarly, gluons transmit the strong force between quarks. The Z and $W^{\pm}$ bosons are responsible for the weak force that is active in radioactivity. The Higgs boson provides a field inside quarks that gives them their mass. The smaller mass of leptons also comes from the Higgs boson although the neutrinos are sometimes considered to be massless.

The particles that are made of quarks are called "hadrons" and if they have an odd number of quarks they are "baryons." The nucleons we are familiar with, protons and neutrons, are baryons. They are each made up of three quarks. A proton has two up and one down quark while a neutron is made of two down and one up quark. These quarks are held together by the strong force transmitted by gluons. The quarks in a nucleon are moving very rapidly and the energy of their motion converted to mass is responsible for most of the mass of the nucleon. The mass of a neutron is slightly larger than the mass of a proton because down quarks interact more strongly with the Higgs field than up quarks do.

Another interesting fact is the conservation of quarks and conservation of leptons. The total number of quarks and the total number of leptons in the universe are constant. However, the study of astrophysics described in the previous chapter shows that there is much more matter in the universe than the type of matter we know about. The Standard Model has not helped us to understand what this different type of matter is.

Summing Up

You now know how physicists describe atoms as having dense nuclei surrounded by electrons in orbits having specific shapes and orientations. You know how a specific configuration is described by a set of quantum numbers and how absorption and emission of photons causes electrons to move from one type of orbit to another. Most importantly you learned how this knowledge led to the invention of the laser with all its many applications. In addition, you learned about the nucleus itself; what it is made of and how it changes through radioactivity, fission and fusion processes. Finally, you learned that all matter and fundamental forces in nature come from elementary particles.

This completes your course in advanced physics. Hopefully you enjoyed learning about the creation in which we live and are stimulated to continue your study of physics at an even more advanced level.

Answers to the Problems

16.1

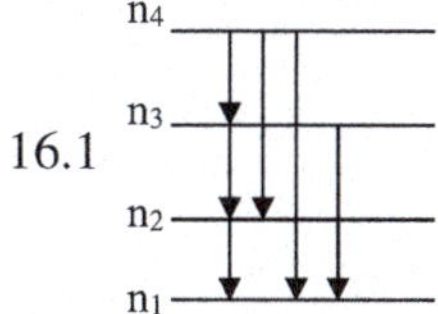

Once the electron got up to the fourth orbit, there are *many* ways it could get down. First, it could jump straight from the fourth orbit back to the first. That would result in one wavelength of light. It could also jump from the fourth to the third (another wavelength of light for a total of 2 so far), from the third to the second (another wavelength for a total of 3), and then from the second to the first (yet another wavelength for a total of 4). However, it could jump from the fourth to the third (we already considered that wavelength), and *then* jump straight from the third to the first (that's a new wavelength, so the total is now 5). Also, it could jump from the fourth to the second (another new one for a total of 6) and then from the second to the first (we already considered that one, however). Thus, there are a total of 6 possible wavelengths of light that can be emitted when the electron decays from the fourth orbit back to the first one.

16.2 The Bohr model can only be used on atoms or ions with just one electron. Thus, the following atoms or ions will work: H, Li^{2+}, Be^{3+}.

16.3

Knowns: $E = -5.45 \times 10^{-19}$ J; $Z = 1$

Unknowns: n, r_n

a. In this problem, we know that $Z = 1$ because we are dealing with a hydrogen atom. We also know the energy. Equation 16.7, therefore, can be used to solve for n. Since the energy was given in Joules, we will have to use R in Joules as well.

$$E = -RZ^2(1/n)^2$$

$$-5.45 \times 10^{-19}\,\text{J} = -(2.18 \times 10^{-18}\,\text{J})(1)^2(1/n)^2$$

$$(1/n)^2 = (-5.45 \times 10^{-19}\,\text{J})/(-2.18 \times 10^{-18}\,\text{J}) = 0.25$$

$$n = 2$$

The electron is in the second Bohr orbit.

b. From Eq. 16.5,

$$r_1 = 5.29 \times 10^{-11}\,\text{m}.$$

and

$$r_2 = 4r_1.$$

Therefore the distance of the electron from the nucleus is

$$r_2 = 4(5.29 \times 10^{-11}\,\text{m}) = 21.16 \times 10^{-11}\,\text{m}$$

16.4

Knowns: Z = 3; $n = 1$
Unknowns: r_1

This is a direct application of Eq. 16.7.

$$r = \left(0.529\,\mathring{\text{A}}\right) \cdot n^2/Z = \left(0.529\,\mathring{\text{A}}\right) \cdot (1)^2/3$$

$$r = 0.176\,\mathring{\text{A}}$$

So we see that a Li^{2+} ion is actually *smaller* than a hydrogen atom.

16.5

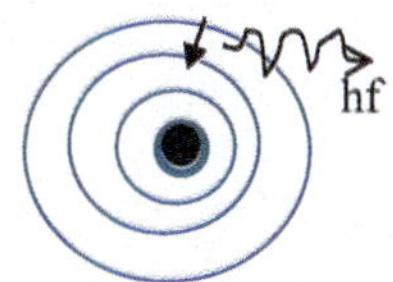

Knowns: $Z = 1$; $n_{\text{initial}} = 3$; $n_{\text{final}} = 2$
Unknowns: λ

$$\Delta E = RZ^2\left[(1/n_{\text{final}})^2 - (1/n_{\text{initial}})^2\right]$$

$$\Delta E = (13.6\,\text{eV})1^2\left[(1/2)^2 - (1/3)^2\right] = 1.89\,\text{eV}$$

This tells us that the electron must emit light with energy of 1.89 eV to make the transition.

From the energy, we can get the frequency

$$E = hf$$

$$f = E/h = 1.89\,\text{eV}/4.14 \times 10^{-15}\,\text{eV} = 4.57 \times 10^{14}\,\text{s}^{-1}$$

Then, we use the speed of light to go from frequency to wavelength.

$$f = c/\lambda$$

$$\lambda = c/f = \left(2.998 \times 10^8\,\text{m/s}\right)/\left(4.57 \times 10^{14}\,\text{s}^{-1}\right) = 656\,\text{nm}$$

The electron must emit light with a wavelength of 656 nm, which is red light.

16.6

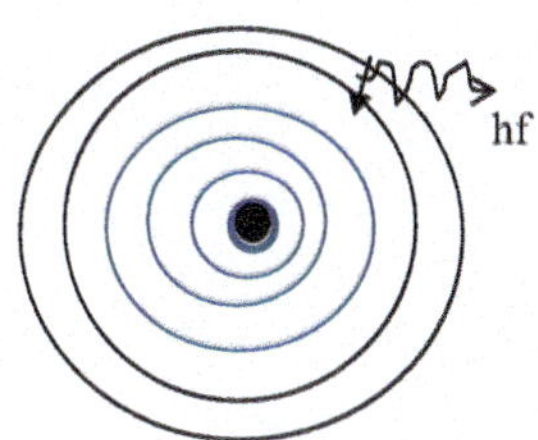

Knowns: $n_{\text{initial}} = 5$; $f = 6.67 \times 10^{14}$ Hz; $Z = 3$
Unknowns: n_{final}

$Z = 3$ because we are dealing with Li^{2+}. In order to find the final orbit, we need the energy of the photon emitted.

$$E = hf$$

$$E = \left(4.14 \times 10^{-15}\,\text{eVs}\right)\left(6.67 \times 10^{14}\,\text{Hz}\right) = 2.76\,\text{eV}$$

Now that we know the energy of the light emitted, we can use Eq. 16.10 to determine n_{final}:

$$\Delta E = RZ^2\left[(1/n_{\text{final}})^2 - (1/n_{\text{initial}})^2\right]$$

$$2.76\,\text{eV} = (13.6\,\text{eV})3^2\left[(1/n_{\text{final}})^2 - (1/5)^2\right]$$

$$(1/n_{\text{final}})^2 = (1/25) + 0.0225$$

$$n_{\text{final}} = 4$$

The electron lands in the fourth Bohr orbit.

16.7

Knowns: ${}^{56}_{26}\text{Fe}$; $m = 55.9349u$
Unknowns: E_{B}/nucleon

Since iron's atomic number is 26, all Fe atoms have 26 protons. The mass number indicates that a ${}^{56}_{26}\text{Fe}$ nucleus has 30 neutrons. The sum of the masses of 26 protons and 30 neutrons is:

$$26 \times (1.0073\,\text{amu}) + 30 \times (1.0087\,\text{amu}) = 56.4508u$$

Since the mass of a ${}^{56}_{26}\text{Fe}$ nucleus is only $55.9349u$, there is a mass deficit of $0.5159u$.

This mass deficit is converted to energy according to

$$\begin{aligned} E = mc^2 &= (0.5159u)\left(1.6605 \times 10^{-27}\,\text{kg/u}\right)\left(2.998 \times 10^{8}\,\text{m/s}\right)^2 \\ &= 7.700 \times 10^{-11}\,\text{J} \end{aligned}$$

The question asks for the binding energy per nucleon, so we must divide this by the total number of nucleons in the nucleus, which is 56.

$$7.700 \times 10^{-11}\,\text{J/56 nucleons} = 1.375 \times 10^{-12}\,\text{J/nucleon}.$$

16.8 ${}^{?}_{?}? \rightarrow {}^{220}_{86}\textbf{Rn} + {}^{4}_{2}\text{He}$

A ^{220}Rn nucleus has 86 protons and 134 neutrons. This nucleus is the result of the nucleus in question losing 2 protons and 2 neutrons. Thus, the original nucleus must have 88 protons and 136 neutrons, which is $^{224}_{88}\mathbf{Ra}$.

16.9

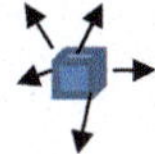

Knowns: $t_{1/2} = 8$ days; $m_i = 10.0$ g; $t = 10$ days
Unknowns: m_f

This one is not so easy, because the elapsed time is not an integral multiple of the half-life.

Thus, we need to get the value for k and then use Eq. 16.13. From Eq. 16.14

$$k = \text{days}\,0.693/(8\,\text{days}) = 0.09\,\text{days}^{-1}$$

Now we can use Eq. 16.13:

$$N = N_o e^{-kt} = 10.0\,\text{g}\,e^{-(0.09/\text{days})(10\,\text{days})} = 4\,\text{g}$$

16.10 $^{27}_{13}\text{Al} + ^{4}_{2}\text{He} \rightarrow _ + ^{1}_{0}n$

To get the sum of the subscripts to equal each other on each side of the equation, the missing product must have 15 protons. To get the sum of the superscripts equal to each other on both sides of the equation, it must have a mass number of 30 as well. Thus, the missing nucleus is $^{30}_{15}$P.

This is two smaller nuclei forming a larger one. Thus, it is fusion.

Study Guide for This Chapter

(The mass of a proton is 1.0073u, and the mass of a neutron is 1.0087u. The speed of light is 2.998×10^8 m/s; 1u = 1.6605×10^{-27} kg; $R = -2.18 \times 10^{-18}$ J = 13.6 eV.)

1. What assumption did Bohr make that led to the success of his theory?
2. In the Bohr model, an electron can be in orbit #1 or orbit #2, but it cannot be anywhere in between. If that's the case, how does the electron jump from orbit #1 to orbit #2?
3. Suppose you excite an atom and you see no light coming from it. Does that mean electrons are not moving up and down in orbitals?

4. Suppose an electron in the Bohr atom absorbs energy and jumps to the $n = 3$ orbit. How many wavelengths of light can the electron possibly emit in getting back to the $n = 1$ orbit?
5. What is the most stable nucleus in Creation?
6. What is an alpha particle? What about a gamma ray? What about a beta particle? Which can pass through the most matter? Which can pass through the least?
7. What happens when a positron collides with an electron? What is the process called?
8. What is the difference between nuclear fission and nuclear fusion? Which nuclei tend to undergo fission? Which tend to undergo fusion?
9. What is the energy of an electron in the $n = 2$ Bohr orbit of a He^+ ion?
10. An electron jumps from the $n = 3$ orbit of a Li^{2+} ion to the $n = 1$ orbit. What is the wavelength of the light emitted?
11. An electron jumps from the fifth Bohr orbit to a lower orbit in a hydrogen atom. If the light emitted has a frequency of 3.16×10^{15} Hz, what orbit did the electron end up in?
12. An electron in a He^+ ion is orbiting the nucleus at a distance of 0.265 Å. What is the electron's energy?
13. The mass of a ${}^{19}_{9}F$ nucleus is $18.9984u$. What is the binding energy per nucleon of the nucleus?
14. The nucleus ${}^{131}_{53}I$ is radioactive and decays by beta emission. This is the nucleus most commonly used in the thyroid "cocktail" which is used to treat thyroid disease. Write a nuclear equation for the beta decay of this isotope.
15. The radioactive isotope ${}^{222}_{86}Rn$ decays into ${}^{218}_{84}Po$. What kind of radioactive decay is this?
16. A ${}^{14}_{7}N$ nucleus is stable but has too much energy. What can it do to release the energy?
17. A nuclear chemist studies an unknown radioactive isotope. The sample of isotope has a mass of 14.0 g when the nuclear chemist begins the study. In 22.2 min, the mass is 13.6 g. What is the half-life of the radioactive isotope?
18. ${}^{27}_{13}Al$ and ${}^{3}_{1}H$ fuse to make ${}^{27}_{12}Mg$ and one other product. What is that other product?

Next Level

19. A baryon is made of an up quark, a strange quark, and a bottom quark. What are its spin, charge in Coulombs, and mass in kilograms?
20. Consider the following four particles: g, ν_τ, Z, and s. Which of these are fermions and which are bosons and why?

Some Final Thoughts

Congratulations on completing this course on advanced physics concepts. We have covered a lot of different topics from the classical description of motion to the weird things that occur to ultra-small objects or objects moving at ultra-fast speeds. We have learned about the nature of gravity and its importance in our Creation. We have shown how electromagnetism and optics are related. We have discussed our current state of knowledge of the matter that makes up our Creation and have described the thermal, mechanical, electrical, and optical properties of this matter.

I hope you enjoyed it and will remember what you learned about optics when you see the use of lasers, or acoustics when you listen to music, or transistors when you use electronic devices, or any of the other many things you learned about how physics effects your everyday life. There is still a lot of research and development going on in the field of physics. Much of this is focused on learning how to use our knowledge of physics for the betterment of mankind. Some of it is trying to extend our knowledge by answering the important unanswered questions that still remain in the field. Much of what you learned was not part of the high school physics course that I took, and future generations of students will study new information not covered in this course.

I spent over forty years as an active researcher in physics. It was one of the most fun and rewarding careers I can imagine. Because physics is an international community, I was able to travel all over the world collaborating with colleagues on research projects and attending international meetings. It is a privilege for me to be able to share with you my enthusiasm for this field through this textbook.

My experience as a physicist learning about the world we live in has given me a great appreciation for Creation. You can't spend time studying Creation without becoming deeply spiritual. I hope that this has been a fun class for you and that it makes you excited to learn even more about our Creation.

Appendix A

Physical Constants

Acceleration of gravity (on earth)	g	9.8 m/s^2
Atomic mass unit	u	1.66×10^{-27} kg
Avogadro's number	N_A	6.02×10^{23} mol^{-1}
Bohr radius	a_0	5.29×10^{-11} m
Boltzmann's constant	K	1.38×10^{-23} J/K
Charge on electron (or proton)	e	1.6×10^{-19} C
Coulomb constant	k	9×10^{9} Nm^2/C^2
Earth mass	E_m	5.98×10^{24} kg
Earth mean radius	E_r	6.38×10^{6} m
Earth distance from sun	D_E	1.51×10^{11} m
Electron mass$^-$	m_e	9.1×10^{-31} kg
Electron volt	eV	1.602×10^{-19} J
Faraday constant	F	96,485 C/mole
Gas constant	R	8.31 J/(mole K)
Gravitational constant	G	6.67×10^{-11} Nm^2kg^{-2}
Moon distance from Earth	D_M	3.84×10^{8} m
Moon mass	M_m	7.36×10^{22} kg
Moon mean radius	M_r	1.74×10^{6} m
Neutron mass	m_n	1.67×10^{-27} kg
Permeability of free space	μ_0	$4\pi \times 10^{-7}$ N/A^2
Permittivity of free space	ε_0	8.85×10^{-12} $C^2/(Nm^2)$
Planck's constant	h	6.63×10^{-34} Js
Proton mass	m_p	1.67×10^{-27} kg
Rydberg constant	R_h	2.18×10^{-18} J

R. C. Powell, *Physics*, Undergraduate Lecture Notes in Physics,
https://doi.org/10.1007/978-3-032-09361-5

Speed of light (in a vacuum)	c	3.0×10^8 m/s
Sun mass	M_S	1.99×10^{30} kg
Sun mean radius	S_r	6.96×10^8 m

Appendix B

<u>Units</u>

Basic units

- Mass—kg
- Length—m
- Time—s
- Volume—L
- Number of molecules—mole

Derived units

- Force—Newton, N = kgm/s^2
- Energy—Joule, J = kgm^2/s^2
- Electrical potential—Volt, V = J/C
- Electrical current—Amp, A = C/s
- Capacitance—Farad, F = C/V

<u>Conversions</u>

Mass

- 1 slug = 14.59 kg
- 1 u = 1.66 × 10^{-27} kg
- 1 metric ton = 1,000 kg

Length

- 1 inch = 2.54 cm
- 1 foot = 12 in
- 1 yard = 3 ft
- 1 km = 0.621 miles
- 1 light year = 9.461 × 10^{15} m
- 1 m = 39.37 in
- 1 mile = 5280 ft

Time

- 1 year = 365 days = 3.16 × 10^7 s
- 1 day = 24 h = 1.44 × 10^3 min
- 1 h = 3600 s

Force

- 1 N = 10^5 dynes = 0.2248 lbs

Energy

- 1 J = 10^7 ergs = 0.738 ft lbs
- 1 cal = 4.184 J
- 1 Btu = 1.054 × 10^3 J
- 1 eV = 1,602 × 10^{-19} J
- 1 kiloWatt hour = 3.60 × 10^6 J

Power

- 1 W = 0.738 ft-lb/sec
- 1 horsepwr = 7.46 × 10^2 W
- 1 Btu/hour = 0.293 W

Pressure

- 1 Pascal = 1.45 × 10^{-4} lb/in^2
- 1 bar = 14.50 lb/in^2
- 1 atm = 1.013 × 10^5 Pa = 14.7 lb/in^2

R. C. Powell, *Physics*, Undergraduate Lecture Notes in Physics,
https://doi.org/10.1007/978-3-032-09361-5

Appendix C

<u>Prefixes</u>

10^{-18} atto (a)	10^{18} exa (E)
10^{-15} femto (f)	10^{15} peta (P)
10^{-12} pico (p)	10^{12} tera (T)
10^{-9} nano (n)	10^{9} giga (G)
10^{-6} micro (μ)	10^{6} mega (M)
10^{-3} milli (m)	10^{3} kilo (k)
10^{-2} centi (c)	10^{2} hecto (h)
10^{-1} deci (d)	10^{1} deka (da)

R. C. Powell, *Physics*, Undergraduate Lecture Notes in Physics,
https://doi.org/10.1007/978-3-032-09361-5

Appendix D

Periodic Table of the Elements

R. C. Powell, *Physics*, Undergraduate Lecture Notes in Physics,
https://doi.org/10.1007/978-3-032-09361-5

Periodic table of all elements in the periodic table with the full names, symbols, atomic number and atomic masses.
Credit Emeka Udenze, Creative Commons Attribution-Share Alike 4.0 International

Periodic Table of Elements

IA	IIA	IIIB	IVB	VB	VIB	VIIB	VIIIB	VIIIB	VIIIB	IB	IIB	IIIA	IVA	VA	VIA	VIIA	VIIIA
1 H Hydrogen 1.01																	2 He Helium 4.00
3 Li Lithium 6.94	4 Be Beryllium 9.01											5 B Boron 10.81	6 C Carbon 12.01	7 N Nitrogen 14.01	8 O Oxygen 16.00	9 F Fluorine 19.00	10 Ne Neon 20.18
11 Na Sodium 22.99	12 Mg Magnesium 24.31											13 Al Aluminum 26.98	14 Si Silicon 28.09	15 P Phosphorus 30.97	16 S Sulfur 32.06	17 Cl Chlorine 35.45	18 Ar Argon 39.95
19 K Potassium 39.10	20 Ca Calcium 40.08	21 Sc Scandium 44.96	22 Ti Titanium 47.87	23 V Vanadium 50.94	24 Cr Chromium 52.00	25 Mn Manganese 54.94	26 Fe Iron 55.85	27 Co Cobalt 58.93	28 Ni Nickel 58.69	29 Cu Copper 63.55	30 Zn Zinc 65.38	31 Ga Gallium 69.72	32 Ge Germanium 72.63	33 As Arsenic 74.92	34 Se Selenium 78.97	35 Br Bromine 79.90	36 Kr Krypton 83.80
37 Rb Rubidium 85.47	38 Sr Strontium 87.62	39 Y Yttrium 88.91	40 Zr Zirconium 91.22	41 Nb Niobium 92.91	42 Mo Molybdenum 95.95	43 Tc Technetium (98)	44 Ru Ruthenium 101.07	45 Rh Rhodium 102.91	46 Pd Palladium 106.42	47 Ag Silver 107.87	48 Cd Cadmium 112.41	49 In Indium 114.82	50 Sn Tin 118.71	51 Sb Antimony 121.76	52 Te Tellurium 127.60	53 I Iodine 126.90	54 Xe Xenon 131.29
55 Cs Cesium 132.91	56 Ba Barium 137.33	57 - 71 Lanthanides	72 Hf Hafnium 178.49	73 Ta Tantalum 180.95	74 W Tungsten 183.84	75 Re Rhenium 186.21	76 Os Osmium 190.23	77 Ir Iridium 192.22	78 Pt Platinum 195.08	79 Au Gold 196.97	80 Hg Mercury 200.59	81 Tl Thallium 204.38	82 Pb Lead 207.20	83 Bi Bismuth 208.98	84 Po Polonium (209)	85 At Astatine (210)	86 Rn Radon (222)
87 Fr Francium (223)	88 Ra Radium (226)	89 - 103 Actinides	104 Rf Rutherfordium (265)	105 Db Dubnium (268)	106 Sg Seaborgium (271)	107 Bh Bohrium (270)	108 Hs Hassium (277)	109 Mt Meitnerium (276)	110 Ds Darmstadtium (281)	111 Rg Roentgenium (280)	112 Cn Copernicium (285)	113 Nh Nihonium (284)	114 Fl Flerovium 289	115 Mc Moscovium (288)	116 Lv Livermorium (293)	117 Ts Tennessine (294)	118 Og Oganesson (294)

57 La Lanthanum 138.91	58 Ce Cerium 140.12	59 Pr Praseodymium 140.91	60 Nd Neodymium 144.24	61 Pm Promethium (145)	62 Sm Samarium 150.36	63 Eu Europium 151.96	64 Gd Gadolinium 157.25	65 Tb Terbium 158.93	66 Dy Dysprosium 162.50	67 Ho Holmium 164.93	68 Er Erbium 167.26	69 Tm Thulium 168.93	70 Yb Ytterbium 173.05	71 Lu Lutetium 174.97
89 Ac Actinium (227)	90 Th Thorium 232.04	91 Pa Protactinium 231.04	92 U Uranium 238.03	93 Np Neptunium (237)	94 Pu Plutonium (244)	95 Am Americium (243)	96 Cm Curium (247)	97 Bk Berkelium (247)	98 Cf Californium (251)	99 Es Einsteinium (252)	100 Fm Fermium (257)	101 Md Mendelevium (258)	102 No Nobelium (259)	103 Lr Lawrencium (262)

Index

R. C. Powell, *Physics*, Undergraduate Lecture Notes in Physics,
https://doi.org/10.1007/978-3-032-09361-5

D

The manufacturer's authorised representative in the EU is Springer Nature Customer Service Centre GmbH, Europaplatz 3, 69115 Heidelberg, Germany. If you have any concerns regarding our products, please contact ProductSafety@springernature.com

Printed and bound by CPI Group (UK) Ltd, Croydon, CR0 4YY
07/07/2026
02160922-0001